Red-winged Blackbird
Agelaius phoeniceus

Savannah Sparrow
Passerculus sandwichensis

Western Harvest Mouse
Reithrodontomys megalotis

Turkey Vulture
Cathartes aura

Great Blue Heron
Ardea herodias

Marsh Hawk
Circus cyaneus

Snowy Egret
Leucophoyx thula

Man
Homo sapiens

California Garter Snake
Thamnophis couchi

Gray Fox
Urocyon cinereoargenteus

Starry Flounder
Platichthys stellatus

Bay Mussel
Mytilus edulis diegensis

Raccoon
Procyon lotor

Sand Bass
Paralabrax nebulifer

*Endangered Species

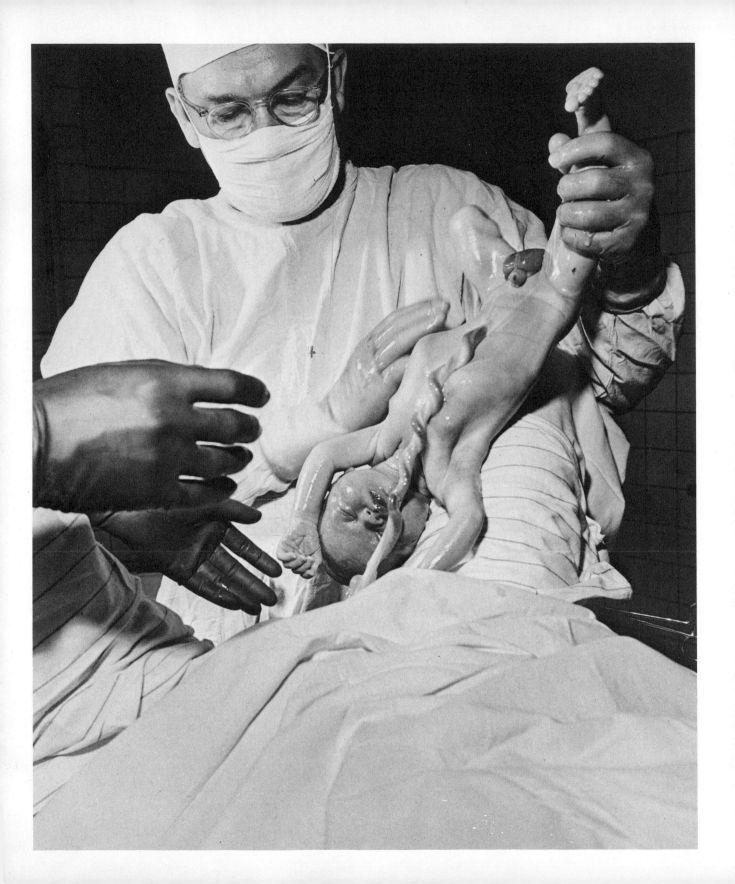

Biology
The World
of Life
Second Edition
Robert A. Wallace
Duke University

Goodyear Publishing Company, Inc.
Santa Monica, California

Library of Congress Cataloging in Publication Data

Wallace, Robert Ardell, 1938–
    Biology, the world of life.

    Bibliography: p. 561
    Includes index.
    1. Biology.   I. Title.
QH308.2.W35 1978   574   77–12384
ISBN 0-87620-109-5

Current printing (last digit):
10 9 8 7 6 5 4 3 2

ISBN: 0–87620–109–5

Library of Congress Catalog Number: 77–12384

Y–1095–2

Designer/Art Director: Don and Debra McQuiston
Supervising Editor: Susan Smith
Scientific Consultant: Gerald Sanders
Photo Research: Barbara Hodder
Illustrator: John Dawson

Printed in the United States of America

The cover is a painting by John Dawson depicting life in a southern
California Salt Marsh. The cover is reproduced on the inside back
cover with the life forms identified.

This book is for Cordia and Holman,
Hershel and Virginia, John and Alma, and
to the strength which derives from deeper roots.

Contents

Preface	*xvii*
1 Biology Yesterday and Today	**1**
The Voyage of the Beagle	**1**
The History of an Idea	**2**
The Beginnings of Biology	3
Photo Essay: Darwin and the	
Voyage of the *Beagle*	**4**
Time and the Intellectual Milieu	14
Darwin's Theory of Evolution	**14**
The Galapagos Islands	15
The Impact of Malthus	16
Natural Selection	16
Descent with Modification	17
Mechanism and Vitalism	20

2 The Beginnings	**23**
The Beginnings of Life	**23**
Miller's Experiments	**25**
The Coacervate Droplet	**26**
The Growth of	
Coacervate Droplets	26
Internal Structure	27
Reproducing Droplets	27
Successful Droplets	28
The Early Cells	**29**
Autotrophy and Heterotrophy	29
An Autotrophy Called	
Photosynthesis	31
Oxygen: A Bane and Blessing	31
The Idea in Science	**32**
Theory	32
The Usual Process	33
What Kinds of Answers Can	
Science Provide?	34
Science and Social Responsibility	**35**
Fads and Trends	37
Exaggeration, Compromise,	
and Understanding	37

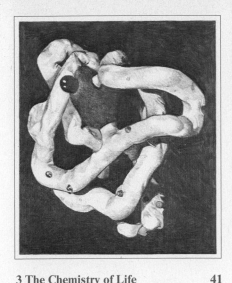

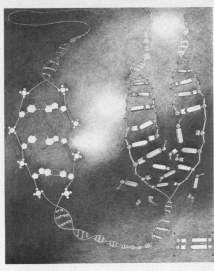

Only the
Sun gives
without
taking.

3 The Chemistry of Life 41
What Biologists Do for a Living 41
Atoms and Elements 42
CHNOPS 43
The Structure of an Atom 43
How Atoms May Vary 43
Isotopes 43
Ions 44
Electron Shells 44
Oxidation and Reduction 44
Electrons and the Properties
  of the Atom 45
BOX 3.1 IONS AND YOUR BEHAVIOR 47
Chemical Bonding 48
Ionic Bonds 48
Covalent Bonds 49
The Magic of Carbon 49
Hydrogen Bonds 51
Chemical Reactions 52
Energy of Activation 53
BOX 3.2 ACIDS AND BASES 54
Enzymes 56
Energy Changes 57
The Molecules of Life 57
Carbohydrates 57
Lipids 60
Proteins 62
Amino Acids 62
Peptides 63
BOX 3.3 UNRAVELING THE
  STRUCTURE OF INSULIN 64
Protein Structure 65

4 Energy 67
The Laws of Thermodynamics 67
ATP: The Energy Currency 68
Photosynthesis 68
BOX 4.1 CHLOROPHYLL 71
The Light Reaction 72
The Dark Reaction 73
Cellular Respiration 75
Glycolysis 75
The Krebs Cycle 78
The Electron-Transport Chain 80

5 Cells 83
Cell Specialization and Control 83
Procaryotic and Eucaryotic Cells 84
How Molecules Move 84
BOX 5.1 THE ELECTRON MICROSCOPE 85
Diffusion 86
Osmosis 86
Active Transport 88
Cell Components 88
Cell Walls 88
The Cell Membrane 90
Mitochondria 91
Golgi Bodies 92
Lysosomes 93
Plastids 93
Vacuoles 94
The Endoplasmic Reticulum 95
Centrioles 95
Cilia and Flagellae 95
The Nucleus 96
Cell Replication 99
Mitosis 99
Meiosis 103
The Double Helix 107
DNA Replication 114
How Chromosomes Work 114
Problems in Cell Biology 120
BOX 5.2 VIRUSES 122

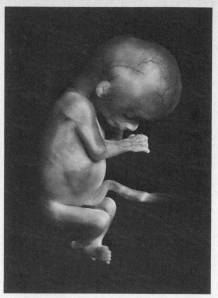

6 Inheritance 125
Mendelian Genetics 125
The Principle of Dominance 127
The Principle of Segregation 127
The Principle of Independent
  Assortment 129
Classical Genetics 131
Sex Determination 133
Sex Linkage in Humans 134
BOX 6.1 THE DISEASE OF ROYALTY 136
Gene Linkage and Crossover 138
Chromosome Mapping 139
Mutations 140
Genic Mutations 140
Chromosome Mutations 140
Population Genetics 141
BOX 6.2 SICKLE-CELL ANEMIA 144
BOX 6.3 THE ATTEMPT TO RECREATE
  LEONARDO DA VINCI 146
Genetic Engineering 147
BOX 6.4 CRIPPLING A MICROBE 148
**Tampering with Life—The
  Recombinant DNA Controversy** 149
BOX 6.5 CLONING INSULIN 150

7 Reproduction 153
Asexual Reproduction 153
Vegetative Fission 153
Budding 154
Sporulation and Vegetative Fission 154
Alternation of Generations 155
The Development of Alternating
  Generations 156
The Advantages of Sexual
  Reproduction 158
**Fertilization in Seed-Bearing
  Plants** 159
Plant Development 161
Animal Fertilization 165
The Menstrual Cycle 171
Conception 174
Contraception 174
BOX 7.1 THE THEORY OF
  RECAPITULATION 175
BOX 7.2 WHY YOUR CHILDREN WILL
  NOT LOOK EXACTLY ALIKE 176
Why Not Conceive? 177
Methods of Contraception 177
Historical Methods 177
The Rhythm Method 178
Coitus Interruptus 179
The Condom 179
The Diaphragm 179
The Cervical Cap 180
Spermicides 180
The Intrauterine Device (IUD) 180
The Birth-Control Pill 180
Sterilization 181
Abortion 183
New Developments in
  Contraception 184

8 Development 187
Types of Eggs 187
Early Development in the Frog 187
The First Cell Divisions 188
Gastrulation 188
Regulation 191
The Flexible Fate 191
Sequencing 192
Stress 192
Induction 193
**Developing Birds and Their
  Membranes** 194
Human Development 196
BOX 8.1 PREGNANCY TESTS 197
BOX 8.2 PRENATAL INFLUENCES 197
The First Trimester 198
The First Month 198
The Second Month 199
The Third Month 201
The Second Trimester 201
The Fourth Month 201
The Fifth Month 201
The Sixth Month 202
The Third Trimester 202
Photo Essay: Human Development 204
Birth 213
Miscarriage 214
Why Have Children? 215

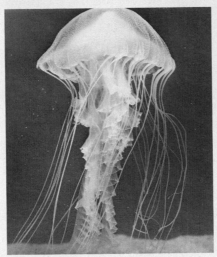

**9 Homeostasis: Controlling the
Internal Environment**   219
Homeostasis   **219**
**Temperature Regulation as an
Example of Homeostasis**   **220**
Cold-Blooded Animals   222
Warm-Blooded Animals   223
The Excretory System   **225**
Solutions to the Water Problem   226
BOX 9.1 SPECIAL SOLUTIONS TO THE
WATER PROBLEM   227
The Human Kidney   230
BOX 9.2 PRESSURE GRADIENTS IN
THE KIDNEY TUBULE   232

10 Systems and Their Control   235
Supportive Systems   **235**
Types of Skeletal Systems   236
Types of Connective Tissue   237
The Human Skeleton   239
Types of Joints   240
Contractile Systems   **241**
Types of Muscle   241
Skeletal Muscles   243
Antagonistic Muscles   *244*
Types of Movement   *244*
Muscle Contraction   244
Contractile Systems in Other
Animals   247
External Respiration   **249**
Circulatory Systems   **252**
BOX 10.1 CLASSIFICATION   254
Types of Circulatory Systems   255
BOX 10.2 BLOOD   256
The Heart   257
BOX 10.3 HEART ATTACK   261
The Lymphatic System   263
Digestive Systems   **263**
Digestive Arrangements   263
The Human Digestive System   267
Digestion in Humans   *270*

11 Hormones and Nerves   273
Hormonal Regulation   **273**
The Human Endocrine System   **275**
How Hormones Work   277
Feedback Systems   278
**The Relationship of Hormonal
and Neural Control**   **279**
Plant Hormones   **279**
Auxins   279
Gibberellin   280
Cytokinins   281
The Nervous System   **281**
The Evolution of the Central
Nervous System   281
The Vertebrate Brain   283
BOX 11.1 THE OCTOPUS   285
The Neuron   **288**
Impulse Pathways   290
The Mechanism of the Impulse   290
BOX 11.2 BOTULISM   293
The Synapse   294
The Human Nervous System   **296**
The Spinal Cord and the
Reflex Arc   296
The Brain   299
The Medulla   *299*
The Cerebellum and Pons   *299*
*The Thalamus and
Reticular System*   *299*
The Hypothalamus   *301*
The Cerebrum   *302*
Hemispheres and Lobes   *302*
The Peripheral Nervous System   305
The Autonomic Nervous System   306
Autonomic Learning   308
Penfield's Mapping   **309**

12 Behavior 313
Ethology and
  Comparative Psychology 313
Instinct 314
Fixed Action Patterns
  and Orientation 314
BOX 12.1 HISTORY OF THE
  INSTINCT IDEA 315
Appetitive and
  Consummatory Behavior 320
Releasers 320
Action-Specific Energy and
  Vacuum Behavior 321
The Behavioral Hierarchy 321
Learning 323
Habituation 323
Classical Conditioning 324
Operant Conditioning 325
How Instinct and Learning
  Can Interact 326
BOX 12.2 REWARD
  AND REINFORCEMENT 327
Rhythms 329
Orientation 331
Photo Essay: The Variety of Life 332
Communication 342
Visual Communication 342
Sound Communication 343

Chemical Communication 345
Why Communicate? 347
Species Recognition 347
Population Recognition 348
Individual Recognition 348
Aggression 349
BOX 12.3 SEXUAL DIMORPHISM 350
Fighting 352
Is Aggression Instinctive? 355
Human Aggression 355
Cooperation 356
Altruism 358
Mindbending 360
BOX 12.4 THE SOCIOBIOLOGY
  ARGUMENT 361
Tobacco 362
Caffeine 363
Marijuana and Hashish 363
Alcohol 365
BOX 12.5 THE HANGOVER 366
Opiates 367
Cocaine 367
Amphetamines 369
Barbiturates 369
Psychedelics 369

13 Communities and Competition 373
Habitat and Niche 373
The Land Environment 374
Biomes 376
Ecosystems and Communities 378
The Water Environment 379
Freshwater Bodies 379
The Oceans 379
BOX 13.1 A DEAD LAKE 381
Photo Essay: The Water
  Environment 384
Photo Essay: The World's Biomes 388
Coastal Areas 398
Island Life 400
Physical Factors 400
Biotic Factors 401
BOX 13.2 GRASS ON THE
  NORTH SLOPE 402
Lessons from Woodpeckers 403
Competition 404
Adaptive Radiation 405
Speciation 407
BOX 13.3 STEREOTYPE
  AND OPPORTUNISM 408
Speciation on Continents 409
Oceans as Islands 409
BOX 13.4 GAINING AN
  ISLAND TOEHOLD 410
The Web of Life 411
Extinction and Us 413
BOX 13.5 THE REDWOODS 414
BOX 13.6 SUBDUE THE EARTH 416

Photo: Life Nature Library *Animal Behavior.*
Photograph by Nina Leen. © 1975 Time, Inc.

14 Populations 419
Population Changes 419
Population Growth 420
Carrying Capacity 421
Reproduction Rate 422
How Many Offspring
  Are Produced? 422
Tapeworms 423
Chimpanzees 423
Birds 426
Humans 428
**Controlling Populations
  Through Mortality** 431
Abiotic Control 432
Biotic Control 433
Predation 434
BOX 14.1 THE GREENPEACE EFFORT 435
Parasitism 436
Competition 438
Disease 439
Death 439
Human Populations 441
Early Man 441
The Energy Pyramid 444

Growth Rate 445
BOX 14.2 STOPPING POPULATION
  GROWTH—THE MATHEMATICAL
  DILEMMA 447
Population Changes Since
  the 1700s 448
BOX 14.3 BIRTH RATES AND
  DEATH RATES 449
After 1850 450
Age Profiles 450
Fertility Rate 451
The Other Effects of ZPG 452
War 454
Intergroup Aggressiveness
  and Commodities 454
War's Effect on Numbers 456
The Evolutionary Implications
  of War 456
BOX 14.4 OUT-GROUP AGGRESSION 457
Commodities and Warfare 458
A Recent Lesson 459
A More Recent Lesson 459
BOX 14.5 DISARMAMENT 462
The Proximal Causes of War 463
The Prospects 463
BOX 14.6 VALUES 465
**New Approaches to the
  Oldest Problems** 466

15 Resources and Energy 469
Food 469
The Distribution of Food 470
Increasing the Production of Food 472
BOX 15.1 HOW HUNGER KILLS 473
Miracle Crops 474
BOX 15.2 THE ASWAN HIGH DAM 475
Domestic Animals as Food 476
BOX 15.3 CLEAR-CUTTING 477
Fishing 478
BOX 15.4 TRIAGE 480
Other Ideas 481
Water 482
Nonrenewable Resources 482
BOX 15.5 DESALINATION 485
BOX 15.6 STRIP MINING 487
BOX 15.7 SHALE OIL 488
Energy 490
Water, Wind, Earth, and Sun 491
Fossil Fuels 492
The Energy of the Atom 494

16 Bioethics and Environment **497**
Pollution **497**
Air Pollution 498
Carbon Monoxide *499*
Nitrogen Oxides *499*
Sulfur Oxides *500*
Hydrocarbons *500*
Particulate Matter *500*
Temperature Inversions *500*
BOX 16.1 THE EFFECTS OF
  AIR POLLUTION 501
The Prospects *503*
BOX 16.2 HIGHWAYS 504
Water Pollution 505
Sewage *505*
BOX 16.3 SEWAGE TREATMENT 506
Chemical Effluents *507*
Heat Pollution *508*
BOX 16.4 A "DEAD SEA" MOVES
  ON NEW YORK 509
BOX 16.5 SPILLED OIL 510
Enforcement *511*
Pesticide Pollution 512

BOX 16.6 PEST CONTROL 513
BOX 16.7 INSECTICIDES 516
BOX 16.8 INTEGRATED CONTROL
  OF INSECT PESTS 516
Biodegradable Pesticides 517
Radiation 517
BOX 16.9 THE BROWN'S
  FERRY INCIDENT 518
BOX 16.10 "THE KYSHTYM
  DISASTER" 518
Nuclear Power Plants *519*
BOX 16.11 RADIATION 521
Heavy Metal Pollution 522
Noise Pollution 523
Hidden Decisions **526**
The Future **528**
Appendix A
The Classification of Organisms 533
Appendix B
Geologic Timetable 543
Appendix C
Evolution 545
Appendix D
Metric Measurements 547
Appendix E
Temperature Conversion Chart 548
Glossary 549
Suggested Readings 561
Photo Acknowledgments 565
Index 571

Preface

I really don't think they were crouching. I think it's just that the windows were high. After all, why would they crouch? It probably only seemed that way, because they held a particular air of mystery for me on those early mornings—and perhaps partly because I could never see them very well, at least not without being obvious. And then there were those rumors about the Mafia living out there! But, the few times I was able to glimpse them, they seemed innocuous enough—reading the paper or just sitting there, in their limousines, resolutely looking ahead, never noticing the other cars, or the people, or the land, on either side.

I often wondered about those people in their sumptuous automobiles as I slowly made my way from my house on Atlantic Beach into the city where I was a visiting professor at Queens College. I suppose my fascination with what I imagined to be the rich or powerful belied my middle-class upbringing, but crawling along on a rough road between the marshy fields on those cold mornings one might search for anything to occupy his mind.

Another thought that came to me from time to time was, why do they keep filling in these marshes? Great machines stood scattered over the land on either side, squat and hard-looking, or standing with their long necks disappearing into the early morning fog. As a biologist I was aware of the importance of the marshes in the ocean's food chain and I resented their destruction.

It dawned on me one morning that no one else seemed to pay any attention to the building activities. Least of all, not those reclining with a kind of heavy presence in their chauffered limousines. Their windows were up and they were removed from the carnage. Whatever was going on out there was of no interest to them. Not them. And then I realized that, in a sense, they *were* being insulted. All of us were. But, of course, *I* was the only one who knew it.

I'm sure everyone on the road on those mornings had his own problems. We were all going to work and had things to do, and who had the time or inclination to worry about marshes? The people in the limousines, it seemed to me, were the most removed from it all. Their car temperature was regulated and they weren't even involved in the physics of dodging chuckholes. So how could those ungainly machines constitute an insult to them?

Being stuck in traffic, even before coffee, breeds philosophy, and it seemed to me that I could see a larger question here. The incessantly irritated and impatient commuters around me had taken no offense at the construction because they simply didn't realize that something more than land was being exploited. But those marshes were part of a critical and fast-disappearing link in a life-chain—and it was a life-chain in which we were all unalterably a part. Rolling up windows, or failing to see, changed nothing. What was being destroyed out there was an important part of our ecosystem—our environment, and, in a sense, part of us.

It seemed to me then, and it seems now, that the most important lesson we can learn is that we are a *part* of the world out there. It is not "our home," nor can we think in terms of its being "them or us." Any effort to "have dominion" over nature can only be self-defeating since we are a *part* of nature. We may be a highly intelligent part, but our newly developed abilities cannot negate our heritage. Rather, what we must do, if we are wise as well as intelligent, is to use our abilities to ensure that the rest of the natural world is sustained. Our reward will be our own sustenance.

And this will be the point of this book—to help to show that while we are a special case in many ways, our species is but one of many in a complex and mysterious world, and that our own best interest lies in coming to understand that world and to live in harmony with it.

About the Book

The format of this book, as you can see, is traditional. We go from the history of biology to the establishment of an elementary molecular background, which will be used primarily in understanding energy and the mechanisms of DNA activity; then from cellular, through systemic and organismic biology, ending with discussions of behavior, populations, and ecology. The underlying theme, or the unifying thread, if you will, involves the twin concepts of evolution and adaptation: how each life form has come to fit into its particular part of the world.

Adherence to tradition, of course, does not justify the appearance of yet another book in an already crowded field, which a few truly excellent books dominate. So what particular or unusual features does this book have to offer to justify its existence? First, and I believe this is important, I think that students might enjoy reading it. I had fun writing it and I'm counting on some of that same spirit to be absorbed by the reader. Life is fascinating, and biology is the study of life, but many of the biology textbooks I have encountered in my years of teaching have somehow managed to be dull. Above all, I hope this book is not dull.

Perhaps a faint sense of honesty, but probably more of what

I consider to be a literary flair, compels me to say something else: this book is biased. Throughout the text I have expressed my own opinion on a number of topics. This was done partly to get it off

my chest, but primarily as a teaching technique—a means of stimulating discussion in these critical times. Hopefully, the opinion is clearly distinguishable from what we like to call "fact," so that the student will be stimulated and not misled at such points. My own training is in behavior and ecology so perhaps the major part of my opinion is interjected in related discussions (if one may consider reproductive and population biology as areas of ecology). Anyway, I'm sure the professor who uses this book will have his or her own biases and will be quick to point out mine to the student.

Primarily, the book is written for students who have little or no training (or perhaps interest) in the sciences. I assume most of the readers will be nonmajors, although several reviewers have suggested the book could be used as the framework for more rigorous courses with the material expanded by the instructor. It was my intention (1) to describe the material as simply as possible, (2) to interject biological viewpoint into questions of philosophy and values—questions to which the student can relate in a very personal sense, and, (3) in the event that some students are turned on by the material and should wish to pursue biology further, to present them with a good, solid background in biology upon which they can base future work.

Acknowledgments
I am grateful to a number of people who helped in developing this book. First I would like to thank the scores of reviewers from a number of biology departments across the country. Their comments were invaluable in the development of content and direction. Secondly, I would like to say it has again been a great pleasure to work with Goodyear Publishing Company. I have greatly enjoyed the camaraderie and professionalism of the group. I am particularly grateful for the encouragement and assistance of Al Goodyear and Jim Levy and for the careful concern, advice, and assistance of Gerald Rafferty, Susan Smith, Don McQuiston, Gene Schwartz, Bob Hollander, Barbara Hodder, Debra McQuiston, John Dawson, and Dennis Jeffrey. And again I find myself in the position of thanking my editor, Clay J. Stratton, a man for all seasons. I greatly appreciate the help of Gerald Sanders and Vernon Avila at San Diego State University, Jack L. King at The University of California at Santa Barbara, and Steven Vogel, Peter Klopfer, and Lee McGeorge at Duke University for counsel during the later stages. I am grateful for the kind welcome of the faculty at Duke and the many fruitful discussions with the graduate students here who often don't realize that they're supposed to take my word for it. Finally, and most of all, I would like to thank Bonni. Flower didn't help.

Biology Yesterday and Today

The door closed quietly. Someone had left the room. Behind, a terrified child, without benefit of anesthesia, lay on the operating table as surgeons huddled and prepared to make the initial incision. A group of beginning medical students watched intently from just outside the bright circle of overhead light. As the operation progressed the eyes of one sixteen-year-old faltered. Finally, no longer able to watch, Charles Darwin left the room.

He had failed again, this time at medicine. What was he to do, this younger son of an English gentleman? Perhaps he would be better suited for the clergy, which was, after all, an admirable calling for one of his social position. So the following year his father enrolled him at Cambridge to study theology—with equally disastrous results. Although young Charles spent three years as a theology student, he seemed to spend more time with beetles than beatitudes. He was not alone in his interests, however. At that time the English countryside was alive with amateur butterfly collectors, rockhounds, and plant fanciers whose position and wealth permitted them to indulge in such hobbies. Even so, Darwin's academic prowess had been so thoroughly unremarkable that at one point his father had told his trifling son, "You care for nothing but shooting, dogs, and rat-catching, and you will be a disgrace to yourself and all your family."

THE VOYAGE OF THE BEAGLE

It must be admitted that at this point in Charles Darwin's career there was little to suggest a mind that was to be regarded as one of the most brilliant and inquisitive in history. In a short time, however, this diffident young man was to hear, through his friend the Reverend John Henslow, of an offer of free passage on a survey ship called the *Beagle*. A naturalist was needed for a voyage that was to last five years. There would be no pay, and the person chosen would have to sleep in a hammock in the cramped chartroom (although he would be permitted to share the captain's table). Armed with Henslow's recommendation, Darwin eagerly applied for the position, and was nearly rejected because of the shape of his nose. Captain Fitzroy, himself only twenty-three, believed that the nose reflected the character of its bearer, and Darwin's nose just didn't show much character.

Darwin's family required some persuasion to accept this "madcap scheme," which they considered scarcely in keeping for a prospective

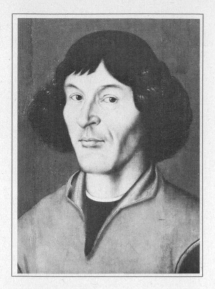

1.1 Copernicus in 1543 proclaimed that the sun, not the earth, was at the center of the solar system. This was a shocking notion for authoritarian and religious society of the time. Some of his ideas were erroneous (he miscalculated the cycles of Mercury and Venus and implied the sun was at the center of the universe) but his basic contributions were great. Although he empirically "proved" his system, he died without his ideas ever being accepted.

member of the clergy. And evidently young Charles had his own trepidations about such a momentous decision. In a letter to his sister Susan he wrote:

. . . Fitzroy says the stormy sea is exaggerated; that if I do not choose to remain with them, I can at any time get home to England; that and that if I like, I shall be left in some healthy, safe and nice country; that I shall always have assistance; that he has many books, all instruments, guns, at my service. . . . There is indeed a tide in the affairs of men, and I have experienced it. Dearest Susan, Goodbye.

Ultimately all the arrangements were made, and in 1831 the H.M.S. *Beagle* set out from Devonport, with Charles Darwin, now the ship's naturalist, gazing, perhaps a bit apprehensively, at the slowly retreating shoreline of his homeland. As the heavy wooden vessel, about the size of a tugboat, creaked and groaned its way across the Atlantic toward South America, young Darwin's worst fears were realized. Shipboard life was tougher than he expected, but worse yet, Darwin tended to get seasick! He made the best of it and the boat continued relentlessly, captained by tough young Fitzroy, perhaps one of the world's best navigators. At long last the *Beagle* reached South America and then headed down the coast on the first leg of its voyage—past the coasts of Brazil and Argentina, weaving through the terrible pounding gales of Cape Horn, to finally turn northward along the desolate coasts of Chile and Peru. Fortunately there were periods of respite when the *Beagle* dropped anchor to put foraging parties ashore. Darwin wasted no time getting to land. But after resting a bit and regaining his land legs he was irresistably drawn deeper into these new places, places which harbored all manner of new and fascinating things. He took copious notes on everything he saw and, following his natural urges, he brought back to the ship all manner of things to the amusement and the occasional dismay of the crew. Since the ship sometimes remained anchored for months he ventured far inland into the wild South American terrain. Darwin, an excellent horseman himself, soon developed the greatest respect for the horsemanship of the rugged gauchos who often accompanied him.

To understand just what this adventure meant to Charles Darwin, and would later mean to the world, we must stop at this point to consider the state of scientific thinking at that time. What was known of the biological world? What prejudices or beliefs did Darwin have? More important, what seeds of ideas? What hunches?

THE HISTORY OF AN IDEA

New scientific ideas have appeared throughout man's history. Many of these ideas have been erroneous, of course, but they often provided at least a framework for the expansion of our knowledge. Nevertheless, new ideas—and even new facts—have not always been met with enthusiasm. In the best of times, when an assumption did not fit what was known about the real world, the assumption was discarded. The notions that remained, then, were likely to be based to some extent on the available facts, and as new information came to light, our body of scientific knowledge expanded. It would be satisfying, somehow, to say that scientific knowledge progressed stepwise from ancient times until the present, knowledge and understanding accumulating all the while until we reached the modern crescendo of scientific expansion, all to the good of man. Alas, this is not the case.

As far back as the early civilizations of Babylonia, Egypt, and Greece, scholars were trying to figure out the nature of the world in which they lived. Some of these early scientists were looking for some unifying concept, some thread of continuity. However, after the fall of these ancient civilizations, scientific curiosity fell into a steady decline. For a period of more than ten centuries, between 200 and 1200 A.D., there were virtually no scientific advances at all. In fact, much of what had been known was forgotten. Although the Greek mathematician Eratosthenes had calculated the circumference of the earth to within fifty miles more than 200 years before the birth of Christ, in 1492 Columbus had trouble convincing anyone except a few Moorish astronomers that the world was round!

During these centuries the writings of a few ancient scholars were preserved in dusty monasteries, and some of the ideas they contained were dutifully mulled over and transcribed from one parchment to another. But instead of active investigation to verify statements or ideas, men turned to religious authority for absolute answers, seeking a kernel of order in a world that was beyond their understanding. As a result, science and religion became hopelessly intertwined, so that "scientific" statements that were otherwise testable became religious doctrine and therefore not to be questioned. Thus, by a twist of fate and politics, the church became on one hand the seat of higher learning and on the other hand a formidable opponent of new ideas. The Polish astronomer Copernicus (1473–1543), who voiced the heretical theory that the earth was not the center of the universe, escaped retribution by dying shortly after his work was published. However, when the great Galileo (1564–1642) produced detailed evidence that the earth did indeed revolve around the sun, his writings were banned in Rome and he was summoned before the Inquisition and forced to recant his belief in the Copernican theory.

Nevertheless, as the world became more predictable, and men began to feel less at the mercy of magical forces and unseen beings, there was a great surge of scientific thinking. In fact, the very year Galileo died marked the birth of one of the greatest scientists of all time—Isaac Newton (1642–1727). By the age of twenty-four Newton had already formulated the idea of universal gravitation, and in 1685 he presented a set of carefully proven laws of motion that were to revolutionize the physical sciences.

The Beginnings of Biology
Until the eighteenth century, science was limited for the most part to topics dealing with inanimate matter—mathematics, astronomy, physics. It was

(*Text continues on page 13.*)

1.2 Galileo Galilei embraced the findings of Copernicus, which stated that the sun stands at the center of the universe with all other bodies moving around it. But this implied, at least to religious minds, that man did not stand at the center of all creation. The old scholar of 69 was forced to recant under dire threats of the Church. He knelt before ten cardinals, robed in their scarlet, and placed the earth back in its immovable position before all the world. It is said that as he arose from his knees he muttered, "Eppur si Muove"—"It moves nevertheless."

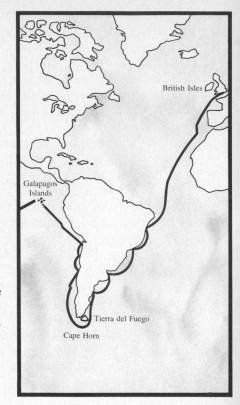

the voyage as a Lamarckian and "save for a few grave doubts," his views, he said, were perfectly orthodox. They were, however, a bit radical for the conservative Captain Fitzroy who, after arguing endlessly with Darwin, concluded the young scientist was something of a heretic.

Darwin's general discontent with Lamarckian explanation was to be shattered the day in 1835 that the *Beagle* sent its anchor clattering downward at Chatham Island in the Galapagos. The jutting lava cliffs suggested that this was a younger world than that known by English scientists and theologians. Yet it abounded with the most peculiar of life forms. Were these also young? Then what was one to think of the story of creation? The newness of everything was to take its toll. Darwin wrote Henslow, "Oh, the degree to which I long to be once again living quietly with not one single novel object near me." To another friend he wrote, "I dare hardly look forward to the future, for I do not know what will become of me."

Darwin and the Voyage of the Beagle

Charles Darwin became a disciple of Sir Charles Lyell while reading Lyell's *Principle of Geology* aboard the *Beagle*. Lyell saw the world as a slowly changing place, not greatly affected by the violent upheavals that were assumed to have wrenched new life forms from the ancient clay. He wrote, "We hear of sudden violent revolutions. I shall take a different course. We are not authorized in the infancy of our Science to recur to extraordinary agents." Darwin was deeply impressed and wrote his cousin, "I am become a jealous disciple. I am tempted to carry parts to a greater extent even than he does." And thus he did.

Whereas Lyell rejected Lamarck, Darwin replaced him. Darwin had apparently begun

□ Through most of the trip Darwin was to be racked with seasickness. The frequent stops along the eastern coast of South America and up its western coast caused the fog-bound Britisher to write ecstatically about the fresh, clear air and dizzying heights. The journey to the Galapagos was highlighted by one of sailing's greatest challenges: passing westward around the Cape at Tierra del Fuego. The sail was distinctly anticlimactic after the Galapagos and Darwin, never a sailor, wrote, "I loathe, I abhor the sea."

□ The *Beagle* was a lumbering square-rigger with high freeboard, as were most of the ships of the time. Such ships are sturdy and remarkably roomy, but the required structural bracing renders them somewhat bulky and very difficult to manage. The *Beagle* was, in its way, a monument to the nautical abilities of young Captain Fitzroy.

□ The *Beagle* at anchor in the Murray Narrows of what is now called the Beagle Channel. The boat was often greeted by friendly peoples such as those of Tierra del Fuego—a peculiar group who had managed to survive until they were decimated by the advantages of civilization.

□ Ona Indians of Tierra del Fuego. These photographs were taken about 1880 by some of the earliest white explorers. The Indians were truly remarkable, often sleeping naked in driving sleet, or going about carrying babies shielded from the elements only by a piece of hardened leather. Their main food was a wild camellike animal called the guanaco as well as shellfish, berries, and fungi. Darwin brought three of these people back, two men and a young girl. Their story is recounted in *Voyage of the Beagle*, a popular account written by Darwin.

☐ Darwin was most unimpressed with the stark and desolate terrain of the Galapagos Islands. Coves such as the above were bountiful, however, so Fitzroy was not reluctant to stay long enough to give Darwin time to take a close look. Darwin wanted to stay even longer, but finally Fitzroy pressed on.

☐ The diversity of life in the Galapagos must be partly a result of the great shielded tide pools such as this one on the northwest shore of Isabella Island (right). The pools are harbors for animals that would be less successful on the surging waves surrounding the islands. There are, in addition, fresh-water pools formed from ''lenses'' sitting atop the heavier salt water of the porous island soil of some of the islands.

□ A variety of habitats are clearly shown in this scene of Bartholomew Bay (left). Sandy, flattened beaches give way before dense lowland vegetation that is not able to succeed in the higher areas. In these places the vegetation is spotty if it exists at all. Discrete ''pockets'' such as these contribute to the development of plants and animals specialized for each type and hence tends to hasten evolutionary processes.

□ Land iguana, *Conolothus subcristatus* (above). It is only distantly related to the marine iguana, being more solidly built with a shorter tail and crest. It eats green vegetation, mainly cactus fruit, munching happily away on cactus spines. The land iguana sleeps in shallow burrows and during the day it scurries back to these shelters if threatened. Sadly, introduced animals are outcompeting this relic and humans continue to hunt it for its meat and skin.

□ Marine iguana, *Amblyrhynchus cristatus*, a four-foot reptile that basks in large groups on the blackened lava (left). Thousands can be seen on Narborough at one time. Darwin called them ''imps of darkness,'' as he watched them sunning on rocky beaches after a night wedged in a lava crevice. When they are hunting they drop into the water, their limbs flattened against their body, propelled by their undulating tail. They graze on the abundant seaweed that sways beneath the surging water. At the sight of a shark they scramble onto the land by pulling themselves up with long, curved claws.

☐ The flightless cormorant, *Nannopterum harrisi* (above), whose population is less than 1,000 (notice the legband). Like many island species of birds and insects, it has completely lost the ability to fly, its wings reduced to useless appendages. Theoretically, the loss of flying ability is an adaptation that decreases the chance that the animal will be blown away while flying. Although this is an endangered species, bodies of these birds are frequently recovered from fishermen's traps and nets.

☐ A Galapagos penguin, *Spheniscus mendiculus* (right), in front of mangrove trees in the littoral zone of an island. Mangroves are peculiar, their stilted roots giving rise to very shallow forests from which one can often see the island's interior as well as the sea's horizon.

☐ One of the thirteen species of Galapagos tortoise, *Testudo elephantopus* (bottom left). The species are so distinct that an experienced observer could tell which island any specimen came from. The arched shell identifies this animal as one that stretches to reach the fruit of *Opuntia* and shrubby plants. This species prefers high, cool areas with dense vegetation.

☐ The giant tree cactus, *Opuntia* (bottom right), is a close relative of the prickly pear of the United States and is particularly abundant along coastal deserts, such as this area on Tower Island. They are often over thirty feet tall and are virtually the sole source of food and water for some populations of tortoises.

□ The Galapagos tortoise, *Geochelone elephantopus* (above). Its lower shell reveals it as a ground forager, unable to reach very high. There are three to four thousand of this particular subspecies left. In years past hundreds of thousands of the hardy tortoises were taken by sailors to await a slow death aboard ship. They were often stacked upside down and left without food or water for up to a year when they were killed for food.

□ While the Galapagos lowlands are arid places, the middle and higher elevations are cooler, wetter, and abound with green vegetation (left). In periods of drought the plants survive by utilizing the moisture of fog. The soil is very fertile, formed by disintegrating lava, and the undergrowth is lush with ferns and shrubs. Epiphytes (plants that grow on other plants) are abundant in these higher areas, an example being the air ferns shown here.

□ A feral goat, a wild descendant of imported herds. Goats have replaced the tortoises as the prime decimator of *Opuntia* during the dry season, sometimes cutting the plant down in order to get at the fruit. (Wild cattle and donkeys also eat the prickly fruit.) The goats were introduced to provide sailors with a food supply. The idea worked well until English buccaneers began to avail themselves of the food supply. Then the Viceroy of Peru decided to eliminate the pirate's food, so he introduced dogs to the islands. The dogs, of course, couldn't catch the goats on the rough lava beds so they turned to other food sources and continue to threaten many defenseless species on the islands. They are particularly fond of young tortoises and eggs as well as young flamingos and both forms of iguanas. Another threat to the tortoises are introduced pigs, which eat young tortoises and eggs as well as ground-nesting birds and sea turtle eggs. Tamer, naturally-occurring rats are dwindling before the more vicious introduced rats, which attack young tortoises and eggs of tortoises and birds.

□ A view of the incredibly rough surface of Isabella Island (opposite page, top). Such terrain led one internationally known scientist to give his impression of each of the islands as "a very beautiful and different hell."

□ The woodpecker finch (bottom left), probing with a twig into a hole, per chance to chase out an insect. This is a remarkable adaptation in which a behavioral pattern clearly replaces a morphological structure. No one knows how the behavior develops but perhaps the birds are born with a tendency to pick up and handle bits of twigs, later learning to forage with them by watching others.

□ Sally lightfoot crabs, *Grapsus grapsus* (bottom right), cling tenaciously to surf-pounded rocks. They hide in the dark recesses of the lava. Occasionally all the crabs scurry from their crevices at once, producing a startling, rasping surge of red contrasted against the dull, black rock.

☐ Blue-footed boobies, *Sula nebouxii* (above), share the islands with their cousins, the red-footed boobies. Boobies may nest deep in volcanic craters, the young unable to fly out until they are three months old. One might wonder about their reaction to their first glimpse of the incredible world awaiting them outside the protection of the crater.

☐ The Galapagos flamingo, *Phoenicopterus* (top right), is probably the most endangered bird in the islands. It is extremely shy and will quickly desert its mud nest if disturbed. There are very few breeding colonies, and people, dogs, and pigs often destroy the eggs and young of nesting birds. A recent census turned up only 317 birds on five islands. Flamingos not only wade, but are excellent swimmers. When swimming, they tuck their long legs under them and paddle only with their feet, resembling great swans with hooked noses.

☐ A displaying male frigate bird (bottom right). The great air sacs are so easily distended that a male may rest his chin on it and doze. Note the great wing span of this species. When disturbed by humans they do not usually defend themselves but simply turn and face the other direction. They are superb fliers, however, and when they launch themselves they seem simply to rise as if by levitation.

one thing to apply physical laws to salts, stones, or stars, but quite another to consider *life* in those terms. After all, man was a living thing, and therefore life must surely have some special *purpose,* some grand design. We just weren't ready to view ourselves as merely another physical phenomenon.

In the 1700s it was universally believed that all species, or "kinds," of living things were created in their present form—in other words, that they had not changed in any way over the millennia. This was certainly the view of the Swedish botanist Carl von Linné (1707–1778), who devised a system of classification for all living organisms and, in his fondness for Latin, even called himself Carolus Linnaeus. Many argued that Linnaeus' designations, which involved lumping the species together on the basis of their similarities, were artificial and based on personal whim. Few, however, considered criticizing his assumption of special creation or his concept that each species was fixed in its original form, never to change.

One small departure was suggested by a French contemporary, George-Louis Leclerc De Buffon (1707–1788). In 1753 Buffon proposed that in addition to those animals that had originated in the Creation, there were also lesser families "conceived by Nature and produced by Time." He explained that changes of this kind were the result of imperfections in the Creator's expression of the ideal—a philosophical, rather than a biological, point of view.

A decade later, however, another doubter appeared on the scene in England—Erasmus Darwin (1731–1802), the grandfather of Charles. Erasmus was a peculiar man—a physician and an amateur naturalist who wrote on the subjects of botany and zoology, often in rhyme. He referred almost incidentally to the relationships among animals, the importance of competition in the formation of species, the effect of environment on changes in species, and the heritability of these changes. There has, of course, been much speculation on the extent to which young Charles may have been influenced by Erasmus. The effect may not have been as great as one might imagine because Charles apparently had little respect for the musings of his grandfather.

Other men were also beginning to toy with the notion of the heritability of change. In France Jean Baptiste de Lamarck (1744–1829), a protégé of Buffon's, boldly suggested that not only had one species given rise to another, but man himself had arisen from other species. A passionate classifier, Lamarck believed that every organism had a relative position on the "scale of nature"—with man, of course, as the highest form of life. He pointed out, however, that the fossil animals found in older layers of rock were simpler forms and contended that they had become higher animals through a gradual progression. In his view there was a "force of life" that caused an organism to generate new structures and organs to meet its biological needs. Such structures then continued to develop through use, with the change acquired in each generation transmitted to succeeding generations. Lamarck cited as an example the

1.3 Isaac Newton, one of the truly important figures in the history of science. Newtonian physics was to dominate the world of the physical sciences until the 20th century. It was Newton who, in 1687, provided final and irrefutable evidence for the Copernican doctrine of the solar system.

long neck of the giraffe, which he maintained had evolved as each generation of giraffes had stretched their necks in an effort to reach the topmost branches of trees and then transmitted genes for this longer neck to their offspring.

At the time, Lamarck's arguments did little to persuade his lecture audiences. With the foment of the French Revolution, there was some lively discussion in intellectual circles, but in society at large the firm conviction that each form of life had arisen through special creation ruled out serious consideration of any other concept.

Time and the Intellectual Milieu

The intellectual climate into which Darwin was born was far more conservative than that in France. The English had been horrified by the French Revolution, and ideas such as those held by the "French atheists" were either dismissed out of hand or viewed with extreme suspicion. Partly for this reason, the church continued to hold strong sway over the sciences, and biologists in England continued to adhere rigidly to the traditional tenets.

Even so, men such as Erasmus Darwin had begun to consider the idea of heritability of change, at least from a philosophical standpoint. It has been suggested that the similarity between these ideas and the concepts formulated later by Charles Darwin merely reflect an inescapable convergence of thinking, given the state of knowledge and a changing intellectual milieu. In other words, perhaps the time was right.

It must be stressed, however, that such views were by no means universal throughout the scientific community. Half a century later, after Darwin had published his *Origin of Species,* he observed:

It has sometimes been said that the success of the "Origin" proved "that the subject was in the air," or "that men's minds were prepared for it." I do not think that this is strictly true, for I occasionally sounded out a few naturalists, and never happened to come across a single one who seemed to doubt about the permanence of species. . . . I tried once or twice to explain to able men what I meant by Natural Selection, but signally failed.

DARWIN'S THEORY OF EVOLUTION

When Charles Darwin set out on his voyage aboard the *Beagle,* it was with the same conviction shared by others of his time that life had originated through special creation and all species were fixed in form. The physical sciences, however, were less hampered by religious tenets, and there were signs of stirring in other diciplines. Charles Lyell (1797–1875), who was only a few years older than Darwin, had published the first volume of his work *Principles of Geology* before the *Beagle* set sail, and Darwin had asked to have the second volume sent to him en route. Lyell had rejected the Biblical thesis that the earth had been created in the year 4004 B.C. and contended that it was actually much older. On this basis he theorized that the earth's features had not appeared suddenly through catastrophic upheavals, as had been thought, but were instead the result of slow, steady processes over an exceedingly long period of time. Lyell had not extended his theory to any speculation about the origin of living organisms, but Darwin immediately seized on the implications. He had dismissed the idea that all species had sprung from simpler forms. He didn't believe that mud around a pond formed frogs and he wasn't prepared to believe that frogs could suddenly begin to produce snakes. But the dimension of time opened up possibilities for some other kind of evolutionary process. Nevertheless, he had formed only the seed of an idea. Darwin had been thoroughly schooled in beliefs of long tradition, and rejection of those beliefs was not going to be easy.

As Darwin continued his investigation of the plant and animal life and fossil beds of South America, he was struck by the differences in living things from one place to the next. The shells awash on the Pacific shores were not at all like those he had found on the Atlantic beaches. He also noticed differences in the birds and mammals as he traveled from one place to another. In some cases the change was marked. In other cases there were gradual changes in form as one type of terrain gave rise to another. Had each of these forms been created separately, Darwin wondered, each with its own subtle characteristics making it different from the others?

The Galapagos Islands

When the *Beagle* finally left the shores of South America, an impatient Captain Fitzroy set the sails of his sturdy boat for a straight run to the Galapagos, a chain of islands that lie about 580 miles off the coast of Ecuador. Little did he or Darwin realize what the results of this brief visit would be. Fitzroy was surging toward the remodelling of world science, and Darwin was ensuring for himself the life of a beleaguered recluse.

When the anchor was lowered in the shallows of the barren islands, Darwin scrambled ashore as usual, and immediately encountered a rather strange assortment of animals. "A little world in itself," he wrote, "with inhabitants such as are found nowhere else." There were lizards three feet long that grazed beneath the turbulent sea—in Darwin's words, "imps of darkness, black as the porous rocks over which they crawl." And there were the incredible giant tortoises for which the islands are named. Even more remarkable, each island had its own type of tortoise, distinguished by characteristics that enabled local fishermen to tell immediately from which island they came.

Darwin also noticed a group of small birds—now called Galapagos finches, or Darwin's finches—drab little birds but definitely finches. There are thirteen species differing primarily in the size and shape of their beaks. Each species uses its specialized beak differently and picks up its food in a specific way. Darwin surmised that these birds must have come originally from the South American mainland, since the volcanic islands of the Galapagos would have been formed later than the continent. But how could these finches differ so markedly from those on the mainland, and why was it that the species on the islands differed so much from each other?

1.4 The Swedish naturalist, Carolus Linnaeus, the founder of the scientific method of naming living things. He apparently conceived of species as unchangeable entities, the products of divine creation. He lived and worked in the century before Darwin and had no idea of the concept of evolution.

1.5 Beef cattle (like the black Angus, opposite page) were developed over long years of breeding only the heaviest, stockiest animals in the herd. The results are heavy, short-legged animals that are rather slow and that do not possess great endurance. However, they turn grain to beef with much greater efficiency than did their forebear, the longhorn (right).

Today, the cattle of less developed countries more closely resemble our longhorn, the famed symbol of the untamed West. One may wonder why the herds of protein-poor countries are composed of such thin, rangy, and low-yield cattle. One reason is that modern American beef cattle would immediately succumb to the rigors of the environment if they were placed on the stark grazing areas to which many forms of less productive native cattle are adapted. The longhorn, you will recall, was a tough, highly adaptable beast, capable of surviving harsh conditions and long cattle drives.

It seemed almost as if one type of finch had been modified into a number of new types. Darwin was unable to dismiss this thought. But it was another twenty years before he was to resolve it in the context of any comprehensive theory.

The Impact of Malthus

After his return to England Darwin had occasion to read an earlier essay by the Reverend Thomas Malthus (1766–1834), which was probably our first clear warning of the danger of overpopulation. In this essay, which appeared in 1798, Malthus pointed out that populations increased in a geometric progression, and if man continued to reproduce at the same rate, he would inevitably outstrip his food supply and create a teeming world full of misery and vice. Darwin gleaned one important idea from this treatise. The reproductive potential of animals is high, but not all animals survive long enough to reproduce; hence it is this difference in survival among members of populations that has kept those populations in check. Darwin calculated that a pair of slow-breeding elephants, for example, could produce 19 million progeny in only 750 years. Yet through the years the total number of elephants seemed to stay about the same. If some members of a population were unable to survive, what factors determined which among them were to be the successful ones? Darwin could only account for this by the operation of some system of natural selection.

Natural Selection

Darwin had long been familiar with the principles of *artificial selection* from his experiences as the son of a country gentleman. He knew that breeders were able, through careful selection of animals for mating, to accentuate desired characteristics in the offspring. By mating only the

offspring of the greatest milk producers, they could develop high-yield dairy cattle. And by permitting only the offspring of good laying hens to breed, they could eventually produce hens that were veritable egg-laying machines. The results of artificial selection could be seen in only a few generations, since individuals with undesirable qualities were kept from transmitting those traits.

Darwin saw natural selection as an analogous process in which the animals that would reproduce were selected, not by the breeder, but by the environment. The process in this case was far less efficient, since individuals with less-desirable characteristics were often able to produce *some* offspring and thus the traits would be maintained in the population for a time. Those individuals with characteristics *totally* out of keeping with their environment would leave no offspring at all; hence their characteristics would disappear from the population. And since those individuals with characteristics which provided some advantage in reproducing could be expected to outreproduce the rest of the population, they would eventually come to predominate.

Descent with Modification

Since this theory of natural selection depended on inequalities among members of a population, the source of the variation had to be accounted for. Why were the individuals different? Darwin proposed that variations appear randomly—that no driving force, no direction, and no design are necessary. He also assumed that among inherited traits, some were better than others. He came to the conclusion that a heritable trait, if it provides some advantage, would be expected to increase the reproductive output of its bearer. If the long neck of the giraffe is inherited and is helpful in acquiring food, then the giraffes with longer necks will be better nourished, and hence more likely to have the energy to leave offspring.

A. Lamarck's Explanation B. Darwin's Explanation

Some of these offspring will have longer necks than others, and will be more successful than their shorter-necked brothers and sisters, thus long necks would become increasingly common in the population. The long-necked giraffes would continually be favored through the generations, as long as the species specialized in eating leaves of trees. The result would be a tendency for any generation to have longer necks than any generation of their predecessors.

It is important to realize that Darwin had no hard proof for this idea. There was no experimental evidence he could offer, since he knew nothing of the field we now call genetics. He was therefore reluctant to present his theory without being able to explain the mechanism by which the process of natural selection actually worked. However, in 1858, when a young biologist working in the area of Indonesia, Alfred Russell Wallace (1823–1913), outlined a similar but far more sketchy theory of natural selection, he provided the impetus for Darwin to pull together his voluminous notes and prepare a paper to be presented to the Linnaean Society of London. In spite of the overwhelming data he had gathered during the twenty years since his voyage, Darwin asked that Wallace be permitted to present his paper first and receive credit for the idea, rather than have any man think he had behaved in a "paltry spirit." Wallace, just as much a man of honor, declined. The outcome was that they presented their papers together, with Darwin speaking first, in keeping with his much more substantial evidence.

Darwin's carefully formulated ideas were greeted with enthusiasm in some quarters, but needless to say, the response was not universal. He was forced to defend his suggestion of "descent with modification" not only against scientists who demanded hard evidence, but also against the attacks of philosophers, theologians, and a general public who were reluctant to part with the idea that they were members of a unique and chosen species.

Darwin himself was poorly equipped for such a battle. He returned to England quite ill after his five-year odyssey, and he never recovered his health. Plagued by anxiety and self-doubt, he had remained in seclusion after his return, rarely venturing outside his house and avoiding everyone but his family and a few close friends. His defense however, was by no means single-handed. Many of the best minds of the time leaped to the defense of the grand idea. Some of these men were brilliant, hard-nosed and combative, savoring the taste of intellectual battle. Certainly no one in their right mind would want to challenge Thomas Huxley in open debate. But the proponents did not join the fray empty-handed. Here, after all, was a unifying concept, one that made sense of it all. It accounted for the observations. It was not to be rejected on any basis other than a better explanation. And there was none.

We will discuss the principles of evolution throughout this book, but for now let us note that the concept of evolutionary descent is generally accepted by the scientific community on the basis that it accounts for the

1.6 Today, biology students associate the name Lamarck with error. Lamarck (above), we know, was *wrong.* But we must keep in mind that, in Darwin's time, he was a major intellectual force. He was an early evolutionist and, furthermore, he believed that man himself had evolved. However, he also believed that the use of organs would ensure their transmission in strengthened form to the next generation. He was most ridiculed in the popular press for asserting that giraffes, by continually stretching for higher leaves, would tend to have offspring with yet longer necks. In time, he explained, all giraffes would have long necks, as we see in column A at left.

Darwin's theory, on the other hand, suggested that from among any population of giraffes, some are born with longer necks than others and that these would be able to reach more leaves. Better-fed animals would then most likely be able to rear their young, and thus longer necks would come to predominate in subsequent generations (column B). We should probably keep in mind that when Darwin set sail on the *Beagle,* he himself was an adherent to Lamarckian biology, at least in broad principle.

1.7 In the chaos that followed the immediate publicity given Darwin's theory of evolution, the satirists (many choosing to remain essentially uninformed about what Darwin really said) had a field day. This cartoon purports to show the transition of worms into apes and gnomes, culminating in a dapper gentleman paying tribute to Darwin. The Victorian press abounded with such cheap shots that were obviously elicited by a deep-seated reluctance to relinquish the notion of man's "specialness."

PUNCH'S ALMANACK FOR 1882. [December 6, 1881.]

MAN · IS · BVT · A · WORM ·

data. Today the major arguments concern the mechanism of its operation. For example, are new species formed through an accumulation of small changes over a long period of time, or are such changes produced only by major upheavals?

Darwin's theory of natural selection has also given rise to a number of other questions. If the species we see around us are the products of evolutionary change, what are their histories? More important, how are they related to each other? And the question that hangs over it all: where does man fit in? Are we really the cousins of other animals? Can we conceivably be related to *them?*

Mechanism and Vitalism

One of the basic questions centered over where it all began. If one species is descended from another, what did the *earliest* ones look like? Did they have much the general form we see in living species, or have they changed radically from some simple ancestor? In other words, how did life begin?

One idea was that life might have originated by some sort of chance combination of chemicals. If even the *possibility* of such a beginning could be demonstrated, then perhaps the nature of "life" could be defined in terms of simple chemical interactions. According to this view, called *mechanism*, a living thing might conceivably be no more than the sum of its chemical reactions.

Taken to its extreme, this mechanistic view implied that if everything were known about the chemical make-up of the members of any group, their behavior could be predicted under any given circumstances. We could know when any individual would feel hungry or sexy, whom he would favor in a football game, and how he would feel about entering the European Common Market. Such thinking was a bit much for those who believed that living things have certain special qualities that are lacking in nonliving things—qualities that are not reducible to mere chemical reactions. According to their view, called *vitalism,* living things uniquely possess some mysterious sort of "life force."

In the absence of any direct evidence of such a life force, the mechanistic approach at least offered a framework for continuing the search for the possible chemical beginnings of life. The working hypothesis was that life could have arisen from the interactions of nonliving chemicals as a form of "spontaneous generation," in which, at some minute point, matter crossed the dividing line between nonlife and LIFE.

The Beginnings

In order to begin our search for how life may have started, we must visualize a different kind of earth. And since no one can give us a first-hand account, we must rely on a measure of imagination. We must also make assumptions, but they must be based on certain facts that we do have available. For example, it has been determined that the atmosphere was different then. Today we have an oxygen-laden atmosphere. Our air contains about 78 percent molecular nitrogen (N_2), 21 percent molecular oxygen (O_2), 0.03 percent carbon dioxide (CO_2), and traces of rare gases such as argon and helium. The Russian scientist A. I. Oparin proposed in 1936 that the atomosphere of the early earth must have contained much more hydrogen (H_2) than it does now. The result was that much of the free oxygen, nitrogen, and carbon dioxide could combine with the hydrogen to form new molecules. So instead of free nitrogen, for example, there was ammonia (NH_3), oxygen joined with hydrogen to form water vapor (H_2O), and of particular importance for the beginnings of life, carbon joined with hydrogen to form methane (CH_4). Methane is simply a hydrocarbon, an organic molecule, and organic molecules are associated with living systems, as we shall see.

We can only guess at what the earth was like in those days. Perhaps the most difficult thing to imagine is that the planet—this planet, our earth—was lifeless. There were not, nor had ever been, ears to hear or eyes to see. There were no crusty lichens or slimy molds. Nothing lurked anywhere. The globe was a lifeless unchronicled ball of matter covered by a very thin layer of hot swirling gas. The earth itself was hot, molten, and volcanic. Then the surface began to cool and solidify. The piercing shrieks and groans and deep rumblings from below the crust startled no one. Heavy billowing clouds, miles thick, surrounded the darkened sphere. The blackness was split by spewing gapes in the earth and thunderous lighting from above. When the water vapor condensed and fell to the sterile earth, it exploded with the heat and was lofted skyward again with a crackling hiss. What could have happened to the molecules of the earth's substance at such a time? How might they have become rearranged to form the progenitors of life?

THE BEGINNINGS OF LIFE

Molecules associated with life are, for the most part, highly complex structures. Often they are long branching chains of atoms, far more

complicated than the simple water, ammonia, and methane, which comprised the early earth's atmosphere. For life to appear, somehow these simpler molecules had to be joined together. The trouble was that the molecules, by themselves, had no tendency to join. They could coexist quite nicely side by side without interacting at all. As time went by, however, changes took place which encouraged them to join.

With the cooling of the earth, pools of water condensed on its surface. More important, as the atmosphere itself cooled there must have been torrential rains of incredible proportion. As the water rushed over the exposed surface it swept away the salts and minerals of the topmost rock. These mineral-laden waters poured into the cracks and crevices of a parched earth, turning pattering streams into pounding rivers. The harsh waters fell into the lowest reaches of the planet's surface, and the oceans were born. It was here, in these violent briny waters, that the earth's myriad chemicals were forced into immediate proximity for long periods.

Still, there was little likelihood of any of the molecules joining together on their own. If they were to join, some outside source would have to provide the energy to make it happen. And there are, in fact, a number of possible sources of such energy. One is the sun. In addition to visible light rays, the sun produces ultraviolet rays, X rays, gamma rays, and various other radiations. Another possible source of the critical energy is heat. The earth's surface was still far from placid. Violent eruptions were occurring, and molten and cooling rock covered much of the earth's surface. Lightning continually streaked earthward from the dense, heavy clouds above, so the critical energy could have been electrical.

But was this enough? Could a mixture of methane, water, and ammonia form complex molecules by the addition of some outside source of energy?

MILLER'S EXPERIMENTS

In 1953 a graduate student at the University of Chicago provided an answer. Stanley L. Miller built an airtight apparatus through which four gases—methane, ammonia, water, and hydrogen—could be circulated past electrodes (Figure 2.2). He permitted the gases to circulate together in his chamber for a week, energizing the mixture electrically, and then he analyzed the contents of the chamber. What he found surpassed his expectations. There were an astounding number of organic compounds. Among these were certain important amino acids, the components of protein, those magical molecules of life.

So that no one would question his experimental method, Miller ran two controls. In one case he sterilized the gas mixture at 130°C for eighteen hours before he subjected it to the electrical power source. The yield of complex molecules was the same as in his first test. In the second he ran the test exactly as before, except for the step of adding electric energy. In this case there was no significant yield of the new molecules. He then felt safe in concluding that the formation of organic molecules was not due to the

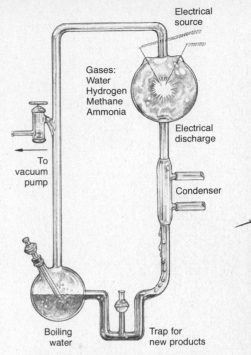

Gases:
Water
Hydrogen
Methane
Ammonia

Electrical source

Electrical discharge

Condenser

To vacuum pump

Boiling water

Trap for new products

2.2 In 1953 a graduate student, Stanley L. Miller, subjected a mixture of methane, ammonia, water, and hydrogen to a series of electrical charges. He imagined this to be a rough duplication of conditions on the primitive earth when the "primordial soup" was subjected to bolts of lightning. The result justified his expectations. After a week, the inorganic molecules had joined to form amino acids.

2.1 It is difficult for us to imagine a primitive earth. Could our beloved earth have been so hostile? From the molecules that comprised the sterile and deadly globe came the peaceful valleys, singing brooks, and swaying grasses. We are left today with only reminders of how terrible it must have been, as we see sharp mountains with deposit lines angling upward from the horizon and as we hear of mass evacuations before predicted volcanic activity. Even today, though, we must remind ourselves that the environment is essentially disruptive and that work must be done to keep it from disorganizing our ordered molecules.

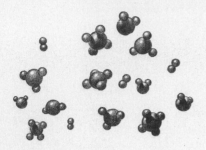

Primal atmospheric gases, hydrogen (H₂), ammonia (NH₃), water (H₂O) and methane (CH₄) along with energy constitute the conditions under which life may have begun.

In the primitive seas, the constant interaction of the primal gases and energy leads to the inevitable synthesis of new, more complex molecules, a few were basic parts of the molecules of life.

Continued synthesis may have brought about the first proteins, carbohydrates, lipids, and nitrogen bases.

presence of living contaminants, but was the result of the fusing of smaller molecules that had been subjected to a source of energy.

Since then, Miller's experiment has been duplicated many times, and even extended. Researchers have subjected methane in hydrogen-laden atmospheres to energy sources and have produced a great variety of organic compounds. Among these are important molecules of life, such as purines, pyrimidines, and sugars, which we will consider later. Other workers, using other gases and a number of types of energy sources, have added to the long list of synthesized organic compounds.

In the face of so much evidence, it now appears that the synthesis of complex organic compounds in the early days of the earth was not only possible, but perhaps *inevitable*. Any large molecules so produced would probably survive for a considerable time, and thus have greater likelihood of entering into other reactions. Why should these large molecules have survived for long periods? Their stability is due to two phenomena: the relative lack of oxygen and the absence of microorganisms. Oxygen can break down complex substances through the processes of oxidation. This is why both iron and a half-eaten apple turn rust colored in air. Their constituents are breaking down in the presence of oxygen. And microorganisms, such as bacteria, can act to degrade, or rot, complex substances. Of course, in earth's present atmosphere, which is high in oxygen and filled with microorganisms, no free complex organic molecule could survive for long. Thus life on earth can never arise again. At least not in this way.

THE COACERVATE DROPLET

So here we are. We now have a system whereby large, complex organic molecules could have been produced when the earth was young. It is apparent however, that the problem of the origin of life is not solved, not even theoretically. It is a long step from acknowledging the existence of organic molecules to accounting for even the simplest forms of life. Any living body is composed of a fantastically ordered arrangement of molecules. The precision of their relationships and interactions is even more astounding than the early biologists dreamed. How, then, could a living system arise from a simple hot chemical soup?

Again we have no firm answers, but we have some ideas. Oparin pointed out that *colloidal* protein molecules, in which the particles are suspended in a gellike state, tend to clump together into increasingly complex masses. These are called *coacervate droplets*. Each droplet consists of an inner cluster of colloidal molecules surrounded by a shell of water. The molecules of this water shell are arranged in a specific manner in relation to the colloid center, so that there is a clear demarcation between the colloidal protein mass and the water in which it is suspended.

The Growth of Coacervate Droplets

The peculiar orientation of the water molecules around the colloidal mass gives this water shell special physical properties. In effect, it acts as a sort of membrane; that is, the water layer acts as a kind of screen, allowing some molecules through and retarding others. The colloidal mass inside has its own peculiar qualities as well, since the complex molecules of which it is composed also tend to arrange themselves in an orderly manner. This comes about through the interaction of a number of forces in the mass, such as positive and negative charges along the molecules. (We will discuss these positive and negative charges in more detail later.) The colloid, as a result of its peculiar arrangement, has a tendency to absorb other molecules from the medium in which it is suspended. However, it does not absorb all molecules with equal vigor. In fact the mechanism is so selective that the colloid may take almost all of certain chemicals from the medium and practically none of others. As a result of this continual absorption of certain kinds of molecules, the droplet grows. In some cases the arrangement of the colloid may become further differentiated or specialized. The molecules at the surface may become more active and rearrange themselves into a membranelike structure just under the water layer. This new membrane is even more selective in what molecules it will allow to pass than was the water "membrane." Thus with increasing selectivity in the kinds of molecules the mass will accept, the structure becomes more and more specific. At this stage the complex droplets have many of the properties of living things and under a microscope may be mistaken for living things. (Even experienced biologists have, on occasion, attempted to classify such structures as species.)

Internal Structure

It is important to realize that the specialized coacervate droplet provides a special environment for any molecule absorbed into its structure. First of all, because of the selectivity of the droplet surface, only certain types of molecules are brought together. Then, once these molecules have been incorporated into a droplet, they are brought into increasingly closer association as a result of the compactness (or high density) of the droplet. Thus certain kinds of reactions become more likely inside the droplet than outside it, and each of these reactions is likely to influence other reactions in new ways. In addition, the catalytic activity of substances such as metal compounds is facilitated by the regular arrangement of the colloidal molecules.

So what we have is a rather highly structured mass of matter that is susceptible to increased orderliness as it grows. And as a result of its orderliness, the likelihood of certain molecular events is increased. In a sense the droplet "regulates," or at least influences, the types of chemical reactions that can take place inside its mass. Now we begin to see that, whereas coacervate droplets are not "alive" as we usually understand the term, they do have many of the qualities we associate with living things.

Reproducing Droplets

If we are going to accept these droplets as precursors of life (or as distant relatives), we must somehow account for their reproduction. Again we rely

2.3 (*continued*)

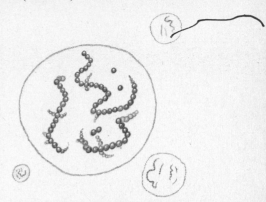

Masses of newly synthesized proteins clumped together, forming clusters, surrounded by a thin shell of water: the coacervate.

Coacervates, in their watery coating grow to critical mass by the intake of more protein, divide and grow again. This phenomenon is suggestive of reproduction, a characteristic of life.

As coacervate systems increase, molecular synthesis continues. Large "active proteins" and primitive nucleic acids are possible additions.

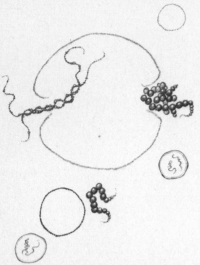

Newer molecules are incorporated into coacervates. More regulated chemical synthesis begins. Nucleic acids begin to determine which proteins will be produced.

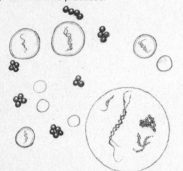

Coacervates, now in an advanced state, occur in populations. Competition for the limited molecules begins.

The earliest cells, complete with membranes, active nucleic acids and enzymes, capture and use the molecules of their surroundings and each other. Competition increases, with some types failing.

on a strong imagination—held in check by circumstantial evidence. To begin with, *most* of these primordial droplets obviously did not reproduce. They must have formed in an astounding array of sizes and compositions, each with its own peculiar structure. Some were more stable than others, and these, of course, had the greatest chance of finally reproducing. Of course, here we are using the term "reproduction" in the broadest sense. Consider one example of how reproduction might have occurred. As certain of the droplets continued to survive, they would have continued growing as new molecules entered their masses. At some point they would have probably fragmented as a direct or an indirect result of their having reached a critical size. Now, if their structures were highly regular and ordered, the resulting fragments would have each been very similar to the original droplet, and each fragment could then begin its own growth.

Reproduction of the droplets could also have occurred in another way. Among the many types of molecules comprising the colloid there were *nucleic acids.* We will discuss the chemical makeup and behavior of these remarkable molecules later. For now, however, bear in mind that they are the critical components in the mechanisms of heredity. It is possible that nucleic acids in a pool of their component parts, called *nucleotides,* could arrange those free nucleotides to correspond to their own images—another form of "reproduction." It is even possible that nucleic acids could form some kind of "code" for molecules other than replicas of themselves, and thus produce such critical molecules as proteins. Just how this might have come about is not at all clear, but if you are interested, you might read the June 1, 1973, volume of *Science.* The point is that there are ways in which certain highly complex coacervate droplets *could* have come to reproduce themselves. The most "successful" of these droplets, of course, would be those which came closest to producing *replicas* of themselves, since they had effectively demonstrated the stability of their own particular structure by their very presence.

Successful Droplets

Through eons of slow accumulation, as the less successful ones disappeared, the more successful droplets could have developed into the first cell-like bodies. With the continual disruption of less satisfactory bodies, as time passed the percentage of more stable forms increased, and the components of the lost droplets were recovered by increasingly efficient early cells. These cells, then, were the result of the most fortuitous of the molecular combinations that had taken place in an untold number of trials through millions of years. The point at which a droplet had become what we can call a cell is arbitrary. There was no sudden and dramatic appearance of life on the earth. No trumpets blared; no flags waved. There were only a few peculiar droplets that were perhaps a little more organized than the rest.

This picture of how life was first generated on earth is by no means universally accepted. Some scientists contend that coacervate droplets probably did not nurture reproducing nucleic acids. Instead, they consider it more likely that free nucleic acids came together and then built a cell around themselves. Others, of course, feel that there is an essential distinction between life and nonlife, and that all life—or at least that of man and the other vertebrates—was created by some higher being. And then there are a host of lesser theories, which need not concern us here. The whole question is an academically fascinating one, however, so let us briefly consider how these early cells might have made their living.

THE EARLY CELLS

As you probably already know, there are only two basic ways an organism can get its energy, or "food." Plants utilize carbon dioxide, water, and sunlight to make their food. They are called *autotrophs*, meaning "self-feeders." Animals do not normally make their own food, but must derive it from other organisms that have made it. They are called *heterotrophs*, meaning "other-feeders."

The prevailing notion is that the earliest cells were heterotrophs—that they utilized nutrients from their environment. Such a life style enabled them to survive without having to develop a food-producing mechanism. But it presented other problems. At first, with so few cells in the rich, warm primordial soup, there was no problem in getting food. As time went by, however, the cell population began to expand drastically, since the cells were actively reproducing themselves generation after generation. And eventually the nutrients on which they subsisted must have fallen into short supply.

There must also have been many different types of cells in those days, since they had sprung from a wide array of coacervate droplets that had formed more or less through various chance combinations. Some types of cells had advantages over the others. Those which were more adept at obtaining food from the environment were, naturally, able to reproduce at a more rapid rate, and it follows that in time these cells would come to be more prevalent. In this ever-changing and complex early world it is also likely that further chance changes took place within certain cells. When such changes impaired their food-absorbing ability, these altered cells perished in the competitive environment. But when the changes produced some slight advantage in competing for food, these altered cells increased in number over the rest. Again, the net result as time went by was that these early cells became increasingly efficient.

Autotrophy and Heterotrophy

The incredibly complex biochemical reactions that take place in cells today are probably a far cry from what went on in the early cells. So we might ask how they could have ever developed such complex systems. What was the impetus? Perhaps such complexity was spawned by the need to find new sources of supply for specific nutrients that had fallen into short supply. We know that certain food molecules were critical for cell reproduction.

2.3 (*continued*)

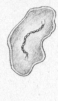

With the forces of natural selection at work, only the most efficient survive. New energy sources must be found and used.

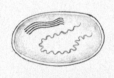

Light, an untapped energy source, is now used as light-trapping chlorophyll molecules are incorporated (left), thus autotrophs begin. Heterotrophs also improve, using oxygen now available from autotrophs, they become more efficient.

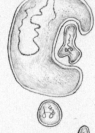

From the first autotrophs and improved heterotrophs, the forces of natural selection mold more efficient descendents. Some of the old persist in limited environments.

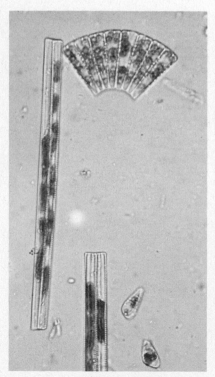

Thus a cell unable to find such nutrients had to either develop some alternative or perish. Among the small chance changes that were continually arising in the cells, then, there could have been some change that enabled a cell to manufacture critical nutrient *A* from a more available nutrient *B*. These cells, which had been held in check by scarce nutrient *A*, were then able to reproduce wildly—that is, until nutrient *B* fell into short supply. At this point perhaps there appeared, by chance, some other slight change that enabled a cell to manufacture *B* from nutrient *C* and thereby to make nutrient *A*. These cells were the ones that survived. Finally, the most viable cells would have been those with the ability to carry out long chains of biochemical reactions. Such biochemical complexity would have set the stage for existing in an increasingly complex and variable world. Any such system, of course, would demand an efficient energy system to drive the mechanism. It is interesting that the chemical ATP, which is a prime source of energy in present cells, is one of the molecules that is synthesized in experiments simulating the conditions of the early earth.

Today all animals are heterotrophs. They no longer absorb their food from a surrounding hot soup, it is true. Instead they eat autotrophs—or other heterotrophs that originally got their food from autotrophs they had consumed. All plants are autotrophs; they are able to make their own food, and as a result they have played a unique role in the history of life. Let's briefly consider how they might have arisen and how they might have affected other life.

An Autotrophy Called Photosynthesis

Basically, autotrophy involves the manufacture of complex molecules out of simpler materials. It is likely that as some of the early cells developed complex chemical pathways for making the nutrients they needed, they were finally able to make carbohydrates from carbon dioxide and water. Carbon dioxide would have been abundant as a by-product of heterotrophic cycles. We can only speculate about how early plants might have manufactured their food, but the process by which plants now do this is called *photosynthesis.* The simplified chemical description of this process is written as

$$6CO_2 + 12H_2O \xrightarrow[\text{chlorophyll}]{\text{sunlight}} C_6H_{12}O_6 + 6O_2 + 6H_2O$$

That is, six molecules of carbon dioxide and twelve molecules of water, in the presence of sunlight for energy and a pigment called chlorophyll, form one molecule of glucose (a sugar), six molecules of oxygen, and six molecules of water. We will discuss the pathway in more specific terms later.

Oxygen: A Bane and Blessing

It is important to note that oxygen is released in the photosynthetic reaction. This is good news for us, since we need oxygen, but it sounded the

2.4 Autotrophs and heterotrophs. There are very few truly carnivorous animals (perhaps a few flies) but cats and sharks come close. Here a lion is about to extract her nutrients from a Thomson's gazelle which has, in turn, robbed grass. The amoeba (top left) cannot make food but must engulf tiny bits of nutrient material, often preferring the more proteinaceous forms. It will eat plants, animals, or bacteria. Lions and amoebas, then, are heterotrophs. Autotrophs are represented by plants. One of the least considered, but most important, groups of plants in our daily lives are the algae. Algae in the sea (bottom left) are enormously productive food producers and some day we may be forced to go directly to the algae instead of eating the fish that eat the algae. Most autotrophs that we know about are the large terrestrial plants such as aspen trees. We live among them, and in the absence of profits to be made, protect and eulogize them, but we often forget we are inextricably bound with them into a cycle of life.

death knell for many of the early heterotrophs. The release of more and more oxygen gradually changed the atmosphere from one full of hydrogen to one rich in oxygen. Now, it is logical to assume that some cells developed a means of utilizing the free oxygen, perhaps in energy-producing reactions. Oxidation, as we shall see, is a much more efficient means of producing energy than the fermentationlike processes the cells had been using. New types of cells with such abilities would have had a great advantage over the old types, and in time they would have come largely to replace them. However, the oxygen that proved a boon for the new kinds of cells probably acted as a poison for many of the old heterotrophic organisms and hastened their disappearance from the earth.

Not all scientists believe that the increase in the earth's oxygen is primarily the result of photosynthesis by plants. Some believe that most of this oxygen has come from the sun's action on the water of the upper atmosphere. We know that as water (H_2O) is split into its component parts oxygen is released. This knowledge may provide some consolation as we watch the destruction of the world's great forests and the slow poisoning of the algae-laden seas.

Whatever its source, the appearance of high levels of oxygen had important effects. As oxygen collected in the upper atmosphere in the form of ozone (O_3), it formed a shield against certain types of the sun's rays. The ultraviolet rays in sunlight had probably provided energy for the activities of early cells, but the disruptive effects of these rays would also have been a threat to their continued survival. In fact, the early cells probably survived only by staying five to ten meters beneath the surface of the water, where the force of the ultraviolet rays was diminished. The development of a protective layer of ozone meant that dry land was now safe to inhabit, and soon the shorelines were invaded by strange primitive life forms. An important effect of the oxygen-rich atmosphere, however, was that life could no longer arise in the way it once had. In other words, as life developed, it destroyed the very mechanism that had produced it.

THE IDEA IN SCIENCE
The foregoing description of the possible beginnings of life is, of course, pure conjecture. Almost every point can be argued, and some key points defy conceptualization, let alone proof. However, the scientific mind has never been marked by a reluctance to launch expansive theories from small scraps of information—and in a way that's good. It provides a rather loose intellectual milieu in which to rigorously test small facets of great theories. And it is only the accumulation of hard data that can ever bring an idea past the state of hypothesis, or even of theory.

Theory
And by the way, what is a *theory?* There are some who discount the concept of evolution by saying, "It's only a theory." However, bear in mind that the status of theory in science is a relatively lofty one. To illustrate, suppose

someone comes up with a good idea, one that explains a host of observed phenomena in nature. At first it is just that, an idea. After it has been carefully described and its premises precisely defined, it may then be a hypothesis that can be subjected to testing. If rigorous, carefully controlled testing substantiates the hypothesis, more and more confidence will be placed in it. If it continues to account for incoming data from the real world, it may finally gain the status of a theory. The theory itself, however, may remain unproven and unprovable.

The reception of a theory—or, for that matter, of a fact—may depend on the intellectual milieu into which it is introduced. The theory of evolution, for example, caused an initial shock in the Victorian world but was assimilated by the intellectual community surprisingly rapidly. One possible reason for its widespread acceptance is that the deeply religious citizens of that era saw in evolution a preexisting *program* for change—a program authored by the Deity. Of course, the theory also had strong opposition at the time and is still met with strong resistance in some quarters today. It is interesting to note, however, that the arguments against evolution are not universally applied. For example, no one has ever seen a hydrogen atom, but detailed descriptions of its behavior may be found in any chemistry book. Moreover, we have been able to use the description of this theoretical atom to make water! In spite of the same kind of abundant evidence for evolutionary theory, people somehow have much stronger feelings about evolution than they do about chemistry.

In science any idea or hypothesis must (ideally) be testable to gain full acceptance. This does not mean that scientists are a breed of eminently rational people, unfettered by personal prejudice. They aren't, but fortunately such sterling character is not actually necessary. Important contributions have stemmed from hunches or from ideas that simply appealed to the prejudices of individual scientists. The fact is that the "scientific method" works in the face of all bias simply because almost any idea is sure to be attacked by someone and if it has weaknesses they are likely to be revealed.

The Usual Process

When a scientist reports the results of an experiment, which is only one way of gaining information, several hurdles confront their acceptance. First, the general method comes into question. Other scientists working in that area of research have some idea whether the data could have been gotten in the way the experiment was conducted. Jane Goodall followed chimpanzees through African forests for several years and through observation and a few simple experiments brought back important new information regarding their behavior. Her findings were enthusiastically accepted. However, any reports based on extended periods of following the great blue whale through its daily activity might be greeted with some skepticism. Whale following is presently impossible.

Second, the scientific audience will question the experimental tech-

2.5 Jane Goodall was among the first, and most publicized, modern researchers to move directly into the field to live among her subjects, the chimpanzees of the Gombe Stream Preserve of East Africa. The chimpanzees avoided her for months but finally came to accept her as she moved among them. Her work involved detailed descriptions of individual animals, their interactions, and their family relationships. Her work has recently been hampered by African political problems. (Photograph by Baron Hugo van Lawick. © National Geographic Society.)

nique. Could the results have been produced only by the experimental conditions as described? Third, the readers will ask whether the conclusion is, in fact, supported by the data. Finally, the experiment will be repeated to see whether the results correspond to those originally reported. Not only do experiments face such rigorous tests, but even simple observations may be subjected to the same skeptical analysis.

Of course, this isn't the only way information gets into circulation. Someone who happened to witness the birth of a volcano—which would be difficult, or at least inadvisable, to duplicate—might provide us with some observations. Nor can it be said that all good scientists formulate theories only after a painstaking analysis of voluminous and detailed evidence. Some of the best science is often dreamed up from the armchair, as it were. The actual testing may be done by more pedestrian scientists. In fact, it has been said that the inspired guess comes first, and that the data are only a dodge to give "proof to the disbelievers." The scientific world may be full of "disbelievers," it is true, but they, in their role of constant skeptic, have undoubtedly saved us from many "great ideas" that just couldn't be made to fit a real and very complex world.

What Kinds of Answers Can Science Provide?

Because science is limited to testable evidence, it can provide answers only to certain kinds of questions. Unfortunately the confusion between science and religion has often blurred the issues, but science cannot, by definition,

2.6 Disagreement among scientists. Here, a protest is being lodged at the spring 1977 meeting of the National Academy of Science. The object of concern was a forum on recombinant DNA research (which we will discuss later). Such disruptions were rather startling when they began in the 1960s, but in general they may have helped to make the scientific community more responsive to public needs.

resolve questions either of faith or of "right" and "wrong." Thus science *cannot* tell us whether there is a God since any such Being, by definition, could choose not to be examined. Science also cannot tell us whether war is good or bad or whether capital punishment or abortion is immoral, since such statements would have to be based on some system of values about what is good or bad. However, science *can* measure certain physiological responses and arbitrarily tell us at some point whether an organism is dead or alive. And science *can* measure the effects of certain conditions on living organisms and tell us the effects of birth and death rates on population size. And scientists *can* tell us what bombs do to children.

SCIENCE AND SOCIAL RESPONSIBILITY

For some time now there has been a movement to make science more responsible to the needs of man and society. Many have wondered, for example, whether a scientist has the moral right to develop a destructive device and at the same time disclaim responsibility for the use to which it is put. However, this question leads to a larger issue. Should scientists determine for themselves what areas of knowledge mankind should or should not have? Should they refuse to develop anything that could cause the wholesale destruction of people or the mindless poisoning of our environment? Or is it the responsibility of the scientist to seek knowledge in any form, so that an informed society is better able to make its own decisions?

One argument for greater personal involvement on the part of scientists is that they are in the best position to realize the full implications of their work. This argument is bolstered by the obvious lack of information among the general public. An all-to-common lament is, "But why haven't they said something? Why haven't they told us?" The truth of the matter is, prominent biologists have covered the country on speaking

tours and have written bestselling books on the state of our environment. Television talk shows have been inundated with alarmed scientists from every discipline. Activist student groups and concerned-citizen groups have sprung up all over the country. Each day the news media bring us stories of environmental disasters. Yet somehow the general populace remains either uninformed or unconcerned.

Fads and Trends

Part of the problem is that *ecology* has become a household word—and like any household word, it has become so familiar that people are sure they know exactly what it means. In fact, the general public has been bombarded to the saturation point with "eco-messages" from every quarter—often in contexts that had little to do with the environment. To make matters worse, the worst offenders among us took up the ecology cry, generally in hope of selling us something. Paper companies touted their tree-planting programs while they stripped mountains bare. Television commercials extolling the virtues and concern of the oil companies flooded our homes while their products did the same to our air and waters.

During the late 1960s ecology was a national fad. Fads are part of the American culture, perhaps the result of a great national energy coupled with a short attention span. But like all fads, the crest of the ecology movement passed, leaving the public with a sense that the environmental crisis was an outmoded topic. This topic is likely, however, to be raised again and again, with increasing urgency, until we realize that the fashions of the well-fed may have rather little to do with the real world.

Exaggeration, Compromise, and Understanding

To some extent the demise of this early effort to arouse public concern for our environment was a predictable response to some rather shocking statements by environmentalists—some valid, others less so. One result was that sober scientists in various fields began to offer rebuttals to the often shrill warnings of the "eco-freaks." They told us that the problem had been exaggerated, or that there actually was no real problem. We were also told that by the time any problem did appear, it could be resolved by improved technology.

Even among the environmentalists there were disputes over both the seriousness of the problem and the source of the danger. Some maintained that, with the possible exception of Southeast Asia, the environment was in no real danger of losing its potential to support increased populations. This was in disagreement with those who felt that the major threat to human existence was man's burgeoning numbers. Others said it was misdirected technology that had gotten us into trouble, and a redirection of priorities could get us out. With such arguments going on among those who were supposed to know—those who had seen the data—it was no wonder that the nonscientist was confused. The confusion, unfortunately, led to still further disenchantment in some quarters. Perhaps no one was right! There

2.7 A few of the bobcats killed at one ranch in Texas. The reasons for killing such an endangered species are varied. Ranchers often cite the bobcat as a chicken killer (while choosing to overlook its role in controlling rodents). Among many cultures, including our own, such killing is often taken to be a symbol of maleness or manhood. If the cats were run down on foot and then throttled with bare hands, perhaps a case could be made for this argument, but they have very little chance of escaping trained hounds and high-powered rifles. Sometimes the reasons given for such killing is economic. Perhaps the hides can be sold. If the demand exists, it will be met, whether the poachers are African tribesmen killing the last of the rhinos or Louisiana citizens spotlighting alligators at night. The sale of such skins is not restricted throughout most of the world and one can find all sorts of spotted-cat skins on the proud backs of status-conscious Europeans any cool Saturday afternoon.

were even claims that the environmentalist movement was not only misdirected, but sinister. One piece of literature pointed out, among other things, that the first Earth Day took place on Lenin's birthday.

Probably there are few, without economic interests directly at stake, who would seek to undermine ecological efforts. After all, what advantage can there be in despoiling one's own habitat? The important questions today do not concern the benefits of a "good" environment. We all want that. Instead, we hear arguments over what comprises a good environment. Which is the more important component of man's habitat, a clean river or an air-conditioned house? Unfortunately, in many cases we can't have it both ways.

We also hear arguments over what methods to employ in regaining or maintaining a good environment. Should we force industries to produce clean effluent at the cost of higher prices and fewer goods for us all? Should we limit the size of families by law? Should we publicly finance those who are running for public office, so that our elected officials are not beholden to the large corporations that now finance their campaigns?

Again, these are questions that science cannot answer. However, the answers will be forthcoming one way or another, even if it is through silence. What science can do is provide us with the information that is available, so that we can make the best decisions possible in an increasingly complex and precarious world. You won't find all the answers in this, or any other, book. But perhaps you will begin to grasp some of the more critical problems and to understand the ramifications of your decisions. We can no longer afford to respond off the tops of our heads; we must all begin to *understand* our world, and ourselves, at the deepest level possible for us.

2.8 The North American continent abounds with powerful, free-ranging rivers. In the early days of the colonization by Europeans the waterways provided open transportation routes. In later years they were not only a source of water for increasingly large settlements but a source of energy as well. Waterwheels, however, have given way to enormous hydroelectric dams, many of which supply power to populations thousands of miles away. The euphemism for blocking the mighty waters is "harnessing." Other bodies of water have provided a focal point for human populations and have spawned great cities—cities that seemed to lose touch with their quiet source of life. Rivers came to be seen as things to be crossed or dammed, or as ready dumps to carry waste downstream, downstream where somebody *else* lives.

The Chemistry of Life

3

Chemistry? Why chemistry?! Often beginning students of biology, particularly those who don't intend to make a profession of the sciences, are aghast at confronting chemistry in what they thought was going to be a biology course. Is chemistry really necessary? The answer is absolutely yes and no!

Biology can be "done" without chemistry. Darwin had only a rudimentary knowledge of the subject, and many modern naturalists do not use it in any way in their work. However, some knowledge of the subject is necessary in order for you to know why you breathe, urinate, and also, perhaps, why your mood changes before a thunderstorm. Some knowledge of chemistry as it pertains to biology may also help you to make critical personal and political decisions in the years ahead. For instance, the pollution-control devices now required on most automobiles trap hydrocarbon molecules, but allow nitrous oxides to pass into our air. Automobile manufacturers argue that if the hydrocarbons were also allowed to pass into the air, they would neutralize many of the nitrous oxide ions by bonding with them to produce relatively harmless compounds. But what are these compounds, and what are the effects of hydrocarbons? Even if you have never had any great yearning to be a scientist, some idea of the properties of these molecules will, if nothing else, make you a more perceptive consumer and a more discerning voter.

WHAT BIOLOGISTS DO FOR A LIVING

To the layman the word *biologist* often conjures up the image of a balding little man whose right shoe squeaks as he pads through aisles of dusty books all written in Latin. Others may visualize butterfly chasers leaping gleefully through the brush with their nets poised, or birdwatchers peering through field glasses in the cold, wet dawn in hope of catching one glimpse of the rare double-breasted seersucker. Perhaps such descriptions do fit some working biologists. But biology is much more than this. Many biologists search for the mysteries of life in clean, well-lit laboratories, amid tinkling glassware. About fifteen years ago it was said that a biologist is one who thinks that molecules are too small to matter, a physicist is one who thinks that molecules are too large to matter, and anyone who disagrees with both of them is a chemist. Perhaps it is true that the biologists whose scope is limited to molecules must periodically be convinced of the existence of the platypus, but no one else has been able to explain to the birdwatcher how a tiny hummingbird has the energy to cross the Gulf of Mexico. The dividing lines between these disciplines will undoubtedly become less distinct as time passes and more and more parts are fitted into the "great puzzle." So far, however, neither the field biologists nor the molecular biologists have been able to tell us, after all, what life is really about. On one hand, observations of the behavior of groundhogs do not provide us with a complete description of the mysterious process we call life. On the other hand, we don't know nearly enough about where the molecular data fit into the big picture. As molecular biologists search for the integral units of life, they are often left sifting electrons through their fingers, "life" having escaped them somewhere along the way. Rest assured that we're not even going to approach such formidable questions as the "nature" of life. We will, however, consider a few of the basic concepts of molecular biology. As you continue to read, you may gain some appreciation for the problems confronting those dealing with larger questions.

Now, here's the plan. We will first learn what an atom is supposed to look like. Then we will explore the nature of a few specific molecules and compounds. In ensuing discussions we will consider the ways these particles work. You should be forewarned that in following chapters we will frequently refer to the basic principles mentioned here.

Before we go on to specifics, one other point should be mentioned. We will often consider some highly detailed and incredibly complex features of many kinds of plants and animals in general terms. The reason is that in spite of their great morphological and ecological differences, all living things fuction biochemically in much the same way. In fact, this similarity is further evidence of the common origin of present life forms.

ATOMS AND ELEMENTS

All matter, whether it comprises an alga, a redwood tree, or a distant star, is composed of *elements*. *Atoms* are the smallest indivisible particles of these elements. Atoms can be further divided into component parts, of course, but when this is done, special qualities of that element are lost. Each element, or type of atom, has its own special chemical behavior by which it is identified.

There are 92 elements that occur in nature. Other elements have been synthesized in the laboratory, so that the total number of known elements is now 103 (at least, was at the time of this writing). Each element is named and designated by letter symbols. For example, hydrogen is designated by the letter H. Nothing is all that simple, however, so we find that sodium is designated by Na, from its Latin name *natrium*. The symbol standing alone usually designates one atom. For example O refers to one atom of oxygen. The symbol 2O refers to two atoms of oxygen. But the symbol O_2 means that two oxygen atoms are joined together into one *molecule* of oxygen, the form in which oxygen usually exists in nature.

CHNOPS

CHNOPS is a "word" you might like to remember to identify the six elements that make up 99 percent of living matter. However, if you were to drop this word casually at your next gathering, it is not likely that everyone will know you are referring to carbon, hydrogen, nitrogen, oxygen, phosphorus, and sulfur. If they do, you are in the company of organic chemists and should leave immediately.

The Structure of an Atom

Now let's consider what an atom is supposed to look like. We have to say "supposed to" because atoms cannot be seen, so all the evidence is circumstantial. It is, however, impressive. People have been imprisoned on less compelling evidence. Atoms, according to theory, are composed of a dense center around which smaller particles spin in orbit. The center, called the *nucleus,* consists of *protons,* which are usually accompanied by *neutrons.* Each proton carries one positive charge, designated by a plus sign. Neutrons, as the name implies, carry no charge. Each electron carries one negative charge, designated by a minus sign. Electrons are much smaller than the protons (about 1/1835 the mass) and are in orbit around the nucleus. An atom of hydrogen, the simplest element, consists of one proton and one electron, with no neutrons (Figure 3.1).

Helium is the next simplest element and is composed of two protons, two neutrons and two electrons. The number of protons in the nucleus is its *atomic number.* Elements are arranged according to their atomic number. Elements differing by one proton may have vastly differing chemical properties. You may have noticed in the figures that the positively charged protons are balanced by the same number of negatively charged electrons so that the atom has a neutral charge. Also, it should be mentioned that these figures are not drawn to scale. The reason is that the distance between the nucleus and its electrons is enormous. You can get some idea of the relative distances if you imagine that the period at the end of this sentence is a nucleus. Its nearest electron would probably be across the street somewhere.

How Atoms May Vary

Isotopes

The atoms of a given element always have the same number of protons in their nuclei. If this were to change, the atom would be a different element. However, in certain elements the number of neutrons in atoms may vary. For example, most oxygen atoms have eight protons and eight neutrons, but there are also some oxygen atoms that have nine neutrons, and some that have ten. These variant forms of the same element are called *isotopes.* There are eight different neutron variants of oxygen, so oxygen has eight isotopes. Some elements have as many as twenty isotopes.

Generally all the isotopes of an element have about the same chemical properties, but in some cases the tiny difference can be important. Some isotopes of uranium, for example, are much less stable than others.

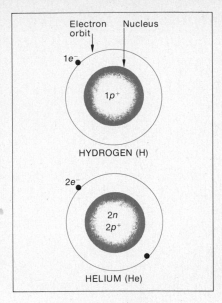

3.1 The hydrogen atom (top) and the helium atom (bottom). Note that hydrogen has one proton (this defines it as hydrogen) and one electron (thus it is electrically balanced). The helium has two protons (making it helium) and two neutrons in its nucleus. The neutrons bear no charges so the atomic number of helium is two, while hydrogen is one. Helium is also balanced in that its two positively-charged protons of the nucleus exist with two negatively charged electrons in orbit around the nucleus.

Ions

An atom may also vary in electric charge. In its ordinary state each atom has the same number of negatively charged electrons as it has positively charged protons; hence, it is electrically balanced. The number of protons, or positive charges, can't change without changing the element. However, an atom can gain or lose electrons, and when this happens, it is no longer electrically balanced. If the atom loses an electron, it is left with a net positive charge, since its protons now outnumber its electrons. If it should happen to gain an extra electron, it would have a net negative charge. Charged particles—atoms that have somehow acquired either a positive or a negative charge—are called *ions*. Ions, as we will see later, function importantly in living systems.

Electron Shells

Electrons are in orbit around the atomic nucleus and are moving at almost unbelievable speed. They do not circle about the nucleus at random, but move in definite energy paths, with each path at a specific distance from the nucleus. The paths of electrons are usually depicted as concentric spheres around the nucleus, called *shells*. The electrons in the outer shells, it turns out, have to orbit much faster—that is, at much higher energy levels—than those in the inner shells.

It is possible for an electron to move from one shell to another within the atom. However, in order to move outward to a higher-energy orbit the electron must be "excited" by some external energy source, such as heat, light, or electricity. When the electron falls back into a lower orbit, the energy that had provided its boost is released. Figure 3.3 gives you an idea of the energy relationships.

OXIDATION AND REDUCTION

If an electron is given enough of an energy boost it may escape from its atom altogether. The loss of electrons by this process is called *oxidation*

3.2 Oxidation is continually taking place, within our bodies and outside, as hydrogen ions (or electrons) are passed to oxygen. The loss of the hydrogen ions is often disruptive, being the basis for such varied processes as fire, rotting, and digestion. Have you noticed the yellowing pages of an old book? The yellowing may be due to slow oxidation unless you've used them to line your pet's cage. The process of oxidation may take place rapidly if the "energy of activation" has been provided, here, perhaps, by the intense heat of a match.

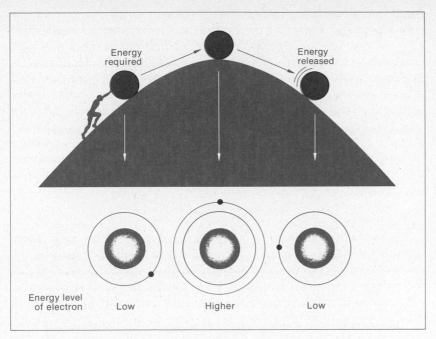

Energy required

Energy released

Energy level of electron   Low   Higher   Low

Electrons may exist in high or low energy states. An electron moving from an orbit near the nucleus to one farther out is moving from a lower to a higher energy level. "Exciting" an electron by raising it to a higher level by the input of energy is analogous to pushing a boulder up a hill. At the higher energy levels they both increase their potential energy, which can be released if they return to their former states.

because in some cases the lost electron is recaptured by an oxygen molecule. In fact, some substances cannot lose electrons unless oxygen is available to accept them. The capture of such electrons is called *reduction*. This simple process of electron transfer—the oxidation of one material and the reduction of another—has tremendous implications for life. In some cases, for example, an electron travels in company with a proton—that is, as a hydrogen atom. Thus oxidation and reduction involve the transfer of hydrogen from one element to another. When oxygen is reduced by two such hydrogen atoms, a water molecule is formed.

3.4 Helium (top) and neon (bottom) are called inert atoms. They are extremely unreactive with other elements because their charges are balanced and their electron shells are filled.

ELECTRONS AND THE PROPERTIES OF THE ATOM

Another important property of atoms is the fact that only a certain number of electrons can occupy any one electron shell. The first (inner) shell will accommodate only two, the second only eight, the third only eighteen, and the fourth only thirty-two. However, no matter how many shells an atom has, the outer shell is the one that is critical. Many of the chemical properties of the atom are determined by this shell. The reason is that the atom will tend to fill this shell with electrons.

When the outer shell of an atom contains its full electron complement the atom is *inert;* it is almost impossible to make it react with any other element. Helium and neon, for example, are inert elements (Figure 3.4) Argon is also an inert element, despite the fact that its outer shell is not full. It works out that if the third or fourth shell of an atom has as many as eight electrons, the atom, actually behaves as if this shell were full. These inert elements that are seeking no electrons are not particularly interesting in

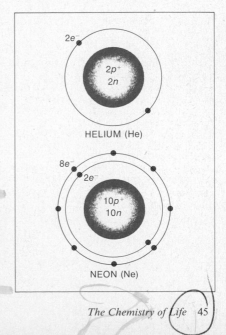

$2e^-$

$2p^+$
$2n$

HELIUM (He)

$8e^-$
$2e^-$

$10p^+$
$10n$

NEON (Ne)

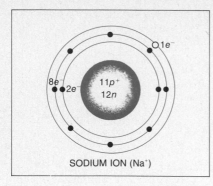

SODIUM ION (Na⁺)

3.5 Since sodium has eleven electrons, with ten existing in its first two shells, the third shell can harbor only one electron. Sodium cannot rectify the situation by adding seven more electrons to the third shell. The resulting charge imbalance would be too great. Instead, it tends to give up its outermost electron. Thus sodium is highly reactive with anything that accepts electrons.

3.6 Argon has an unfilled third shell, but an atom with as many as eight electrons in the third shell behaves as if that shell were full. Oxygen, on the other hand, is reactive because it has only six electrons in that shell.

Table 3.1   Some Elements of Living Matter

Element	Symbol	Atomic number	Electrons in each shell				Percentage of earth's crust
			1	2	3	4	
Hydrogen	H	1	1	0	0	0	0.10
Carbon	C	6	2	4	0	0	0.03
Nitrogen	N	7	2	5	0	0	Trace
Oxygen	O	8	2	6	0	0	46.60
Sodium	Na	11	2	8	1	0	2.90
Magnesium	Mg	12	2	8	2	0	2.10
Phosphorus	P	15	2	8	5	0	0.10
Sulfur	S	16	2	8	6	0	0.05
Chlorine	Cl	17	2	8	7	0	0.05
Potassium	K	19	2	8	8	1	2.60
Calcium	Ca	20	2	8	8	2	3.60
Iron	Fe	26	2	8	8	8	5.00

terms of life processes. With elements, as with people, the most fascinating are the dissatisfied ones.

Why is the number of electrons in the outer shell so important? Let's consider the case of oxygen. A look at Table 3.1 tells us that oxygen has an atomic number of 8; that is, it has eight protons and therefore needs eight electrons. However, since the inner shell will require two of these electrons, this leaves the outer shell with six—two short of its shell requirement. This is why oxygen so readily accepts electrons from other atoms.

The situation is different for sodium. Sodium has an atomic number of 11, so it has two electrons in its first shell, eight in the second shell, but only one in the third shell. Since it isn't feasible for a sodium atom to gain another seven electrons, it tends instead to lose its single outer electron. Of course, this ionizes the sodium, giving it a slight positive charge. Sodium ionized in this way is then written Na⁺. For the same reason magnesium, which has twelve electrons, often appears as Mg⁺⁺ (do you see why?). The fact that chlorine has an atomic number of 17 may give you a clue to the reason salt (NaCl) is formed so easily. Work it out.

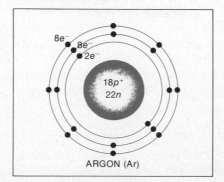

ARGON (Ar)

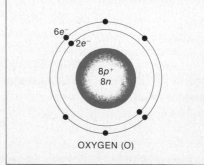

OXYGEN (O)

Box 3.1   Ions and Your Behavior

One day, let us say, not long ago, a preschool child in San Francisco was throwing a tantrum. The day was overcast and foggy. In New York the weather was bright and sunny, but a postman felt a migraine coming on and had to sit down. In Alabama a high school senior taking his college entrance exams was doing so surprisingly well on the math problems that he barely noticed a tropical storm battering the school. In each of these cases, strangely enough, the weather may well have been partly responsible for the behavior of the people.

We know that weather may depress or cheer us, make us pensive or exhuberant, but we are only now learning that our moods may be mediated by the chemistry of the air, in particular, by the ions in the atmosphere. For example, negative ions may make us "smarter." In 1938, a group of entering freshmen at Massachusetts State College (now the University of Massachusetts) was taking I.Q. tests just as a storm was approaching. As the storm grew, their scores rose astoundingly (still the highest in the school's history), but after the storm, scores dropped to 10 percent below average. Now, University of Pennsylvania researchers have found that negatively charged ions in the air promote healing, relieve asthma and hay fever and, in general, make spirits rise. But positively charged ions, it seems, cause fatigue, dizziness, and headaches. Falling barometric pressure, by the way, generates positive ions. These findings are tentative at present, but the idea is interesting.

CHEMICAL BONDING

When atoms join together to form molecules, they do so in specific ways. Moreover, since the atoms of each element have their own peculiar bonding properties, only certain kinds of molecules can form. The qualities of the constituent elements, however, may suggest very little about the nature of the molecule they form. For instance, sodium is a deadly poison, and so is chlorine, but together they form table salt.

Ionic Bonds

Let's take a closer look at how table salt is formed. We know that sodium has two electrons in its first shell, eight in its second, and only one in its third. Therefore sodium has a tendency to give up its outer electron (at the expense of becoming ionized) in order to meet its outer-shell requirements. We also know that chlorine has two electrons in the first shell, eight in its second, and seven in its third. Hence it can fill its outer shell requirement rather easily by picking up a single electron. Now, when a strong electron donor such as sodium encounters a strong electron acceptor such as chlorine, there is a quick switch of an electron from one atom to the other. This satisfies the outer-shell requirements of both atoms, but it leaves the sodium, which has lost an electron, with a positive charge (Na^+). And of course the chlorine, with its additional electron, now has a negative charge (Cl^-).

It is probably no surprise to you to learn that opposites attract. The two ions, Na^+ and Cl^-, are held together by their opposite charges, and the result is sodium chloride, NaCl. *Ionic bonding*, then, involves the transfer

3.7 Chlorine has a vacancy in its third shell so it has a high affinity for sodium. When the electron in sodium's third shell is accepted by chlorine, its charge goes with it, thus chlorine becomes negative and sodium becomes positively charged. Since they bear opposite charges, then, the two deadly poisons combine to make table salt.

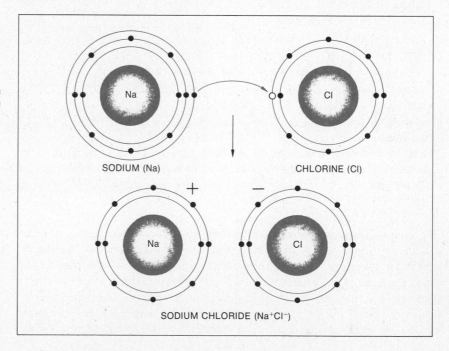

SODIUM (Na)   CHLORINE (Cl)

SODIUM CHLORIDE (Na^+Cl^-)

of an electron from one atom to another and the joining together of the resulting oppositely charged ions as a molecule. In some situations ions with opposite charges can exist near each other without forming molecules. For example, when salt is poured into water the Na and Cl atoms dissociate and remain as separate ions in the water (Figure 3.8). This is because water molecules have unevenly distributed charges so they cluster around the oppositely charged ions and tend to segregate them. We'll learn more about this when we get to hydrogen bonding.

Covalent Bonds

In another type of bonding atoms do not exchange electrons, but instead share them. Let's consider an example. Oxygen, as we have seen, needs two electrons to satisfy its outer shell. One way it can acquire them is by sharing electrons with two hydrogen atoms through *covalent bonding*. This bond, which forms a water molecule (H_2O), satisfies the shell requirements of both the oxygen atom and the two hydrogen atoms (Figure 3.9).

When a covalent bond is formed between two different kinds of atoms, the shared electron is usually attracted more stongly to one element than to the other. In the case of water, for instance, the shared electrons are pulled more strongly toward the oxygen nucleus than they are toward the two hydrogen nuclei. As you see in Figure 3.9, the result is an asymmetrical distribution of charge on the water molecule. In some kinds of covalently bonded molecules this tendency for electrons to be more attracted to one kind of atom is pronounced, and it others no asymmetric charge results at all. Thus there is actually no clear line of distinction between ionic and covalent bonding. To summarize, in ionic bonding an electron from one atom is completely captured by the other atom; in covalent bonding the electron may be almost, but not quite, captured, or it may be shared about equally between the two atoms.

A word should be said here about notation, how chemical structures are written. Bonding is often shown by a line from one chemical symbol to the other. For example, when two hydrogen atoms join together the resulting hydrogen molecule may be written as H—H. When an oxygen atom joins with two hydrogens the bonds may be shown as H—O—H. When two oxygen atoms join together, each of them is seeking two electrons, so their two shared electrons form a *double bond*, O=O. Look back at Table 3.1. Why can molecular nitrogen by written N≡N?

The Magic of Carbon

We have not discussed carbon in any of our examples of bonding. A glance at Table 3.1 shows why. Carbon has six protons and six electrons. We know that it has two electrons in its inner shell, so it must have only four in its outer shell. In order to satisfy its shell requirements, then, it must either give up these four electrons or gain four more. Either of these changes, however, would throw the charge too far out of balance. So the only way carbon can resolve this dilemma is by sharing its four outer electrons with

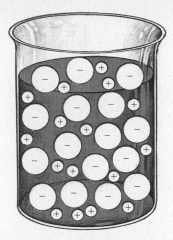

3.8 When NaCl is placed in water, the salt molecule separates into its component parts. Thus sodium and chlorine move apart and behave as independent molecules. Water has this effect on salt because of the peculiar charge distribution on the water molecule.

3.9 Covalent bonds (top) form as two hydrogen atoms move in to share their electron with an oxygen atom. The "unbalanced" structure (bottom) of a water molecule results in a stronger negative charge on one side of the molecule.

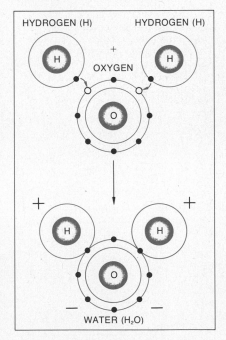

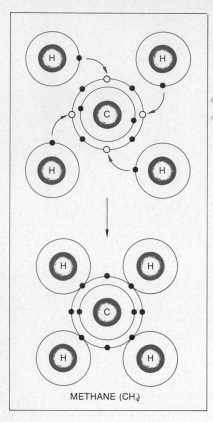

3.10 Methane is the simplest hydrocarbon, composed of one carbon and four hydrogen atoms covalently bonded. Note that by sharing their electrons, the hydrogen atoms fill the shell requirements of the carbon.

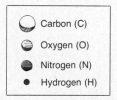

○ Carbon (C)

◐ Oxygen (O)

● Nitrogen (N)

• Hydrogen (H)

other atoms through covalent bonding. For this reason, when carbon appears in a molecule, it is usually found sharing four electrons with other atoms.

Moreover, because carbon can meet its shell requirements in two ways—by either a gain or a loss of four electrons—it can combine with either electronegative or electropositive elements. It can share the electrons provided by oxygen, nitrogen, sulfur, or phosphorus atoms, or it can provide the electrons sought by hydrogen ions (which have lost their electrons).

Because of its peculiar bonding properties, carbon serves as the backbone for a wide variety of molecules. For example, if a carbon atom simply adds four hydrogens, the result is methane (Figure 3.10). Chemists would ordinarily represent this as CH_4, or:

$$\begin{array}{c} H \\ | \\ H-C-H \\ | \\ H \end{array}$$

or

Actually, the four hydrogens are not at right angles, but they protrude in such a way that they are as far from each other as possible. Thus, the structure is more accurately represented as:

We will illustrate molecules in the simpler fashion, so just keep in mind that it is a simplification. If two or more carbon atoms attach to each other, we may get a variety of larger molecules:

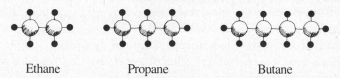

Ethane   Propane   Butane

The molecules produced by any combination of hydrogen and carbon atoms are called *hydrocarbons*. And as you might guess, it is possible to build some very long hydrocarbon chains. Note that in these kinds of molecules the electron-shell requirements of every atom are satisfied; each carbon atom has four other atoms with which to share electrons and each hydrogen atom is satisfied through a single covalent bond with a carbon atom.

Carbon chains may also branch out to form more complex arrangements:

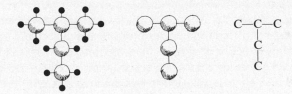

And their ends may join so that they form rings:

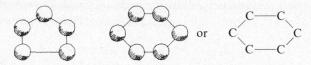

Sometimes the notation for such structures is simplified to show just the carbon backbone.

There are several very important *functional groups* which can attach to hydrocarbons. These include the *hydroxyl groups,* —OH, which can form alcohols:

The *amino groups,* —NH$_2$, which can form amines:

The *carboxyl groups,* —COOH, which can form acids:

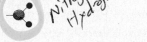

These functional groups, once they have formed, tend to behave as if they were a single atom. They move together as a group when a larger molecule of which they were a part becomes dissociated. When they exist as unattached charged particles they are sometimes referred to as *free radicals.*

Because of carbon's peculiar bonding abilities, it permits the formation of complex molecules, which then interact in still more complex ways to implement life's processes. Thus the chemistry of carbon is, in a sense, the chemistry of life.

Hydrogen Bonds
There are a number of other, weaker types of bonds that may hold atoms together. The most important of these is the *hydrogen bond.* This bond forms only between a few small electronegative atoms such as oxygen and nitrogen. Strangely enough hydrogen bonds are important to life just

because of their weakness. It takes very little energy to form them, and they are easily broken.

A hydrogen bond is formed when two electronegative atoms share a single hydrogen atom—that is, when the hydrogen atom forms a bridge between them. As an example, water molecules may be held together by hydrogen bonds. The bonding takes place because the oxygen atoms draw their electrons in close about them and, as a result, are somewhat electronegative. This means that the hydrogen atoms (the protons from which the electrons were drawn) are left with a slight positive charge. They are therefore attracted to the nearest available electronegative particle, which is likely to be the oxygen atom of the next water molecule. Thus water molecules are joined in a loose, constantly changing latticework.

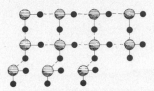

Hydrogen bonds last for an incredibly brief time—about 10^{-11} ($\frac{1}{100,000,000,000}$) seconds. But in their short existence they bestow upon water the unusual qualities of being very fluid and, at the same time, relatively stable. These qualities are critical to life in many ways. Have you ever noticed how long it takes for water to boil as the liquid seems to resist being changed to a gas? As the bonds quickly break and reform they impart a remarkable stability to the water. Hydrogen bonding also makes water resistant to change from its liquid form to its solid form, ice. This stability, or resistance to change, enables us to breathe in a wide range of temperatures by retarding the evaporation of moisture from our lungs, an important factor for terrestrial animals. The unequal charge distribution of water molecules, which permits hydrogen bonding in the first place, also gives water vital qualities as a solvent. The solvent properties of water tend to cause complex molecules to break down, allowing their components to interact in new ways. Thus all biochemical reactions in our bodies take place in water. Also, because of the fluid nature of water, caused by the continually breaking hydrogen bonds, it is able to pass through our tiniest capillaries and to seep deep into the earth's crust to reach the roots of the largest trees.

CHEMICAL REACTIONS

Chemicals, as we know, can interact—so let's take a brief look at how such reactions take place. Basically, chemical reactions involve simply the breaking of bonds and the formation of new bonds. Thus our task here will not be too difficult.

We have already mentioned chemical reactions such as the type represented by

$$A + B \rightarrow AB$$

An example of this sort of reaction is the formation of hydrochloric acid from hydrogen and chloride ions:

$$H^+ + Cl^- \rightarrow HCl$$

Here one ion each of hydrogen and chlorine form one molecule of hydrochloric acid. In many cases, however, we have to designate the number of constituents in the product. For example, the ionized hydrogen and chlorine would ordinarily be in molecular form themselves, as H_2 and Cl_2, so we would get two molecules of hydrochloric acid:

$$H_2 + Cl_2 \rightarrow 2HCl$$

In the formation of water we are also breaking the bonds of hydrogen molecules (H_2) and oxygen molecules (O_2) and forming new bonds in the water molecules. Thus

$$2H_2 + O_2 \rightarrow 2H_2O$$

That is, two molecules of molecular hydrogen and one molecule of molecular oxygen form two molecules of water. Notice that the same number of hydrogen and oxygen *atoms* appear on both sides of the arrow, so that the equation is balanced.

Many reactions of this kind are reversible. For example, we can also break water down into hydrogen and oxygen:

$$2H_2O \rightarrow 2H_2 + O_2$$

We sometimes write reversible reactions with a two-headed arrow:

$$2H_2O \leftrightarrow 2H_2 + O_2$$

In another type of chemical reaction the constituents may simply switch partners. Instead of two chemicals interacting to form a single substance, they may simply exchange components and come out as two completely different molecules. If we combine hydrochloric acid (HCl) with sodium hydroxide (NaOH), for instance, we get water (H_2O) and table salt (NaCl):

$$HCl + NaOH \rightarrow H_2O + NaCl$$

The hydrogen atom from the HCl and the hydroxyl group ($-OH$) from the NaOH have joined as water and the remaining atoms have formed table salt.

Energy of Activation
Some chemical reactions take place spontaneously; that is, the atoms themselves contain whatever energy is necessary to make them react when

Box 3.2 Acids and Bases

Acids are recognizable by their sour taste. An acid is any substance that donates positive hydrogen ions—H^+—to a solution. Since H^+ is hydrogen that has lost its electron, it is simply a proton. Thus acids can be considered proton donors. Hydrochloric acid, HCl, is a "strong" acid because it has a strong tendency to dissociate, or to release H^+ ions into its environment. It cannot be neutralized by the weak negative chlorine ions, Cl^-, which are released by the same action.

Bases taste bitter and flat. They reduce the amount of H^+ in a solution. In other words, bases are proton acceptors. The ion OH^- is a powerful base because it will readily capture free H^+ ions to form H_2O. For this reason, when an acid and a base combine, they form a compound plus water. For example,

$$NaOH + HCl \rightarrow H_2O + NaCl$$

The acidity of a substance is measured on the pH scale. The term pH means "hydrogen power." The reference point on this scale is 7; thus substances with a pH value below 7 are acids and those with a pH value above 7 are bases. A substance with a pH of 7 is neutral. The scale is based on the concentration of H^+ in one liter of water. For example, pH 3 means 10^{-3} moles of H^+ ions per liter of water, and pH 8 means there are 10^{-8} moles per liter. A *mole* is the number of grams of a compound equal to its weight. Almost all biological processes take place in a pH environment of 6 to 8, with a few important exceptions such as digestive processes in the stomach (which occur at about pH 2). The pH level in living systems is sometimes regulated by "buffers," acid–base pairs that serve to "soak up" small amounts of excess acid or base ions.

pH range showing relative number or H^+ and OH^- ions in solution.

Strongest Acids	10,000,000,000,000	H^+ ions	← at pH 1 →	10	OH^- ions	Concentrated Nitric Hydrochloric, Sulfuric Acids (stomach)
	1,000,000,000,000	H^+ ions	← at pH 2 →	100	OH^- ions	
	100,000,000,000	H^+ ions	← at pH 3 →	1,000	OH^- ions	
Weak Acids	10,000,000,000	H^+ ions	← at pH 4 →	10,000	OH^- ions	
	1,000,000,000	H^+ ions	← at pH 5 →	100,000	OH^- ions	
	100,000,000	H^+ ions	← at pH 6 →	1,000,000	OH^- ions	Blood and Tissues of Organisms
Neutral	10,000,000	H^+ ions	← at pH 7 →	10,000,000	OH^- ions	
	1,000,000	H^+ ions	← at pH 8 →	100,000,000	OH^- ions	
Weak Bases	100,000	H^+ ions	← at pH 9 →	1,000,000,000	OH^- ions	
	10,000	H^+ ions	← at pH 10 →	10,000,000,000	OH^- ions	
	1,000	H^+ ions	← at pH 11 →	100,000,000,000	OH^- ions	
	100	H^+ ions	← at pH 12 →	1,000,000,000,000	OH^- ions	Concentrated Sodium Hydroxide
Strongest Bases	10	H^+ ions	← at pH 13 →	10,000,000,000,000	OH^- ions	
	1	H^+ ion	← at pH 14 →	100,000,000,000,000	OH^- ions	

3.11 The molecules of nitroglycerin in dynamite will rearrange themselves to a lower energy level when the proper energy of activation is provided. In the case of nitroglycerin that energy is usually provided by a concussion which forces the molecules closer together and causes them to interact. Strangely enough, heat *may* not do the trick here. I have seen old-timers in Alaska carry burning dynamite for a torch as they walked into dark mining tunnels.

they come in contact with each other. Reactions of this kind have been known to cause explosions as amateur chemists randomly mix chemicals. Other reactions need some extra energy boost to set them off. For example, a rock at the top of a hill may need a shove before it begins to release its energy by tumbling down. And a match can burn, but it can exist quite nicely without ever bursting into flame. To initiate the oxidation process you have to add the heat produced by friction. Once the reaction has started, it will continue, at least long enough for one to light whatever one is smoking. The energy boost required to initiate a given chemical reaction is called its *energy of activation.*

In the case of the match, why should heat be able to provide the critical energy of activation? What does heat do? Heat causes an increase in the kinetic energy of the molecules; that is, it increases the speed of their random movements. The more active the molecules are, the greater the likelihood that they will bump into each other, and it is in such encounters that molecular bonds are altered.

Another way of increasing the likelihood of molecular contact (or chemical reaction) is to increase the concentrations of the two reactants. Or the mixture of reactants can be compressed to force the molecules closer together. Thus an increase in pressure has the same net effect as increasing the concentrations.

All these methods are fine for initiating interactions in laboratory

3.12 How an enzyme is believed to work. The enzyme, usually a complex protein, has an area that roughly fits the substrate upon which it will act. In this example the enzyme is separating the substrate into its component parts, but in other cases enzymes may join smaller molecules to form larger ones. It is believed that the enzyme doesn't precisely fit the substrate configuration so that when the enzyme/substrate complex is formed, the substrate is bent out of shape, or stressed, a bit. The stress exposes certain parts of the molecular substrate to other reactants in the medium, and the substrate molecule, which would otherwise have been quite stable, reacts. After the reaction, the substrate breaks away from the enzyme, leaving it unchanged and ready to initiate the same reaction with new substrates. Each type of enzyme reacts with only one kind of substrate.

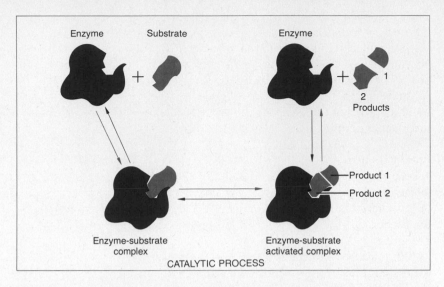

chemicals, but how are chemical reactions activated in a living organism? The application of heat or pressure or drastic increases in the reactant concentrations in living things would be disruptive to say the least. As you may have guessed, there is another way chemical reactions may be initiated—one more compatible with the delicate mechanisms of life.

Enzymes

Suppose you tried to impress your little brother by mixing hydrogen and oxygen in hope of forming water. He probably won't be very impressed because nothing would happen. However, if you added a tiny piece of platinum to the mixture, the hydrogen and oxygen would react violently enough to blow his hat off. And, interestingly enough, afterwards you could recover your platinum. The platinum served as a *catalyst,* and a characteristic of most catalysts is that they trigger or speed up reactions between other elements without actualy entering into the reaction themselves. They do not change the type of reaction that can take place, or the direction of the reaction. They simply make it possible for the reaction to take place by lowering the energy of activation. It is through this process of *catalysis*—chemical reactions initiated by catalysts—that the "cold" chemistry of life can take place.

The special types of catalysts found within living things are called *enzymes* (identified by one memorable student as "naval officers"). All enzymes are proteins. Proteins, as we shall see later, are large, highly complex molecules, and it is this complexity that provides for the specialized action of different enzymes. Figure 3.12 illustrates the process by which an enzyme is believed to work. Note that the reactants are not an exact fit with the enzyme molecule, so that for an instant certain bonds in the reactants are placed under a bit of stress. The resultant weakening of these bonds renders the reactant susceptible to attack by other agents, such as the functional groups of another molecule or free H^+ or OH^- ions. Once the new molecule has formed, it detaches and diffuses away from the enzyme molecule. This whole process takes place so quickly that a single enzyme may go through several million such cycles a minute in a dazzling display of efficiency. The synthesizing enzymes are capable of joining two substrate molecules. This is what happens, for example, in the process of building proteins, but enzymes can also act to break down molecules into their component parts, as in the process of digestion.

Energy Changes

All chemical reactions are accompanied by a change from one energy level to another. That is, the product of a chemical reaction will have either more or less energy than the starting material. If the product has more energy, the process is called *endergonic;* if it has less energy, the process is called *exergonic.*

But just as a boulder cannot roll up a hill without a push, neither can endergonic chemical reactions take place without an expenditure of energy. Thus in an endergonic reaction:

$$A + B \xrightarrow{\text{energy}} AB$$

However, once you have gotten your boulder up the hill, it has considerable potential energy, and if it should roll back down, that energy would be released. Similarly, when the product of a chemical reaction has less energy than the starting material, that extra energy must be released during the reaction:

$$AB \xrightarrow[\text{energy}]{} A + B$$

Of course, if the boulder happened to hit you on its way down, it might cause certain changes in your body. In this case the boulder's energy would be doing work. We'll see what kind of work can result from exergonic reactions when we discuss photosynthesis and respiration. Before we consider energy relationships in living things, however, let's take a look at a few key kinds of substances which enter into those reactions. Their names are bandied about rather freely, especially by people on diets, but have you wondered what a carbohydrate really is? Or a fat, or protein, or nucleic acid? Now, at last, you can know.

THE MOLECULES OF LIFE

Carbohydrates

Carbohydrates are molecules that contain carbon, hydrogen and oxygen in a proportion of one carbon to two hydrogens to one oxygen atom (CH_2O). They are important because they are the principle energy source for most living things and because they serve as the basic material from which many other kinds of molecules are built. Carbohydrates may be very simple, like

3.13 Glyceraldehyde, glucose, and fructose are important simple sugars. They are called simple sugars because they are not combined with other sugars. Note that glucose and fructose may exist as straight chains or as rings. Interestingly, fructose formed from glucose in the male reproductive tract is an energy source for sperm. Glucose is an important energy source in many kinds of organisms. In humans it is stored in the form of glycogen in a number of places including the liver and the muscles.

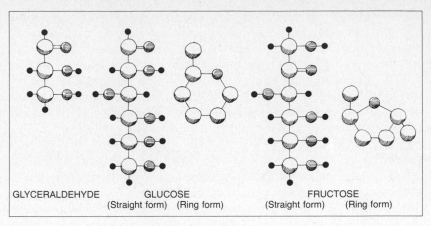

GLYCERALDEHYDE GLUCOSE (Straight form) (Ring form) FRUCTOSE (Straight form) (Ring form)

3.14 Sucrose, or table sugar, is a double sugar composed of both glucose and fructose. In the formation of sucrose, glucose loses an OH group that joins with an H from fructose to form water. The two simple sugars then join at the point at which these elements were lost. The reaction does not proceed easily and in living systems an energy source must be provided. When the elements of water are chemically added to sucrose, it breaks down into glucose and fructose, its component parts. Reactions involving loss of water are called condensation or dehydration; those in which water is added are called hydration, and they result in a chemical breaking apart known as hydrolysis.

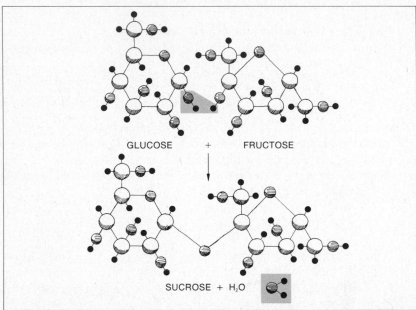

GLUCOSE   +   FRUCTOSE

SUCROSE + H₂O

glyceraldehyde, or they may be more complex, like glucose and fructose. These longer carbohydrates may exist as either chains or rings.

The simpler carbohydrate molecules are called *monosaccharides,* or simple sugars (Figure 3.13). Two simple sugars may combine to form a *disaccharide,* or many of them may join together to form various *polysaccharides.* For example, when the elements of water are removed from glucose and fructose, these sugars may combine to form sucrose, or table sugar (Figure 3.14). The process of removing water is called *condensation* or *dehydration.* When water is chemically added to condensation products, a process called *hydrolysis,* the product is broken down back into its component substances. You can see why hydrolysis is an important process in digestion.

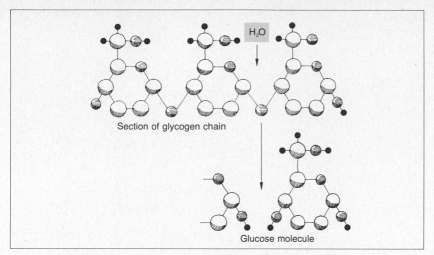

Section of glycogen chain

Glucose molecule

3.15 Glucose units can be joined together in long chains to form polysaccharides. Glucose units may also form cellulose, which is a rather resilient molecule found, among other places, in the cell walls of plants. Cellulose is partly responsible for the great strength of wood. Glucose units may form starch, which is a storage form for this simple sugar in plants or it may be found as glycogen, the usual form in which glucose is stored in animals. In comparing this with the next figure, note the difference in the linkages between glucose units in glycogen and starch as compared to cellulose.

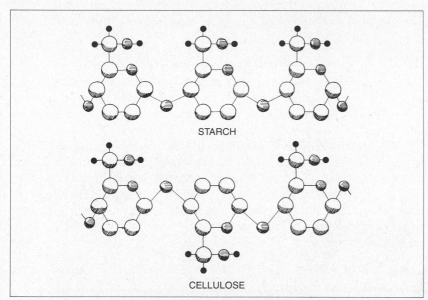

STARCH

CELLULOSE

3.16 Segments of starch and cellulose molecules. Starch is the form in which many animals store glucose units. (Notice the relationship between a link in this chain and an isolated glucose molecule.) The links in starch are easily broken by the digestive processes of many animals, but the links holding the units of cellulose together resist disruption by the digestive enzymes of all but a few species, such as protozoa that live in the digestive tract of termites. Since cellulose is a component of wood, we don't normally eat wood (except those people who use flour with "added fiber" and that fiber turns out to be sawdust—as was discovered in at least one case in 1977). Compare the kinds and arrangements of atoms in starch and cellulose.

Glucose molecules may combine through condensation to form long branching chains of *glycogen* (Figure 3.15). Glycogen, which is a starch, is the principal form in which glucose is stored in the body, chiefly in the liver. When your body is in need of glucose, the glycogen is disrupted by hydrolysis and the component glucose molecules are released into your bloodstream.

Glucose units may also be joined by a somewhat different sort of linkage to form cellulose, a relatively insoluble polysaccharide that is common in plants. Our digestive systems can readily break down starches into their component parts, but the particular bonding in cellulose makes it fairly resistant not only to hydrolysis, but also to most digestive enzymes.

3.17 Molting of the desert locust. As the insect grows and finds its old exoskeleton too confining, it takes a firm grip on a plant and, using enzymes to split its old exoskeleton, it wriggles free, leaving a ghostly apparition of its former self. (Of course, it can always return to view that shell and see for itself how undeveloped it once was, just as an author might treat a first book.) Its soft body will grow another exoskeleton that will be composed primarily of glucose units and various salts. These substances form a covering called chitin, a material that is indigestible by most organisms.

This is why people don't eat much wood and are rarely seen grazing. Cows and termites have some help: they serve as hosts to certain one-celled organisms that live in their digestive tracts and can break down cellulose into sugars.

The hard covering, or exoskeleton, of insects is also composed of units of sugar bonded in a way that makes them indigestible. I was once following a young man down a street in Austin, Texas, during one of the area's regular cricket invasions when he suddenly scooped a handful of the hapless insects from a wall and stuffed them into his mouth. I suppose I could have explained to him the nature of sugar bondings, but I walked into a trash can instead.

Lipids

Like carbohydrates, *lipids* are composed of carbon, hydrogen, and oxygen. Lipids, however, have far less oxygen. Also, lipids are insoluble in water.

Fats are only one type of lipid, although they are perhaps the best known. There are a number of kinds of fats, and these are of interest for a variety of reasons. Most important, however, fats and their fatty acid components are the chief forms of energy storage in the body. They contain twice as much energy for their weight as do carbohydrates or proteins. You would think from this that fat people ought to be bursting with strength and energy. You'll see as we go along why they aren't.

Each molecule of fat consists of two types of compounds, the alcohol *glycerol* and *fatty acids*. Glycerol is a chain of three carbon atoms, each with a hydroxyl (OH) group. In the formation of fats, the hydroxyl groups of the glycerol join with the hydrogens of the carboxyl (COOH) units of

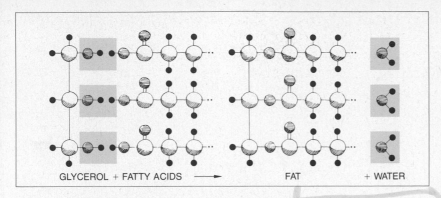

GLYCEROL + FATTY ACIDS ⟶ FAT + WATER

fatty acids through the process of condensation (the removal of water). This process is outlined for you in Figure 3.18.

It should be pointed out that the fatty acid chains are highly variable. They are usually sixteen to eighteen carbon units long, but they may be much shorter or longer. The three fatty acid components also don't have to be all the same length and fats may vary in the way the carbon atoms are bonded. For example, in some fats the carbon chains are fully saturated with hydrogen atoms. In other fats, however, certain carbon atoms may be joined by a double bond instead of filling their shell requirements by adding hydrogen atoms (see Figure 3.19). The former type of fat is called *saturated.* Guess what the latter type is called.

There are other important lipids besides fats. For example, there are the *waxes,* with their long saturated fatty acid chains. And then there are

SATURATED FAT

UNSATURATED FAT

3.18 A fat molecule is composed of a glycerol molecule joined to fatty acids. Three molecules of water are lost in the bonding of three fatty acid molecules to a single molecule of glycerol. The fatty acids form long hydrocarbon chains the ends of which link with the glycerol molecule. The key questions that determine the nature of the fat are: how long are the chains, and are the fatty acids saturated or unsaturated (as we see in the next figure).

Fat molecules store a great deal of energy, but it is not easily retrievable. Marathon runners normally expend their available glucose after two hours of running and then must switch to fat metabolism. For some reason the shift is considered traumatic, and perhaps more easily accomplished by women than men.

3.19 Unsaturated and saturated fats. In recent years a generally overfed American populace has become increasingly concerned about the qualities of all the foods available to them. For example, are our fats saturated or unsaturated? Have you heard advertisers use the term "polyunsaturated?" Househusbands or housewives may not know that unsaturated simply means that carbon atoms are linked together by covalent bonds of two pairs of shared electrons—in other words, by double bonds. Carbon atoms linked by double bonds are called unsaturated because they are able to form new bonds with other atoms. Unsaturated fats are generally oily, like olive oil, while saturated fats, like lard, are usually solid at body temperature.

3.20 Cholesterol is a steroid. All steroids are composed of our basic interlocking rings, but a variety of side groups may be attached to those rings. Cholesterol is abundant in foods such as milk, butter, and eggs, and it may be synthesized in the liver as well. Cholesterol circulates in the bloodstream and if present at high levels, it may be deposited in the aorta, the artery leaving the heart, possibly causing arteriosclerosis and heart attacks. As it circulates through the bloodstream it may be picked up by specialized cells of the sex and adrenal glands and altered to form the steroid sex hormones. No one yet knows how these hormones work their wonders. So if steroids don't kill us outright, they at least seem to keep us in a lot of trouble.

the *phospholipids*. In phospholipids the third hydroxyl group of glycerol is attached to a molecule that contains phosphorus. Phospholipids are important in the structure of living membranes, as we shall see.

Another important class of lipids is the *steroids*. The steroids are fatlike substances that have a structure quite unlike that of the other lipids we have considered. Whereas the other lipids are formed from the bonding of fatty acids and alcohols, steroids are formed from four interlocking carbon rings with a variety of side groups attached (Figure 3.20). Some people develop a deep and abiding affection for steroids when they learn that sex hormones are included in this group.

Proteins

The general structure of proteins is simple enough to describe, but this simplicity is deceptive. Proteins are actually immense molecules with incredibly complex structures. Let us not be deterred, however, as we plunge ahead into new realms of knowledge.

We will, of necessity, be dealing largely in generalities. Also, our generalities will be based on many assumptions, since we know the actual structures of relatively few proteins. Fortunately, however, proteins are composed of smaller building blocks that we do know something about.

Amino Acids

Proteins are made up of chains of smaller nitrogen-containing molecules called *amino acids*. As you may have guessed from the name, amino acids contain an amine group (NH_2) and an acid group (COOH). Both of these functional groups are attached to the same carbon. The carbon chain itself comes in a variety of forms, and for the sake of simplicity, we'll just call it *R,* for radical. For a molecule to qualify as an amino acid, then, it need only have the configuration

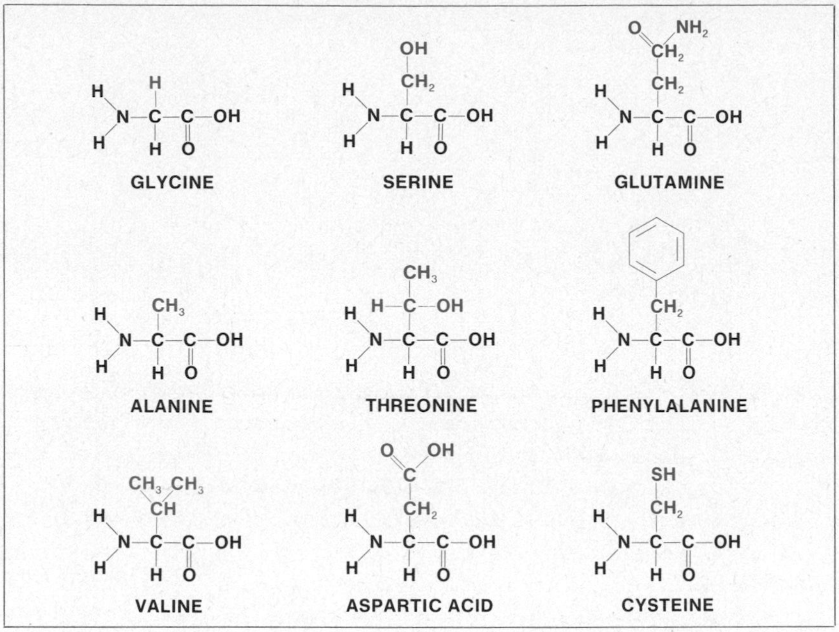

GLYCINE

SERINE

GLUTAMINE

ALANINE

THREONINE

PHENYLALANINE

VALINE

ASPARTIC ACID

CYSTEINE

3.21 Some of the amino acids found in proteins. Note the variety of structures among amino acids, but also note that they all have two things in common: an amino group and an acid, or carboxyl, group. There are only twenty amino acids in all, and the body can manufacture twelve of these from other amino acids. It is essential that the other eight, however, be provided in the diet, so they are called the "essential" amino acids. It seems paradoxical that the amino acid element that is probably available from the fewest sources is nitrogen. And yet it is this same nitrogen that provides us with such problems with its disposal as we will see in our discussion of the kidney. The importance of amino acids, of course, lies in their role in the formation of proteins.

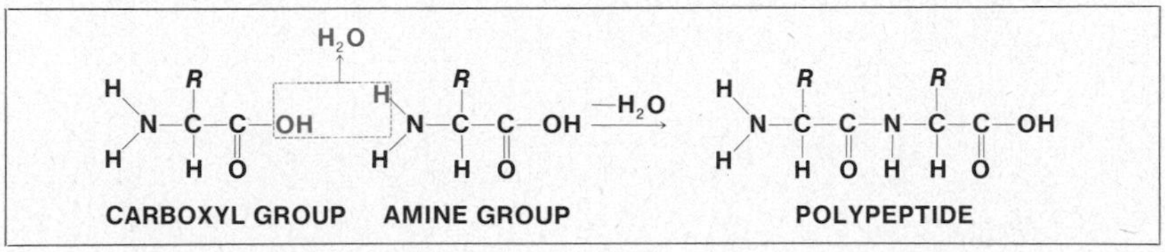

CARBOXYL GROUP   AMINE GROUP

POLYPEPTIDE

3.22 The carboxyl group of one amino acid can join with the amine group of the next amino acid by each contributing to the formation of water. The result is a peptide bond. Polypeptides, then, and their larger cousins, proteins, are chains of amino acids joined by peptide bonds.

Remember that the COOH may be written

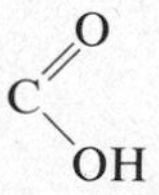

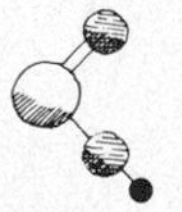

There are many known amino acids, but only twenty appear in proteins. Some of these are shown in Figure 3.21. As you can see, the carbon group *R* determines the name of the compound.

Peptides

Amino acids join together to form long *peptide chains* (polypeptides), and one or more of these then make up the protein. Polypeptide chains may consist of as many as 300 amino acids. The amino acids need not be of the same type, but they are always linked together in the same ways; the OH part of the carboxyl group joins with the H of an amine group to form water, as shown in Figure 3.22. The resulting linkage is called a *peptide bond.*

Box 3.3 Unraveling the Structure of Insulin

In 1954, after ten years of painstaking research, Frederick Sanger and his colleagues at Cambridge University in England were able to tell us the full sequence of the amino acids in insulin. This work, for which Sanger was awarded a Nobel Prize, consisted in breaking the insulin molecule into smaller pieces of protein and then analyzing the pieces. First he subjected the entire molecule to strong acid to break it down to its component amino acids. He was then able to identify each of these amino acids by a process called chromatography. This involves filtering the product through a column of an adsorbent material for which different molecules have different affinities. Each amino acid thus adheres to the material at a different rate and so is stopped at a specific point as it moves down the tube. By comparing his results with the known adsorbing point of each acid, Sanger determined which amino acids were present in insulin. He was also able to learn something about their relative amounts.

To investigate the sequence of amino acids in the molecules Sanger used more specific hydrolyzing agents to break certain bonds in the long chain. Pepsin, for example, hydrolyzes bonds only between tyrosin or phenylalanine and other amino acids. To determine which end of the segment was free and which was attached, he added a chemical (dinitrofluorobenzene) that attaches to any terminal amino group and turns the compound a bright yellow.

After analyzing many such short segments, Sanger was eventually able to work out their structures, and where the terminal amino acids of these groups were identical, he could overlap them to reconstruct

longer chains. For example, an overlap of the segments

A—C—X—N—O—F
           N—O—F—G—O—I

yielded the longer segment

A—C—X—N—O—F—G—O—I

Later he was able to determine how the two component molecules of insulin were joined by bonds between sulfur atoms of cysteine molecules. Thus, after years of tracking down small bits of information, patiently, one step at a time, Sanger was eventually able to put all the pieces together and tell us for the first time the nature of a relatively simple protein.

It is the sequence of amino acids in the chain that determines the biological character of the protein molecule. A change in the position of a single amino acid in a sequence of thousands can alter or destroy the activity of the protein. You can imagine the incredible number of proteins possible through the variable sequencing of amino acids. You can also imagine the high degree of specificity that is possible in such a system. It turns out, in fact, that although all proteins utilize only twenty amino acids as building blocks, almost every species has proteins that are peculiar to it alone.

Protein Structure

A protein molecule may be described in terms of its amino acid sequence, which constitutes its *primary structure*. As protein chains form, however, they also take on a *secondary structure*. As a result of the hydrogen bonds that develop between every fourth amino acid, the protein molecule spirals into a helical shape, like a circular staircase. Its helical arrangement imparts its secondary structure. Proteins that have structural functions in the body—those that protect the body or keep it from falling apart—usually form ropelike fibers or sheets. Hair, nails, cartilage, bones, and tendons all contain such fibrous proteins. These proteins usually have only primary and secondary arrangements.

But not all proteins are fibrous. Some fold back on themselves in a *tertiary structure* to form dense clumps, or globules (see Figure 3.23). Enzymes, for example, are globular proteins with complex tertiary structures. A protein maintains its tertiary structure by means of hydrogen bonds, hydrophobic (water-repelling) forces, and disulfide linkages (bonds between cysteine amino acids along the chain). Bonds of this kind are weak and can be broken easily by such things as heat or acid. Once the tertiary bonds are broken, the molecule can become fibrous (or "denatured"). This is what happens when you cook the protein-rich white of an egg. Such changes are irreversible, as you know if you have ever tried to unfry an egg.

In addition to their tertiary structures, two or more globular proteins may twine around each other to form a *quaternary structure*. Hemoglobin, the blood's oxygen-carrying molecule, has such a configuration.

Now that you have some idea of the basic chemistry of life, let's turn our attention to how such molecules function in living systems.

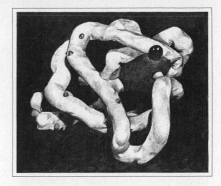

3.23 The tertiary folding of a hypothetical protein. *R* groups would extend outward in every direction, filling the internal areas and building a roughly globular structure. Globular proteins, then, are actually quite dense. The tertiary formation is maintained mostly by weak hydrogen bonds, by repelling interactions between different parts of the chain, and by linkages between sulfur groups. All these can be broken quite easily, for example by heat, and the former structure can never be regained. The process is called *denaturation* and, in general, it involves unfolding tightly coiled peptide chains so that they form a random configuration. Of course, denatured proteins lose their biological activity.

Energy

4

If we are ever to understand the nature of the "real world," we must learn more about the behavior of energy. Energy exists in a variety of forms, at different levels, and it ebbs and flows in countless directions throughout the universe. Moreover, energy can be trapped or stored, and then released to do work at some later time. Molecules, physical things, can be rearranged so that they hold more energy. Then, when they are converted back to their former structure, that energy is released. A person holding a jar of nitroglycerine would probably be obsessed by a single hope: that the molecules will not rearrange themselves to a lower energy level.

Potential energy is stored energy; it is doing no work. When potential energy is released to do work, it then becomes *kinetic energy*. The jar of nitroglycerine and the rock at the top of the hill have potential energy which can be released to do work.

Both potential and kinetic energy can take many different forms. A car battery has potential electrical energy. When that energy is released to turn over a starter, it becomes mechanical energy. As the parts of the starter move, friction causes some of the initial energy from the battery to be dissipated as heat energy. Thus we see that not only can energy exist in different forms, but it can also be converted from one form to another.

THE LAWS OF THERMODYNAMICS

There are a few simple laws that govern the behavior of all energy, regardless of its form. According to the *first law of thermodynamics,* energy can be neither created nor destroyed, but can only be changed from one form to another; hence the total amount of energy in a system remains the same, regardless of the physical or chemical changes it may undergo. This means that the energy that is released as our boulder topples down the hill must be directly proportional to the amount of energy we had to exert to get it up the hill. It also means that the energy that is "lost" as the heat of friction is not actually lost; it is merely unavailable for use in that system. However, in any energy transformation, according to the *second law of thermodynamics,* energy tends to dissipate. Thus after each energy transformation, the available energy in a system will be slightly lower than before. As a result, according to the second law of thermodynamics, all processes act so as to "increase entropy." *Entropy* is a measure of disorder, so this means that a system becomes increasingly disorganized as its available energy decreases.

4.1 A rock perched on a ledge has potential energy. So does a stick of dynamite, a car battery, and a peculiar molecule called ATP (which we will encounter shortly). Potential energy is simply the ability to do work. The term indicates that energy can be stored. The hibernating skunk behaves as if it knows this. So does the bird stuffing itself in the spring before its long migration northward. The principle of energy storage, particularly at the cellular level, makes a logical and fascinating story.

"Only the Sun gives without taking," the hip saying of the 60s once scrawled on the men's room wall of Schultz's Biergarten in Austin, Texas, is true to the extent that the sun probably receives nothing from the earth. The author probably intended to make a social comment in the 30 seconds or so he had available, but it makes sense biologically as well. The sun produces its radiation when hydrogen is converted to helium (another question could center over where the hydrogen came from). Once that energy reaches the earth, however, it may enter the biological system—a place where there are no free lunches. When that energy is passed along, it is transferred at a cost. It moves in countless ways through reciprocal and interdependent living systems.

With a little imagination, you can see that these energy laws have some very interesting implications for living systems. Any organism must expend energy simply to remain organized. And since life can only continue in highly organized states and our environment is essentially disruptive, living systems must have both a source of energy and some efficient means of converting it from one form to another.

In the biological world the ultimate source of all energy is the sun. This energy is captured, changed, shifted about, and filtered through living systems, where it is used to reduce entropy. The sun's energy begins its useful progression in green plants, where it is stored in rather high-energy molecules. Each stage in life's energy cycles extracts its share of energy from those molecules, until finally all that is left are low-energy molecules of carbon dioxide and water. Let us now consider a few of the chemical reactions that result in your having the energy to do whatever it is you would rather be doing.

ATP: THE ENERGY CURRENCY

The standard energy currency of cells is a fairly simple molecule called *adenosine triphosphate* or ATP. ATP is referred to as "energy currency" because it must be "spent" for work to be done. It is spent by breaking one of its high-energy bonds, thus converting the molecule to adenosine diphosphate, or ADP, plus inorganic phosphate, or P_i. In other words,

$$ATP \xrightarrow{\text{energy}} ADP + P_i$$

Now, as you know, in order to have money to spend, work must be done (by someone). So it is with the chemistry of life. In order to produce ATP, energy must be expended somewhere along the line. That energy is used to put a molecule of inorganic phosphate back onto an ADP molecule, a place where it really doesn't "want" to go. It has to be forced back onto the ADP molecule by the expenditure of energy. Thus

$$ADP + P_i \xrightarrow{\text{energy}} ATP$$

But if ATP provides the energy necessary for life's functions, and energy is necessary for the reconstruction of ATP, we are left with a problem. Where does the energy necessary to put the phosphate on the ADP come from? Unlikely as it seems, that energy comes from the sun. What we will see now is how green plants use sunlight to form ATP, which they then use to make glucose, and how animals, by eating plants, use the plants' stored glucose to form ATP of their own.

PHOTOSYNTHESIS

Photosynthesis is a remarkable process that takes place in green plants and some other organisms. In fact, green plants get their color from a pigment

Only the
Sun gives
without
taking.

4.3 ATP is actually adenylic acid (comprised of adenine, ribose, and a phosphate) attached to two more phosphates.

The last two phosphates are attached by high energy bonds. When these bonds are broken, large amounts of energy are released, energy that is available to do work.

called *chlorophyll,* which drives the photosynthetic mechanism by capturing the energy in light. There are two distinct steps in the photosynthetic process, one that requires light (the light reaction) and one that doesn't (the dark reaction). The net result of this vital drama is that carbon dioxide and water are converted to glucose and oxygen:

$$6CO_2 + 12H_2O + \text{sunlight} \xrightarrow[\text{enzymes}]{\text{chlorophyll}} \underset{\text{glucose}}{C_6H_{12}O_6} + 6O_2 + 6H_2O$$

Many of the details of this process remain cloaked in mystery to this day, especially some of the dark reactions. However, more is known than you would probably care to have described to you, so we will consider photosynthesis only in rather general terms.

Box 4.1 Chlorophyll

Chlorophyll is a pigment molecule found in the chloroplasts of all green plants. The green color you see in such plants is actually caused by light rays from the center portion of the spectrum, which are reflected back to your eye by the molecular structure of the chlorophyll. The colors you do not see—the reds and purples at either end of the spectrum—are the ones that are absorbed. This scheme is not coincidental. The red and violet light rays are the ones that contain the highest concentration of light energy, energy that the chlorophyll uses in the vital process of converting light energy from the sun into usable energy for all living things.

Chlorophyll is usually a mixture of two different compounds known as chlorophyll A and chlorophyll B:

$$C_{55}H_{72}O_5N_4Mg$$
Chlorophyll A

and

$$C_{55}H_{70}O_6N_4Mg$$
Chlorophyll B

Most green plants have more chlorophyll A, although a few, including some algae, ordinarily utilize chlorophyll B or some related compound instead.

The chlorophyll molecules are located in the chloroplasts of green plants. Each cell may contain as many as eighty of these small bodies, which hold the chlorophyll molecules stacked in piles within their shelflike grana. In this way the largest possible reactant surface is made available for the important chemical reactions in which these pigment molecules participate.

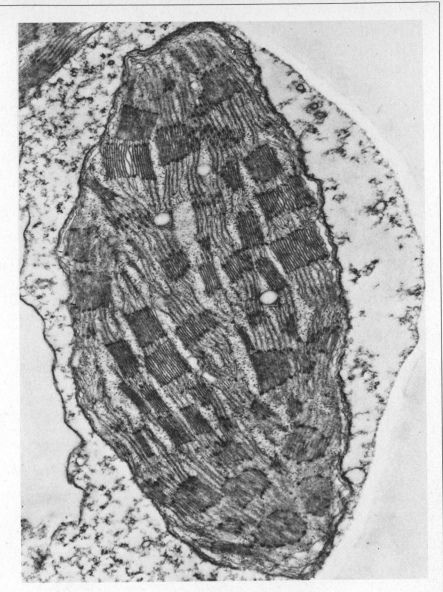

CHLOROPHYLL A

The Light Reaction

Photosynthesis begins when light strikes a chlorophyll molecule and initiates the light reaction. The chlorophyll is activated by the light energy, so that two of its electrons move to a higher energy state and are then captured by an electron acceptor. The chlorophyll replaces its lost electrons by literally tearing electrons from a water molecule, and this disruption causes the water molecule to fall apart, releasing the oxygen as a gas. This entire process takes place within the system designated as photochemical system II (Figure 4.4).

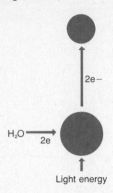

From the electron acceptor in photochemical system II, the electrons are passed through an electron-transport chain. The chain consists of a series of electron-acceptor molecules, each at a lower energy level than the last, so that the electrons progressively lose their energy as they are passed from one molecule to the next. Some of this energy is used to make ATP from ADP and P_i.

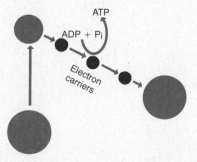

Finally the electrons are passed along to another pigment system at the bottom of photochemical system I (designated as I because it is believed to have evolved first). The fact that the first system occurs second is a bit confusing, but it's probably the least of your worries. Then light energy again boosts the electrons to another electron acceptor, this one at an even higher energy level than the last. From this acceptor the electrons are passed along to a nucleotide of adenine called $NADP_{ox}$ (since it is in oxidized form), thus reducing it to $NADP_{red}$. You will notice in Figure 4.4 that the electrons accepted by $NADP_{ox}$ are still at a high energy level.

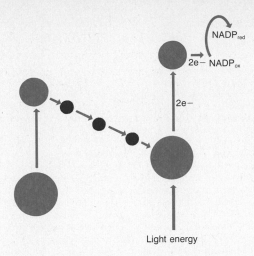

Light energy

The light reaction is sometimes called the Z reaction because someone thought its diagram formed a Z (actually, it forms an N unless you turn the book sideways).

The Dark Reaction

The dark reaction is the process by which plants actually manufacture food from water and carbon dioxide. It is called the dark reaction because light is not necessary to drive this part of the mechanism. The energy in this case comes from the ATP and high-energy electrons in the $NADP_{red}$ that were produced in the light reaction. The major biochemical pathway in this reaction, also called the *Calvin cycle,* is traced in Figure 4.4.

The dark reaction consists of fourteen or fifteen complex steps, which we need not go into here. But, basically what happens is that a carbon dioxide molecule is added to a five-carbon sugar molecule that is already present in the cell. This five-carbon sugar is *ribulose diphosphate,* or RuDP:

$$CH_2\text{—}O\text{—}P$$
$$|$$
$$C\text{=}O$$
$$|$$
$$CHOH$$
$$|$$
$$CHOH$$
$$|$$
$$CH_2\text{—}O\text{—}P$$

The additional carbon results in a very unstable six-carbon sugar, which quickly splits into two three-carbon sugars. Then there is a complex series of events in which the bonds between carbon atoms are broken and reformed, eventually producing two molecules of phosphoglyceraldehyde, or PGAL. The PGAL is the end product of photosynthesis. PGAL may then enter into a number of metabolic pathways. If the plant is in need of

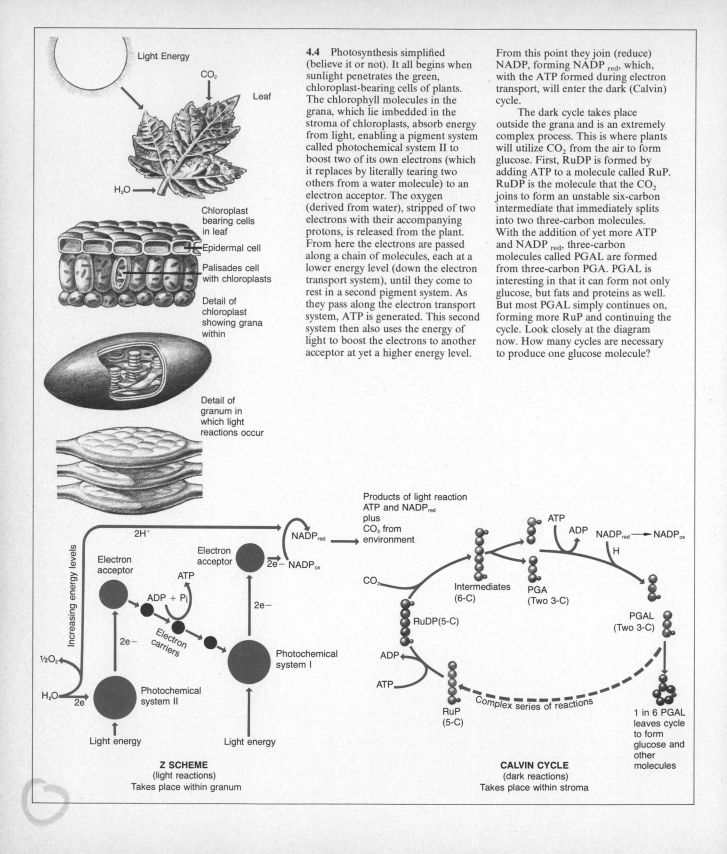

4.4 Photosynthesis simplified (believe it or not). It all begins when sunlight penetrates the green, chloroplast-bearing cells of plants. The chlorophyll molecules in the grana, which lie imbedded in the stroma of chloroplasts, absorb energy from light, enabling a pigment system called photochemical system II to boost two of its own electrons (which it replaces by literally tearing two others from a water molecule) to an electron acceptor. The oxygen (derived from water), stripped of two electrons with their accompanying protons, is released from the plant. From here the electrons are passed along a chain of molecules, each at a lower energy level (down the electron transport system), until they come to rest in a second pigment system. As they pass along the electron transport system, ATP is generated. This second system then also uses the energy of light to boost the electrons to another acceptor at yet a higher energy level.

From this point they join (reduce) NADP, forming $NADP_{red}$, which, with the ATP formed during electron transport, will enter the dark (Calvin) cycle.

The dark cycle takes place outside the grana and is an extremely complex process. This is where plants will utilize CO_2 from the air to form glucose. First, RuDP is formed by adding ATP to a molecule called RuP. RuDP is the molecule that the CO_2 joins to form an unstable six-carbon intermediate that immediately splits into two three-carbon molecules. With the addition of yet more ATP and $NADP_{red}$, three-carbon molecules called PGAL are formed from three-carbon PGA. PGAL is interesting in that it can form not only glucose, but fats and proteins as well. But most PGAL simply continues on, forming more RuP and continuing the cycle. Look closely at the diagram now. How many cycles are necessary to produce one glucose molecule?

Light Energy

CO_2

Leaf

H_2O

Chloroplast bearing cells in leaf

Epidermal cell

Palisades cell with chloroplasts

Detail of chloroplast showing grana within

Detail of granum in which light reactions occur

Products of light reaction ATP and $NADP_{red}$ plus CO_2 from environment

2H⁺

Electron acceptor

Electron acceptor

$NADP_{red}$

$2e-$ $NADP_{ox}$

ATP

ADP + Pᵢ

$2e-$

Electron carriers

$2e-$

Photochemical system I

Increasing energy levels

$2e-$

½O_2

H_2O

2e

Photochemical system II

Light energy

Light energy

Z SCHEME
(light reactions)
Takes place within granum

CO_2

RuDP(5-C)

ADP

ATP

RuP (5-C)

Intermediates (6-C)

ATP

ADP

PGA (Two 3-C)

$NADP_{red}$ → $NADP_{ox}$

H

PGAL (Two 3-C)

Complex series of reactions

1 in 6 PGAL leaves cycle to form glucose and other molecules

CALVIN CYCLE
(dark reactions)
Takes place within stroma

food, it can be used as is. The plant may also convert it into food stores in the form of glucose or starch. Or it may convert the PGAL to proteins or fats. Most of the PGAL, however, is changed back to RuDP, by the expenditure of energy from ATP, so that the cycle begins again.

Following is a discussion of how the food made by plants can be utilized to provide energy for other living things.

CELLULAR RESPIRATION

The glucose manufactured in photosynthesis is now ready to be "cashed in" for ATP molecules to drive the reactions of living cells. The plant may use its own glucose, of course, but that glucose is also available to animals that eat the plants, and then to other animals that eat those animals. Plant glucose is one of the things that the grazing steer is after—as is the man who eats the steak. The process of trading glucose for ATP is referred to as *cellular respiration*, since it takes place within the cell. Cellular respiration takes place in three stages: *glycolysis,* the *Krebs cycle,* and the *electron-transport chain.*

Briefly, what happens during respiration is that the glucose molecule is split, the hydrogen atoms (both protons and electrons) are transferred from the carbon atoms to oxygen atoms, and the energy released as a result of this transfer is used to convert ADP to ATP. Let's look at these steps in more detail.

Glycolysis

The process of glycolysis takes place in the cell fluid, or *cytoplasm.* At this stage the six-carbon glucose molecule is split up and changed into two molecules of a three-carbon substance called *pyruvic acid.*

In the formation of pyruvic acid the glucose is oxidized by the transfer of hydrogen to two nucleotides of adenine called NAD. The NAD_{ox} molecules thus become NAD_{red}. Also, during this process two ATPs are produced, as shown below. If no oxygen is present, these two ATPs are all the energy an organism can get from a glucose molecule. Not much. However, yeasts and other *anaerobic* organisms, those that do not utilize oxygen in their respiration, derive all their ATPs from glycolysis.

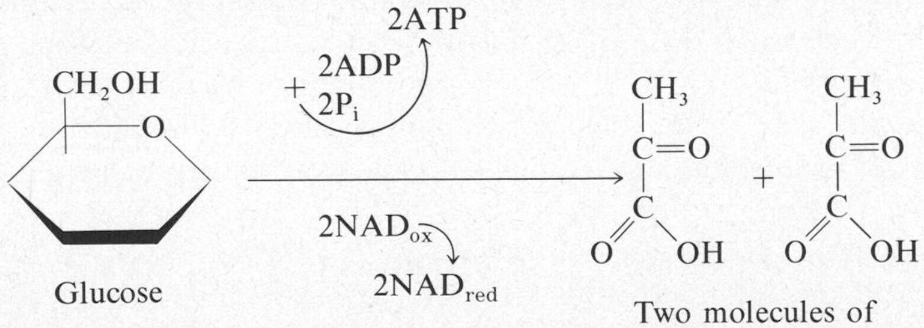

Glucose   →   Two molecules of pyruvic acid

Pyruvic acid is not the end of the line in glycolysis. In yeasts, the respiratory products are carbon dioxide and ethanol, or spirit alcohol. The

4.5 Pyruvic acid is not the end of the line in glycolysis. In yeasts, glucose is converted to ethanol (the alcohol of spirits) and carbon dioxide.

Yeast cells bloom on the skin of the grapes. When the juice of these grapes is left in airtight containers, the glucose of the fruit is changed to ethanol. Notice that CO_2 is released in the formation of ethanol, so you should somehow allow for the escape of gas to prevent your "experiment" from blowing up.

NAD_{red} gives up its two added hydrogens to form the alcohol and is then free to return as NAD_{ox} to pick up two more hydrogen atoms from a new glucose molecule. We have known about the practical aspects of this process for some time, since early man, confronted with all the great puzzles of the universe, busied himself at once with learning how to make booze.

In animals the product of oxygenless respiration is not alcohol, but lactic acid. It is not surprising that anaerobic animals, such as internal parasites, are fairly sluggish. They just don't have the ATPs to be frisky. When *aerobic* animals—those that require oxygen for respiration—are unable to get enough oxygen during prolonged exercise, some of their energy may be derived from glycolysis (since aerobic and anaerobic

4.6 The person who builds up his cardiovascular system through exercise such as jogging is better able to bring oxygen to his muscles as his circulatory system develops along with his muscles. He is thus more likely to stave off exhaustion as pyruvic acid enters the Krebs cycle instead of forming fatigue-inducing lactic acid.

processes can occur simultaneously in these animals). In this case lactic acid accumulates in the muscles, and they weaken with fatigue. If enough lactic acid accumulates, the muscles will fail to contract, or contraction will be sufficiently impaired to affect coordination. This is why a runner who is not in condition is usually gasping for air at the end of a race and may stagger across the finish line. A conditioned runner is likely to have developed his cardiovascular system along with his muscles, so that it will provide him with sufficient oxygen to prevent the buildup of lactic acid.

In mammals lactic acid is removed from the muscles by the blood system and carried to the liver, where it is reconstituted to glucose and stored as glycogen. This transformation back to glucose is uphill, of course, and thus requires an expenditure of energy in the liver. However, this is no great hardship, since the energy level of lactic acid is still fairly high. In other words, the boost to glucose is a short one.

In aerobic animals, the pyruvic acid formed in glycolysis has a

4.7 Respiration, tracing glucose through glycolysis, the Krebs cycle, and electron transport. Glucose enters our diagram at upper left. It is immediately converted to two three-carbon pyruvic acids, and two ATPs are formed that are converted to acetyl CoA as CO_2 is lost and NAD is reduced. The acetyl CoA is then ready to enter the mitochondrion.

When the two-carbon acetyl CoA crosses the mitochondrial membrane, which is studded with F_1 particles containing phosphorylating enzymes, it loses its CoA and joins oxaloacetic acid, forming a six-carbon molecule of citric acid. The molecule then continues through a series of changes, occasionally releasing CO_2 (which is lost) or electrons (which reduce the coenzymes FAD or NAD). One ATP is formed in the Krebs cycle. (Where?) When the cycle is complete oxaloacetic acid is formed, ready to join incoming acetyl CoA and repeat the process.

The reduced NAD and FAD unload their electrons at the electron transport system, the electrons being transferred to lower and lower energy levels, releasing energy at each step. This energy goes to form ATP by some unknown mechanism. When the electrons reach the lowest energy level, these tired particles are given over to oxygen, forming metabolic water. In all, then, thirty-eight molecules of ATP have been produced, two in glycolysis, the rest in the Krebs cycle and, particularly, the electron transport chain.

continuing and critical role. It goes through a series of reactions in which almost all the energy it holds is extracted from it. That energy goes to make new ATPs, many more than can be formed in glycolysis. Let's take a closer look as the energy-loaded electrons of pyruvic acid are passed along until they end up as mere components of humble water.

The Krebs Cycle

We know that in the absence of oxygen a glucose molecule provides an aerobic organism with only two molecules of ATP to carry out its functions. But if oxygen is present, it's a whole new ball game. The pyruvic acid will go through a series of reactions that generates thirty-six more ATPs. In aerobic respiration the two pyruvic acid molecules formed from the glucose are shunted into the *Krebs cycle,* named for its British discoverer Sir Hans Krebs. Eventually the high-energy pyruvic acid is broken down completely to the simple end products carbon dioxide and water, having released most of its energy along the way. Both the Krebs cycle and the next stage of respiration, the electron-transport chain, take place in tiny structures within the cell called *mitochondria.* (You'll have a chance to look at these in the next chapter.)

If oxygen is present the three-carbon pyruvic acid molecules produced by glycolysis oxidize—with NAD molecules again accepting two hydrogens—and lose one carbon in the form of carbon dioxide. The remaining two-carbon fragment, acetic acid, then unites with a protein substance called *coenzyme A,* or CoA, to form *acetyl-CoA:*

$$\underset{O}{\overset{\overset{\textstyle CH_3}{\overset{\textstyle |}{\underset{\textstyle \parallel}{C}}}}{}}\diagdown CoA$$

It is actually the acetyl-CoA that enters the Krebs cycle. It should be stressed that the body doesn't get all its energy from glucose. Fats and some amino acids also convert to acetyl-CoA, although they do so by a less direct route, and they too enter the Krebs cycle.

In the Krebs cycle the acetyl group separates from its CoA enzyme and combines with a four-carbon molecule called *oxaloacetic acid.* The result is the six-carbon compound citric acid, the common acid of lemons and limes. The citric acid is then oxidized, which results in the splitting off of two carbons in the form of carbon dioxide, and the remainder is again oxaloacetic acid, which is recycled to pick up another acetyl group from the next acetyl-CoA molecule. The cycle continues as long as acetyl-CoA is available. This process is illustrated in Figure 4.7. As you can see, the stored energy in the acetyl group has been used to convert three molecules of NAD_{ox} to NAD_{red}, one molecule of FAD_{ox} (another electron acceptor) to FAD_{red}, and one molecule of ADP to ATP. The glucose molecule has by this time been completely oxidized. Its products, however, will go on to provide the cell with more energy as they are passed along to the electron-transport chain.

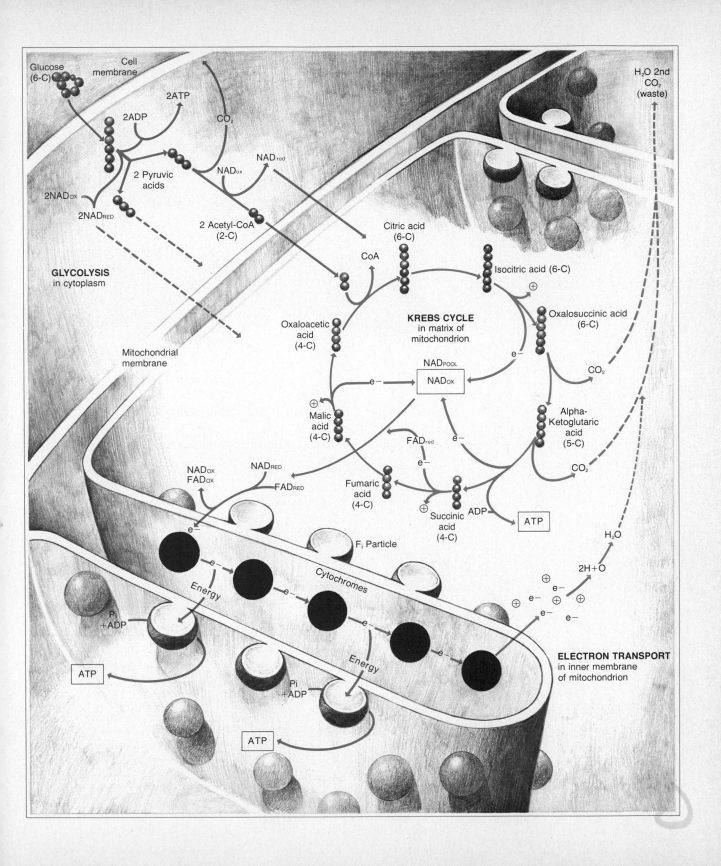

4.8 The basic cycle of life on earth involves two tiny cell inclusions: the chloroplast and the mitochondrion. The chloroplast first captures the energy of the sun's radiation and builds glucose. The mitochondrion then utilizes this glucose to form ATP, the chemical that stores energy in the cell. The CO_2 and H_2O formed as by-products of the mitochondrion's activity are returned to the chloroplast to be used in the formation of new glucose.

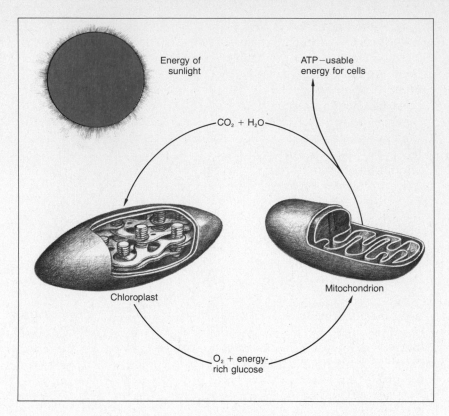

Energy of sunlight

ATP—usable energy for cells

$CO_2 + H_2O$

Chloroplast

Mitochondrion

O_2 + energy-rich glucose

Note that the electron acceptors NAD and FAD are all nucleotides, and each of these substances readily accepts and releases two electrons at once. Do you see a certain parsimonious efficiency in a system that makes multiple use of nucleotides in a continuous oxidation and reduction process? What do you suppose is the advantage of a cycling system in living things?

The Electron-Transport Chain

The last stage of aerobic respiration is the *electron-transport chain*. In this process the electrons that had been accepted by the NAD and FAD molecules are now passed along to a series of electron carriers similar to those referred to in photosynthesis. Here again each acceptor is at a lower energy level than the preceding one, a system that permits the energy of the electrons to be freed in a stepwise process. The energy that is released is used to form more ATP (Figure 4.7).

At the end of the chain the electrons join with protons, and the resulting hydrogen atoms combine with oxygen to produce water. In the series shown in Figure 4.7 the electrons are passed along in pairs from NAD, and three molecules of ATP are formed. However, if the electrons are instead passed along from FAD, which has a slightly lower energy level, only two ATPs are formed.

Let's stop at this point to retrace the respiratory reactions. During glycolysis the six-carbon sugar glucose is split into 2 three-carbon molecules. These molecules are then rearranged as pyruvic acid, and in the process a small quantity of ATP is made available for energy. In the absence of oxygen the pyruvic acid is converted in yeasts to carbon dioxide and alcohol and in aerobic animals to lactic acid. In the presence of oxygen in animals the pyruvic acid is instead converted to acetic acid, which combines with coenzyme A. The resulting acetyl-CoA then enters the Krebs cycle, where carbon dioxide is produced, along with some ATP, and electrons are transferred to FAD and NAD molecules. The FAD_{red} and NAD_{red} then release these electrons to the electron-transport chain, and as the electrons travel down the chain to lower energy levels, ATP is generated at each step. The remaining electrons are finally deposited in the "electron dump," oxygen, to form water. The energy left over after these reactions appears as metabolic heat. That's why the person next to you is so warm.

When we trace the energy of the sun through living systems, we can envision a cycle like that shown in Figure 4.8. Energy from the sun is utilized by the chloroplasts in plant cells to make oxygen and glucose, and some of the sun's energy is stored in the glucose. A number of things can happen to this glucose, but eventually some of it reaches the cells of aerobic animals. The mitochondria in these cells transform certain constituents of the glucose to carbon dioxide and water, and in so doing they release the stored energy of the glucose to make ATP, a chemical which provides usable energy to the body.

All this takes place in cells. So let's have a closer look at these incredible structures.

4.9 The tiny kangaroo rat is a desert-dwelling rodent. It is a teetotaler in that it doesn't drink—anything. It derives its water from the food it eats and as a by-product of metabolic activity. It conserves what water it gets by depositing very dry excrement and by hiding in its burrow during the heat of the day, creeping forth in the cool of the desert evening for its meals. Its large eyes enable it to see well enough at night to find food and, hopefully, to avoid predators.

There are a number of animals that do not drink, including some insects, but this is not to say they don't need water; they simply need very little.

Cells

In the mid-1600s Robert Hooke was appointed Curator of Experiments for the prestigious Royal Society of London. At the group's weekly meetings Hooke was to provide some sort of demonstration for them, something of scientific interest for their entertainment and information. This was a difficult task, since the society comprised the elite of the scientific community—a rather jaded lot, to be sure. As it happened, however, Hooke was interested in the recently developed microscope and managed to employ this fascinating new device most effectively in several of his demonstrations. One night, in an attempt to find something new to examine under his microscope, he cut a very thin slice of cork with his own sharpened penknife. When he peered through the microscope, he was somewhat surprised at what he saw. Hooke wrote at the time that the cork seemed to be composed of tiny "boxes" or "cells"—and a new field of science was born.

Many years were to pass before the scientific world would grasp the extreme importance of Hooke's "little boxes." In fact it was not until the nineteenth century that a comprehensive cell theory was advanced. The idea that all living things—from oak trees to squids, from tigers to men—are made up of cells is primarily due to the work of two Germans, the botanist Matthias Jakob Schleiden and the zoologist Theodor Schwann.

Today *cytology,* the study of the cell, has been enhanced by an array of sophisticated techniques and by the invention of the electron microscope. A cell, we now know, is the smallest unit that can exist as an independent organism. It has, at some point in its development, everything necessary for life. A cell that is part of an organism cannot ordinarily survive independently, but many kinds of isolated cells have lived and reproduced under carefully regulated laboratory conditions. Also, some organisms, such as the protozoa, consist of only a single cell. Here we are concerned primarily with the cells of multicellular animals, but many of the principles we will discuss apply to protozoa.

CELL SPECIALIZATION AND CONTROL

Cells come in a variety of sizes and shapes, each with its own particular limits and abilities. Not only do the cells of various organisms differ, but the cells of a single organism may be quite different from each other. The

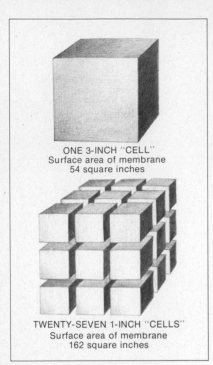

ONE 3-INCH "CELL"
Surface area of membrane
54 square inches

TWENTY-SEVEN 1-INCH "CELLS"
Surface area of membrane
162 square inches

5.1 The cellular structure of living things essentially breaks bodies into a number of parts. Each part has its own surface and hence cellularity increases the total area of cell membranes in living things. Since cell membranes are living regulatory structures, the result is increased control over what enters and leaves the body's cytoplasm. Cellularity also has another advantage, that of permitting specialization as cells differentiate according to specific roles. Note that in the illustration both "cells" have the same volume, but the subdivided one has a far greater surface area.

tiny boxlike cells beneath the skin of your hand are not much like the nerve cells that extend all the way from the base of your spine to your foot.

Cells that differ in appearance may also behave quite differently, as would be expected. Some are highly irritable—that is, they respond quickly to environmental changes. Some are contractile, others secrete fluids, and still others have long tails and can swim. Such variety indicates one great advantage of cellularity that might have made the condition worth developing through evolution. That advantage is *specialization*. An organism living in a complex environment would be better equipped to meet its many varied demands if it consisted of a number of kinds of cells, each with its own special abilities. Think of the advantage of having some cells that can react to light, others that can distinguish particular types of pain, and then others that can contract and move you toward the light and away from the pain. As different kinds of cells developed and then became increasingly coordinated with each other, it is easy to see how the organism they comprised could come to fit more precisely into its part of a competitive world.

Another important advantage to an organism in having many cells, instead of simply being one large cell, is greater *control* over interactions between the organism and the environment. The environment, as you know, is likely to be disruptive; the successful organism is one that is best able to minimize its entropic effects. This greater control is made possible by the increased membrane area provided by many cells. The membranes that bound all cells are more than sacks that hold the fluids in. They are living, responding structures that regulate what moves into the cell from the environment and also what moves out from it. Hence the more membrane area an organism possesses, the greater its control over its internal environment. As you can see from Figure 5.1, a 200-pound cell would have less regulatory membrane surface than a 200-pound organism composed of many cells. This is one reason why there is no such thing as a 200-pound cell.

PROCARYOTIC AND EUCARYOTIC CELLS

Before we become involved with the description and behavior of cells, you should first understand what kinds of cells we will be talking about. We will not be discussing the bacteria and blue-green algae. These are one-celled organisms that differ in several basic ways from all other cells. They are called *procaryotic cells* to differentiate them from the *eucaryotic cells,* which are our primary concern. The differences between the two types of cells are summarized in Table 5.1. The terms are defined later in the chapter.

HOW MOLECULES MOVE

In the regulation of complex systems of cells, one problem is moving things around in those sytems. So now let's see how molecules can move into, out of, and through cells.

Box 5.1 The Electron Microscope

The limits of the light microscope, the one with which you are probably most familiar, lie in its power of resolution. *Resolution* refers to the smallest distance by which two points can be separated and still be distinguished as separate points. If they are separated by a distance smaller than the *resolving power* of the microscope, they will be seen as a single point. The resolving power of the electron microscope is far greater than that of the light microscope. As a result, objects magnified to the same degree can be seen more clearly, and with the greater magnification afforded by the electron microscope, objects can now be seen that could not be seen at all with the light microscope.

With the light microscope the image is seen as contrasting light and dark areas as light passes through the specimen from below. The problem of resolution is therefore compounded by the inherent limitation of light diffraction, the bending of light rays as they pass through an object. The length of the light wave itself (that is, the color of the light) is a critical factor here. As the wavelength increases, so does the angle of diffraction, and its resolution is thus reduced. Under ideal conditions the best light microscopes have a resolution of about 0.25 microns with ordinary white light.

The electron microscope, which became available in the 1930s, works on an entirely different principle. Excited electrons are drawn from a heated filament and are then directed, or focused, by magnetic fields. As the electrons pass through an object, they are deflected or absorbed to various degrees by differences in the density of the matter. They then produce an image on a coated screen or special film below the object. Electrons are easily deflected, so the specimen must be cut exceedingly thin and the operation must be carried out in a vacuum.

The wavelength of an electron beam is only about 0.05 Ångstroms (an Ångstrom is a hundred-millionth of a centimeter) so, theoretically, resolutions of this power are possible. A major disadvantage is that heavy-metal stains must be employed, and these are usually highly poisonous to living things. Because of this and the necessity for extremely thin sections, there is no way living material can be examined. Thus there is no way to know how much of what we see only with an electron microscope has resulted from the method of preparation.

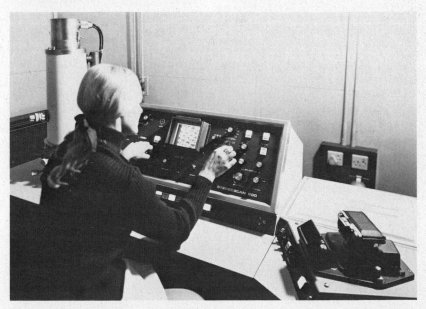

The scanning electron microscope, a specialized device that shows remarkable depth and perspective. At top left is a scanning electron micrograph taken through this microscope of cells at the surface of a mistletoe leaf. At right is a conventional electron micrograph of part of a cell from an onion root tip. The nucleus is at top (note the pores). The cytoplasm lies below the nucleus. The cell was opened by very rapid freezing, which caused it to fracture along stress lines.

5.2 Molecules of perfume escaping from a bottle will soon permeate an entire room. At first the molecules are concentrated near the bottle, but if evaporation is complete, in time their random movements will eventually disperse them evenly throughout the room. They shift from their center of concentration because they collide with greater frequency where they are more dense. They are less likely to collide where they are rarer and hence they change direction less often as they move away from other molecules.

Table 5.1 Differences Between Procaryotic and Eucaryotic Cells

Structure	Procaryotic cells	Eucaryotic cells
Nuclear membrane	Absent	Present
Inclusion bodies (mitochondria, Golgi body, etc.)	Absent	Present
Flagellae	Lack 9-2 structure	Have 9-2 structure
Chlorophyll	Not in chloroplasts	In chloroplasts
Cell wall	Contains amino acids and muramic sugars	Does not contain amino acids and muramic sugars (when wall is present)
Chromosomes	Composed entirely of nucleic acids	Composed of nucleic acids and proteins

Diffusion

The molecules of any substance—gas, liquid, or solid—are constantly in motion and continually bumping into each other and rebounding, thus changing their direction of movement. The principle becomes apparent if you open a vial of perfume (or if you're the baser sort, try hydrogen sulfide) in a corner of a room. Soon the smell will be detectable in all parts of the room. If the solution is allowed to evaporate completely, all parts of the room will eventually smell the same. Until that happens, however, there will be a *concentration gradient,* with a greater concentration of perfume molecules nearer the vial.

The reason for the shift of molecules away from their center of concentration is that their movement is random and they bump into each other with greater frequency where they are more densely packed. When they collide, of course, they rebound and change direction. Their chances of collision are lessened when they are moving outward toward areas of lesser concentration. As a result, this is what they tend to do. The *net* movement of these molecules is outward, away from the area of concentration. Their individual movements will be random, and by chance collisions with other molecules they will diffuse randomly from areas of higher concentration to areas of lower concentration.

Molecules may also be transported by their own random motions through the fluid environments in which living cells exist. While certain kinds of molecules may encounter barriers in the form of membranes or other structures, other types of molecules can pass with relative impunity through such obstacles. In some cases the difference depends simply on the size of the molecules in relation to the size of the pores in the membrane; smaller molecules pass through more easily. A few substances, such as carbon dioxide, oxygen and water, are able to diffuse rather freely through almost any cell membrane, while proteins usually cannot.

Osmosis

Battlefield nurses in World War II were taught to bathe exposed viscera wounds only with a salt solution, not with plain water. If you can muster the courage to place a drop of your own blood into water and watch the results under a microscope, you will see why. The blood cells will quickly swell and rupture, releasing their contents and leaving only colorless, limp "ghosts" of themselves. Plain water can have the same effect on other living cells.

Osmosis is the net movement of water through a membrane separating two solutions—a membrane that is more permeable to water than to whatever is in solution. Remember that the membrane separates both water molecules and molecules of the solute. Suppose we have two sugar solutions separated by a membrane, and that one solution has a higher *concentration* of sugar than the other (see Figure 5.3). Now we can discuss the movement of the water molecules and the sugar molecules separately. If the membrane is more permeable to water than it is to sugar, then the sugar molecules can go through the membrane only with difficulty, or not at all. Water molecules, however, pass through easily and will follow the law of diffusion: they will move from the side of their higher concentration to the side of their lower concentration. The lower concentration of water, of course, is on the side with the higher sugar concentration. Therefore water from one side will pass through the membrane to the other side where its concentration is lower.

The result is that water will *rise* on the side that started with the highest sugar concentration until the concentrations are equalized, or until the weight of the greater volume forces it back through the membrane at the same rate it enters. (Some recent data indicate that the movement of water through the membrane may be due to other factors besides simple diffusion, but you will be pleased to learn that these concepts are beyond the scope of this book.)

Cytoplasm, or cell fluid, is an aqueous (water-based) solution of a large number of types of molecules. When a cell is immersed in pure water or in a solution with fewer solute molecules than the cytoplasm, there is a tendency for water to move into the cell, causing it to swell and possibly rupture. This is what happened to the blood cells we mentioned above. When the fluid outside the cell has a higher concentration of solute molecules (and therefore a lower concentration of water), the effect is the opposite. If a cell is placed into a more concentrated solution, the water from the cell moves outward, causing the cell to dehydrate. However, a salt concentration of about 0.85 percent would have no effect on the volume of a human cell. In this case the number of particles in solution would be about the same both inside and outside the cell. Note that the important factor here is the concentration of the solution, not the type. We have discussed the concentrations of both solutes and water. In order to understand how osmosis affects cells you must clearly differentiate these two concentrations in your mind. You may, or may not, like to know that different types of cells vary widely in their permeability to water. A human blood cell is highly permeable to water. In contrast, a one-celled amoeba is able to live at the bottom of fresh-water ponds partly because it forcibly

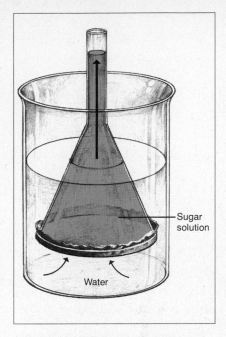

Sugar solution

Water

5.3 Osmosis will cause water molecules to move through a semipermeable membrane from a place where the water is in higher concentration to where it is lower. In the illustration the inverted funnel is immersed in pure water, but inside the funnel the water is diluted by sugar. Note that we're speaking of sugar diluting water (instead of the other way around) because the membrane covering the funnel will not allow the sugar molecules to pass, hence we are interested only in the movement of the water molecules. The resulting movement in this case will cause the sugar solution to become more dilute, thus raising its level in the funnel.

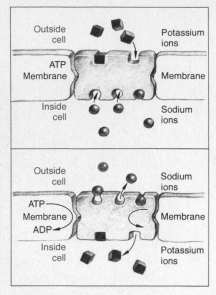

5.4 Cell membranes are living, responding structures that can actively move particles across themselves. In some cases, such movement must be accomplished against an osmotic gradient, and thus it requires energy (probably from ATP). Some substances move easily across the membrane on their own. These include water, oxygen, and carbon dioxide. Carbon dioxide is concentrated inside cells (where it is produced) and oxygen is concentrated outside. These, then, simply diffuse along their concentration gradients. Movement against a concentration gradient is called active transport. No one knows how it is accomplished, but apparently certain carrier molecules pick up the molecules to be transported at one side of the membrane, diffuse across, and release them on the other side. Another hypothesis says that the carrier molecules simply revolve in place (as shown), picking up molecules on one side of the membrane, turning, and releasing it on the other.

expells excess water that enters it, but also partly because its membrane is rather resistant to water.

Active Transport

In some cases a living cell can move molecules across its membrane from a place of lower concentration to a place of higher concentration. Such a chore is obviously "uphill," in that it goes against the normal tendency of the molecules to move in the opposite direction. As an example of such movement, consider cells that contain far less sodium than does their surrounding medium. Membranes are permeable to sodium to some degree, and so some sodium manages to diffuse in. As fast as it moves in, however, the cells pump it back out. The mechanism of the "pump" is not completely understood, but it is generally assumed that some sort of carrier molecule is involved—a membrane molecule that picks up sodium from inside the cell reverses its position in the membrane and releases it on the other side (Figure 5.4). A variety of molecules are subject to active transport, and some, such as potassium, may be carried the other way—into the cell.

CELL COMPONENTS

We will now consider some of the components of cells. The treatment is of necessity a bit artificial because of the complexities of cell composition. Just keep in mind that neither the descriptions nor illustrations are likely to depict the situation for all cells, nor for any cell all the time.

Cells are living things, so they change. In fact, the inside of a living cell is a seething, roiling mass of viscous, grainy fluid with tiny bodies of every description—bodies that are constantly growing, extending, moving, multiplying, appearing and disappearing. We know that cells must maintain and repair themselves, and that they must divide and grow. We also know that the cell's sensitive cytoplasm responds to specific environmental stimuli, and as those stimuli change, so does the behavior of the cell. Thus the vigorous panorama we see inside living cells should not be entirely unexpected. It should serve to remind us, however, that the goings-on within cells are incredibly complex, and that we cannot really hope to understand everything that is happening there. At least not yet.

Cell Walls

Only plant cells have cell walls; animal cells do not. Whereas cell membranes are living things, cell walls are dead, rather inflexible, and highly permeable products of cellulose and related compounds, which exist in a mat of fibers (Figure 5.6).

The cell wall of plants is commercially important in a number of ways. It is the tough cell wall that makes plant tissue useful as a building material. It is for cellulose that paper companies have leveled vast areas of American forests in response to our carefully cultivated demand for "disposable" commodities. Cellulose is also a major component of celluloid, rayon,

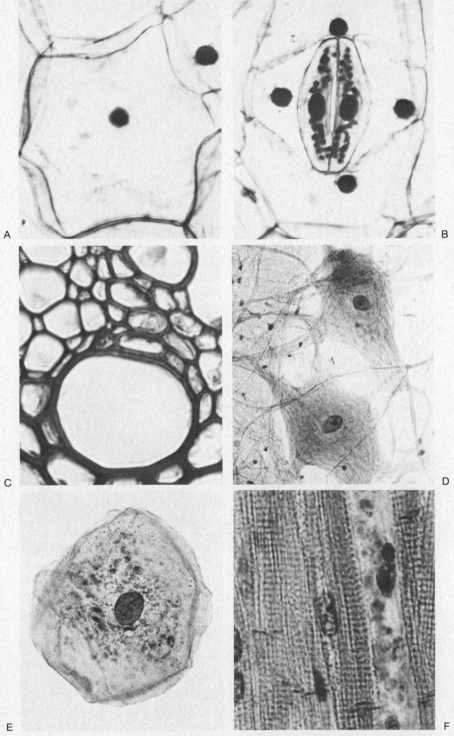

5.5 Cell variety. A. The epithelial ("skin") cells from the underside of the wandering Jew plant *(Zebrina).* B. Guard cells from the underside of the *Zebrina* plant. The nuclei and chloroplasts are stained darkly. The guard cells, when open, allow the passage of gases in and out of the leaf. C. Conducting cell of a corn plant. Note the heavy wall and the thinner walls of the surrounding pith cells. Because such cells serve as conducting tubes they are dead, the cytoplasm gone with only the cell wall remaining. D. Human spinal cord cells. E. Simple squamous epithelium from a human. These are very flat and delicate cells that can be scraped from inside your cheek with a toothpick—a behavior considered tacky outside of class. F. Heart muscle. Note that distinct cells are not visible. This is because the heart is not composed of distinct cells, as we shall see later.

5.6 A micrograph of a plant cell wall and its accompanying illustration. As plant cells divide, a thin layer of sticky material forms between the daughter cells. Basically it is composed of polysaccharides and acts to hold cells together. Directly under this substance, and against the membranes, the cells lay down their cell walls composed of cellulose. The molecules of the walls are arranged in sheets at right angles to each other to give them unusual strength and durability. There are small openings in these walls, usually opposite a similar hole in an adjacent cell. Through these openings pass small strands of cytoplasm that aid in communication between the cells. Young cells have thin, flexible walls to permit their growth, but older cells are bound by heavy, rigid walls, and, in some cases, a second cell wall may be constructed for added strength. Obviously, such strength is needed to keep tall trees from collapsing of their own weight.

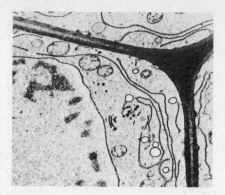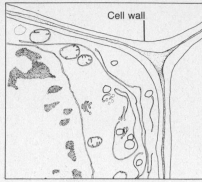

cotton, and hemp. (Hemp, which is the Cannabis plant, was supplied in the United States by the cultivation of *Cannabis sativa* or marijuana. This is the same plant that is presently giving Kansas law-enforcement officers fits.) The lignin found in cell walls was once considered a totally useless by-product of paper manufacturing and presented a disposal problem. Researchers were hired to find a way to alter it so that it could be sold to us somehow, and the clever fellows managed to find a use for it in the manufacture of synthetic rubber, vanillin, and adhesives.

The Cell Membrane

All cells are bounded by a delicate structure called the cell membrane or plasma membrane. The membrane is so thin that its presence was postulated on circumstantial evidence alone until the advent of the electron microscope. In 1940 J. F. Danielli, a biologist at King's College in London, reported that the cell membrane consists of a rather wide area lying between two thinner, denser areas. The outer layers are believed to be protein and the light inner layer phospholipids and steroids. Together these three layers comprise the *unit membrane*. The molecules of the inner layer are apparently oriented parallel to the cell's radius, while those of the two outer layers lie perpendicular to the cell's radius (Figure 5.7).

Danielli's description was based on an elaborate series of permeability studies. In recent years more direct methods, such as electron microscopy and X-ray diffraction techniques, have been employed. The newer evidence at first seemed to confirm Danielli's model, but later reports provided disconcerting discrepancies. It now seems that membranes are generally composed of lipids and proteins, but the arrangement of these molecules may vary widely. For example, membrane thicknesses may be quite different from cell to cell, and the ratios of different lipids may vary from one type of cell to another. The most likely explanation of these differences is that there are different structural arrangements of the various lipids and proteins among cells. For example, perhaps in some cells the sandwich is reversed (in effect, making the lipid the "bread"), or maybe in some cases the proteins are globular instead of extended. A great many alternative descriptions have been suggested, but at this point the basic

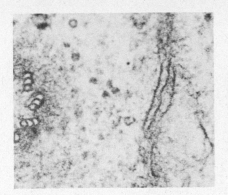
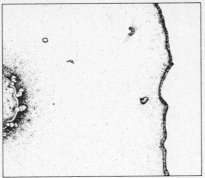
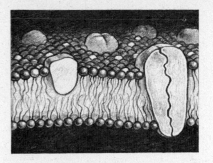

5.7 A micrograph (left) with its accompanying illustration, showing the cell membrane. A hypothetical model of the cell membrane is shown at right. There is some argument about precisely how cell membranes are constructed. Phospholipids have water-soluble heads and water-insoluble tails. Essentially phospholipid molecules are arranged so that their hydrophobic (water-fearing), fatty acid tails are directed inward. The phospholipids lie intimately against globular proteins, some of which are enzymes, while others are engaged in active transport or are supporting structures.

model of the unit membrane still appears to be correct in its essentials.

You will recall that cell membranes are selectively permeable; that is, they permit certain molecules to cross through and prohibit the passage of others. Thus this membrane regulates the constituents of the cell's interior, helping to protect it from an essentially hostile environment. Not all particles entering or leaving a cell must pass through the membranous material, however. Electron microscopy has revealed that the membrane surface is marked with many tiny pores through which some materials may pass. The specific roles of such pores are not well understood at present.

Mitochondria

Mitochondria are tiny structures, about the size of bacteria, which occur in almost all types of eucaryotic cells. They appear in a variety of shapes— round, elongate, and even threadlike. In some cells they seem to squirm around and move through the cytoplasm and in other cells they are fairly stationary. In either case, however, they tend to aggregate in places where work is going on—places where the greatest amount of energy is required. Not only are they found in the most active parts of cells, but they also occur in greater numbers in more active types of cells. There are more mitochondria in liver cells, for instance, than there are in less active cells such as cartilage. The concentration of mitochondria at places where work is being done is not coincidental; they are the "powerhouses" of the cell. You may remember that it is in the mitochondria of aerobic animals where cellular respiration takes place, so it is here that ATP, CO_2, and H_2O are generated. (Glycolysis primarily takes place in the cytoplasm.)

A mitochondrion is composed of two membranes, with the inner membrane folded into shelflike *cristae* (Figure 5.8). The mitochondrion is filled with a fluid *matrix* in which other reactions necessary to the synthesis of ATP occur. When ATP is produced, it is released into the cytoplasm at the surface of the mitochondrion. The mitochondria of some protozoa and insects have tubelike structures rather than cristae, but the advantage is the same: they increase the surface area.

Some interesting hypotheses have been advanced to account for the evolutionary development of mitochondria in cells, but the question has

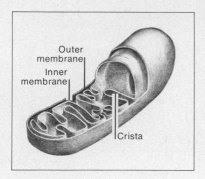

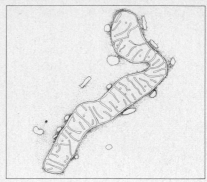

5.8 A cutaway view, micrograph, and graphic representation of mitochondria. Mitochondria are double-membraned organelles in which oxygen is used in a reaction that produces vast amounts of energy. That energy is tied up in the energy-rich molecules of ATP. Mitochondria are about the size of some bacteria and there is some argument about how they came to be in cells. Perhaps mitochondria once "infected" cells—bacterialike. You will find mitochondria concentrated in those cells and those parts of cells where the greatest work is going on. They are also concentrated at the ends of nerve cells, and assist in transferring impulses along the nervous system. These cell bodies are peculiar in that they have their own DNA, and as the cell divides they undergo their own replication.

not yet been resolved. One idea is that they pinched off from the cell membrane. As the story goes, the first cells were probably no larger than the mitochondria themselves are now. But as cells grew, their volume increased more than their surface area (in a sphere, as surface area is squared, volume is cubed). Cells developing inpocketings in their membranes became more efficient since this provided greater surface area. Some of these pinched off and gradually became modified into mitochondria. Another hypothesis is that mitochondria were originally bacterialike organisms that moved into the interior of cells as parasites. One point that may support such an idea is the mitochondria have their own DNA, or hereditary material, and undergo their own replication in coordination with the cell's division.

Golgi Bodies

In 1898 Camillo Golgi, an Italian cytologist, was experimenting with some cell-staining procedures and discovered that when he used certain stains, such as silver nitrate, peculiar "bodies" became apparent in the cells. The "reticular apparatuses" he described had never been noticed before, but when other workers looked for them, they turned up in a variety of cells. However, because they could not be seen in living cells, there was a great argument over whether they really existed or were just artifacts or debris produced by the staining process itself.

The electron microscope resolved the debate. Indeed, these strange bodies did exist. It was found that they had a characteristic and identifiable structure no matter what kind of cell they were found in. In every case they appeared as a group of tiny flattened vesicles, lying roughly parallel to each other, in the cell's cytoplasm (Figure 5.9).

Even after their existence was confirmed, an argument continued over their function. What did they do for a living? In spite of a great amount of research on the function of these strange bodies, we're still not sure. But there is some evidence that they serve as a sort of packaging center for the cell. Enzymes and other proteins, as well as certain carbohydrate molecules, are collected in these bodies, and here they are enclosed in "sacks" so that they are held apart from the rest of the cell. In some cases, after these

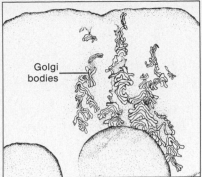

Golgi
bodies

5.9 Golgi bodies apparently serve as "packaging centers" for the cell. They look like flattened sacs of membranes pressed against each other and seem to function in packaging the substances that are formed on the endoplasmic reticulum. Some complex molecules, such as sugars and proteins, may be assembled at the sites of Golgi bodies, molecules that are found later at the surface of the cell membrane. Golgi bodies may also function as a meeting place for the substances that will make cell walls in plants. Their precise role in the life of the cell is not completely understood, but, as you can see, it is obviously a complex one. Whereas animal cells may have only 10 to 20 Golgi bodies, plant cells may contain several hundred.

molecules are packaged they are transferred to the cell membrane, where they are excreted out of the cell.

Lysosomes

The mysterious *lysosomes* are roughly spherical structures about the size of mitochondria. Their insignificant appearance, however, belies the rather startling role they play in the history of a cell. It is believed that lysosomes are packets of digestive enzymes that are synthesized by the cell and packaged by the Golgi bodies. If these digestive enzymes were floating free in the cell's cytoplasm, the cell itself would be digested. In fact, Christian de Duve, who first discovered the lysosomes, called them "suicide bags." The dramatic description is not entirely unwarranted, since they actually serve the function of destroying the cell that bears them. In some cases the destruction of certain cells is beneficial to the organism. For example, a superfluous cell could be an old one that is not functioning well, or one in a part of the body that was undergoing reduction as part of a certain developmental process.

Lysosomes may also act to selectively dispose of mitochondria, red blood cells, or bacteria. Fragments of these structures have been found within these puzzling sacs. Malfunctions of lysosomes have been found in connection with a number of human diseases, including cancer.

Plastids

Plastids are found in plant cells, but not in animal cells. They are similar in some respects to mitochondria. For example, they are double-membraned, and they also have their own DNA. Strangely, their DNA may differ somewhat from that of the cell. Plastids come in a greater range of sizes than do mitochondria. However, the most significant difference between these two structures lies in their roles: mitochondria make free energy available from food and special plastids make food from solar energy. Plastids, then, are the bodies in which you would expect to find chlorophyll.

There are two major types of plastids—those that are colored and those that are white or colorless. The best known of the colored plastids are

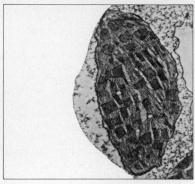

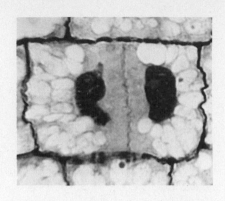

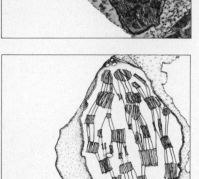

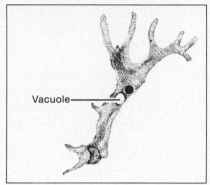

Vacuole

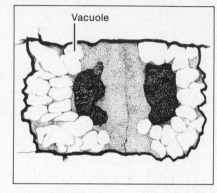

Vacuole

5.10 Plastids (left) are found in plant cells and, in some respects, are similar to mitochondria. They contain a peculiar DNA, which is quite different from that contained in the chromosome. Some plastids contain chlorophyll and are called chloroplasts.

5.11 Vacuole in an amoeba (center), a protozoa. Vacuoles often function in expelling wastes by contracting and forcing wastes out through the membrane. Others may contain engulfed food.

5.12 Vacuoles are particularly prominent in plant cells (right) where they appear as great fluid-filled sacs, pressing everything else against the outer edges of the cell. Some are active in metabolism; others are simply storage vessels.

the green chlorophyll-containing *chloroplasts*. Other chloroplasts may contain yellow or carrot-colored pigments, which, logically enough, are called *carotenoids*. The colorless plastids are storage bodies in which fats, starches, or proteins may be kept until needed by the cell.

Vacuoles

Vacuoles are fluid-filled sacks found in the cells of both plants and animals. It is in plant cells, however, where they reach their greatest development; in fact, they may be the most conspicuous body in the cytoplasm. There are a number of types of vacuoles, each with a different function. Some, for example, are highly active in the cell's metabolism, and others are simply storage vessels.

In plants the vacuoles are filled with a "cell sap," a fluid whose volume increases through osmosis. The pressure of the swelling vacuole forces the cell membrane against the cell wall and imparts firmness to the plant tissue. A plant wilts when there is not enough fluid to keep its huge vacuoles filled. Besides water, the vacuole sap may contain sugars, organic acids, proteins, and pigments. The acid component is what gives oranges and lemons their tart taste.

Protozoa may contain *contractile vacuoles,* which enable them to squeeze waste material or excess water out through the cell membrane. Such vacuoles may appear and disappear in response to the organism's needs or they may be relatively permanent structures—a useful feature in our identification of these tiny organisms.

The Endoplasmic Reticulum

The *endoplasmic reticulum* (ER) varies widely in appearance in different cells (Figure 5.13). It may exist as elongate vesicles or as long tubelike structures. In any case, however, it always appears as spaces surrounded by membranes. Some of these membranes are studded with tiny granular particles called *ribosomes.* Those with ribosomes are called *rough ER,* and those without, *smooth ER.* Ribosomes may also be found in the cytoplasm, independently of the ER. We'll discuss these fascinating little structures later. For now, though, you will have to settle for knowing that they are the sites of protein synthesis. They appear to be almost round, and in rapidly growing cells they may occur in small clusters of five to ten.

In early electron-microscope studies it was found that the ER somehow had a close association with the cell nucleus. There is now an abundance of evidence of a connection between the ER and the outer portion of the nuclear membrane. Since the nuclear membrane differs in important ways from other cell membranes, this finding has several implications. For example, the nuclear membrane may not be as "special" as it was once thought to be; it may be simply an extension of the general membrane component of the cell. Because of its large surface area, the ER could conceivably have the function of expediting interactions between the cell's nucleus and its cytoplasm by providing an expanded interface between them.

The ER membrane may be confluent not only with the nuclear membrane, but with the cell membrane as well (see Figure 5.21). This means that there may be an open channel from the nucleus to the outside of the cell. The cell might thus be able to transport products manufactured in the nucleus directly to the outside. But more important, with such a connection, the nucleus might be able to react more quickly to changes in the cell's environment.

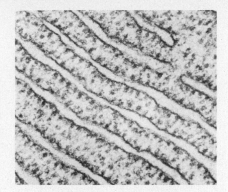

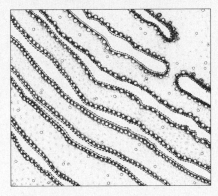

5.13 Rough endoplasmic reticulum (ER). The ER may be an extension of the cell membrane that communicates directly with the nuclear membrane. Some ER has ribosomes attached and is possibly involved in producing certain proteins. There are a number of enzymes attached to the vast ER membrane. The ER may also act to break up the cell into compartments so that different ongoing chemical processes do not greatly interfere with each other.

Centrioles

Centrioles are small, dark cylindrical bodies just outside the nucleus in an area of specialized cytoplasm. They are normally found in the cells of animals, algae, and fungi; they are rarer, however, in the cells of higher plants.

Centrioles have a characteristic structure of nine fibrils (or little fibers) running along their length, just below the surface (Figure 5.14). Usually they are paired, with each centriole lying at right angles to the other.

Cilia and Flagellae

Cilia and *flagellae* are hairlike projections from certain plant and animal cells. They are not distinguishable from each other except in terms of

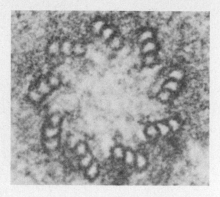

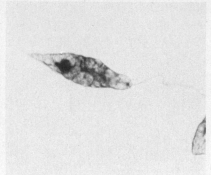

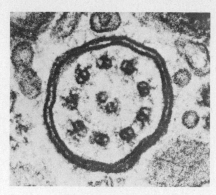

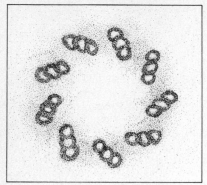

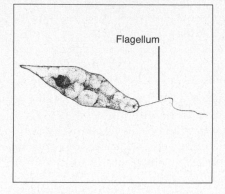

Flagellum

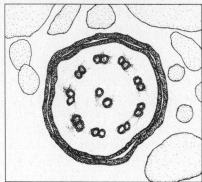

5.14 Cross section of a centriole (left). Centrioles are small cylindrical bodies just outside the nuclear membrane. They characteristically contain nine fibers and usually appear in pairs that lie at right angles to each other in a T-configuration.

5.15 Some protozoa, like the euglena (center), are propelled by whiplike, thrashing appendages called flagellae. Flagellae are not essentially different from the shorter appendages called cilia, except for their length.

5.16 A cilium (right) and flagellum have a characteristic appearance in cross section. They are composed of eleven tiny fibers, nine of which are arranged in a circle, as are those of a centriole. The other two lie parallel to these inside the circle.

length; cilia are about 10 to 20 microns long while flagellae run up to thousands of microns long. Both cilia and flagellae function either in moving the cell or in moving substances across the cell surface. For example, a sperm cell swims by beating its whiplike flagellum, and upward sweeping motions of cilia clean your breathing passages (unless you have killed them by smoking).

The cilium or flagellum is actually an outpocketing or extension of the cell membrane and contains a cytoplasmic matrix. The rather consistent arrangement of such structures has generated much speculation about their origin. There are usually eleven groups of fibrils extending the length of the appendage, and these are arranged in a very particular order. Nine of the fibrils lie just under the surface of the structure, just as in the centrioles, but there are two additional fibrils running along the center (Figure 5.16). Partly for this reason, it is believed that each cilium develops as the extension of one of a pair of centrioles.

The Nucleus

The physical prominence of the nucleus among the cell's various structures is matched by the role it plays in the cell's history. This large, fairly obvious body is intimately involved with processes such as the cell's internal regulation and reproduction.

The nucleus is surrounded by a membrane; this membrane differs

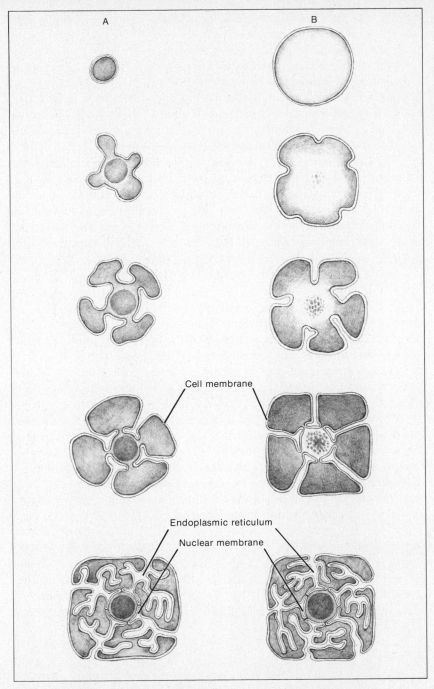

A B

Cell membrane

Endoplasmic reticulum
Nuclear membrane

5.17 Two opposing theories that seek to explain the presence of the nucleus within the cell. In A, the nucleus is the primary structure and the ER and cell membrane are added secondarily as extensions of the nuclear membrane. In B, the primal structure is the cell. The cell constituents organize themselves into a nucleus as the ER develops by inpocketings of the cell membrane.

from the cell membrane in that it is doubled (Figure 5.17). The inner and outer parts of the nuclear membrane pinch together in certain places over the nuclear surface to form scattered thin-walled areas called *nuclear pores*.

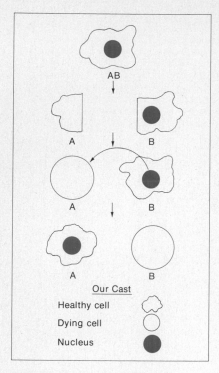

5.18 One of the classic experiments showing the importance of the nucleus involved cutting a living amoeba (a single-celled animal) in half, leaving one half with the entire nucleus. The deprived half died, the nucleated half survived. Then when the nucleus was removed from the surviving half and placed into the dying half, the latter recovered and continued a normal existence.

5.19 How much can the activity of the hereditary units called genes be influenced by the environment? We know that coat color in many animals, for example, is determined genetically. However, in the varying hare the genetic component that dictates hair color is directly controlled by temperature. When it grows colder, the cells that had been making brown hairs change and white hairs appear (left). In spring the sequence is reversed (right). Of course the result is a prey animal that is difficult to see in either summer or winter.

Although they are not actually pores, since there is no pore opening, they provide areas of high permeability that probably facilitate the flow of molecules between the nucleus and cytoplasm.

A number of relatively straightforward experiments have been done to show how critical the nucleus is to the cell's welfare. For example, if an amoeba is cut in half, the half with the nucleus continues to function as before. The half without the nucleus is unable to move about or capture food, and it soon dies. However, if the nucleus is surgically restored to this half within a few days, it recovers and is as good as new (Figure 5.18).

This doesn't mean that the nucleus has absolute control over the cytoplasm. As you might expect, biological stories usually aren't that simple. There are indications that the cytoplasm also has certain influences on the behavior of the nucleus. In addition, other environmental factors may influence the nucleus. For example, a change in the environmental temperature causes the nuclei of hair-producing cells to turn the hare's coat from one color to another as the seasons change (Figure 5.19).

Inside the nucleus is a fluid matrix in which a number of different types of bodies are suspended. When cells are stained in certain ways, a netlike structure becomes visible within the nucleus. This material is called *chromatin* because of its affinity for the stains (*chromatin* is from the Greek word for "color"). At a certain period in the cell's cycle the chromatin shortens and thickens and forms *chromosomes.* Chromosomes occur in a variety of simple rodlike, curved, or forked shapes. Each species, however, is characterized by a specific number of chromosomes. In normal human body cells there are exactly 46.

Chromosomes are composed of protein and a molecule called DNA, which we will consider shortly. It is these structures that are the sites of *genes.* You are undoubtedly aware that genes are the determiners of hereditary characteristics. They determine the characteristics of the cells, and hence of the organism. Basically, genes operate by determining what kinds of proteins the cell will produce. This is an extremely critical role, as we will see. But we're getting ahead of our story.

Another structure found in the nucleus is the *nucleolus*. Each nucleus usually contains only one nucleolus, but it may contain more. The nucleolus is an oasis of a certain type of molecule called RNA, which will play an important role in another story—how protein is made.

Now, to pull together what we have learned about the cell and its constituents, Figure 5.20 depicts the general structures of an animal cell and Figure 5.21, a typical plant cell. Keep in mind that these are stylized representations and that few real cells would look like these. However, most real cells, at some time in their life cycle, would include the various bodies shown.

CELL REPLICATION

It may not surprise you to hear that when carrot cells reproduce, they form more carrot cells. And oyster cells form more oyster cells. An oyster cell is very good at being an oyster cell, so when it reproduces, its best bet lies in making a very close approximation of itself. If a good functioning oyster cell were to generate new kinds of cells, there would be no guarantee that those new cells would work—and in fact, they probably wouldn't. So there is an advantage in having cells form new cells which are very similar to themselves. Such accurate duplication is assured by the exacting mechanism of cell reproduction.

Let's see, now, how cells go about reproducing themselves so that highly regular offspring result. Later we will learn something about the process at the molecular level.

Mitosis

The process of cell reproduction is called *mitosis*. In general terms, this is a process in which each cell replicates its chromosomes, along with some cytoplasm, and then divides into two new cells. Each new cell receives one of the replicated chromosomes, with the result that it has a full chromosome complement identical to that of the parent cell. As the cell divides, the cytoplasm is distributed about equally between the "daughter" cells. Some of the inclusion bodies, such as the mitochondria, also reproduce themselves at cell division. Others, such as ribosomes, are synthesized anew by the daughter cells. The result is new cells which are *about* the same size, have *about* the same inclusion bodies in the cytoplasm, and have *identical* chromosomal complements.

For purposes of discussion the process of mitosis is divided into four stages called *prophase, metaphase, anaphase,* and *telophase.* When the cell is not dividing it is said to be at *interphase.* This was once called the "resting" phase of a cell, but we now know that this description is inaccurate. At interphase many of the cell's components must double prior to the next mitotic division. The distinction between one phase and another is in a sense arbitrary, since there is a continuous progression from one phase to the next. Those who find comfort in labeling sometimes add further breakdowns, such as late prophase or early anaphase.

5.20 There is no typical animal cell, but this has not stopped us from showing you one. What is depicted here is a cell as it might appear with stop-action photography. The cytoplasm is normally a boiling mass of shifting composition. At no time could all these inclusion bodies be seen at once. Each body is usually discovered only under very special staining conditions. Nevertheless all the structures you see here are believed to exist at some time in the cell's history. The cell's shape, of course, is highly variable and will differ within types as well as from one type of cell to another.

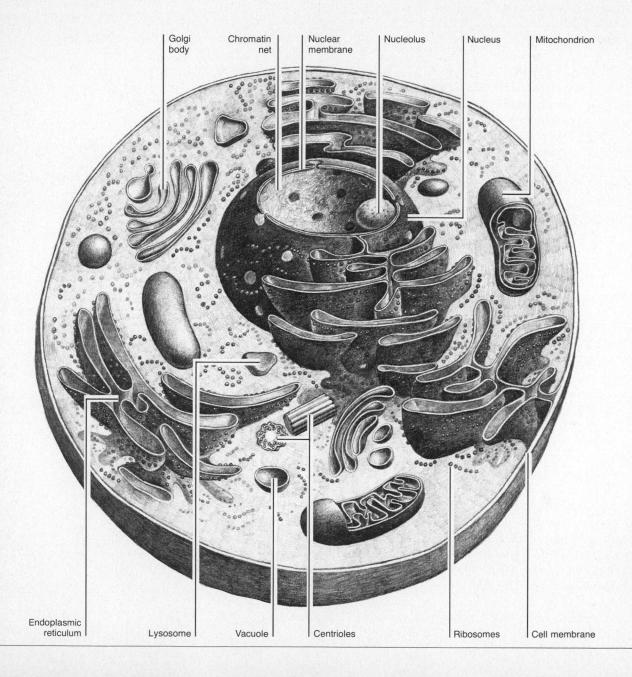

Golgi body · Chromatin net · Nuclear membrane · Nucleolus · Nucleus · Mitochondrion

Endoplasmic reticulum · Lysosome · Vacuole · Centrioles · Ribosomes · Cell membrane

5.21   This plant cell is also a composite of what we are able to piece together about cells. Some of the differences in animal and plant cells should now be apparent. Notice the heavy cell wall here, and the enormous vacuoles. Since plant cells are generally less active, they have fewer mitochondria. But since they are green and manufacture their own food, they have chloroplasts. Chromosomes are not yet visible in either the animal or the plant nuclei. In conducting cells of plants the cytoplasm disappears, leaving only the rigid cell walls to form tubelike structures through which food and water will pass.

Endoplasmic reticulum　　Nucleolus　　Nuclear membrane　　Chromatin net　　Nucleus　　Chloroplast

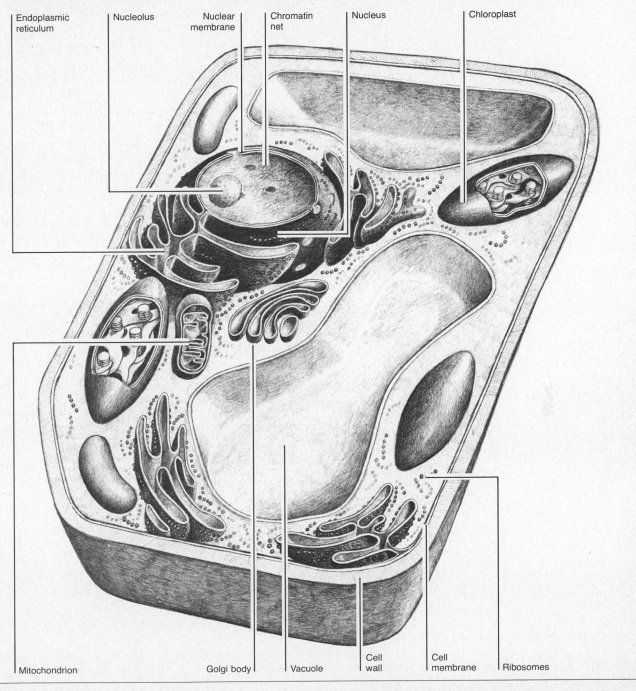

Mitochondrion　　Golgi body　　Vacuole　　Cell wall　　Cell membrane　　Ribosomes

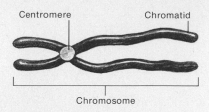

Centromere   Chromatid

Chromosome

5.22 Chromosomes become apparent in the nuclei of cells for the first time at the beginning of prophase. The diffuse chromatin net has condensed and become twisted into strands. It is not always apparent but chromosomes are composed of two parts called chromatids, which are joined at some point along their length to form chromosomes. When two chromatids that are joined in such a way become separated, each former chromatid is then called a chromosome. The chromatids usually lie pressed close together but when they are distinguishable, each part extending from the centromere is called an arm. Thus in this figure, there is one chromosome, two chromatids, and four arms.

At interphase the chromosomal material of the cell may be difficult to distinguish, since it is dispersed throughout the nucleus in fine netlike strands, which have a granular appearance. At the beginning of prophase, however, the chromatin becomes more visible as it begins to gather together and thicken into long twisted strands. The strands may now be called chromosomes. During interphase the genetic component of the cell has doubled, so that now each chromosome is made up of paired strands called *chromatids,* which are joined at some point along their length by a centromere (Figure 5.22).

The animal cell at this stage generally becomes rounder as the cytoplasm becomes heavier and more viscous. Other changes are occurring as well. The centromeres become markedly conspicuous, and the chromosomes continue to shorten and thicken as they move toward the nuclear membrane. At about this time a peculiar *spindle* appears in the cytoplasm—a structure consisting of tiny tubules that are seen as thin lines radiating across the cell from opposite poles. It is about then that the nuclear membrane begins to vanish. As it disappears, the chromosomes move toward the center of the cell and line up across the spindle. This is now the metaphase stage.

Anaphase begins with the division of the centromeres and the separation of the chromosomes into their separate chromatids, each now a single chromosome. Subsequently each of these single chromosomes moves along the spindle lines to opposite poles of the cell. It seems at this point as if the lines of the spindle have attached to the centromeres and are pulling the chromosome apart. The result is that the chromosomes that converge at each end of the spindle are identical. The clumping together of the identical groups of chromosomes at each spindle pole marks the beginning of the telophase stage.

The chromosomes of each group now begin to stretch out and unwind as they revert to their own interphase condition. In animal cells, as the chromosomes become longer and thinner, a nuclear membrane begins to form around each group of chromosomes. The cytoplasm of what was the parent cell is divided as the membrane between the two groups of chromosomes pinches in and finally closes together, dividing what are now two separate cells. In plants, with their rigid cell walls, there is no pinching in. Instead a new cell wall begins to form in the cytoplasm between the divided chromosomes. Each daughter cell may at first be somewhat smaller than the mother cell because it has less cytoplasm, but such differences are quickly reduced by water intake and the generation of new inclusion bodies. The daughter cells may also be of different sizes, since the cytoplasmic division may not be precise, but each daughter cell has an identical genetic component. The chromosomes in each new cell will then replicate themselves from nuclear material as preparations for the next division occur. See Figure 5.23 (which appears on pp. 104–105) for illustrations of mitosis.

Before we take a look at how all this is done at the molecular level, let's consider another process—one that seems similar to mitosis in its procedure, but that has important differences with exceedingly critical consequences.

Meiosis

The description of meiosis should answer some puzzling questions for you if you have thought much at all about the developmental process, especially if you are aware that eggs and sperm are simply specialized cells. At fertilization the nucleus of an egg and a sperm are joined, and from the subsequent divisons of the fertilized egg, a new organism is formed. Because of the regularity of the mitosis process, each new cell of the developing individual is genetically identical to its parent cell. This means that all the cells of the developing organism have their full complement of chromosomes. Since eggs and sperm are both cells, when they combine at fertilization, why don't the resulting offspring have twice as many chromosomes as their parents?

We know that the number of chromosomes is identical in every cell of the body, since all the cells of any organism spring from that first fertilized egg. We also know that chromosome number is important, since every species has its specific number. For the union of an egg and a sperm to produce the correct number of chromosomes in the cells of the offspring, each of them must have exactly half the chromosome complement of other cells, so that the *sum* of their chromosomes equals the full complement. As it turns out, each egg and sperm has exactly half the number which is characteristic of the species. When the egg and sperm are joined, the result is a primal cell with the correct number of chromosomes. Then through millions of divisions of that cell (in conjunction with cell specialization, which we'll worry about later), a new individual is formed.

Now, egg and sperm cells start out with the full complement of chromosomes. Somewhere along the way half the chromosomes are lost. The process by which certain cells in the *gonads* (ovaries or testes) lose half their chromosomes on the way to becoming eggs and sperm is called *meiosis*.

An important point should be noted here. As you see in Figure 5.24, chromosomes come in *pairs*. For every long chromosome, there is another identical long chromosome in that nucleus. This relationship prevails because the egg and sperm contribute identical kinds of chromosomes to the new individual. Thus, for any pair of chromosomes in the body, one can be considered the descendant of a paternal chromosome (from the father's sperm) and the other the descendant of a maternal chromosome (from the mother's egg). The result is a contribution of each parent for every genetically determined characteristic of the offspring (with special exceptions, as we shall see). So, since chromosomes come in pairs—23 pairs in humans, 9 pairs in cabbages, 47 pairs in goldfish—if we have a single gene for eye color from one parent, we will have another gene for eye color from the other parent (although not necessarily for the same color).

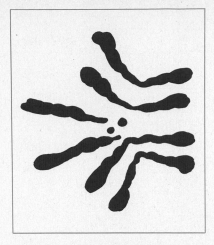

5.24 Pairing in chromosomes of a female *Drosophila*. Note that each chromosome in a cell has its counterpart, one having come from the mother, the other from the father fly.

5.23 Mitosis in an onion root tip. A. Interphase: The chromatin is diffuse, spread thinly throughout the nucleus. The nucleolus is prominent. B. Early prophase: The nuclear material begins to shorten and thicken. C. Late prophase: The nucleolus is disappearing as the chromosomes are becoming distinct. D. Metaphase: The distinct and identifiable chromosomes are lined up across the spindle. E. Early anaphase: The centromeres have divided so that former chromatids are now chromosomes. These single-stranded chromosomes are moving to opposite poles of the spindle. F. Late anaphase: The chromosomes are beginning to aggregate at their ends of the spindle. G. Early telophase: Since this is a plant cell, the cell wall is beginning to form. H. Late telophase: New nuclei are becoming evident. I. Daughter cells have formed and have entered interphase.

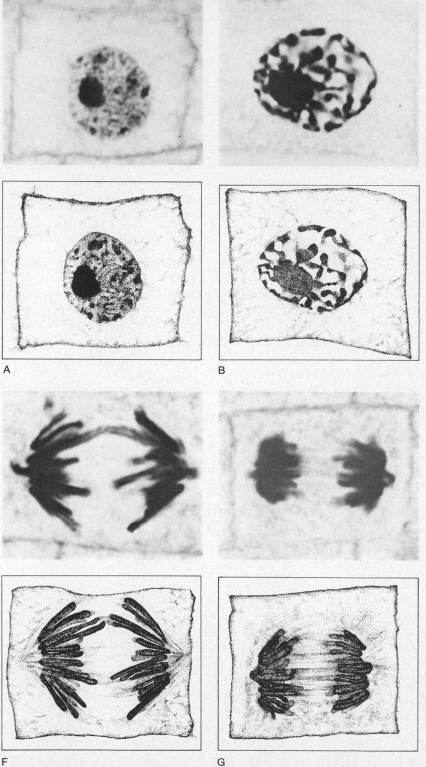

A

B

F

G

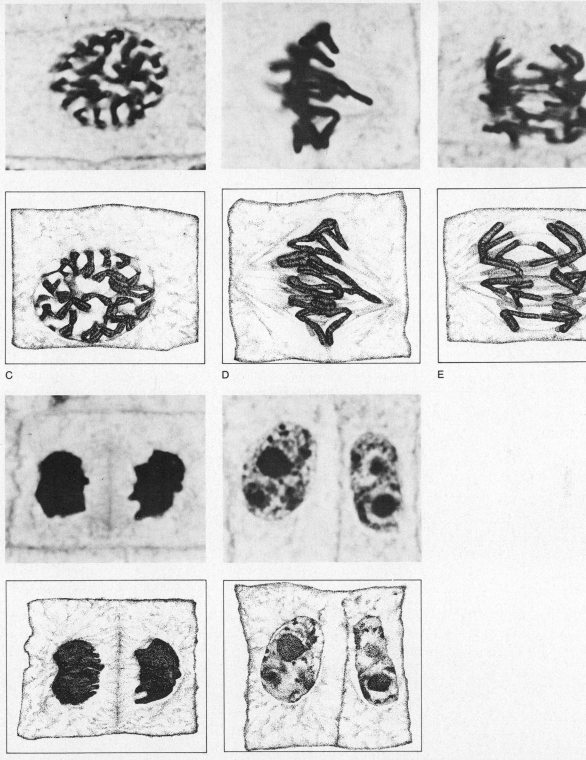

C

D

E

H

I

Now let's see how meiosis works. As we follow the development of a cell in the gonad through its meiotic divisions as it becomes an egg or a sperm, it may seem at first that we are retracing mitosis. But the sharp-eyed, critical student will quickly see important differences.

First of all, the meiotic process is divided into two parts, meiosis I and meiosis II (Figure 5.26). Before meiosis begins there is a replication of the genetic material in the interphase, just as in mitosis. The chromatin net then shortens and thickens to form distinct chromosomes, and the spindle begins to appear as the nuclear membrane begins to disappear. It all sounds familiar so far.

In early prophase I each pair of chromosomes comes together. Each chromosome consists of two identical chromatids, so each chromosomal group is made up of four chromatids at this point. As the chromosome pairs line up side by side, an important—and somewhat surprising—event may take place. The chromosomes may exchange parts. This exchange is important in providing variation among offspring. Other ways of providing this variation and the importance of variation in evolution are discussed in later sections.

In metaphase I, as you would expect if you have managed to remember our discussion of mitosis, the chromosomes line up across the developing spindle. In meiosis, however, the chromosome *pairs* separate, with one member of each pair moving to opposite spindle poles, marking anaphase I. In this case the centromeres do not divide, as they do in mitosis. Thus the chromatids of each chromosome now travel together to the spindle poles. At telophase I, the end of the first stage of meiosis, the chromosomes begin to elongate and revert to their netlike or grainy condition, and a nuclear membrane may form as the cell enters interphase II. Each nucleus now contains only half the number of chromosomes of the parent cell, although each chromosome is two-stranded. Not only are there fewer chromosomes in each cell, but they are probably qualitatively different from any of the original chromosomes as a result of crossovers.

Interphase II differs from the interphase preceding meiosis I in an important way: the chromosomes do not replicate themselves. Aside from this difference, the second meiotic division proceeds in the usual way. In prophase II the chromosomes shorten and thicken, and the nuclear membrane (if it is present) begins to disappear. At metaphase II the chromosomes line up across the equatorial plate as the spindle fibers again appear. Now, at anaphase II, in contrast to anaphase I, the centromere breaks apart, and the chromosomes separate into their chromatids, which move toward opposite poles. At telophase II a nuclear membrane forms around each set of chromosomes. The cytoplasm is divided as new cell membranes appear, thus dividing each group into a separate cell. So where there was one cell with a full complement of chromosomes, there are now four cells, each with half the chromosomal complement. You can take a closer look at the difference between meiosis and mitosis in Figure 5.27.

If the meiotic process has been taking place in a testicle, each of four resulting cells will undergo certain changes, separate, grow a tail, and become a sperm. If the meiosis was in an ovary, however, three of the resulting cells would have degenerated into tiny, nonfunctional *polar bodies,* with only one cell surviving to become an egg. It is not known what determines which of the four cells this will be.

There are two major take-home lessons from the story of meiosis. One is that meiosis halves the chromosome number of eggs and sperm. Second, meiosis provides a means of shuffling and reorganizing chromosomes, thus increasing variation in the offspring. This reorganization takes place in crossovers; in the apparently random lineup of chromosomes at metaphase I, so that paternal and maternal chromosomes are mixed by the time a cell enters meiosis II; and in the apparently random selection of polar bodies in egg formation. You may come to have an abiding appreciation for this process when you realize that it is because of meiosis that you don't look like your little brother.

THE DOUBLE HELIX

As more and more information on the behavior of chromosomes was produced, a growing question was, What are they? What are they made of? Chemical analysis showed that the chromosomes themselves were composed of protein and an acid called deoxyribonucleic acid, DNA. By 1950 it was pretty well accepted that the hereditary material itself was the DNA component of the chromosomes. But how was the DNA arranged in the chromosome? And how did it duplicate itself at mitosis? And in particular, how could it direct protein synthesis? Several American and British groups were attempting to answer these questions, and the prevailing notion was that there were three chains in each DNA molecule. Linus Pauling, the great American chemist, and Rosalind Franklin of England both favored this idea but disagreed over whether the nitrogen bases stuck out of the molecule or projected inward.

Then a brilliant and highly confident young American postdoctoral fellow ran across an ebullient and talkative English graduate student in biophysics at Cambridge University. The results of what transpired were described by James Watson in his controversial best-seller, *The Double Helix.*

The Englishman, Francis Crick, had noted that when the DNA portion of chromosomes was photographed by X-ray diffraction techniques, a peculiar image appeared—one which could only be produced by molecules arranged in a helical, or spiral, pattern. Watson's biological intuition told him that only two strands were involved. After all, reproduction usually involved only two parents and there are paired chromosomes and pairs of genes. Then, on the basis of X-ray diffraction information provided by Rosalind Franklin and others, after careful consideration of a wide range of information, and after building several huge DNA replicas in their laboratories, Watson and Crick stated in 1953 that DNA is composed of two chains coiled around each other as a double

5.25 James Watson (top) and Francis Crick (bottom), who, with a little help from their friends, won the race with the great Linus Pauling to discover the structure of DNA. They unmasked the molecule as a double helix after using such sophisticated techniques as X-ray crystallography and such mundane tools as ball and stick models. This fascinating story has been recounted in Watson's controversial book, *The Double Helix.* Crick continued to work on the puzzle and made enormously valuable contributions to the later discovery of the genetic code—the three-nucleotide sequence. Progress continues, however, and now even the revered (and very tidy) nucleotide vocabulary is under attack. It may be soon revised since it seems the story may not be as simple as has been assumed.

5.26 Meiosis in the anther of a lily. The meiotic process is divided into two parts: Meiosis I and Meiosis II. Keep in mind that both the mitotic and meiotic process are a bit different in animals because no cell wall forms; instead, the cells pinch inward to meet the new cell membrane that is being laid down between them.

Meiosis I A. Prophase I: The chromosomes have synapsed, forming tetrads. B. Metaphase I: The tetrads have lined up across the spindle. C. Anaphase I. The centromeres do *not* divide, but the homologous chromosomes separate and move to opposite poles. D. Early telophase I: Chromosomes cluster at the poles. E. Late telophase I: New nuclei are forming at each pole.

Meiosis II F. Early prophase II. The diffuse chromatin is beginning to condense. G. Metaphase II: Distinct chromosomes line up across the spindle. H. Anaphase II. The centromeres divide. Single-stranded chromosomes move to opposite poles. I. Telophase II. New nuclei and new cell walls are beginning to form, resulting in four haploid cells.

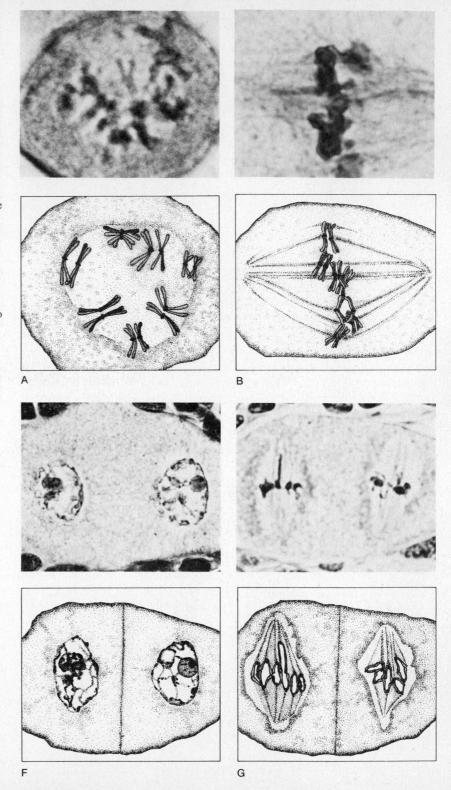

A

B

F

G

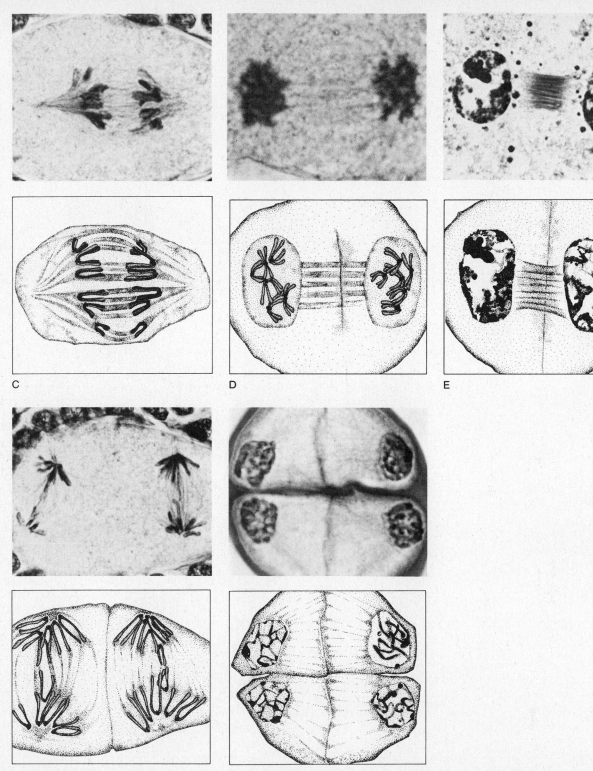

C

D

E

H

I

5.27 A graphic outline of mitosis and meiosis in animal cells. Compare the two processes carefully. For convenience we have made up an animal with *two* pairs of chromosomes. Also to make things conceptually easier, paternal and maternal chromosomes are identified by their darkness. It doesn't matter which you want to call paternal, but it is important to notice the relative number of chromosomes from each parent that makes it into the daughter cells. Notice in particular the different behavior of the chromosomes in metaphase and anaphase of the two processes. Also keep in mind that meiosis is essentially two divisions. Finally, compare the chromosome number in mitotic telophase with that of telophase in meiosis II. Keep in mind that chromosomes are most easily counted by counting centromeres.

MITOSIS

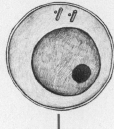

1. INTERPHASE
Chromosomes not visible; division preparations in progress.

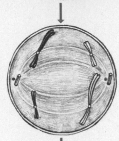

2. PROPHASE
Centrioles migrated to opposite sides; spindle forms; chromosomes become visible as they shorten; nuclear membrane, nucleolus fade.

3. METAPHASE
Chromosomes aligned vertically on cell equator. Note attachment of spindle fibers from centromere to centrioles.

4. ANAPHASE
Centromeres divide; single-stranded chromosomes move toward centriole regions.

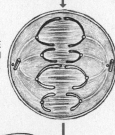

6. DAUGHTER CELLS
Two cells of identical genetic (DNA) quality; continuity of genetic information preserved by mitotic process.

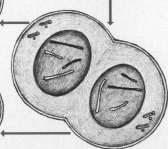

These cells may divide again after growth and DNA replication has occurred.

5. TELOPHASE
Cytoplasm divides; chromosomes fade; nuclear membrane, nucleolus reappear; centrioles replicate (reverse of prophase).

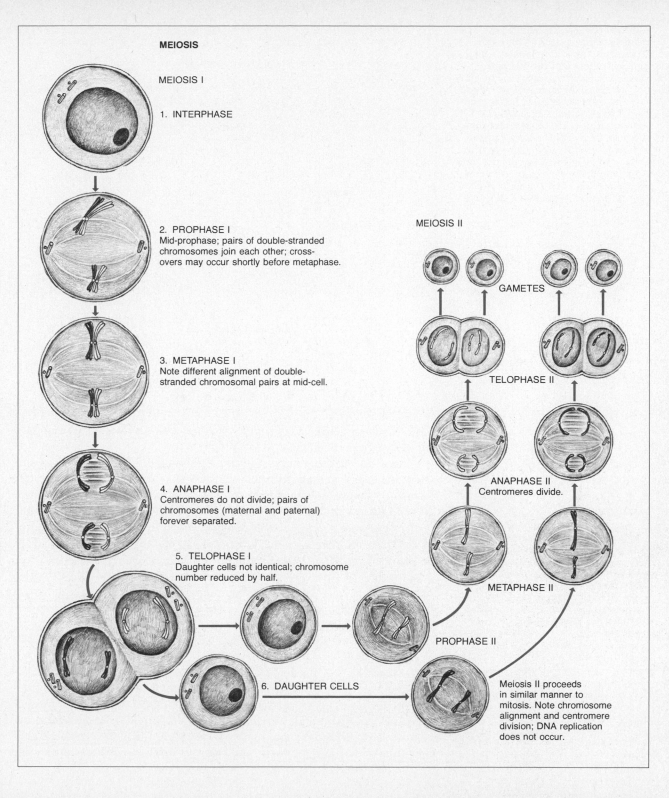

MEIOSIS

MEIOSIS I

1. INTERPHASE

2. PROPHASE I
Mid-prophase; pairs of double-stranded chromosomes join each other; cross-overs may occur shortly before metaphase.

3. METAPHASE I
Note different alignment of double-stranded chromosomal pairs at mid-cell.

4. ANAPHASE I
Centromeres do not divide; pairs of chromosomes (maternal and paternal) forever separated.

5. TELOPHASE I
Daughter cells not identical; chromosome number reduced by half.

6. DAUGHTER CELLS

MEIOSIS II

GAMETES

TELOPHASE II

ANAPHASE II
Centromeres divide.

METAPHASE II

PROPHASE II

Meiosis II proceeds in similar manner to mitosis. Note chromosome alignment and centromere division; DNA replication does not occur.

5.28 The four nucleotides of DNA. The basic building blocks are a sugar (2-deoxy-D-ribose), a phosphate group, and any of four nitrogen-containing bases. The smaller bases have one ring and are called pyrimidines (thymine and cytosine). The larger bases, purines, have two rings (adenine and guanine). In building their model, Watson and Crick assumed the two pyrimidines would be too short to reach each other from their sugar-phosphate backbone and that the two purines, reaching toward each other, would overlap. Thus they decided that a purine must be paired with a pyrimidine.

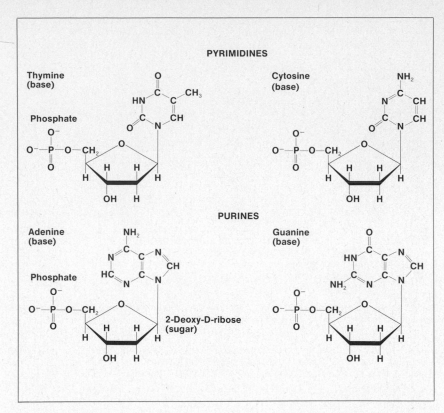

5.29 DNA replication. As the two strands of DNA unwind, their bases are exposed to the nuclear fluid that contains a large number of free bases—thymine (T), adenine (A), cytosine (C), and guanine (G)—sugars (S), and phosphates (P). The exposed bases on the chains, then, join with their free-floating complementary bases from the nucleus. As these free bases fall into line along the strand of DNA, they become connected by a precise alignment of sugars and phosphates. Thus each strand manufactures a strand that is identical to the one from which it had separated and the familiar double helix is reestablished.

At left the molecular structure of DNA is shown. At right we see a model of DNA replication as geometric figures represent each base and the specificity of base pairing becomes obvious.

helix. You can get the idea if you visualize a ladder twisted so that the rungs remain perpendicular to the sides.

According to their model, the two longitudinal sides of the ladder are alternating sugar and phosphate molecules. The "rungs" of the ladder are formed by four nitrogenous bases—the purines adenine (A) and guanine (G) and the pyrimidines thymine (T) and cytosine (C). These bases, with their sugar and phosphate components, are called *nucleotides* (Figure 5.28). There is one base per sugar-phosphate, and two bases join together to form each rung. The bases extend inward from the ladder's sides and are held together by weak hydrogen bonds. It was found that two of the larger bases (purines, with two nitrogen-containing rings) were too large to fit between the sides of the ladder, and two pyrimidines, with their single nitrogen-containing rings, were too short to reach across. The arrangement could work only if a pyrimidine joined with a purine to form the rungs (T to A and C to G).

According to Watson and Crick's model, the nucleotides along one of the strands of the double helix could follow any sequence. For example, one might "read" along a strand and find ATTCGTAACGCGT in one segment. Also, the strands were found to be very long, so that the necessary complexity for life could be provided by the variation in sequence and the long numbers of nucleotides in any chain. If the DNA from a single cell

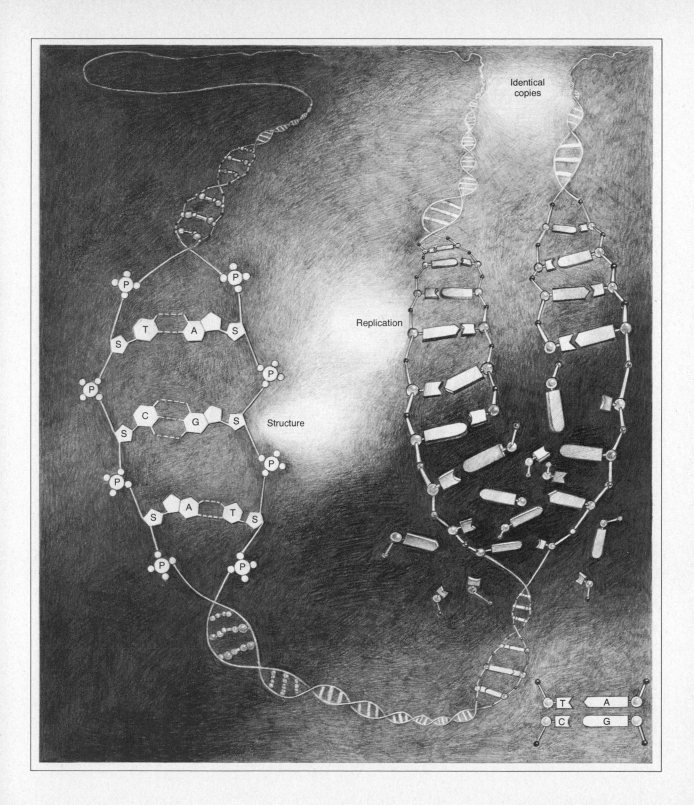

Identical
copies

Replication

Structure

T A
C G

were extended in a straight line, it would be about five feet long! There are about 10 billion nucleotide pairs in man's 46 chromosomes. Think of the number of possible sequences!

Now, if the sequence in one strand of DNA is known, the sequence of the other strand can be deduced. The purine adenine (A) will always be bound to the pyrimidine thymine (T) of the other strand. And wherever cytosine (C) exists along one chain, it will be attached to guanine (G) on the other strand. So what would be the nucleotide sequence of the complementary segment of the strand described above?

DNA Replication

As you recall from our discussion of mitosis, after the cell has divided each chromosome replicates itself, so that the chromosomal component is doubled before the next division starts. But how does the chromosome, with its intertwining strands, do this?

Actually, Watson and Crick's model of DNA structure suggested the mechanism—conceptually a simple one. When the time comes for the chromosome to replicate itself, its DNA strands separate. The weak hydrogen bonds holding them together are broken, and the molecule opens up. Purines and pyrimidines are now exposed to the nuclear sap, which is loaded with many kinds of molecules. From among these free-floating molecules, adenines, guanines, cytosines, and thymines attach to the exposed bases of the open DNA strand according to their base-pairing formula (T to A, C to G). The new and the old strand of bases then bind together by forming weak hydrogen bonds along their length. Thus, step by step, an identical DNA molecule is produced from each strand of the original, using the raw materials from the nucleus. Each new DNA strand will become a chromatid at the next cell division (p. 113).

HOW CHROMOSOMES WORK

Chromosomes function not only in reproductive processes by the action of eggs and sperm, but in all the other cells of the body as well. After all, skin cells have nuclei; so do nerve cells and muscle cells and almost all the other cells that comprise organisms. And in those nuclei are chromosomes. In fact, it is those chromosomes that direct a developing cell to become skin or muscle or nerve. Now let's find out how they do it.

At the same time we can answer another question. You may have been astonished—or at least mildly interested—to learn that practically all the cells in your body have the same chromosomal component. After all, those cells all arose as a result of mitotic divisions of that first fertilized egg. Since we know that mitosis results in an equal distribution of chromosomes, how can those same chromosomes cause a cell to become muscle in one case and nerve in another?

Before we launch into this discussion, you might note that cells become specialized into muscle or nerve or whatever as a direct result of differences in their component enzymes. Enzymes are biocatalysts, you

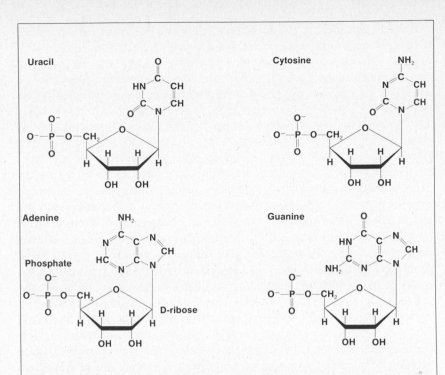

Uracil

Cytosine

Adenine

Phosphate

D-ribose

Guanine

5.30 DNA manufactures mRNA through a process similar to its own replication. However, as mRNA is made, thymine is replaced by another nucleotide called uracil. As long mRNA molecules break away from their DNA template, they move into the cytoplasm to direct protein formation. The DNA, then, establishes the sequence of bases along the mRNA. Obviously, entire DNA strands do not unwind when forming mRNA. If they did, since each cell contains the same chromosome complement, they would all make the same enzymes and no specialization could occur. One of the enduring questions in biology is, How is the manufacture of mRNA directed? How does one part of a chromosome come to actively make mRNA while other parts lie dormant, their DNA strands tightly interwound? And what causes long dormant segments to suddenly become active?

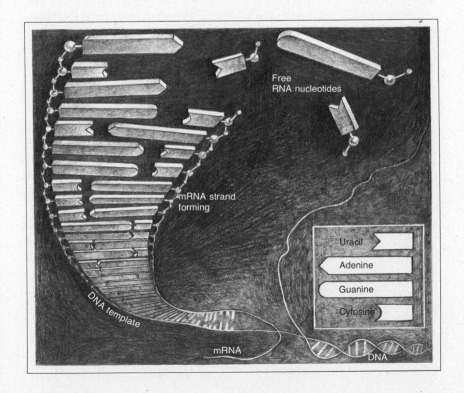

Free RNA nucleotides

mRNA strand forming

DNA template

mRNA

DNA

Uracil

Adenine

Guanine

Cytosine

will recall, and their role is to facilitate certain very specific biochemical events. It it through the precise summation of such events that cells become specialized. So what we are really asking here is how do chromosomes make those peculiar proteins called enzymes? How do events in the nucleus alter the developmental pathway of the rest of the cell?

The first thing that happens is that the chromosomes unwind, just as they did in chromatid replication in mitosis. As they unwind, the hydrogen bonds between purines and pyrimidines are broken, exposing these nucleotide bases to the nuclear fluid, and free nucleotides attach to the exposed bases along the DNA strands. As before, free cytosine attaches to exposed guanine, free guanine to exposed cytosine, and free adenine to the exposed thymine. However, where adenine lies exposed along a DNA molecule, it is attached, not to free thymine, but to *uracil,* another nucleotide in the cytoplasm. Here, then, the DNA strands are not building complementary strands of DNA, as they do when they replicate themselves in mitosis. They are building another substance—*ribonucleic acid,* RNA. RNA has one more oxygen atom in the sugar component of its nucleotides, and in place of thymine it has the nucleotide uracil (Table 5.2). Also, most RNA exists as a single strand, and not as a helix.

So, as the DNA unwinds, RNA is manufactured along its length. Sections of this RNA will then separate from the DNA strand and move from the nucleus into the cytoplasm, each one carrying a message encoded in its nucleotide sequence. This RNA is called, appropriately enough, *messenger RNA,* or *m*RNA (Figure 5.30).

The *m*RNA, then, is a rather long, and, as we will see, somewhat unstable molecule which contains guanine, cytosine, adenine, and uracil and the sequencing of its nucleotide bases is very precisely directed by the DNA inside the nucleus. Before we see what this very specific *m*RNA does in the cytoplasm, let's learn something about how its message is carried— the mysterious genetic code. The genetic alphabet consists of only four nucleotide bases, yet the messages are sufficiently precise to code for every kind of protein that makes up the highly complex structures of living organisms. How is this complexity achieved with only a four-unit code?

Actually, we already have enough information to figure out the answer to this question. Although proteins are very large and complex, they are made up of amino acids—of which there are only twenty kinds. So the problem reduces to simple mathematics. If there are only four nucleotide bases, obviously one base cannot code for each amino acid, since this could produce only four amino acids. A two-base combination for each amino acid would provide four times as many possible combinations (4 × 4 = 16), but this is still not enough.

So, early in the search for the genetic code it was postulated that a three-base combination was necessary to code for each amino acid. At first this was just a guess; the verifying data came later. It was reasoned, however, that with four nucleotide bases, a combination of any three as a

Table 5.2 The Genetic Code, or the "Language of Life." There are 64 Triplet Combinations. Each Triplet Is Shown with the Amino Acid for Which It Codes.

		Second letter				
First letter		U	C	A	G	Third letter
U		UUU UUC } PHE UUA UUG } LEU	UCU UCC UCA UCG } SER	UAU UAC } TYR UAA stop UAG stop	UGU UGC } CYS UGA stop UGG TRP	U C A G
C		CUU CUC CUA CUG } LEU	CCU CCC CCA CCG } PRO	CAU CAC } HIS CAA CAG } GLN	CGU CGC CGA CGG } ARG	U C A G
A		AUU AUC } ILE AUA AUG MET (start)	ACU ACC ACA ACG } THR	AAU AAC } ASN AAA AAG } LYS	AGU AGC } SER AGA AGG } ARG	U C A G
G		GUU GUC GUA GUG } VAL	GCU GCC GCA GCG } ALA	GAU GAC } ASP GAA GAG } GLU	GGU GGC GGA GGG } GLY	U C A G

Source: F. H. C. Crick, "The Genetic Code: III." Copyright © 1966 by Scientific American, Inc. All rights reserved.

Phenylalanine	Serine	Tyrosine	Cysteine
Leucine	Proline	Histidine	Tryptophan
Leucine	Threonine	Glutamine	Arginine
Isoleucine	Alanine	Asparagine	Serine
Methionine		Lysine	Arginine
Valine		Aspartic Acid	Glycine
		Glutamine	

code for a particular amino acid would provide sixty-four possible combinations (4 × 4 × 4 = 64). But there are only twenty amino acids. So, as you may have guessed, some nucleotide triplets are coded for the same amino acid.

Of the sixty-four nucleotide triplets, called *codons,* it is known that sixty-one code for specific amino acids. But what about the other three, UAA, UAG and UGA? These particular codons, interestingly enough, serve as punctuation marks in the language of life; they signal the end of a given message. The codon AUG apparently doubles as an initiator, indicating the start of a message.

Now that we have some idea of the nature of the genetic code, let's see how it operates to produce the astonishing array of proteins found in living things. When the newly formed *m*RNA leaves the nucleus and enters the cytoplasm, it diffuses through the cytoplasm until it encounters one of the cell's many ribosomes. It joins temporarily with the ribosome to form a

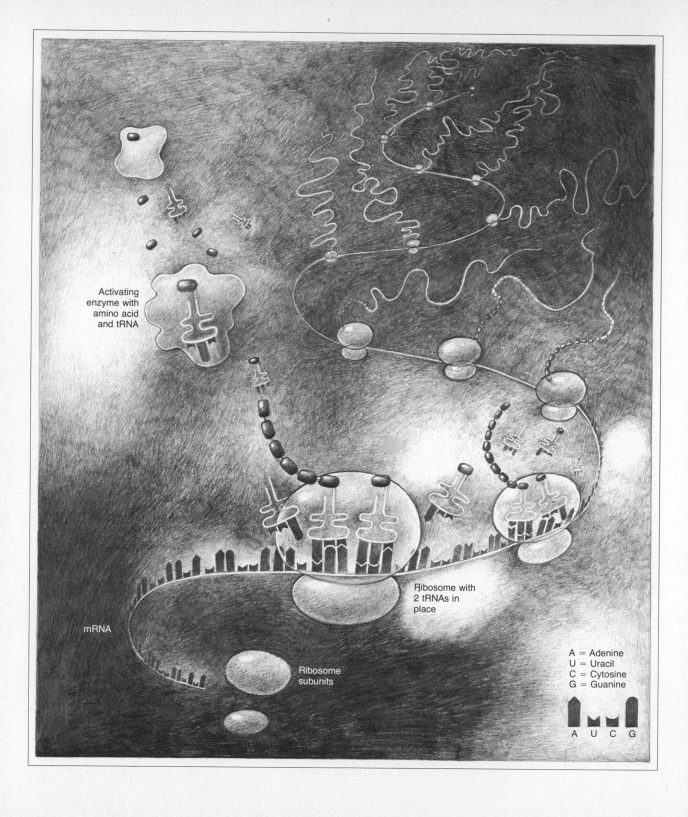

Activating
enzyme with
amino acid
and tRNA

Ribosome with
2 tRNAs in
place

mRNA

Ribosome
subunits

A = Adenine
U = Uracil
C = Cytosine
G = Guanine

A   U   C   G

ribosome-mRNA complex, the site of protein formation. The function of the ribosomes is to decode the message that was inscribed by the DNA. The amino acids from which the new protein will be formed are escorted to the ribosome-mRNA site by another RNA nucleotide called *transfer RNA*, or *t*RNA. (There is also a *ribosomal* RNA in the ribosome; we don't know yet exactly what it does.)

There are at least as many kinds of *t*RNA molecules as there are amino acids, and each kind of *t*RNA molecule can attach only to one type of amino acid. The specificity of the *t*RNA molecule is determined by the three nucleotides that are found at one end. For example, if these three bases all happen to be adenine, an AAA sequence, then the *only* amino acid that the *t*RNA molecule can attach to is one called *phenylalanine*. Now, when a *t*RNA with this nucleotide arrangement arrives at the *m*RNA-ribosome site with its phenylalanine, where do you think the phenylalanine will be inserted on the ribosome-*m*RNA complex? Of course! Only at a place on the *m*RNA where there are three complementary uracils (UUU). So both the kind of amino acid and its position in the developing protein are dictated by the DNA.

Figure 5.31 shows the process by which a new protein is synthesized. The ribosome attaches to one end of the *m*RNA strand and moves along its length, stopping to read off each nucleotide triplet to the *t*RNAs, which have diffused to the *m*RNA site with their amino acids. When it pauses at a triplet, a *t*RNA molecule with the proper complementary sequence drifts to that position, briefly plugs into the *m*RNA codon, and then moves away, leaving its amino acid behind. Meanwhile, the ribosome has reached the next codon site, and another *t*RNA falls into place, leaving its amino acid behind the first one. Thus amino acids are aligned, one by one, in proper sequence for the formation of a specific protein. When all the slots in the sequence have been filled with the appropriate amino acids, and these have attached to each other to form polypeptides or proteins, the protein moves away into the cytoplasm, and the *m*RNA begins to break down.

The resulting protein may be a structural material or it may be an enzyme, depending on the *m*RNA sequence along which it was constructed. If it is an enzyme, it will participate in the biochemical activities of that cell, and hence will help to determine its nature—what the cell looks like and what it does (perhaps the same things one might worry about a blind date). We know that the chromosomes are identical in every cell, but how can the same DNA component result in cells of entirely different types?

We know that different cells are specialized by different enzymes. We also know that enzyme production is directed by the DNA of the chromosomes through *m*RNA strands formed along its length, which then move into the cytoplasm to direct protein formation. Since proteins, even within a single cell or organism, are of many different kinds, we can assume that many kinds of *m*RNA are formed along a single DNA chain. Thus it is clear that a specific *m*RNA is produced only along a certain section of the

5.31 Protein synthesis. Messenger RNA is assembled along parts of separated strands of DNA. The rather large mRNA molecules then move through the nucleus membrane into the cytoplasm. In the cytoplasm they are joined by ribosomes that "read" the base sequence of the molecule. A sequence of three bases (a triplet) is read at once, and each triplet codes for a particular amino acid. As a triplet is read, a smaller RNA molecule, transfer RNA, or tRNA, moves to the site with an amino acid in tow. The tRNA has an *anticodon* sequence that fits the triplet's *codon*. When the appropriate tRNA locks onto the codon it releases its amino acid to a developing chain of amino acids and then unlocks and moves away. Any tRNA molecule can carry only one kind of amino acid. After the mRNA has directed a number of amino acid sequences, it breaks apart. Long sequences of amino acids form proteins, of course, and some of these are enzymes. Enzymes determine the nature and behavior of the cell.

long DNA molecule. It would be inappropriate for DNA to continually synthesize all of the varieties of *m*RNA, since this would mean all of the cell's proteins would be produced. So only certain segments in each DNA molecule are active in *m*RNA production. In other words, parts of the chromosomes are "shut down" while other parts are busy forming *m*RNA, with the working segment depending on which kinds of proteins are to be synthesized.

This explanation of cell differences, as you will immediately see, simply places the question of control one step further back. We are now faced with the question of how the active section of a chromosome is determined. At this point the data become rather varied and complex, but there is some evidence that the position of the cell in the body may be critical. The idea, in such an explanation, is that the milieu of a cell may have an effect on its chromosomes.

PROBLEMS IN CELL BIOLOGY

The sight of complex flowcharts and remarkable electron-microscope photographs of cell processes often leads us to believe that the great questions have been answered, and all we need to do now is to tie up a few loose ends. Nothing could be further from the truth. In fact, we have barely touched the "hem of the garment" in many areas of cell biology.

As an example of our ignorance, we are unable to control aging at almost any level. What causes cells to lose their tone? Why do fat cells under the eyes give way as we grow older? We just don't know! The only consolation for spiteful agers is that the day will also come for the ebullient youths around them—unless a method of cellular control is developed. In our youth-worshipping culture we are presented with a vast array of moisturizers, hormone creams, lotions, and exotic "aging remedies." The search goes on. Meanwhile, we can really say little more than "lead a good life"—and we don't even know what *that* is.

In spite of what may have seemed great assurance in the foregoing discussions, we actually know very little about such matters as mitochondrial replication, spindle formation, cell-environment interaction, cell differentiation (what makes cells different), cell growth, chromosome structure, or the permeability of membranes. We know only a precious few of the simplest enzymatic pathways.

Cancer is certainly one of our most dreaded and puzzling cellular problems and continues to absorb a great deal of research attention throughout the world.

Some cancer is the result of an uncontrolled mitosis on the part of certain cells. Cell division, as we know, is a tightly regulated procedure in the living organism, from the time of fertilization onward. Once the body has stopped growing, cell division must be carefully geared to replace only the cells that have worn out or died. In some cases, however, cell division begins to lurch out of control, and produce new cells with different qualities. These new kinds of cells may spread rapidly as cancer.

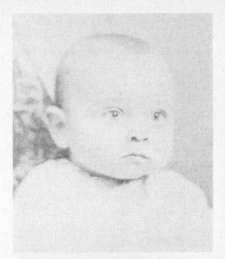

5.32 Changes in the behavior of chromosomes are largely responsible for the aging process. Putting glasses and hair on the infant at the upper left would not have caused anyone to exclaim, "Ah ha! Harry Truman!" But as the infant aged, the trappings of maturity gave way to the signs of old age. We don't know yet what causes areas of activity along chromosomes to shift as one ages, but the knowledge might well unlock the secrets of eternal youth. Would a perpetually young population be good or bad?

Box 5.2 Viruses

Viruses are remarkable little organisms, about 1/20,000 the size of bacteria, that reproduce entirely inside cells. Viruses are unique among living things—if, in fact, they are alive. Outside of a living cell they are completely inert. In some cases they look like crystals, and as lifeless as ordinary table salt. As soon as a virus enters a living cell, however, it begins to take over the metabolic machinery of that cell. It reorganizes the cell's processes so that the cell begins to engage in producing more viruses instead of continuing its normal activities. Finally, when the cell ruptures, new viruses are released to take over new cells.

Single or double strands of DNA or RNA are usually found inside the virus. Outside, it is coated with a protein layer that protects its nucleic acid from the cell's normal enzymes. This protein layer also gives the virus a special affinity for certain kinds of host cells. Some viruses, for example, have an affinity for bacteria. When such a bacteriophage, or bacteria-eater, accidentally bumps into a proper host cell, it attaches by its tail through covalent bonding. An enzyme of the virus digests a hole in the cell wall, and the virus then contracts its "head" and releases its nucleic acid contents into the bacterium.

Not all viruses have "tails" as we see in the prickly measles virus at top. Much of our information about viruses, however, has come from studies on such types as the coliphage virus (at bottom), which attacks a common intestinal bacterium.

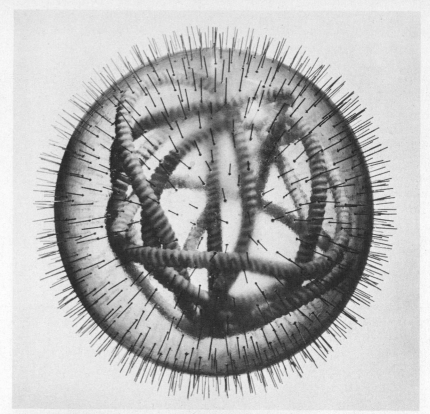

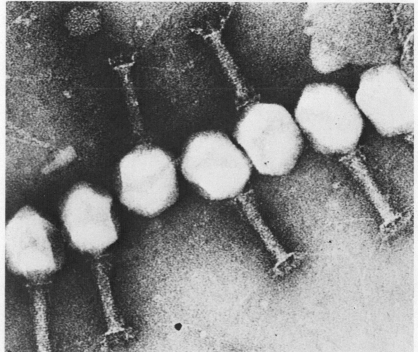

Viruses, certain chemicals, irradiation, and automobile exhaust products have all been linked with cancer, but in some cases the disease seems to appear spontaneously, perhaps through some small coding mistake in DNA replication. We may not be able to prevent cancer until we have a complete understanding of chromosome structure and the process of mitosis. In the meantime, techniques of virus control and antibody synthesis may enable us to control the symptoms.

Little by little we are accumulating bits of information about the nature and processes of cells, but there seem to be few shortcuts. Our hope of ever understanding these matters lies, now, in the slow accumulation of hard-won data by diligent researchers. Perhaps we have finally begun to ask some of the right questions. But with all the areas in which we still know too little even to formulate questions, much of the knowledge we need is likely to come from "pure" research—research that is not prompted by a specific utilitarian goal, but is done simply to "find out." The breaking of the genetic code and the description of DNA were the results of pure research. Thus it may prove scientifically tragic that government funding of pure research has been cut so drastically in our country. The federal funding of the sciences is being channeled into heart and cancer research at present. These are, of course, worthy causes, but the fact is often overlooked that those people involved in solving very specific problems for mankind must draw on the wealth of background information provided by researchers who simply wanted to know.

Inheritance

Occasionally you will hear someone just back from a trip to Mexico or Spain boasting of his bullfighting experiences. You might be disillusioned to learn that he hadn't fought any bulls at all—that his "bullfight" was with a heifer. At certain times of the year guests are invited to the ranches where the valiant "brave bulls" are bred to watch the testing of the cows—to determine which stock will be used for breeding. As part of the festivities guests may then be invited to try their hand at caping a young cow. The breeders are interested in the cows because it is axiomatic in such circles that a fighting bull gets its strength from the sire, but its courage from the cow. The origin of such beliefs is lost in tradition, but it is evident that man utilized the principles of genetics long before he became interested in their basis.

You will recall that Darwin knew something about artificial selection from his familiarity with livestock breeding. The offspring of heavier animals, he knew, tend toward heaviness; heifers which were daughters of high-yield milk producers tended to produce more milk. The principle seemed clear enough to animal breeders: some traits are inherited. However, in Darwin's time it was believed that the traits of both parents blended in the offspring, just as two colors of ink do when they are mixed; hence these characteristics could never show up in "pure" form in subsequent generations. In other words, a black horse and a white horse would produce a gray colt, and when these gray horses were mated, they would continue to produce gray offspring, but the original black or white could never reappear.

Although such thinking was firmly entrenched at Darwin's time, it ignored a body of contradictory evidence. For example, as far back as the first century Plutarch had noted that some children have qualities obviously inherited from their mothers and fathers, but that others often display the characteristics of their grandparents which had been missing in their parents. Also, there was no satisfactory explanation for the common observation that when black and honey-colored cocker spaniels are crossed the pups are not usually a blend, but instead are either black or honey-colored.

MENDELIAN GENETICS

It is interesting that the answers to many of the questions about heredity which plagued Darwin were actually published during his lifetime by a

6.1 Gregor Mendel was a well-trained, insightful, persevering, and very lucky abbot. The results of his years of breeding pea plants were published only seven years after Darwin's *Origin of Species*. It is unfortunate that he published in Darwin's shadow because his findings could have provided the missing mechanism for natural selection. There is some evidence that Darwin himself encountered Mendel's work, but failed, perhaps, to grasp its importance. In any case, Mendel's writings were found in Darwin's library.

monk named Gregor Johann Mendel (1822–1884), who was a member of an Augustinian order in Brünn, Moravia (now Czechoslovakia). Unfortunately, Mendel's work was given so little notice at the time that Darwin either never learned of it or its implications escaped him.

The Abbot Gregor Mendel was a rather remarkable person in his own right, although he is usually depicted as a kindly old man of the cloth who, while puttering around in his monastery garden, somehow stumbled on important laws governing genetic transmission. Our penchant for building myths to embellish the memory of already worthy people has led us astray again. Early in his life Mendel began training himself, and he became a rather competent naturalist. To support himself during those years he worked as a substitute high school science teacher. The professors at the school, noting his unusual abilities, suggested that he take the rigorous qualifying examination and become a regular member of the faculty. Mendel took the test and did surprisingly well, but he failed to qualify. The conditions for teaching high school were rigorous indeed.

The monastic order to which Mendel belonged, confident of his abilities, sent him in 1851 to the University of Vienna for two years of concentrated study in science and mathematics. When Mendel returned to the monastery, he began his plant-hybridization studies in earnest. He was especially noted for his high level of intelligence and his vigor, and he applied these qualities to the study of plant breeding. He developed new varieties of fruits and vegetables, kept abreast of the latest developments in his field, and became active in community affairs.

In 1865, while Darwin was still puzzling over the enigma of heredity, Mendel presented a single paper at a meeting of the Brünn Natural Science Society. The paper was published in the society's proceedings the following year. Apparently, however, no one understood what Mendel was even talking about, and the work for which he was to go down in history was met with a resounding silence.

Actually, Mendel said three things:

1. When parents differ in one characteristic, their offspring will be *hybrids* for that particular characteristic; however, the offspring may resemble one parent of the other instead of showing a blend of both traits. This means that the trait of one parent must be dominant over the trait of the other parent. This phenomenon he called the principle of *dominance*.

2. When a hybrid reproduces, its gametes (eggs or sperm) will be of two types. Half will carry the dominant trait provided by one parent and half will carry the recessive trait provided by the other parent. This he called the principle of *segregation*.

3. When parents differ in two or more characteristics, the occurence of any characteristic in the next generation will be independent of the occurence of any other characteristic, so that any combination of the parental components may appear in the offspring. This is called the principle of *independent assortment*.

Now, at this point you are probably mystified as was Mendel's audience, but let us continue. Perhaps the best way to clarify things it to see exactly how Mendel arrived at his conclusions.

The Principle of Dominance

To begin with, Mendel based his information on a very carefully planned series of experiments and, more important, on a statistical analysis of the results. The use of mathematics to describe biological phenomena was a new concept. Clearly, Mendel's two years at the University of Vienna had not been wasted.

The careful planning that went into Mendel's work is reflected in his selection of the common garden pea as his experimental subject. There were several advantages in this choice. Pea plants were readily available and fairly easy to grow, and Mendel had already developed some thirty-four pure strains. These strains differed from each other in very pronounced ways, so that there could be no problem in identifying the results of a given experiment. Mendel chose seven different pairs of characteristics to work with:

1. Seed form—round or wrinkled
2. Color of seed contents—yellow or green
3. Color of seed coat—white or gray
4. Color of unripe seed pods—green or yellow
5. Shape of ripe seed pods—inflated or constricted between seeds
6. Length of stem—short (9 to 18 inches) or long (6 to 7 feet)
7. Position of flowers—axial (along the stem) or terminal (at the end of the stem)

Mendel's approach, a novel one at that time, was to cross two true-breeding strains that differed in only one characteristic, such as seed color. Peas ordinarily self-fertilize, so for this cross the pollen was transferred by hand. He called this original parent generation P_1 and designated their first-generation offspring the F_1 (first filial) generation. When the F_1 plants were allowed to self-pollinate, so that they crossed with each other at random, the offspring resulting from this cross were called the F_2 generation, and so on.

Now, when Mendel crossed his original P_1 plants, he found that the characteristics of the two parents didn't blend, as prevailing theory said they should. When plants with yellow seeds were crossed with plants that had green seeds, their F_1 offspring did not have yellow-green seeds. Instead, all of them had yellow seeds. Mendel termed the trait which appeared in the F_1 generation the *dominant* trait and the one which had failed to appear the *recessive* trait. But he was now left with a vexing question. What had happened to the recessive trait? It had been passed along through countless generations so it couldn't have just disappeared.

The Principle of Segregation

In the next stage of his experiments Mendel allowed the F_1 plants to self-pollinate, and lo and behold, the missing recessive trait reappeared in

Table 6.1 Mendel's Pea Plant Experiment

Dominant form	No. in F_2 generation	Recessive form	No. in F_2 generation	Total examined	Ratio
Round seeds	5,474	Wrinkled seeds	1,850	7,324	2.96:1
Yellow seeds	6,022	Green seeds	2,001	8,023	3.01:1
Gray seed coats	705	White seed coats	224	929	3.15:1
Green pods	428	Yellow pods	152	580	2.82:1
Inflated pods	882	Constricted pods	299	1,181	2.95:1
Long stems	787	Short stems	277	1,064	2.84:1
Axial flowers	651	Terminal flowers	207	858	3.14:1

their F_2 offspring! Also, the *ratio* of F_2 plants with recessive traits to those with dominant traits was fairly constant, regardless of the particular characteristic involved. The exact results for each of the seven pairs of characteristics Mendel used in his tests are shown in Table 6.1.

Finally, in the third year of the experiments, Mendel allowed the F_2 plants to self-pollinate. He found that all those with recessive traits produced only recessive F_3 offspring. However, the F_2 plants that showed a dominant trait produced both types of F_3 offspring. One-third of them produced *only* dominant offspring. The other two-thirds produced both dominant and recessive offspring, but they produced three times as many offspring with the dominant trait as they did with the recessive trait. In other words, all the F_2 recessives and one-third of the F_2 dominants bred true. The other two-thirds of these dominants produced mixed off-spring—but in the same 3:1 ratio of dominant to recessive as in the plants their F_1 parents had produced.

Let's consider a specific example of a cross between two pure lines—plants that breed true when they self-pollinate and are identical in all characteristics except one. In this case suppose one strain always produces round seeds and the other always produces wrinkled seeds. We have already seen that the round form is dominant, so we'll designate that form as R and the recessive wrinkled form as r. Since both plants are true-breeding, we can assume that in each plant the genetic components (from their own parents) are identical for their respective traits. That is, in one plant the components are RR and in the other they are rr. Such plants are said to be *homozygous* for that particular characteristic, meaning that all the gametes they produce will be the same.

Figure 6.2 shows what happens when these two plants are crossed. Remember that they are identical in all characteristics except one, so only the differing characteristic is diagrammed. Since half the gene component in the F_1 generation comes from each parent, the genetic composition, or *genotype,* of any F_1 plant would have to be *Rr.* Hence these plants would be *heterozygous* for the characteristic in question. Now, when these plants produce gametes (eggs and pollen), each gamete will carry either an *R* component or an *r* component, and the combination that occurs when two gametes come together at fertilization will determine the genetic characteristic of the resulting F_2 plant.

The results we should expect in the F_2 generation are diagrammed in a form called a *Punnett square.* As you can see in Figure 6.2, the physical characteristics, or *phenotypes,* of the F_2 seeds occur in a 3:1 ratio. However, a moment (or several moments) of reflection will show that the F_3 ratio Mendel obtained in the third year of his experiments could have been gotten only if one-fourth of the F_2 plants were homozygous dominant for the characteristic, one-half were heterozygous, and one-fourth were homozygous recessive. So the genotype of this F_2 generation is described as 1:2:1.

These results led Mendel to conclude that every heritable characteristic is determined by two components, one from each parent. (Before you start feeling smug about the fact that you knew this, remember that it was Mendel who told you.) Thus if two homozygous parents were represented by *RR* and *rr* (as for seed form), their offspring would have to be heterozygous *Rr,* since one component would have come from each parent. And when this *Rr* individual produced gametes, half would contain an *R* component and the other half would contain the *r* component. A cross between two *Rr* individuals could then be expected to yield offspring with the genotypes *RR, rR, Rr,* and *rr,* or a genotype ratio of 1:2:1. What would be the genotypes and the phenotypes of the offspring if an *Rr* crossed instead with a homozygous recessive?

Mendel's experiments with monohybrid crosses enabled him to show that the genetic element for each characteristic, now called a *gene,* consists of two components (or *alleles*), one contributed by each parent. However, only one of these two components may be expressed in physical form. Hence an individual carrying both components may show no sign of the other parental characteristic at all. The principles of dominance and segregation implied by these results ruled out (in Mendel's mind, at least) the earlier notion that heritable characteristics were always blended in the offspring.

The Principle of Independent Assortment

In his next set of experiments Mendel crossed *dihybrids,* true-breeding plants that differ in two characteristics. In one such experiment plants with *round (R) yellow (Y)* seeds were crossed with a strain that had *wrinkled (r) green (y)* seeds. We saw in Table 6.1 that round and yellow are both dominant, so we shouldn't be astounded to learn that all the F_1 plants had round yellow seeds. Since one P_1 parent was *RR YY* and the other was *rr yy,* all the F_1 plants would have to be *Rr Yy,* or heterozygous for both characteristics. When these plants were then allowed to self-pollinate, we

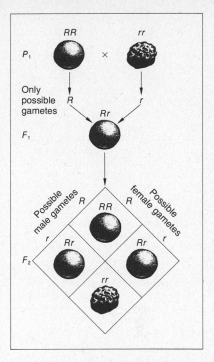

6.2 Arriving at the expected ratio in the F_2 generation of plants that differ in only one character (here, seed shape). The parent (P_1) generation is a cross between a pure-breeding plant with round seeds (*RR*) and a pure-breeding plant with wrinkled seeds (*rr*). Obviously, a plant that had received an *R* gene for round seeds from each parent could only pass along *R* genes to its progeny, and the same for a plant receiving only *r* genes from its parents. The F_1, then, must be *Rr* since one gene for seed shape must come from each parent. If two individuals from the F_1 are crossed, the expected ratio of their offspring can be calculated by use of a Punnett square. Remember, though, that these are only *expected* results and this ratio would probably be approached only if a great number of crosses were made.

6.3 In this case we have the expected F_2 ratio from a dihybrid cross (a cross between plants that differ in only two characters—here, seed shape and seed color). One pure-breeding parent in the P_1 generation bore both dominant traits, the other, both recessive traits. Obviously they could pass along only dominant and recessive genes, respectively. This means the F_1 is necessarily heterozygous for both traits. At meiosis, however, the traits assort independently. This means that when the F_1 is crossed with itself we must account for all possible combinations in order to determine the F_2 ratio. Only one possible combination is labeled. If you were to work this out, you would get nine genetic combinations. Calculate the genotypes and phenotypes and then check page 133 to see how wrong you were. Again, keep in mind that we're only dealing with probabilities. This means that all the F_2 *could* be *rryy* (double recessives), but the odds are very much against it.

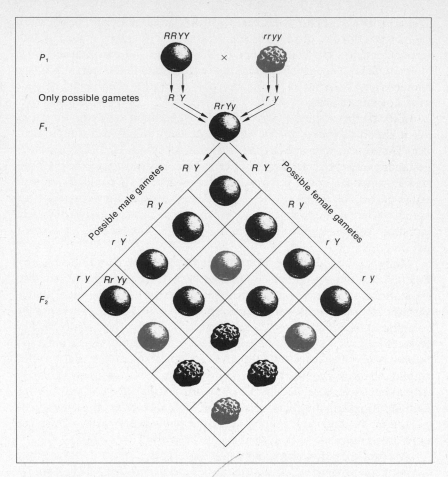

would expect a random and independent assortment of the various characteristics to produce the F_2 generation shown in Figure 6.3

It was apparent to Mendel from the results of this dihybrid cross that the segregation of genes for one characteristic was not affected by the segregation of genes for the other characteristic. Thus he was able to deduce that genetic combinations follow the principle of independent assortment. Now, it should be pointed out here that if you were to run a single experiment to test Mendel's results, you might end up with all wrinkled green seeds in the F_2 generation—just as when you toss a coin ten times, it *might* come up tails each time. You could expect the results shown in Figure 6.3 only if you ran enough tests for the law of averages to apply.

Before we leave the subject of Mendelian genetics, you should know that there are the inevitable exceptions. Not all characteristics are expressed in dominant-recessive form. In some cases a characteristic may appear in the offspring in a form intermediate between the characteristics of each parent. For example, four-o'clocks may have either red or white flowers. A cross between the two produces pink-flowered plants. A situation in which one characteristic is not dominant over the other is called *intermediate inheritance*. As before, one component, or allele, of the gene is contributed by each parent, but in this case neither allele gives way to the other. Instead they both exert an influence on the physical characteristic determined by that gene. This means, of course, that the pink-flowered plant is heterozygous for flower color. Could a cross between two such plants produce red or white flowered plants again? What kinds of gametes would these plants produce?

There are also other ways in which intermediate effects arise, and such variations within species may have a number of parameters. For example, people do not come in only two sizes, large or small. They come in a wide range of heights, weights, and shapes, just as they differ widely in bone size, nose shape, and any number of other physical characteristics. Such variation can be accounted for if we assume that certain traits are coded for by more than one gene.

Suppose some characteristic such as the length of bill in a woodpecker were coded for by four separate genes, with "longness" indicated by a capital letter and "shortness" indicated by a lowercase letter. Then a bird with the genotype *AA BB CC DD* would be expected to have the longest bill, provided environmental factors such as diet were conducive. A bird with the genotype *Aa BB Cc DD* would have a somewhat shorter bill, and one with the genotype *Aa bb cC dd* would have a still shorter bill. Now, since there is a bill size that is "best" for a woodpecker, we would expect to find more birds with bills in the intermediate range than at either size extreme. Thus if the bill length of a population of woodpeckers were plotted on a graph, we might expect something approaching a "normal" curve (Figure 6.4).

CLASSICAL GENETICS

It is unfortunate that those biologists who did read Mendel's reports were skeptical of his findings. It is also unfortunate that no one apparently saw enough significance in them to bother repeating his laborious experiments. However, six years after Mendel died in obscurity, his work was rediscovered by three separate researchers who independently arrived at essentially the same conclusions. At this point the field of genetics burst into an era of dizzying expansion. Although Mendel didn't live to see this happen, his experiments provided much of the basis for what is now called post-Mendelian, or classical genetics.

Thus far it had been assumed that the chromosomes of any organism were identical, and that meiosis was simply a way of dividing them up so that the gametes would receive the same number. It was also believed that during meiosis the maternal and paternal chromosomes separated and each set traveled together at anaphase, so that any gamete would contain chromosomes from one parent or the other, but not both. In the early 1900s, however, it was found that chromosomes are not identical, but come in a variety of sizes and shapes, and that they also come in pairs—two of

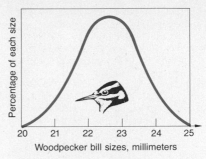

6.4 Woodpecker bills come in a number of sizes and shapes. To the untrained eye, they may all look about the same, but there *are* differences. Variation exists within populations, so some bills are long, some short, but the majority are of an intermediate length. In the population shown in this figure, there are few birds with bills 21 millimeters long and virtually none with bills of 20 millimeters. Nor do any have bills longer than 25 millimeters. Practically every woodpecker you see, though, has a bill within the 22 to 24 millimeter range. Why do you suppose this is? When a population shows a continuous distribution like this, instead of all bills being of 2 or 3 sizes only, you can assume that the trait is controlled for by a combination of several genes. Of course, a bird genetically disposed to have a long bill might not achieve that status if he has been subjected to a poor diet.

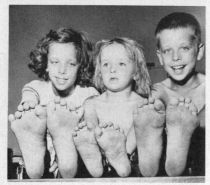

6.5 Polydactyly, the condition of having more than five digits on the hands or feet is a dominant trait in humans. In the top photograph a man exhibits six fingers on each hand. In the lower photograph two girls and their brother show variation in the condition. The girl at left has six toes on both feet, her sister has six toes on one foot and her brother has the normal complement of five toes on each foot.

Table 6.2 Some human traits that are inherited

	Dominant	Recessive
Hair, skin, etc.	dark hair	blond hair
	nonred hair	red hair
	curly hair	straight hair
	abundant body hair	little body hair
	early baldness (dominant in males)	normal
	white forelock	self-color
	piebald (skin and hair spotted with white)	self-color
	pigmented skin, hair, eyes	albinism
	black skin (two pairs of genes, dominance incomplete)	white skin
	ichthyosis (scaly skin)	normal
	epidermis bulbosa (sensitiveness to slight abrasions)	normal
	normal	absence of sweat glands
Eyes and features	brown	blue or gray
	hazel or green	blue or gray
	"Mongolian fold"	no fold
	congenital cataract	normal
	nearsightedness	normal
	farsightedness	normal
	astigmatism	normal
	free ear lobes	attached ear lobes
	broad lips	thin lips
	large eyes	small eyes
	long eyelashes	short eyelashes
	high, narrow bridge of nose	low, broad bridge
Skeleton and muscles	short stature (many genes)	tall stature
	dwarfism (achondroplasia)	normal
	midget (ateliosis)	normal
	polydactyly (more than five digits)	normal
	syndactyly (webbed fingers or toes)	normal
	brachydactyly (short digits)	normal
	progressive muscular atrophy	normal
Some conditions affecting systems	hereditary edema (Milroy's disease)	normal
	hypertension	normal
	normal	hemophilia (sex-linked)
	normal	sickle-cell anemia
	normal	diabetes mellitus
	tasters (PTC)	nontasters (PTC)
	normal	congenital deafness
	normal	amaurotic idiocy
	migraine	normal
	normal	dementia praecox (several pairs of genes)
	normal	phenylketonuria (PKU)

Source: Adapted from C. Villee, *Biology,* 4th ed. (Philadelphia: Saunders, 1962), and Evelyn Morholt, Paul F. Brandwein, and Alexander Joseph, *A Sourcebook for the Biological Sciences* (New York: Harcourt, Brace & World, 1966).

each type. Chromosomes that comprise the pair are called *homologs*. In addition, it was found that meiosis divides chromosomes not only quantitatively but also qualitatively, so that each gamete receives both the same *number* and *kind* of chromosomes.

A young graduate student at Columbia University, Walter S. Sutton, began a study to try to relate Mendel's findings to what was known about meiosis. His results were surprising. For example, he found that when chromosome pairs lined up at meiosis, there was no way to determine which way a paternal or maternal chromosome would go at anaphase. Some maternal chromosomes moved to one pole and some to the other, but there was no way to predict their direction. Now, at this time it was not known that chromosomes were the carriers of genes. The independent assortment of chromosomes, however, confirmed a hunch Sutton had that hereditary units are associated with chromosomes. Also, since there were far more hereditary characteristics than there were chromosomes, he reasoned that each chromosome must be responsible for many genetic traits.

In 1910, after Sutton's report, an innocuous little insect entered the world of genetics. Thomas Hunt Morgan, also at Columbia, began a program of breeding experiments with the fruit fly (or vinegar fly), *Drosophila melanogaster*. These little flies are ideal subjects for genetic studies. They are easy to maintain in the laboratory; they mature in twelve days and reproduce prolifically; and they have a variety of features that can be readily identified. They also have giant chromosomes in their salivary glands, on which specific areas, or loci, are readily distinguishable. In addition, they have only four pairs of chromosomes. They have little economic importance, and they don't seem to bother man much where the two species come in contact. In fact, it has been said that God must have invented *Drosophila melanogaster* just for Thomas Hunt Morgan.

Sex Determination

One of the first things Morgan did was attempt to cause genetic changes, or mutations, in his flies. He subjected them to cold, heat, X rays, chemicals, and radioactive matter. At first he found nothing, but then he noticed among a group of normal red-eyed flies that there was a single male with strange *white* eyes. Morgan was convinced that this was a mutation (a subject we will take up later). He carefully nurtured his little white-eyed specimen and crossed it with several of its red-eyed sisters. In these crosses he found that the F_1 generation were all red-eyed, just as you would expect if white eyes were recessive. Simple enough so far—but then the story became a bit more complicated.

When the F_1 flies were allowed to interbreed, the result was 3,470 red-eyed and 782 white-eyed flies. This is not the simple 3:1 ratio you might have expected according to Mendelian principles. Also, there was another troubling factor. All the 782 white-eyed flies were males. The red-eyed flies, however, were of both sexes—2,459 females and 1,011

Answer to Figure 6.3.
You should get:

GENOTYPE	PHENOTYPE
1 RRYY	9 Round Yellow
2 RRYy	3 Round green
1 RRyy	3 wrinkled Yellow
2 RrYY	1 wrinkled green
4 RrYy	
2 Rryy	
1 rrYY	
2 rrYy	
1 rryy	

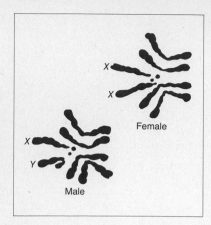

6.6 The chromosomes of male and female *Drosophila*. Note that each chromosome of one sex is represented in the genotype of the other, except for the twisted *Y* chromosome of the male. Because there are few chromosomes and they are rather easily identifiable, *Drosophila* were chosen for intensive research.

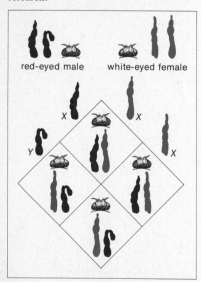

red-eyed male   white-eyed female

6.7 In *Drosophila* the gene for white eyes (a recessive gene) exists on the *X* chromosome. In females the gene can be masked by a dominant gene for red on the other *X* chromosome. But males carry no such dominant gene on their *Y* chromosome, so if a male's *X* should bear the gene for white eyes, it will be expressed.

males. At this point you may have come up with the obvious answer: only males are white-eyed. Alas, when the original white-eyed male was crossed with its F_1 red-eyed daughters, the result was 129 red-eyed females, 132 red-eyed males, 86 white-eyed males—and 88 white-eyed females! If you have a penchant for puzzles, can spare a few minutes, and think you are smart, try to figure out how this could have happened. In case you need a clue, keep in mind that sex is determined by chromosomes. Meanwhile, the rest of us will trudge along.

Morgan surmised that, although chromosomes may segregate independently, the genetic determiners (or genes) may not. He then surmised that the sex-determining factor and the eye-color factor are somehow linked together. Hence, as one goes at anaphase, so must go the other.

By this time Morgan knew something about the chromosomes that determine the gender of an offspring. He knew that of the four pairs of chromosomes in *Drosophila*, only one pair was responsible for gender. He even knew which ones were the *sex chromosomes* and which were the *autosomes*, the chromosomes responsible for other characteristics. In females the sex chromosomes are a rod-shaped pair of chromosomes, for some reason called *X chromosomes*. In males the sex chromosomes consist of only one *X* chromosome and a J-shaped chromosome called a *Y chromosome* (Figure 6.6). The *X* chromosomes are usually larger and contain far more genes than the *Y* chromosome. This fact will be critical in understanding certain of the effects of sex linkage in heredity.

In the production of sperm in males the *X* and *Y* chromosomes line up together at meiosis, so that half the gametes will contain an *X* and half will contain a *Y*. Since females have only *X* chromosomes, the eggs will always contain an *X*. If a *Y*-bearing sperm reaches the egg first and fertilizes it, the offspring will be *XY*, and thus male. If an *X*-bearing sperm fertilizes the egg, the offspring will be *XX*, or female. This is the case in many animals, including mammals, but in other organisms the situation may be different. For example, in birds the sex chromosomes are identical in males *(ZZ)* and different in females *(ZW)*.

Now, regarding the question of eye color in *Drosophila*, Morgan showed that the genetic allele for eye color is located only on the *X* chromosome, and white eye color is recessive. Therefore females will have white eyes only if both eye-color alleles are for white eyes. In males, however, when a white-eye factor turns up on the *X* chromosome, there is no dominant counterpart on the *Y* chromosome, so the white eye color will be expressed (Figure 6.7).

Sex Linkage in Humans

There have been several interesting applications of the findings regarding sex inheritance in the past few years. For example, the cells in women have tiny specks in their nuclei called *Barr bodies*. Also, white blood cells in women show a small "drumstick" attached to the nucleus (Figure 6.9).

MALE					FEMALE				

MALE

1   2   3   4   5

6   7   8   9   10

11   12   13   14   15

16   17   18   19   20

21   22   X Y

FEMALE

1   2   3   4   5

6   7   8   9   10

11   12   13   14   15

16   17   18   19   20

21   22   X X

6.8 Human chromosomes of normal male and female. Note that the pairs are identifiable and numbered. Thus a disease called trisomy 21 is understood to be the presence of three chromosomes of the type labeled as 21. Notice that the chromosomes are identical in males and females except for the sex chromosomes. Also note the small size of the *Y* chromosome. Considering the large number of chromosomes held in common by the sexes, one might wonder how the sexes could come to differ in so many ways.

Cells from normal men lack both the Barr body and the drumstick. You may remember a stir at a recent meeting of the Olympic games when a European woman who showed remarkable strength in the field events was found to be genetically male, to lack the Barr bodies and drumsticks in her cell nuclei. For certain athletic competitions, sex tests are now mandatory.

Color blindness in humans is caused by a recessive allele on the *X* chromosome. Thus a woman who is heterozygous for this condition will show no symptoms of color blindness. Among her children, however, she can expect one son in two to be color blind and one daughter in two to be carriers of color blindness like herself. In her sons the recessive gene on the *X* chromosome is not overridden because the tiny *Y* chromosome bears no allele for color discrimination. Figure 6.10 shows the expected occurrences of color blindness in this woman's grandchildren, the F_2 generation. There are a number of other phenomena associated with sex-linked genes that have great impact on people's lives. For example, muscular dystrophy is caused by a recessive gene on an *X* chromosome. The effects of this abnormality are so pronounced, however, that males carrying the gene usually die in their teens before reproducing. As a result, females rarely have the disease, since the probability of receiving the allele from both parents is very low. However, they can carry a single gene for the condition, so that when they reproduce they may pass it along to their offspring. Another such disorder is hemophilia, which is also a recessive characteristic carried on an *X* chromosome (Box 6.1).

In other instances, there may be an imbalance of *X* and *Y* chromosomes. For example, certain females may have cells with no Barr bodies or drumsticks. In such cases there is only one *X* chromosome instead of two.

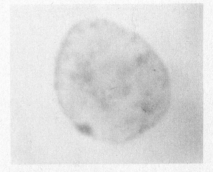

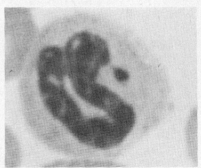

6.9 Electron micrographs of the drumstick (top) and Barr body (bottom) from cells of a woman. Drumsticks are found in white blood cells. The sex of an individual can be identified according to these telltale structures, normally lacking in males.

Box 6.1 The Disease of Royalty

Because it was the practice of ruling monarchs to consolidate their empires through marriage alliances, hemophilia was transmitted throughout the royal families of Europe. Hemophilia is a sex-linked recessive condition in which the blood does not clot properly, so that any small injury can result in severe bleeding, and if the bleeding cannot be stopped, in death. Hence it has sometimes been called the "bleeder's disease."

Hemophilia has been traced back as far as Queen Victoria, who was born in 1819. One of her sons,

Leopold, Duke of Albany, died of the disease at the age of thirty-one. Apparently at least two of Victoria's daughters were carriers, since several of their descendents were hemophilic. Hemophilia played an important historical role in Russia during the reign of Nikolas II, the last Czar. The Czarevich, Alexis, was hemophilic, and his mother, the Czarina, was convinced that the only one who could save her son's life was the monk Rasputin—known as the "mad monk." Through this hold over the reigning family, Rasputin became the real power behind the disintegrating throne.

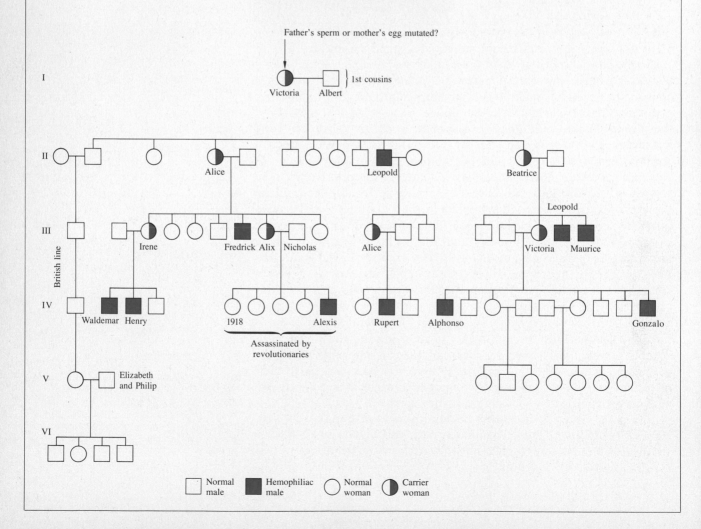

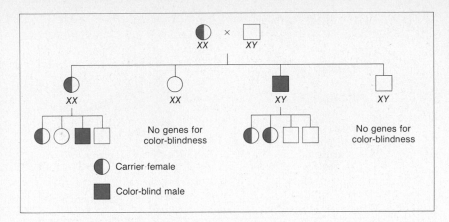

6.10 How color blindness is inherited. The original mating is between a carrier female and a normal male. Spouses (not shown) are presumed to be free of the color-blindness gene.

Carrier female

Color-blind male

No genes for color-blindness

No genes for color-blindness

The physical characteristics that accompany this condition are female genitalia, but underdeveloped breasts and tiny ovaries—and for some reason, short stature. The presence of an extra X chromosome in males (XXY) also results in certain anomalies. Such males usually have sparse body hair, some breast development, and cells that show both Barr bodies and drumsticks.

There has been some discussion of the observation of an XYY condition in certain criminals. Of course, there is no information on its frequency in the general population of noncriminals. Nevertheless, several legal appeals have been entered on behalf of men convicted of murder and other violent crimes based on the argument that these men had been "genetically driven" to their crimes and could therefore not be judged accountable for them.

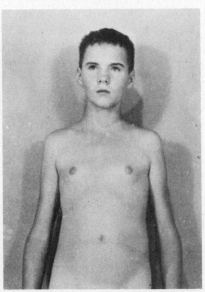

6.11 Down's syndrome (or *mongolism*) is evident in the child at left. The eyefold and open mouth are characteristic. The condition is caused by an extra chromosome in the twenty-first pair (all human chromosomes bear a number). This is not a sex-linked trait, so it appears in both males and females. Such individuals are mentally deficient and short of build, often with heart abnormalities. These children are born more frequently to older women. Klinefelter's syndrome, at right, is caused by an extra X chromosome in males (XXY). Cells may show Barr bodies and drumsticks. Such individuals may be tall, lacking in some body hair and showing slight breast development.

6.12 Curved wings and normal wings in *Drosophila*. The curved wing is one of the obvious traits, due to a single gene mutation, which makes this small insect so easy for geneticists to work with. The curved wing trait is recessive, as is black body color, the normal color being grey. It was discovered later that the genes for wing shape and body color lie on the same chromosome and thus cannot obey the law of independent assortment.

6.13 Because some genes are linked on the same chromosome, you would expect these results from the cross shown below. Actually the linkage may be disrupted by crossovers.

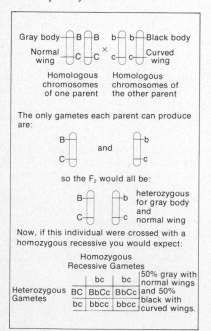

Gene Linkage and Crossover

Recall from our discussion of Mendel's work that the assortment of chromosomes is independent. Thus the assortment of genes on one chromosome of a pair would have to be independent of those on the homologous chromosome. If two genes are on the *same* chromosome, however, it is likely that they will be passed together to the gamete. The occurrence of two or more hereditary units on the same chromosome is called *gene linkage*.

In *Drosophila*, for example, the recessive alleles for a black body (*b*) and a curved wing (*c*) are carried on the same chromosome. The dominant alleles are for a gray body (*B*) and a normal wing (*C*). So if we diagram a cross between a homozygous normal fly, *BB CC*, and a homozygous recessive fly, *bb cc*, we will get the results shown in Figure 6.13. The only gametes the gray, normal-winged fly can produce are *BC* and the only gametes the black, curved-wing fly can produce are *bc*. Hence their F_1 offspring would all be heterozygous for both characteristics—*Bb Cc*—and would, of course, show gray bodies and normal wings.

Now, if this *Bb Cc* fly were crossed with a homozygous recessive, *bb cc*, you would expect half of their offspring to be gray with normal wings and half to be black with curved wings (this is worked out for you in the Punnet square). In all likelihood, however, this cross would produce about 37 percent gray, normal-winged flies; about 37 percent black, curved-wing flies; 13 percent *black*, normal-winged flies; and 13 percent gray, *curved*-wing flies. We would have to have two conditions to account for this: (1) the genes for both body color and wing shape would have to be on the same chromosome, with their alleles on homologous chromosomes, and (2) the chromosomes would have to be able somehow to exchange genes.

You can probably believe condition 1 easily enough on the basis of our previous discussions, so let's consider how condition 2 might exist. Recall that in meiosis the homologous chromosomes line up during metaphase I, and as anaphase begins, the homologs of each pair move toward opposite poles. While they are lined up side by side in prophase, however, the homologs may stick together at various places and then exchange parts of their chromatids. The opposite chromatids of the homolog cross each other and then break off at their point of contact. Each broken part then fuses with the opposite portion of its homolog (Figure 6.14).

You might wonder how the chromatids manage to break at *exactly* the same place on each one, so that one chromosome does not end up with more genes than the other chromosome. The geneticists investigating the phenomenon of crossover are asking the same question—but with no answer yet. We do know, however, the basic mechanism by which crossover occurs, and also that some crossover takes place at virtually every meiosis.

CHROMOSOME MAPPING

If genes are arranged along the chromosomes and specific alleles are precisely exchanged through crossovers, then the genes for certain characteristics must lie at specific points along each chromosome. But how do we know which genes lie where? Finding the answer involves "mapping" the chromosome—a technique that is based on information from observations of crossovers.

Thomas Morgan and A. H. Sturtevant, one of the brilliant young graduate students attracted to Morgan's laboratory in those days, hypothesized that if genes were arranged linearly along the chromosome, then those lying closer together would undergo crossovers less often than those lying farther apart. The idea assumed randomness, or at least a large degree of variability, in the point at which chromosomes would break. Genes lying closer together would thus have a greater *probability* of being passed along as a unit. Stated another way, the percentage of crossover is proportional to the distance between two genes on a chromosome—so *percent crossover* is the number of crossovers between two genes per 100 prophase opportunities.

As an example, suppose two characteristics, which we can call *A* and *B,* show 26 percent crossover. We can assign 26 crossover units to the distance between the two genes. Then if some characteristic *C* turns out in breeding experiments to have 9 percent crossover with *B* and 17 percent crossover with *A,* it would be located between *A* and *B* at a point 9 units from *B* and 17 units from *A.* After the information from many such experiments has been compiled, a chromosome map can be constructed that indicates the position along the chromosome of the genes which code for certain characteristics. Figure 6.15 shows chromosome maps developed by this technique for chromosomes in the salivary gland of *Drosophila melanogaster.*

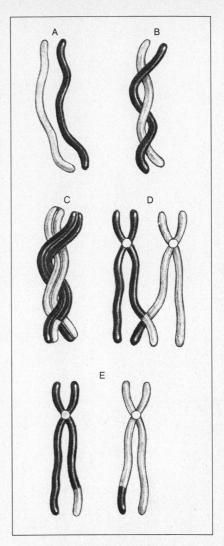

6.14 In meiosis, homologous chromosomes synapse or join together in early prophase (see A and B). During this time, genetic components may be exchanged. Following this, each chromosome appears to divide longitudinally, forming chromatids (C). This is followed by a separation of homologous chromosomes with the crossover of chromatids as they separate (D). The result is a pair of chromosomes that have exchanged genetic information (E).

```
           26
  |_____|
  A                B


        17      9
  |_____|_____|
  A        C     B
```

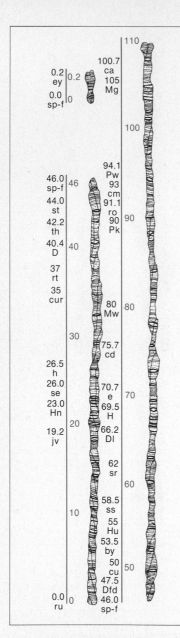

6.15 Map of chromosomes from the salivary gland of *Drosophila melanogaster.* The units are shown in the numbered column at right. The relative positions of specific genes (such as one called jv) are shown by the numbers above the gene designation.

MUTATIONS

Mutations, or changes in the hereditary material, may be caused by X rays, chemicals, ultraviolet radiation, and a number of other agents. In fact, as you are sitting peacefully reading this fascinating account, your body is being bombarded with tiny subatomic particles from outer space—radiation. If one of these particles should strike you right in the gonad, and alter a nucleotide in a developing gamete, the structure of that gamete could be changed. Recall that a small change in a nucleotide sequence could result in an entirely new codon. Hence if that gamete happens to enter into the reproductive process, whatever change has taken place will be passed along to the next generation, unless the change is lethal or otherwise prevents reproduction, or unless it is reversed through a second mutation, an unlikely event.

Genic Mutations

There are different kinds of mutations. One is associated with alterations in the genes and is called *genic mutation.* Genic mutations occur during the duplication of DNA. They may involve the omission of nucleotide units, the repetition of nucleotide units, or mistakes in bonding between these units. Or they may involve the substitution of one nucleotide for another—either the substitution of one purine or pyrimidine for another or the substitution of a purine for a pyrimidine or a pyrimidine for a purine.

Not all such changes result in a change in the characteristic. For example, the sequence GAC codes for the insertion of aspartic acid into a protein, and a change from GAC to GAU causes no alteration in the code. Also, other mutations sometimes occur that compensate for the effect of the first change and thus conceal it.

It should be apparent that the addition or deletion of a single nucleotide could change the entire amino acid product, since the codons are read in groups of three nucleotides. For example, ACG-GCC-GGA codes for threonine, alanine, and glycine. If the first G should be repeated, the sequence would instead read ACG-GGC-CGG-A . . . and the amino acids that would be inserted into the protein would be threonine, glycine, and arginine. Thus this one added nucleotide could change every amino acid from that point on in the sequence.

Chromosome Mutations

A second class of mutations involves some physical change in the structure of chromosomes. These, logically enough, are called *chromosomal mutations.* Such changes may occur when the chromosomes are moving about or undergoing changes—for example, during meiosis. At such times a part of a chromosome may be broken off and lost (a *deletion*), or it may break off and reattach at its other end (an *inversion*), or a segment may be duplicated, resulting in a doubling in the number of certain genes (a *duplication*). The most serious of these mishaps is a deletion, since the absence of a large number of genes is likely to result in death.

Of course, not all mutations are bad. In fact, mutations are the source of much of the variation we see between individuals in a population. Among these variations, certain charcteristics will enable an individual to fit into the environment better than other individuals. Thus these better-equipped individuals will be more likely to survive and reproduce as the environment exerts its effect on the population. This phenomenon of natural selection is one we will discuss in more detail later.

In some cases mutations may have little effect on the phenotype of the individual. For example, a slight difference in hair color as the result of a mutation might scarcely be discernible. Some mutations may also be highly beneficial in some environments, but not in others. A change that produces blue eyes makes them more sensitive to light, but this sensitivity ceases to be an advantage in environments subject to glaring sun. In general, however, random changes are not likely to be helpful to a population that has become finely attuned to its habitat through countless generations. Suppose you raised the hood on a sportscar, and knowing nothing about motors, you made an adjustment. It may be that the change you made was just the one necessary to make the car run even better, but the odds are a lot greater that you screwed up something. The odds with random mutations are about the same.

POPULATION GENETICS

Brown eyes are dominant over blue eyes in humans. So why doesn't everyone have brown eyes by now?

The answer is not a simple one. In fact, at the beginning of the century many geneticists raised this point as an argument against the validity of Mendel's principles. A fuller understanding of the implications of Mendel's findings, however, reveals that the principles he described can, in fact, readily account for such seeming paradoxes.

To approach such questions we must begin by considering genes in populations. We have already been using the term *population*, and you may have taken it to mean a group of individuals. You were not wrong, but at this point we can give it a more precise meaning. In biology, a *population* designates a group of interbreeding or potentially interbreeding individuals. With this definition in mind, let's now consider the ratios of different alleles for a specific characteristic, such as eye color, in a population.

First, imagine a population of only two individuals in which one sex is homozygous for dominant trait *A* and the other sex is homozygous for recessive trait *a,* as shown in Figure 6.17. Now, we know that all their F_1 offspring will be heterozygous for that characteristic. The Punnett square then shows us that in the F_2 generation three out of four individuals will show the dominant trait and only one will show the recessive trait. So it might appear that we are on the way to eliminating the recessive gene from the population. However, if you now plot the F_3, F_4, F_5, and so on, generations, you will find that the proportion of dominant and recessive genes in the population has not changed at all. In fact, look again at the F_1

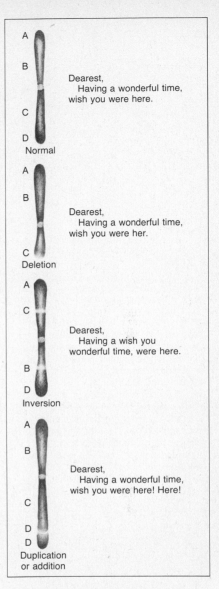

6.16 Chromosomal mutations and their epistolary counterparts, showing the normal message with deletions, inversions, and duplications (or additions) to indicate how such changes can garble the message.

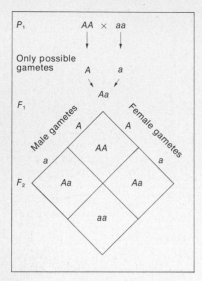

6.17 A demonstration of the Hardy-Weinberg principle. One of the parents bear the dominant trait and one the recessive, but in the F_1 only the dominant phenotype exists and in the F_2 the recessive is outnumbered 3 to 1. So it would seem that the recessive is disappearing from the population. However, an examination of the genotype shows that the ratio of dominant to recessive genes is precisely the same in each generation. The ratio will remain constant through the generations if mating is random, unless it is disrupted by some exterior influence such as immigration, selection, or drift.

and F_2 generations. There is a 1:1 ratio of the two alleles even at these stages.

A population of two is unrealistic, but it serves to point out how the phenomenon occurs in larger populations. We can show by means of Punnett squares that the frequency of alleles for any characteristic will remain unchanged in a population through any number of generations—unless this frequency is altered by some outside influence.

This interesting fact was pointed out independently by G. H. Hardy, a British mathematician, and W. Weinberg, a German physician. The principle they formulated is called the *Hardy-Weinberg law*. According to this law: *In the absence of forces that change gene ratios in populations, when random mating is permitted the frequencies of each allele (as found in the second generation) will tend to remain constant through the following generations.* We will consider forces that change gene ratios in populations later, but for now let's consider the basis for the law.

First of all, note the ratios of the combinations AA, Aa, and aa in the F_2 generation, which is our first opportunity to see all the possible combinations. We see in the F_2 that one-fourth of the population is AA, one-half is Aa, and one-fourth is aa. So, to determine what fraction of the F_3 generation will be offspring, say, of AA and Aa crosses, we simply multiply $\frac{1}{4} \times \frac{1}{2}$ and get $\frac{1}{8}$. In the third generation, then, assuming random mating, we can expect one-eighth of the population to have the genotypes resulting from this cross. What part of this generation will be the offspring of AA and AA crosses? Of Aa and aa crosses?

Now, there are four possible types of crosses that could produce AA individuals in the F_3 generation. You can work out for yourself what they are or conserve your energy and read the answer on page 145. For any of these four mating possibilities the boxes can be filled in accordingly. Each of these combinations will produce one-eighth of the F_3 generation, so what part of the population will be heterozygous, or Aa? Simple enough: $\frac{1}{8} + \frac{1}{8} + \frac{1}{8} + \frac{1}{8} = \frac{1}{2}$. And what was the fraction of Aa individuals in the F_2 population? One-half. If you work out the results for all possible combinations for the F_3 generation, or for any combination in any generation past the F_3 generation, you will see that the ratio of genetic components in a population remains stable. This is why we continue to have blue eyes in our population, even though they are a recessive trait.

Now, if this is the case, how can gene ratios in populations change? After all, evolution is nothing more than that—changes in gene ratios. If there were some great advantage in having brown eyes that gave brown-eyed people a better opportunity to survive and reproduce, then the ratio of brown eyes to blue eyes could shift. It could also change if only people with eyes of a certain color were considered sexually attractive, or if mutations were more likely to occur in one direction than another—say, from a to A rather than from A to a. If the population were small, shifts in gene frequency could occur rapidly, since in a small group only a few accidental deaths could result in a great change in the overall ratios. Also, gene frequencies could change as a result of the flow between populations through immigration and emigration. Any of these things can occur in human populations, so the Hardy-Weinberg model is to some extent an artificial one. However, it helps to explain the relative constancy we see in populations around us.

Now for those burning with a fierce love of mathematics, let us consider the mathematical statement of the Hardy-Weinberg law. The law is stated as

$$(p + q)^2 = 1$$

which can be written in expanded form as

$$(p + q) \times (p + q) = 1$$

$$\text{or}$$

$$p^2 + 2pq + q^2 = 1$$

where

p = frequency of the dominant allele (A in our example)
q = frequency of the recessive allele (a in our example)
p^2 = frequency of the homozygous dominant genotype (AA)
$2pq$ = frequency of the heterozygous genotype (Aa)
q^2 = frequency of the homozygous recessive genotype (aa)
1 = a population in genetic equilibrium (one in which the above conditions are met)

The Hardy-Weinberg law has very specific and important implications for us. For example, if we know the prevalence of the trait of albinism, the absence of pigment, in the population, we can predict, within limits, the probability that a couple will have an albino baby. Here's how this would work. Normal skin and eye pigment in humans is dominant over the albino condition a. The genotype aa occurs in about 1 in every 20,000 people. According to the Hardy-Weinberg equation, this frequency would be given by q^2, so the frequency of genotype aa is

$$q^2 = \frac{1}{20,000}$$

and the frequency of a single allele for this trait is thus

$$q = \sqrt{\frac{1}{20,000}} = \frac{1}{141}$$

The frequency of the dominant allele, A, would then be

$$p = 1 - q$$

$$\text{or} \quad p = 1 - \frac{1}{141} = \frac{140}{141}$$

6.18 The English mathematician G. H. Hardy. Hardy and the German physician W. Weinberg, working independently, discovered in 1908 what has come to be called the Hardy-Weinberg principle. The principle states that gene frequencies in a population remain stable over long periods of time unless the frequencies are shifted through exterior circumstances. The principle explains the continuance of recessive traits in a population, a problem that plagued many geneticists at the turn of the century.

Box 6.2 Sickle-cell Anemia

Sickle-cell anemia is a recessive condition characterized by fragile red blood cells that collapse into a sickle shape. These cells are unable to carry oxygen and may clog the blood vessels, so that persons who have the disease suffer painful symptoms and usually die at an early age. The cause of sickle-cell anemia has been traced to a mutation in a single gene, an alteration of one nucleotide in a triplet. The result is that in one of the four complex polypeptides that comprise hemoglobin, a single amino acid is changed—and so is the capability of the entire molecule.

In persons homozygous for this condition two of the four polypeptides are affected. These people do not normally survive long. In heterozygous persons only one polypeptide is affected, and the red blood cells are able to function under normal oxygen requirements. Under conditions in which the body demands additional oxygen, as during strenuous exercise or at high altitudes, these blood cells are likely to collapse. Nevertheless, the heterozygous condition also carries a compensatory advantage in certain environments. Persons who are heterozygous for sickle-cell anemia are more resistant to malaria than those with normal hemoglobin. Thus sickle-cell anemia is maintained in malarial areas from Africa to India.

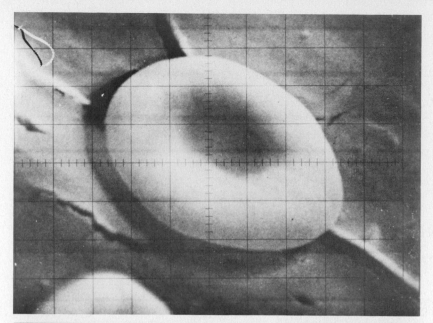

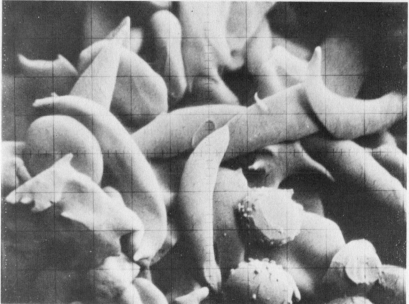

At top is a normal red blood cell, at bottom are sickled cells showing the ragged and collapsed configuration that is associated with their inability to carry oxygen. In the heterozygote condition, the disease is not quite so severe. Many cells are normal, or nearly so, so that the individual is able to live a shortwinded life.

6.19 An albino with his dark-skinned brothers at a feast in the Coral Islands, about 100 miles east of New Guinea. Albinism is a recessive trait and therefore is more likely to appear in small isolated groups. Do you see why this should be the case?

The heterozygous condition, *Aa*, would therefore occur in the population with a frequency of

$$2pq = 2 \times \frac{140}{141} \times \frac{1}{140}$$

$$= \frac{1}{70} \quad \text{or 1.4 percent}$$

Since 1.4 percent of 20,000 is 280, this means that only 280 people in every 20,000 will be carrying a recessive allele for albinism. Hence, in the absence of a family history of this characteristic in either parent, the chance that any couple will have an albino child is very slim.

In contrast, among American blacks a disease called sickle-cell anemia, produced by the recessive condition *ss*, appears in about one person in 400. Using the Hardy-Weinberg equation, we can see that the heterozygous carrier condition (*Ss*) can be expected in almost one person in every ten in this population. With this kind of information governments and health organizations can establish priorities in dealing with health problems. Whereas the bureaucratic or political mind might view one in 400 as too low a ratio to warrant urgent attention, one in ten is a ratio that certainly cannot be ignored. However, decisions about what steps to take in such a situation are exceedingly difficult. For example, should all the available funds be spent on efforts to overcome only the symptoms of the

Answer to the question in text on p. 142.

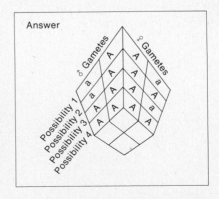

Box 6.3 The Attempt to Recreate Leonardo da Vinci

Leonardo da Vinci was the illegitimate son of Piero, a notary from the town of Vinci, and a peasant girl named Caterina. The third wife of Leonardo's father later bore a son named Bartolommeo who idolized Leonardo although he was forty-five years younger. After the death of the legendary Leonardo, Bartolommeo attempted an amazing experiment. He studied every detail of his father's relationship with Caterina. Then Bartolommeo, himself a notary by family tradition, returned to Vinci and found another peasant wench who seemed similar to Caterina, according to all Bartolommeo knew. He married her and she bore him a son whom they called Piero. Strangely, the child actually looked like Leonardo and was brought up with all the encouragement to follow in the great man's footsteps. Stranger yet, the boy became an accomplished artist and was becoming a talented sculptor. However, he died young, thus ending the experiment.

disease, or should extremely rigorous controls be exercised over marriage—the only way we now know to reduce the occurrence of the disease? Or is there some possibility of controlling the problem at a level somewhere between these extremes? In any case, the number of individuals whom we now know must be carriers for this condition serve as an impetus for some sort of immediate action.

GENETIC ENGINEERING

In a rare disease called phenylketonuria (PKU), the body lacks an enzyme needed to break down a substance called phenylalanine, which is a product of protein metabolism. The phenylalanine thus accumulates in the blood and causes mental retardation. We know that this condition arises from a single gene alteration that inhibits the production of a single enzyme—the one that breaks down phenylalanine. We also know which gene is responsible for the enzyme and the change in this gene that produced phenylketonuria. It is not beyond the realm of possibility, then, that a means will be found to "correct" the aberrant gene by adjusting the nucleotide sequence in the chromosomes. Such programmed alterations in chromosomes fall into a branch of research called *genetic engineering,* a field whose possibilities, though remote, are vast.

Cloning is also being considered as a means of genetic engineering. A clone is a group of cells that have been produced by mitotic divisions of a single parent and are therefore genetically identical. A group of protozoa that are descended from divisions and subsequent divisions of a single cell are considered a clone. In cloning experiments single carrot cells excised from adult carrots have been grown in laboratories and, through divisions, have produced normal adult carrots.

These results have naturally led to the speculation that cells cloned from Uncle Albert might mean that Uncle Albert would always be with us (that might be good news or bad news). As a person grew older, a cell might be excised and grown in a suitable medium, eventually producing another individual with the same kinds of cells as the old one. Would this in fact be the same person? The cloned individual would be genetically identical with the old one, but what about all the experiences that made Uncle Albert the character he was? And would the new Uncle Albert remember he owed you money?

The possibilities of cloning assume more practical dimensions if we pass over the issues of personality and immortality and consider the idea that certain qualities other than physical characteristics may be the result of purely genetic combinations. The first studies of genius, in fact, were prompted by the observation that qualities of genius seemed to appear in certain families with a much higher frequency than could be accounted for by pure chance. What if we could increase the frequency of such qualities in our population by reproducing certain genetic combinations through cloning? Suppose, for example, we had been able to clone cells from

Box 6.4 Crippling a Microbe

At one of the conferences on the risks of recombinant studies, geneticist Roy Curtiss III (right) volunteered to produce a weakened mutant—one that could not survive outside the laboratory.

Given the go-ahead, he first produced an *E. coli* with a defective gene that prevented it from manufacturing its protective coat. The material that the gene normally made had to be provided artificially. But microbes mutate on their own, so soon some were back to producing the normal gene. Curtiss then deleted another gene necessary to coat production. But he was outfoxed by the crafty germs; they reproduced anyway. Dennis Pereira, a graduate student working with Curtiss, found that they were manufacturing a sticky substance called colanic acid which acted as a kind of coat. So Curtiss and Pereira produced a microbe that couldn't make colanic acid. Finally they had a germ that depended on scientists for its livelihood. As an unexpected bonus, this new bug was sensitive to ultraviolet light. Ordinary sunlight would kill it.

One problem remained. Even dying *E. coli* can conjugate with normal *E. coli,* so an escaped germ might be able to pass its dangerous gene to a healthy colleague. Curtiss, however, altered the gene of the dependent bacteria that makes thymine. Thymine, therefore, had to be supplied and, without it, how could DNA be made? Perhaps now the bug was helpless enough.

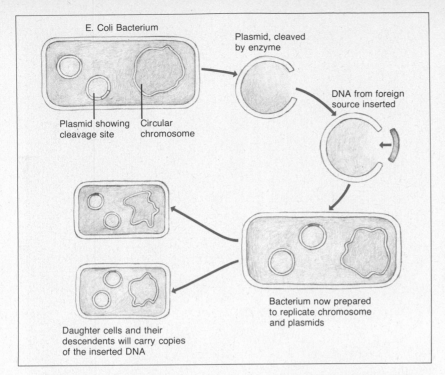

E. Coli Bacterium

Plasmid, cleaved
by enzyme

DNA from foreign
source inserted

Plasmid showing
cleavage site

Circular
chromosome

Bacterium now prepared
to replicate chromosome
and plasmids

Daughter cells and their
descendents will carry copies
of the inserted DNA

6.20 How to mutate bacteria. Most of *E. coli*'s DNA is found in its large circular chromosome. However, other DNA is found in several small circles called *plasmids*. The plasmids are first separated out from ruptured bacteria and subjected to a disruptive enzyme that breaks the circle at specific points. The ends are ragged, or "sticky," being composed of a single strand of DNA, which is ready to accept new nucleotides. These opened loops are then mixed with genes from any number of sources. Another enzyme called *ligase* cements the foreign gene into place in the open plasmid, resealing the loop. The new loops are placed in a cold solution of calcium chloride in which normal *E. coli* have been placed. When the solution is suddenly heated, the membranes of the *E. coli* become permeable and the new plasmid slips in to join the normal DNA complement. The enzymes produced by the altered *E. coli* will be directed not only by its own genes, but also by genes of another species.

Leonardo da Vinci. (If you don't immediately see the implications here, you should learn more about the man.)

The concept of genetic engineering, of course, has many ramifications. How do we determine which characteristics will be of benefit to future generations? And if certain genetic characteristics should be reduced or screened out of the population, who should decide whether your particular genetic components should or should not be reproduced? Do you see any other problems associated with genetic engineering?

TAMPERING WITH LIFE – THE RECOMBINANT DNA CONTROVERSY

This discussion is dramatically titled, but unjustifiably so, since we tamper with life casually, on a daily basis. What we are really discussing is a new research trend toward genetic engineering in bacteria. In particular, our subject is *Escherichia coli*, a common, harmless, intestinal bacteria whose role in our lives is not really understood.

We have been intensively studying *E. coli* for about thirty years, but we have identified only about a third of its 3,000 to 4,000 genes. We know even fewer of the human's hundreds of thousands of genes, but one way to find out what any segment of a human chromosome does is to insert it into the genetic complement of a bacterium and see what it does there. The idea is to transplant one gene at a time into *E. coli* (Figure 6.20), then as the bacterium reproduces through replicating itself, billions of those transplanted genes could be produced in a single day, permitting a variety of large-scale experiments. Eventually, all the genes in human chromosomes

6.21 A technician demonstrates the handling of mutated bacteria under highly controlled conditions. The bacteria are continuously contained in their controlled environment and experiments are conducted by manipulating them with rubber gloves, which are actually extensions of the enclosure.

Box 6.5 Cloning Insulin

Most of the objection to the recombinant-DNA experiments centers over their potential risks. The benefits are hazily described in unspecific terms such as "helping us to understand the nature of the gene." Recently, however, scientists at the University of California at San Francisco have indicated an advantage of such experiments in more specific terms. Using recombinant-DNA techniques, they have inserted the gene that makes insulin in rats into the genetic component of bacteria. This is the first step in what could be mass production of the hormone that is so critical to diabetics.

Traditionally, insulin has been expensive because only minute amounts could be painstakingly extracted from the pancreas of slaughtered livestock. Furthermore some people couldn't accept insulin from other species. Now, however, with the recombinant-DNA techniques, it may be possible to make insulin in large amounts. As the altered bacteria reproduce, they will continually duplicate their own genetic complement *plus* the new chromosome part that has been added. Thus, if that part codes for insulin, the hormone's production would be limited only by the rate of reproduction of the bacteria. Although the gene has been inserted it remains to be seen whether the bacteria will, in fact, make insulin and whether human genes can be inserted as easily. Theoretically, though, it *should* work!*

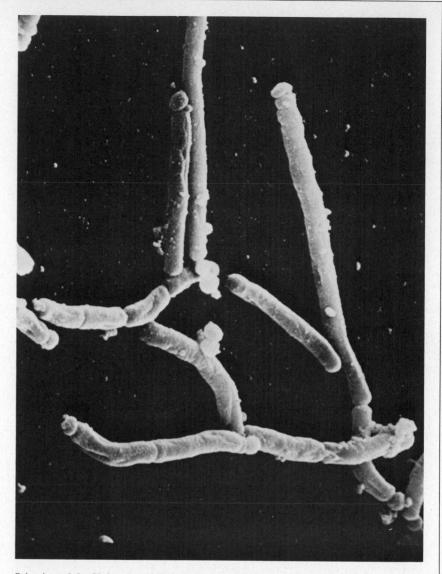

Scientists of the University of California, San Francisco, have successfully cloned the gene for insulin, which means that for the first time anywhere researchers have been able to place an insulin gene into bacteria and make it reproduce itself in large quantities, using recombinant DNA techniques.

Shown here are bacteria used in experiments with the rat insulin gene, as seen through a scanning EM. These bacteria are a special strain of *E. coli* called chi-1776 that have been bred to survive only in a laboratory environment. These cells have abnormal cell walls and have multiple defects so they can only exist in the laboratory under highly specific conditions, not in the natural environment.

*At presstime investigators at the University of California, San Francisco, have announced they are able to use bacteria to produce the human hormone somatostatin, which appears to regulate growth hormones and other hormones including glucagon, which raises the blood sugar.

could be precisely located and studied. The benefits would be enormous. Perhaps the greatest dream of geneticists is to find a way to control or stop cancer. With such a possibility on the horizon, why don't we proceed? Why have city councils met to determine whether to allow such experimentation within the city limits? Why have Senate hearings convened to regulate such experiments? Why did often-arrogant and thorny scientists self-impose a ban on further experimentation. Obviously, there is a catch somewhere.

The first experiment was to have been designed to insert a monkey virus, SV40, into *E. coli*. SV40 is harmless enough in monkeys, but if inserted into other laboratory animals it can produce cancer. It can also turn laboratory cultures of human cells cancerous. It is important to learn how SV40, which contains very few genes itself, can cause cancer. We need to know the mechanism. But what if one of these mutant bacteria were to escape into the air and be swallowed by a human? Since *E. coli* thrives in the human gut, would cancer appear in the unsuspecting host?

The issue is being vigorously debated and the argument has unfortunately become rather heated. Scientists have dropped their usual professional deference toward their colleagues and have even publicly described each other in fecal terms. Critics have likened the experiments to Hitler's efforts to build a master race.

But it is about as difficult to stop research as it is an oil pipeline. So special laboratories are being built to conduct such studies with the safeguards applied according to the risk. For experiments that potentially present the greatest hazard, the precautions are extreme, while concerned scientists ask, How extreme is enough? At present, the plans are to proceed using only weakened *E. coli* so that, should any escape bearing their deadly genes, they would quickly die.

———

Reproduction

The story of reproduction is an interesting one—at least it should be, judging from the attention we give to its mechanisms. In this chapter we will consider the process of reproduction in a variety of organisms. However, we will focus particularly on reproduction in our own species, partly because it's more interesting to most of us, but also because it is important for us to have some knowledge of the subject if we hope to limit our population.

When we think of reproduction we tend, as humans, to think of sexual reproduction, the fusion of male and female gametes. And, as we look around us, it seems that most other living things are sexual reproducers—the trees in our forests, the fish in our rivers, the dogs in our yards, the cockroaches on our floors. However, not all organisms reproduce sexually. It stands to reason that reproduction that involves two individuals must be a more complicated process than one that is done by an individual acting alone. It may also be more interesting. But let us begin with some of the principles of *asexual reproduction.*

7.1 Pronghorn antelope. This buck and doe will stay together for a brief mating period, after which they will separate. In some species the bond may be shorter, just long enough for copulation. In other species the bond may last a lifetime.

ASEXUAL REPRODUCTION
Vegetative Fission

Have you ever wondered how a flatworm reproduces? We'll tell you anyway. The flatworm can reproduce either sexually or by a process called *vegetative fission* (Figure 7.2). The flatworm simply pinches itself in two about midway along its length, and cells then move into the severed area and regenerate the missing parts of each half. The result is two new flatworms.

It is apparent that the cells that rebuild the missing structures in such cases must undergo some sort of fundamental reorganization. After all, these cells had already specialized and were functioning elsewhere in a different role. In some cases, such as regeneration in flatworms or wound healing in higher animals, cell development is reversible; cells have the ability to despecialize, or revert to a more primitive state, and then initiate a new development pathway.

If a flatworm is cut in half, each half will regenerate the missing parts, but when it is cut into smaller pieces, there is a point beyond which a regenerated individual will no longer develop normally. Still smaller pieces are unable to regenerate at all, and the pieces simply die.

7.2 Planarian flatworms reproduce through vegetative fission. Basically, the body pinches in two roughly halfway down its length. Each end then regenerates the missing parts. The tail end, of course, must grow new nerve centers and eyes. The planarian normally reproduces this way, but the process is the same if it is cut in half (left).

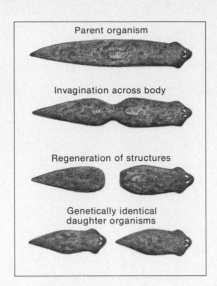

Parent organism

Invagination across body

Regeneration of structures

Genetically identical daughter organisms

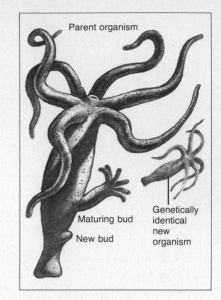

Parent organism

Maturing bud

New bud

Genetically identical new organism

7.3 The *Hydra* can reproduce by a variety of mechanisms, one of which is asexual budding. In this process, buds appear along the trunk of the body. They grow into tubelike structures complete with incipient arms. Finally the structure, resembling a tiny adult, breaks off and swims off alone (right).

Reproduction through regeneration is possible in some species that do not normally reproduce by such means. For example, starfish reproduce sexually, but not long ago vengeful oystermen chopped up starfish that they found in their dredges because the starfish were eating the oysters (an example of man's usual reaction to competing predators). In many cases, however, a new starfish grew from each part the oystermen threw overboard. As a result, the starfish population flourished, and the oyster catches dwindled in response to the increased level of starfish predation.

Budding

Another method of asexual reproduction is *budding,* in which a new organism is produced as an outgrowth of the parent organism. The new organism is well developed by the time it splits off from the parent as a new individual. Numerous plants reproduce in a similar way, as do many of the lower animals. One example is the tiny hydra, a freshwater relative of the sea anemone (Figure 7.3).

Sporulation and Vegetative Fission

In *sporulation* the new organism also develops from a bud of the parent organism. However, in this case the new organism does not remain attached to the parent until it is well developed. Instead, it is shed as a rudimentary cell—a *spore*—that then grows into a new individual apart from the parent. Some spores are capable of surviving in a dormant stage for a considerable time until conditions are favorable for them to begin to mature.

Since sporulation requires so small an expenditure of energy by the

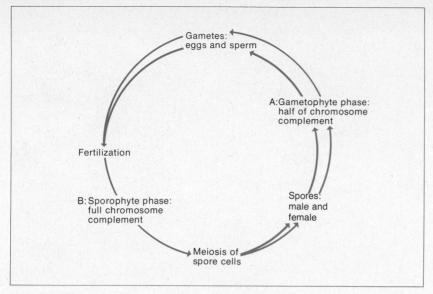

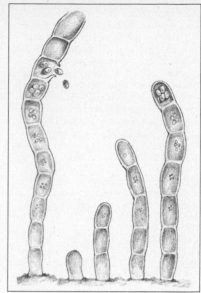

parent organism, you might imagine that spores would be produced in great numbers. And, in fact, this is usually the case. Spore-producers manufacture prodigious numbers of the tiny particles.

In *vegetative fission* the new individual leaves the "parent" at some developmental state between that in sporulation and budding. The "offspring" then is off on its own at a relatively undifferentiated stage of its development and new cellular organization follows.

ALTERNATION OF GENERATIONS

There are a great many species of animals and plants that have both sexual and asexual phases in their reproductive cycle. Often the phases alternate in what is called an *alternation of generations*. In some plant cycles spores grow into organisms that release, not more spores, but eggs and sperm. When a sperm and an egg unite, the fertilized egg will grow into an adult individual that looks entirely different from the one that had grown from the spores. This organism, when it matures, will then produce spores. The genetic basis for this difference in the adult organisms is illustrated in Figure 7.4. The spores are produced by meiosis, hence they have only half the complement of chromosomes characteristic of the other phase, and grow into individuals of generation A, which then release eggs and/or sperm. Of course, both eggs and sperm also have only "half" the number of chromosomes. When these join, generation B is produced, which has the full chromosomal complement, and which, when mature, releases spores to initiate generation A again. Gamete-producing generations are called *gametophytic;* spore-producing, *sporophytic*.

Nearly all higher plants, most algae, and some animals go through an alternation of generations. Depending on the species, however, one

7.4 Some plants show a pronounced alternation of generations. In the sporophyte generation spores grow into adults, which then release eggs and sperm. The sporophyte generation begins when the eggs and sperm are joined in fertilization. These will then grow into adults that produce spores. The spores have only half the "normal" complement of chromosomes, but when they join, they produce plants with the usual diploid number (left).

7.5 In certain simple plants, nuclei may begin to divide in maturing cells. Those nuclei, of course, are genetically identical to the parent. They are released when the cell wall ruptures. Those reaching hospitable areas will begin to grow into new versions of the parent. Because of the simple reproductive mechanism, there is very little variation among the offspring. What would be the advantages and disadvantages of low variation (right)?

7.6 Many North American woodlands are enhanced by the presence of delicate ferns. In the spring one may find them only blushing with the light green that will be their color, as they stand protectively curled, tentatively rising from the forest floor. Later they will radiate outward, full-grown, giving the forest a greenhouse atmosphere, and then one may find those peculiar brown dots on the "leaf" undersides. These are the spore-producing organs that will give rise to the next kind of generation. Ferns are particularly abundant on sodden forest floors, such as are found in rain forests, and are often seen growing from the damp, rotting logs characteristic of such areas.

generation may be much more prominent than the other. For example, mosses have a long gametophyte generation (Figure 7.7), whereas ferns have a much longer sporophyte generation (Figure 7.8). As you may have noticed on your treks into the woods, at certain times ferns have small dark spots on the underside of their fronds. These contain the *sporangia,* the spore-producing structures of the plant.

The Development of Alternating Generations

In species that show an alternation of generations it has been theorized that the sexual or gametophytic phase has been developed in response to drastic cyclical periods in the life cycle. ("Drastic" might be defined as harsh, variable, or unstable.) Under such conditions the variability produced in sexual reproduction might be highly advantageous. The asexual phase, then, might be expected under more optimal, stable, or sheltered conditions. It has been demonstrated, in many species, that sexual reproduction does, indeed, increase in periods of stress. In earlier days steel girders were placed around surviving trees in burned-over areas. The idea was that when the tree grew and came to exert pressure against the girder, the stress would result in greater seed production.

Many species of animals, after experiencing conditions which reduced their numbers, undergo a rapid surge of reproductive activity (which might conceivably be a response to either stress or low numbers). In humans, the baby booms that usually follow great wars have been attributed to stress

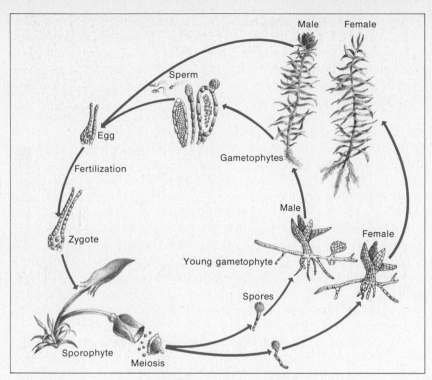

Male   Female

Sperm

Egg

Fertilization

Gametophytes

Zygote

Male

Young gametophyte   Female

Spores

Sporophyte   Meiosis

7.7 The life cycle of a typical moss. The
dominant phase is the haploid
gametophyte, which can be either male or
female and which makes up the familiar
velvety moss mat. Gametophytes can
reproduce vegetatively or can produce
gametes. The fertilized egg becomes the
diploid sporophyte, a small fruiting body
that grows out of the body of the female
gametophyte. It is in this phase, then, that
meiosis occurs. Dispersed spores develop
into either male or female gametophytes.

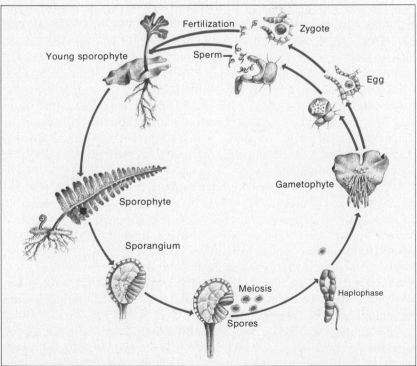

Fertilization   Zygote

Young sporophyte

Sperm

Egg

Sporophyte

Gametophyte

Sporangium

Haplophase

Meiosis

Spores

7.8 The life cycle of a typical fern is
essentially similar to that of the moss, but
it is the sporophyte that is dominant. The
sporophyte at first grows out of the
gametophyte, but then develops roots and
leaves and becomes independent. The
haploid gametophyte remains a relatively
insignificant, fragile, and ephemeral
structure. A single gametophyte usually
produces both eggs and sperm.

(but the possibility of high randiness in returning servicemen cannot be discounted). It is argued that no sound case can be made at this time for a general increase in sexual reproduction as a response to stress. However, this is not to say that the development of the sexual phase in the alternation of generations may not have been a response to such conditions. The idea is interesting, anyway.

The Advantages of Sexual Reproduction

Before going into a discussion of mechanisms of sexual reproduction we might ask, since asexual reproduction is so simple and efficient, why do some animals rely on sexual methods alone? After all, sexually-reproducing forms must normally produce enormous numbers of gametes among which only a few will ever enter into fertilization. So there is great waste in sexual reproduction. Also, there is a risk that some individuals will never be able to interact with another individual so that reproduction may take place. A male bobcat sometimes can't find a female bobcat. Actually, the probability of breeding failure for members of some species is very high. So, if you can reproduce all by yourself, you eliminate the risks associated with sexual reproduction. But sexual reproduction is a common biological phenomenon. Why should this be?

We can begin to understand the adaptiveness of sexual reproduction when we recall that meiotic processes yield gametes, which carry different kinds of genes, and that there is normally no way to predict which of these gametes will eventually enter into fertilization. Also, in most species, mating behavior is a rather indiscriminate process: "mating" animals that respond in specific ways to the opposite sex of their species are likely to choose any number of individuals from among the population. In "nonmating" animals, reproduction may be an even more random process. For example, some fish simply release their eggs and sperm into the water at spawning time, and the eggs and sperm mix. In some cases, the water may be filled with the eggs and sperm of many individuals. There is, of course, no way to predict or control which sperm will fertilize which egg. Both meiosis and randomness in mating, therefore, tend to increase *variation* in a population.

Variation may be an unimportant quality in a homogeneous and unchanging environment; however, in a heterogeneous or unstable environment the degree of variation in a population may be a critical factor. For example, under one set of conditions a certain genotype may be favored while another meets with little success. When conditions change, however, the second genotype may be favored while the first suffers a loss in numbers. Thus with a genetically variable population the interplay of environment and genotype results in an increased likelihood of survival for some members of the group.

Sexual reproduction has other benefits as well. As mutations continue to appear in a population, some will result in an advantage to their bearers. When an individual bearing a new "good gene" mates, by some happy chance, with an individual bearing another adaptive mutation, the combination may produce offspring with a marked competitive advantage. And in time that particular combination may become predominate in the population. In contrast, in asexual reproduction each daughter organism will be exactly like the parent, with no possibility of different gene combinations. Random mutations might arise in a single individual, but there is no mechanism by which these genes might eventually combine with other new genes in the population, to give rise to still other new combinations.

FERTILIZATION IN SEED-BEARING PLANTS

The seed-bearing plants are a highly diverse group in which the sporophyte generation dominates. In fact, the gametophyte generation is so diminished that it exists entirely within the tissues of the sporophyte. Many plants, therefore, propagate by sexual reproduction—that is, through the union of a male gamete (a sperm) with a female gamete (an egg) to form a *zygote,* the primal cell from which the new plant will develop. This union is called *fertilization.* In some plants, such as the willow, the male and female structures are on two different trees. In other plants, such as the conifers (cone-bearing trees), they occur on different parts of the same tree. In flowering plants, however, both structures are usually in the same flower. These structures are the male and female sporangia (since this is the *sporophyte* generation). The male sporangia are the *anthers,* found at the tips of the filaments in the center of the flower (Figure 7.9). They produce spores, which then divide once or twice to become grains of *pollen,* the male gametophyte generation. The female sporangium is the *ovule,* located just above the base of the flower. Certain cells in the ovule undergo meiosis, and these divide a few times to form eight-celled female gametophytes.

Pollen is transferred in a variety of ways to the receptive female part of the plant, the *stigma.* At this point the pollen grain grows a long hollow tube that burrows down toward the female gametophyte. The cells of the pollen grain, the sperm, then move down through the tube, and one of them fertilizes the egg. The fertilized egg, now called a zygote, is the first cell of the new sporophyte generation.

Actually, two sperm enter the female gametophyte. One fertilizes the egg that has developed from one of the eight cells of the female gametophyte. Meanwhile, two of the other female gametophyte cells have fused, and a second sperm joins with them to form a cell with three times the chromosome complement of the gametes. This cell will form the *endosperm* tissue around the zygote, which will nourish it as it develops inside the seed. This endosperm tissue is the nutrient substance you see inside a kernel of corn or a grain of wheat. The seed, protected by a hard seed coat, is then released by the plant to await the environmental conditions that will enable the zygote to begin its growth. This waiting period varies tremendously with different kinds of seeds. The seeds of the mangrove may start to germinate while the fruit is still on the tree, whereas Indian lotus seeds have been known to lie dormant for as long as 400 years.

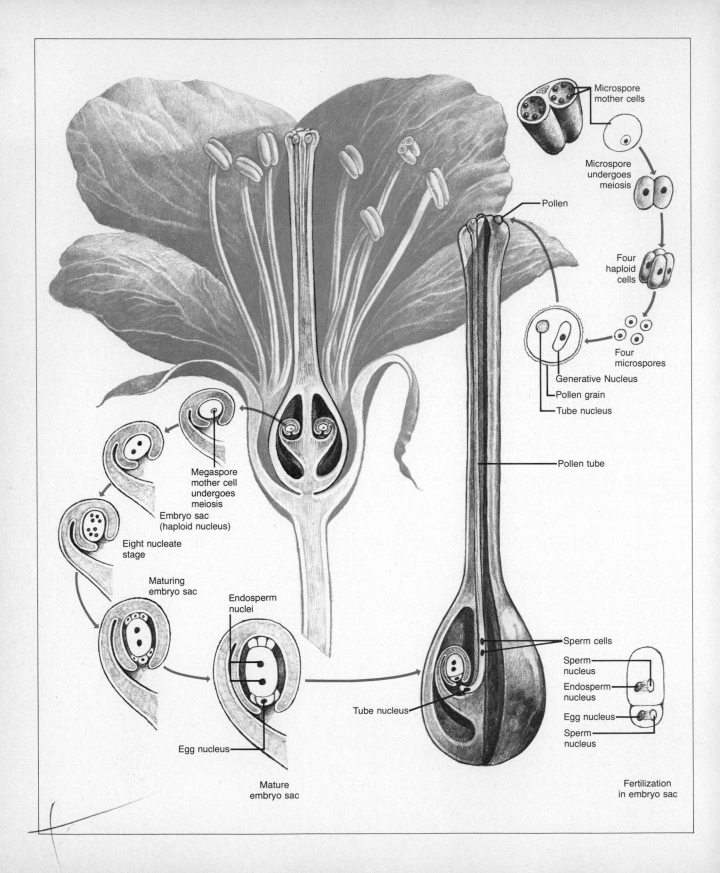

Microspore mother cells

Microspore undergoes meiosis

Four haploid cells

Four microspores

Generative Nucleus

Pollen grain

Tube nucleus

Pollen

Pollen tube

Megaspore mother cell undergoes meiosis

Embryo sac (haploid nucleus)

Eight nucleate stage

Maturing embryo sac

Endosperm nuclei

Egg nucleus

Mature embryo sac

Tube nucleus

Sperm cells

Sperm nucleus

Endosperm nucleus

Egg nucleus

Sperm nucleus

Fertilization in embryo sac

Plant Development

Plant development differs from animal development in two important ways. First, because of the rigid walls of plant cells, new cells cannot develop in the interior tissue of plants, as they can in animal tissue. Instead, plants have to grow by adding new cells to the outside of the established cells. Second, whereas most animals stop growing when they reach maturity, plants continue to grow throughout their lives.

Before the time comes for the seed to *germinate,* or begin growth, the endosperm nucleus goes through a period of rapid division and then the zygote itself begins to divide. As shown in Figure 7.10, the first division of the zygote is uneven, so that one daughter cell is larger than the other. One cell continues dividing to form a chain or stalk of cells; the other cell divides in all directions, producing a globular embryo at the end of the stalk. With progressive divisions the cells in the embryo begin to specialize. Some will eventually produce the *epidermis,* or protective skin; others will produce the *vascular* tissue, which transports nutrients; and others still, will produce the *cortex,* or structural tissue. The structural tissue supports the vascular cells. As each type of structure becomes differentiated, the globule flattens and elongates to form the embryo leaves. These leaves are called *cotyledons.*

There are two parts of the plant that remain permanently embryonic. It is in these areas that future growth will take place. The two growth areas, called *apical meristems,* are at the root tips and the stem tips. The cells of these areas will continue mitosis throughout the life of the plant. As the roots and stems lengthen, the older meristematic cells are left behind to mature and specialize. Figure 7.12 shows the structure of a typical root apical meristem. The apical meristem of the stem differs slightly from that of the root; the stem tip lacks a protective cap and may have leaves. Also, the stem apex is divided by *nodes,* the future sites of new leaf formation.

The stem tip is not protected by a cap but it lies safely snuggled deep in a bud. Thus the small, mitotically active cells behind the tip can carry out their complex functions with some degree of protection from disruptive influences. Whereas it's relatively easy to look at a root tip and name the various areas with great authority, the stem presents more problems. One area may overlap with the next and some areas are surprisingly extensive. For example, the area of elongation may cover several inches.

In trees and shrubs of the earth's temperate zones, the *terminal bud* (Figure 7.13) opens in the spring to give rise to a new section of stem. Then in late summer or fall, as growth ceases, a new terminal bud forms. This means that one can determine how much the stem grew in any year by measuring the distance between the scars. If the weather was unfavorable in a certain year, the distance will be small; for example, an apple tree stem may grow only a half centimeter. But in a good year a willow stem may grow eight feet.

The lateral buds on stems (Figure 7.13) either do nothing, or they grow into branches. If the terminal bud has been removed, branches are

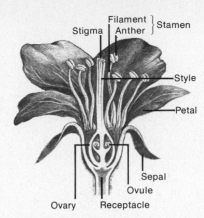

7.9 Fertilization in a flowering plant (opposite page). The overall structures are shown in the figure above. The male gametophyte, or pollen, after landing on the stigma, sends a pollen tube down through the style until it reaches the female gametophyte in the ovule, which is now composed of seven cells, the largest harboring two endosperm nuclei. Notice that there are two sperm nuclei (sperm cells) traveling down the tube, which is maintained by the tube nucleus. One of the sperm nuclei fertilizes the egg (the egg is only one of the seven cells of the female gametophyte), forming a zygote. The other sperm nucleus joins the two endosperm nuclei and this 3n cell will then form the endosperm, the food store of angiosperm seeds.

7.10 Development of *Capsella,* or shepherd's purse. Note the uneven first division. The larger daughter cell will grow and ultimately become the basal cell. The smaller cell will continue dividing linearly for a few divisions until the terminal cell begins forming smaller cells that will produce a globular embryo. The embryonic tissue will grow increasingly specialized until, as we see in part G, several distinct tissues have formed. Note the lack of development in the basal cell.

7.11 A micrograph of a developing *Capsella.* This is a stained specimen and the nuclei show as dark spots in each cell. The micrograph indicates some of the problems with relying on the light microscope to study cell detail. The cell walls are indistinct and many of the organelles are difficult to distinguish. However, compare the photograph with the artist's rendering in Figure 7.10G.

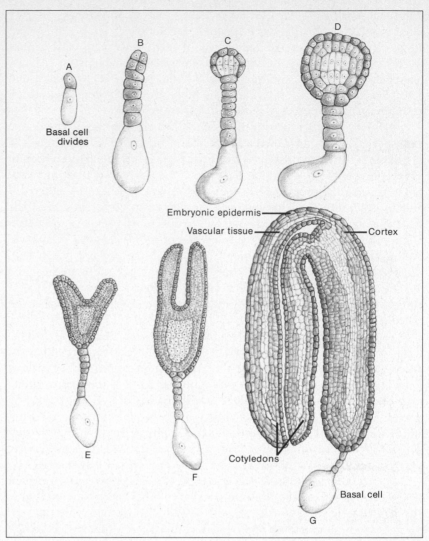

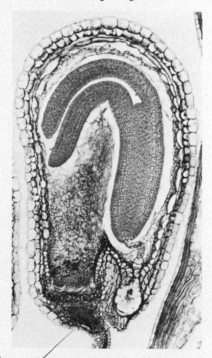

more likly to form. Pruning plants, then, causes them to become more shrublike.

Just as apical meristems account for growth in length, *lateral meristems* contribute to growth in girth. Lateral meristems are also called *cambia* (singular, *cambium*). Growth in diameter presents special problems. For example, the vascular system must be enlarged by adding cells to the *xylem,* which carries water up from the roots, and to the *phloem,* which carries food down from the leaves. Since these areas lie deep within the plant, how can new cells be added without rupturing the plant's surface? This is possible because there are two kinds of cambium. The *vascular cambium* produces new xylem and phloem and the *cork cambium*

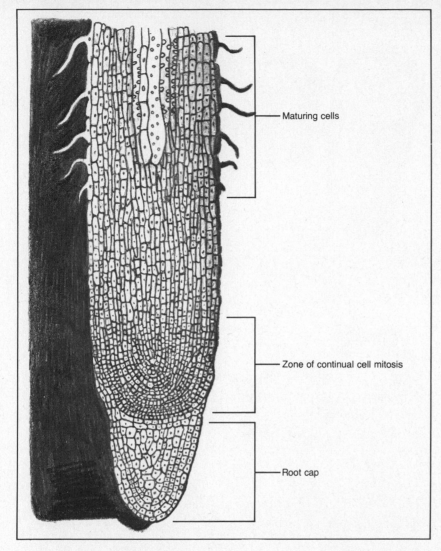

Maturing cells

Zone of continual cell mitosis

Root cap

7.12 The development of the root. The tip is shielded by relatively large cells comprising the root cap. Such protection is necessary as the delicate root pushes its way between soil particles. The area it protects is the site of intense mitotic activity, hence these cells, not yet mature, are smaller. As they mature, they increase their cytoplasm and inclusion bodies, eventually elongating. The most exterior of these may send out delicate root "hairs" through which water is readily absorbed.

7.13 The growth area of a stem. Note that you can read much of the plant's history by the distance between bud scale scars, the distance being greater for good years, when the stem has been able to grow rapidly.

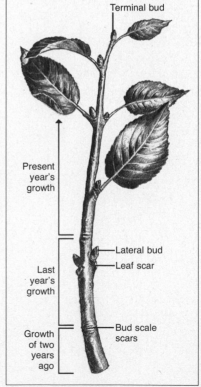

Terminal bud

Present year's growth

Last year's growth

Lateral bud

Leaf scar

Bud scale scars

Growth of two years ago

produces new cork, a protective tissue that functions as an epidermis or skin. The vascular cambium lies between the xylem and phloem, and as it produces new cells, those toward the inside of the stem form xylem and those toward the outside form phloem (Figure 7.14).

Plants contain many other types of cells with specialized functions. The fundamental tissue, for example, includes a variety of cell types, such as the thick-walled *sclerenchyma* and *collenchyma,* which function as structural supports (Figure 7.15). Other types of cells, called *parenchyma,* are thin-walled and have large vacuoles. Some of these cells contain chloroplasts.

The leaves of a flowering plant are the sites of photosynthesis, but

7.14 Section from the stem of a dicotyledonous plant. Note that the woody xylem shows two years growth, the winters being recorded in the areas of smaller cells. The vascular cambium forms xylem on its inner surface, phloem on its outer.

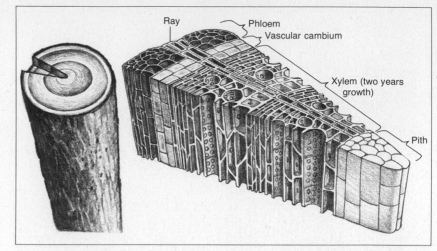

7.15 Sclerenchyma (left) and collenchyma (right). Sclerenchyma cells usually die at maturity, their thickened walls imparting a hard, tough quality to wood. The cells shown contain cytoplasm and so are still alive. Collenchyma is formed in the outer layers of the stem and in leaves. The cells are elongated with thick walls that lend support to young stems and leaves. These cells are dead.

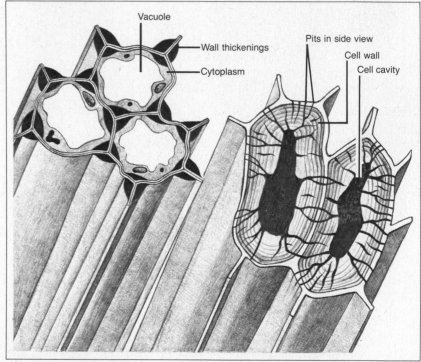

during its early development, before the leaves are fully exposed, the plant must get its food from some other source. In plants such as peas and beans, food is absorbed from reserves in the cotyledons or "seed leaves." The first true leaves will arise from a region at the base of the cotyledons known as the epicotyl (Figure 7.16). As the plant grows, the role of food production will be taken over by the leaves. Leaves occur in a variety of sizes and shapes, as anyone knows, but in Figure 7.17 we have a diagram of a stylized leaf.

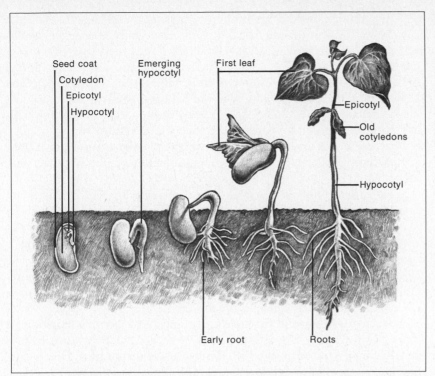

7.16 The germination and early development of the common garden bean. Note the distinct regions of the newly planted bean. The cotyledon, or "seed leaves," serve as food reserves, but these do not become true leaves. The true leaves arise from the epicotyl, a structure at the base of the cotyledons. The hypocotyl will grow first and establish a root system, raising the "bean" out of the ground and thus forming the stem. As the halves of the bean separate the leaves will emerge from between them, leaving the withered and now useless cotyledons hanging below. The plant can now actively begin making food in its rapidly growing leaves.

Seed coat
Cotyledon
Epicotyl
Hypocotyl
Emerging hypocotyl
First leaf
Epicotyl
Old cotyledons
Hypocotyl
Early root
Roots

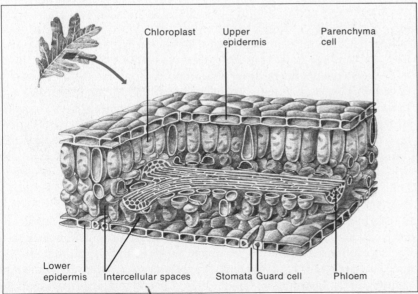

7.17 A leaf cross section. The epidermis is found on the upper and lower surfaces and protects the more delicate inner cells. It is usually one cell-layer thick. In the lower epidermis are openings called stomata through which carbon dioxide enters and the gaseous by-products of photosynthesis escape. The passage of gases through the stoma is regulated by the opening and closing of the guard cells. Most of the internal tissue is comprised of parenchyma cells, which contain chloroplasts.

Chloroplast
Upper epidermis
Parenchyma cell
Lower epidermis
Intercellular spaces
Stomata Guard cell
Phloem

ANIMAL FERTILIZATION

In animals that reproduce sexually, fertilization may be achieved in a variety of ways. It may be accomplished *externally* (Figure 7.18), or *internally,* as is the case in the group of animals that has hair and nurses its young—the *mammals.*

7.18 Insemination in frogs. The male clasps the female from the back, gripping her behind her forelegs. As she releases her eggs he pours sperm over them, and the eggs are then on their own, protected only by their gelatinous coating. This method of insemination is somewhat more intimate than simply releasing eggs and sperm randomly into water, but less "personal" and less efficient than the male injecting sperm directly into the reproductive tract of the female. Other frogs have much more unusual and interesting ways of getting their gametes together, such as pouring them down a leaf funnel over a body of water.

In mammals, fertilization is internal and is accomplished by the act of *copulation*. In most mammals copulation occurs only at specific periods when the female is physiologically and behaviorally receptive (in *estrus* or *heat*). One advantage of such timing is that it may increase the likelihood of young being born at the most opportune time of year, such as when food is most abundant. The receptivity of the female mammal is advertised by odors associated with vaginal discharge in many species, including dogs and horses. In certain primates, receptivity is communicated by a vaginal odor acting in coordination with a reddened and enlarged area around the rump. Receptive female chimpanzees are sometimes referred to as "pink ladies."

In humans, theoretically, there is no period when the female is not sexually receptive or when she is not sexually attractive to males, your personal experience notwithstanding. Women "stand" physiologically ready to copulate at almost any time and there is no particular signal that is involved in attracting males. Anthropologists have suggested that the continual receptivity of women may have evolved as a way to entice larger and stronger males to remain affiliated with the female that may be bearing his children. In such an association, the female might have greater reproductive success with the male's assistance, and the male might leave more offspring, as well, by assisting his weaker reproductive partner who is usually smaller and burdened with the responsibility of nursing the young. His affiliation might have been, and still might be in certain cultures, advantageous in terms of both food and protection. In time, a particular bonding, feeling, or attraction might have developed between reproductive partners that strengthened the union established on a purely physical basis. Perhaps this kind of bond provided the basis for what is called "love" or perhaps such bonding *is* love. It's hard to say since no one has been able to come up with an acceptable definition of the phenomenon. But it is known that the human often develops and, in fact, seems to have a need for strong attachments with another human. Such attraction, as we all know, may be long-term or short-term. Long-term relationships are sometimes formalized as "marriage," a system that in our culture discourages the breaking of the association. It should be pointed out that other species may build long-term bonds. For example, geese often mate for life, and a member of a pair might never mate again if one is lost.

In humans there is no specific point at which a male and female are likely to initiate copulation. The events preceding copulation may vary widely, even between established sexual partners. In certain other animals specific signals always precede the copulatory act, such as the female chimpanzee's "presentation" of her livid rear end to a male (although the blasé males may or may not mount the soliciting temptress). Under normal conditions the physical prerequisites of copulation or *intercourse* in humans are simply an erect *penis* and a lubricated *vagina*. These conditions are usually achieved during the course of precopulatory sexual behavior, or *foreplay*. Foreplay, as well as the copulatory act itself, varies widely in its expression between cultures, from one individual to the next, and even between established sexual partners from one time to the next.

The human *penis* is supplied with a great number of blood vessels, some of which open into large blood chambers. During sexual arousal, blood is restricted from leaving the penis by constriction of certain sphincters (rings of muscles) around the vessels at the base of the penis. Thus the penis becomes filled with blood. The result is that it grows stiff and its length increases, sometimes to a surprising degree—although, perhaps, more surprising to some than others. The penis stands erect at such times, and is said to be in the *tumescent* condition by the more delicate among us.

During foreplay, cells in the vaginal wall of the woman usually secrete a lubricating substance that aids in the insertion of the penis. The *labia minum* and *clitoris* may become enlarged and reddened due to the rush of blood to those areas. The nipples may also harden and enlarge. If the foreplay has been sufficiently extensive, other changes may also occur, such as a mottling of the shoulders and chest and an enlargement of the breasts.

Copulation begins when the erect male penis is inserted into the female vagina. After a few or many pelvic thrusts (depending on the species, and in the case of humans, depending also on the individual, the degree of arousal, and one's schedule), the male ejaculates.

Ejaculation takes place in two stages. First there is a contraction of the entire genital tract, including the epididymis, sperm ducts, and seminal vesicles. Fluid from the seminal vesicles flows into the upper urethra (urinary tract), where repeated contractions of the prostate gland mix it with prostate fluid. As the urethra becomes filled with semen the urethral bulb at the base of the penis expands to accommodate the influx. The second stage begins as a sphincter muscle closes off the urethra at the bladder. The urethral sphincter at the base of the prostate then relaxes, allowing semen to move into the urethral bulb and into the penis. This is followed by contractions of the muscles in the area behind the testicles, which pump the accumulated fluid out through the penis. Several regular contractions may occur, but most of the semen is expelled in the first contraction. The semen may eject from the penis with considerable force in the first contractions and it is in this first fluid that the greater part of the sperm is concentrated.

Ejaculation is ordinarily accompanied by *climax* or *orgasm,* which is the culmination of the copulatory act. Accompanying physiological changes include an increase in blood pressure and a quickened pulse. The postclimactic period is marked by a feeling of relaxation, sometimes to the point of drowsiness. The sphincter muscles in the male relax and the penis again becomes flaccid or limp.

Orgasm in women is much less well documented. In fact, the very existence of the female orgasm was argued until 1966, when a pioneer study of human sexual response was published by the physician-psychol-

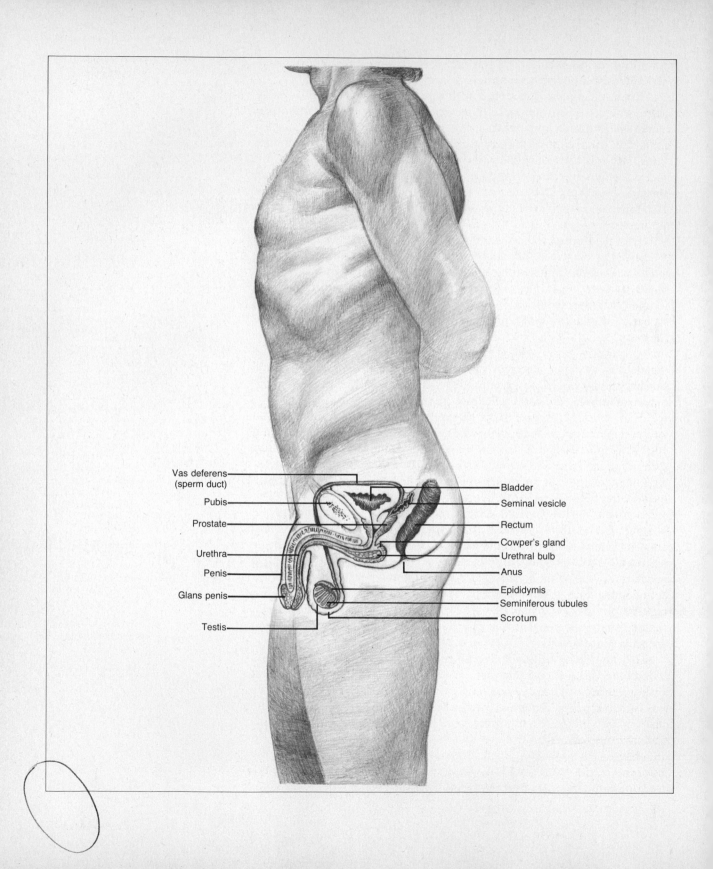

Vas deferens
(sperm duct)

Pubis

Prostate

Urethra

Penis

Glans penis

Testis

Bladder

Seminal vesicle

Rectum

Cowper's gland

Urethral bulb

Anus

Epididymis

Seminiferous tubules

Scrotum

ogist team of Masters and Johnson. On the basis of many carefully monitored observations of various sexual activities, they demonstrated that the female orgasm differs little from the male orgasm. However, it is not so universal a response, and there is a much wider range of individual differences. The response is rarely or never achieved in some women, while others may reach orgasm frequently and easily. The inability to reach orgasm, however, has no effect on the ability to reproduce.

It is not known whether females of other species experience climax. However, Jane Goodall, in her continuing study of the wild chimpanzees of the Gombe Stream Preserve, has witnessed apparent nymphomania in one adolescent female, a behavior that would be difficult to explain in the absence of sensory reinforcement. Also, physiological tests on females of other species have demonstrated behavioral responses in copulation that are similar to those occurring in human females at orgasm.

The topic of copulation has proven a particularly thorny issue in many cultures, including our own. It has become interwoven with religious, legal, and social restrictions, which are imposed to different degrees from time to time, even within the same culture. In our own culture morality has come to refer primarily to sexual matters. To many in our society an "immoral" person is one who is sexually promiscuous, *period.*

To some extent this attitude stems from the concept that childbearing is the only basis for copulation. It has been argued that, in nature, copulation is practiced solely for the purpose of procreation, and that copulation for other purposes is "unnatural," and therefore wrong. Logic aside, the premise is erroneous. Actually, copulation has many functions in nature. Among rhesus monkeys, subordinate males often present their rears to dominant males as an act of submission, an act that has the effect of assuaging aggression on the part of the dominant individual. The dominant male may follow up with a demonstration of his authority by mounting the subordinate and even making several pelvic thrusts, although there is no actual penetration. Among chimpanzees a pink lady may copulate with every adult male in the group, as well as some adolescent ones. Since only one male is needed to impregnate the female, the act of copulation is believed to be a means of developing social bonds within the group. Among baboons the females copulate during the earlier phase of their estrus period with males of any rank, but at the time they are most likely to conceive, they will accept only high-ranking males. Baboons are also highly social, and in this case, too, the earlier acts of copulation may serve to reinforce group bonds.

Humans are highly complex and can only be viewed with what must be a high degree of subjectivity by other humans. It is therefore difficult to define the bonding quality of copulation in our own species. It seems likely on an intuitive basis, however, that copulation may be an important bond builder. If this is so, it would help to explain our caution regarding what seems, on the surface, to be a rather mechanical process. In other words, it may not be advantageous to build bonds indiscriminately.

7.19 The male reproductive system includes the testes, scrotum, seminal vesicles, accessory glands, various ducts, and the penis. The *testes* are the glandular organs in which sperm are produced. Before birth the testes descend from the abdominal cavity into a pouch behind the penis called the *scrotum.* One testicle (usually the left) hangs lower than the other so the two cannot be crushed together. There are two testes (or testicles), which contain hundreds of coiled *seminiferous tubules* in which the sperm develop. These tubules lead into the *epididymis,* which appears to be a large, short tube, but is actually a highly coiled, very thin tube about 20 feet long. The epididymis acts as a filter to remove any cell debris or pigment accompanying the sperm and to propel the sperm into the sperm duct. If more sperm are produced than can be utilized, the epididymis may "digest" some of them and return their nutrients to the body. Also the epididymis acts to increase the fertility of the sperm as they pass through it. The *sperm duct,* or *vas deferens,* transports the sperm over the pelvic bone to the back of the urinary bladder where it joins the sperm duct from the other testis. The *seminal vesicle* is a tightly coiled tube that lies in a pouch at about the point of juncture of the seminal ducts. Fluid from the seminal vesicle is discharged at ejaculation. The *prostate gland* lies just below the bladder and secretes *seminal fluid,* which together with fluid from the seminal vesicle, activates and maintains the sperm. Below the prostate gland the *Cowper's gland* or *bulbourethra,* also contributes a fluid to the semen. The *penis* is composed of three long areas of spongy tissue bound together by fibrous tissue. Two of these spongy *cavernous* areas are attached to the base of the Cowper's gland and at the penis tip, to form the *glans penis.* This structure may be covered by the foreskin, the fold that is removed at circumcision.

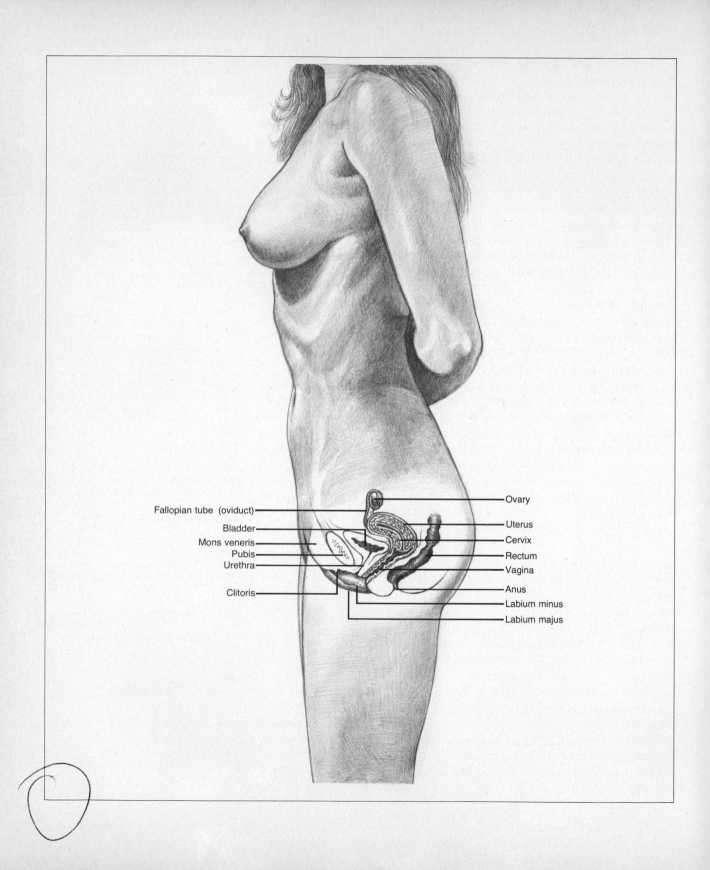

Fallopian tube (oviduct)

Bladder

Mons veneris

Pubis

Urethra

Clitoris

Ovary

Uterus

Cervix

Rectum

Vagina

Anus

Labium minus

Labium majus

An indication that, in humans, copulation may have functions other than procreation is provided by evidence that whereas women may be prepared to copulate at any time, pregnancy can occur only at a specific time in the *menstrual cycle.*

The Menstrual Cycle

The menstrual cycle begins in females at puberty, ordinarily around the age of twelve to fourteen, although the onset has become progressively earlier in American girls, probably as a result of increasingly good health. The cycles will continue for about the next thirty to forty years of the woman's life, usually interrupted only by pregnancies, and terminated finally by menopause.

Menstruation is the result of the shedding of the blood-rich *endometrium,* the lining of the uterus, when the egg released from the ovary has failed to be fertilized. In preparation for receiving a fertilized egg, the endometrium becomes engorged with blood, blood carrying life-sustaining food and oxygen. But if the egg reaches the uterus unfertilized, the endometrial preparations have been for nil and the uterus reverts to its previous condition. The old endometrial lining breaks loose and slides out the vaginal opening, accompanied by blood from ruptured vessels.

An egg is released from an ovary about every 27 to 30 days. If it is not fertilized, menstruation will begin about 14 to 16 days after that and the period will last from 3 to 6 days. However, there is wide variation in the length of both the cycle and the menstrual period among women and any pattern is considered normal as long as it recurs on a rather regular basis. Figure 7.21 describes the relationship of the ovarian, uterine, and hormonal cycles from the start of one cycle to its termination. *Hormones* are "chemical messengers," produced by various glands in the body, that act to regulate the course of events in other parts of the body.

At the end of a menstrual period, the cycle begins anew as deep within an ovary other follicles are developing. These are rounded bodies, each containing an egg which protrudes inward from the follicle wall into a large fluid-filled space. As the follicles mature, they grow larger and migrate to the surface of the ovary. Their development is stimulated by rising blood levels of a hormone called simply the *follicle-stimulating hormone* (FSH), interacting with another hormone called the *luteinizing hormone* (LH). Both of these hormones are produced in a small but rather incredible structure called the *pituitary gland,* which is located at the base of the brain. (We'll look more closely at this important gland later.)

As the follicles mature, they begin to secrete a hormone of their own called *estrogen* that induces a thickening of the uterus lining, the endometrium. For some reason, one bubblelike follicle begins to outgrow the others, and as soon as its ascendancy is established, the others stop developing. This larger follicle then moves to a point just below the surface of the ovary. In response to the pituitary's luteinizing hormone, the follicle ruptures on about the thirteenth or fourteenth day after the follicle-stimulating hormone had initiated its growth. The egg is released, leaving

7.20 The reproductive system of the human female includes the ovaries, oviducts, uterus, vagina, and external genitals. The *ovaries* are located at the sides of the abdominal cavity and are supported by ligaments. Eggs released from the ovaries move into the tubelike *oviducts* that extend to join to either side of the *uterus,* where the embryo develops. The oviduct is lined with cilia that sweep the egg toward the uterus. The uterus is pear-shaped, muscular, and thick-walled, with the lower end opening into the vagina. The inner lining or *endometrium* of the uterus is glandular and is the site where the fertilized egg will develop. The lower part of the uterus, just above the vagina, is called the *cervix.* The vagina is a muscular tube, 3 or 4 inches long, that leads from the exterior to the cervix. The vaginal muscles are thin-walled and the epithelial lining may have a corrugated appearance. The external genitals, or *vulva,* are comprised of the *mons veneris,* a fatty mound that becomes covered with hair at puberty. The *labia majum* are thick folds that lie on either side of the vagina and are also covered with hair. The *labia minum* are smaller folds of skin that lie between the labia majum and the vaginal orifice. Where the labia minum meet above the vaginal orifice is the *clitoris,* a small erectile organ that is highly sensitive and that corresponds developmentally to the penis of the male. The urethra (urinary tract) opens between the clitoris and the vagina. There may be a mucous membrane stretched across the vaginal opening, called the *hymen.*

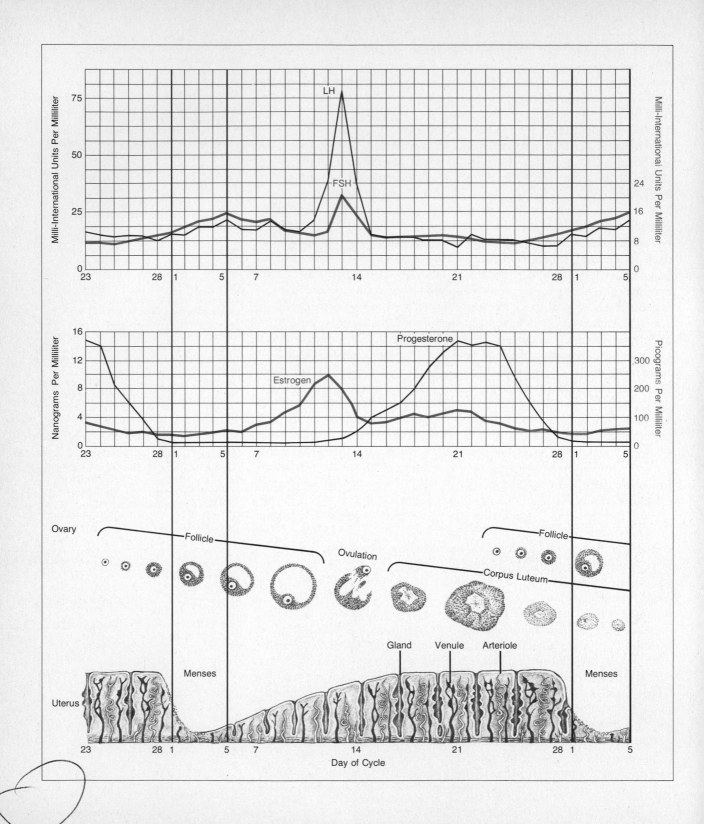

behind the old yellowed body of the follicle, now called the *corpus luteum.* However, the corpus luteum is far from being a corpse, for it begins to secrete another hormone, *progesterone,* which stimulates the uterus to prepare for a fertilized egg. The endometrium thickens and becomes engorged with tiny blood vessels that will carry food and oxygen to the embryo in the event that fertilization occurs. The interaction of the progesterone and estrogen also inhibits the production of the pituitary's follicle-stimulating hormone at this time, to prevent the start of a new cycle. These inhibiting properties of estrogen and progesterone are the basis for their use in birth-control pills, which we will discuss later.

If the copulation has been successful (depending on your point of view), and the egg that reaches the uterus is fertilized, the zygote will implant in the uterine wall and begin its growth, taking its necessities from its mother's blood in parasitic fashion. (Although in some cases this may be only the beginning of a long parasitic existence.) However, when fertilization does not occur, the corpus luteum stops its secretion of progesterone, and the swollen, life-sustaining uterus lining begins to shrink and break down. The connecting blood vessels rupture, and this blood carries the bits of cell tissue out of the uterus and through the vaginal opening during menstruation. The pressure of blood-engorged tissue often causes discomfort and cramping as materials build up and are then dislodged. In some cases the changes in hormone balance at this point cause accompanying emotional depression or despondency. As soon as the uterus has returned to its original state, the pituitary stimulates the development of more follicles, and the cycle, well, the cycle begins again.

7.21 The relationship of female reproductive hormones and the events in the ovary and uterus during the menstrual cycle. Note that FSH and LH are released simultaneously from the anterior pituitary, but that LH is much more abundant. The maturing follicles are, in the meantime, producing estrogen, but progesterone will not begin to rise until after ovulation, when the corpus luteum is formed. The sharp rise in FSH, LH, and estrogen marks the rupture of the egg, but the uterine endometrium is halfway through its period of development by this time.

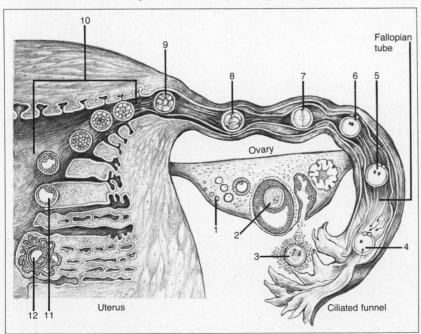

7.22 (1) As the eggs in the ovary develop, one mysteriously gains ascendancy and the growth of the others is halted. (2) The egg develops in a fluid-filled sac that ruptures, and (3) the egg is allowed to enter the Fallopian tube (or oviduct). (4) As it proceeds down the tube it is fertilized and (5-9) development proceeds until (10); by the time the embryo reaches the uterus it has become a blastula. (11) The blastula adheres to the highly developed endometrium of the uterus and (12) begins to develop tissues and membranes of its own that will enable it to draw nutrients and oxygen from its mother's blood.

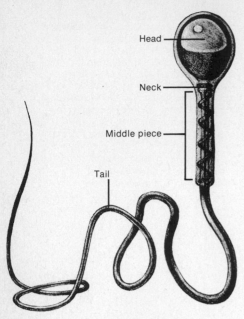

7.23 The human sperm. The head contains an enzyme that breaks down the mucous plug often blocking the cervix. The tail is basically a flagellum that propels the sperm through the female reproductive tract. The sperm does not live long, partly because it carries only a small food supply and partly because of the hazards that await it in the female tract. Many sperm may apparently reach an egg at once, all adhering to the egg surface, but only one will join its chromosomal complement with that of the egg. This union takes place far up in the Fallopian tube, but just how the sperm manage to find their way there remains a mystery.

Conception

Although only one egg is produced during each menstrual cycle, the semen released at each ejaculation may contain millions of sperm, each with its own unique genetic makeup. There is no way to predict which sperm will fertilize the egg. During copulation the sperm are usually ejaculated into the upper reaches of the vagina, near the cervix. Each sperm cell has a whiplike "tail" that propels it along at a rate of about ½ centimeter per minute (Figure 7.23). Surprisingly enough, however, sperm may reach the cervix only 1½ to 3 minutes after ejaculation. This is probably because their own propulsion is further assisted by ciliary action and currents that are set up by contractions of the vaginal muscles.

The sheer number of the sperm in each ejaculation is an adaptation to the hazards they face in making their way to the egg. Some may die as a result of their own physiological changes ("natural causes"). Others may be rendered immobile by the chemical environment of the vaginal tract, or devoured by the woman's roaming white blood cells, which treat them as if they were foreign bacteria. The semen in which the sperm are carried is alkaline, or basic. The fluids in the vagina, however, are slightly acid, and sufficient exposure causes the sperm to become inactive. When the sperm are first activated at the time of ejaculation, they thrash violently and spasmodically, and some make it into the friendly alkaline environment of the uterus. Then, through ways that remain a mystery, a few manage to find they way through the uterus, into the tiny pore of the oviduct (also called the *Fallopian tube*), and up the tube nearly to the ovary, where they encounter the egg on its way down. Conception actually takes place, then, not in the uterus, but far up in the oviduct.

Several sperm may reach the egg at once, but for some reason only one will penetrate the egg's membrane and effect fertilization. The mechanism that prevents fertilization by all but one of the sperm is still unknown.

Once conception has taken place, the zygote will continue down through the oviduct to be received by the ready uterus. As soon as the zygote implants itself in the uterine wall, it will begin to receive sustenance as oxygen and nutrients filter through from the blood of the mother. Then, given the conditions for its continued development, in another 266 days a new individual will join the population.

There has never been another individual quite like this one, and there won't be others to follow. Not exactly like this one. The genetic combination cannot be duplicated and neither can the experiences that will be superimposed on that constitution. In some cases its existence will be brief, the time counted in days or even minutes. In other cases it will join the world's population, where it may become anything from a Sikh to a Republican. Whatever it will call itself, though, it will be unique.

CONTRACEPTION

Now let us consider a problem at the other end of the reproductive spectrum: the avoidance of parenthood.

Box 7.1 The Theory of Recapitulation

Human embryos, in the course of their development, go through stages when aspects of their appearance very strongly resemble characteristics found in the embryos of other species of animals. For example, human embryos develop tails, which usually disappear before birth. (In a small percentage of births, a rudimentary tail must be surgically removed.) Human embryos also develop gill-like pouches, which disappear or are altered to produce other structures such as the ear canal. Gill pouches of shark and man are compared in the art below. Below that you can see the strong similarities in the circulatory systems of shark and human embryos. (Adapted from Simpson, Pittendrigh, and Tiffany, 1957, Harcourt, p. 353.)

In its earliest stages the human embryo is very similar to a starfish embryo. In later stages it bears strong resemblances to fish, amphibian, and reptile embryos. Finally, however, it develops characteristics that are unmistakably human. E. E. Haeckel (1834–1919) summed up this phenomenon, which had been noted earlier, in his statement: "Ontogeny recapitulates phylogeny." This statement is misleading since it assumes that human embryos go through stages in which they resemble adults of lower life forms. Instead, the phenomenon makes more sense to us when we understand that the developmental mechanism of our fishlike ancestors could have been inherited by us. This helps to explain why it is that early embryos of many modern species are so similar. But the parallels cannot be drawn out too specifically, because even in these early embryos characteristics are found that are specific to only one species.

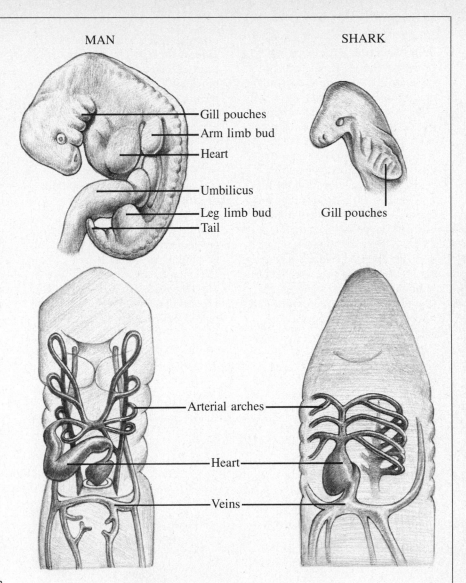

MAN   SHARK

Gill pouches
Arm limb bud
Heart
Umbilicus
Leg limb bud
Tail
Gill pouches

Arterial arches
Heart
Veins

Box 7.2   Why Your Children Will Not Look Exactly Alike

In the developing female embryo, there are about 7 million eggs in the ovaries by the 20th week. By puberty many of these have been lost (though the means of selection is unknown) so that there are by then only about 400,000 eggs. At about the age of sexual maturity (usually around 12 to 16) there may only be 300,000 eggs. Of these, only one will escape a ruptured follicle every 28 days. This goes on for about 40 years, so a total of about 400 eggs are produced. Each woman could actually produce a maximum of about 40 children in a lifetime (the record is 69). So, of millions of potential ova, only 25 to 40 can contribute their genes to the next generation. If no type of developing germ cell has an advantage, the probability of any one developing to maturity is remote and the probability of ova with identical sets of genes approaches nil.

Males, after puberty, normally produce millions of sperm each day and the production may continue throughout life. The chromosomes of each sperm and egg have undergone meiosis and crossing over and there is almost no likelihood of two identical sperm being produced. Of millions of sperm in each ejaculation, only one will fertilize an egg. One reason so many sperm are necessary is that the female reproductive tract is essentially a hostile environment and most sperm will quickly perish there. Also, a mucous plug may block the cervix, and enzymes carried by sperm are necessary to break it down to open the way for other sperm. These same enzymes act to break down the follicle cells surrounding the egg so that a sperm can get through to fertilize it. If no particular type of sperm (in terms of genetic constituents) has an advantage, you can see that the likelihood of identical individuals being produced from the same parents is very slim. In fact, the odds of your having children with identical genetic constitutions have been calculated at about 1 in 14 trillion.

Why Not Conceive?

Why would anyone want to avoid bringing a baby into the world? You might immediately think of a reason. But actually the rationales are varied. In fact, more couples are making the decision to remain childless or to adopt children that are already here. Adoption as a way of becoming a parent has no risk in terms of the health of either the wife or the child, but there is an element of "chance" in the genetic makeup of the child. Also, as an ego trip, it has less going for it. Couples who have no children cite reasons such as personal freedom, financial well-being, fewer worries with resulting better health, and more time to be with each other. They are, strangely, sometimes labeled as "selfish" by people who want children and who, therefore, have them.

One of the most frequently cited reasons for avoiding reproduction involves overpopulation. There are, by almost any measure, too many humans on our limited earth. Worse yet, it seems our numbers are increasing to the point that our population can only be stabilized by very drastic methods. It is doubtful that many couples are aware of, or concerned enough about, overpopulation to willingly forego having children. However, one of the reasons for childbearing has been that society seemed to expect, and even demand, it. In recent years, as more Americans become aware of the population problem, it has become more socially acceptable to forego children. Thus, with the relaxation of public pressure and the generation of more public discussion of alternatives, perhaps some couples forego children because of overpopulation without really knowing it. Even those who have children are having fewer, for now—at least in the United States (which has about 6 percent of the world's people). We cannot assume that Americans are copulating less, so we must assume contraception is being more effectively practiced. How?

Methods of Contraception

Contraception, or avoidance of pregnancy, can be achieved in many ways. Some means are used by the female, some by the male, and some by both partners. Unfortunately, their common bond is that none, except abstinence, is 100 percent effective (Table 7.1).

Historical Methods

Historically, conception has been prevented by a number of means. The ancient Egyptians blocked the cervix with leaves, cotton, or cloth. The encouragement of homosexuality among the ancient Greeks probably reduced the rate of conception in that culture. In the middle ages *condoms* to fit over the penis were fashioned from such materials as linen and fish skins. Douching has long been practiced as a means of flushing sperm out of the vagina. The finger has also been used to remove semen after it has developed a stringy quality through being exposed to the vaginal environment. The Old Testament refers to *coitus interruptus,* or the removing of the penis just prior to ejaculation. The most effective method, of course, has always been total abstention, a method which has met with

Table 7.1 Summary of Contraceptive Methods

Method	User	Effectiveness Rating	Advantages	Disadvantages
Rhythm	Male and female	Poor to fair	No cost; acceptable to Roman Catholic Church	Requires significant motivation, cooperation, and intelligence; useless with irregular cycles and during postpartum period
Douche	Female	Poor	Inexpensive	Inconvenient; possibly irritating
Coitus interruptus	Male	Fair	No cost or preparation	Frustration
Condom	Male	Very good	Easy to use; helps to prevent veneral disease	Continual expense; interruption of sexual activity and possible impairment of gratification
Diaphragm with cream or jelly	Female	Very good	No side effects; minor continual cost of jelly and small initial cost of diaphragm	Repeated insertion and removal; possible aesthetic objections
Cervical cap	Female	Very good	Can be worn 2–3 weeks without removal; no cost except for initial fitting and purchase	Does not fit all women; potential difficulties with insertion
Vaginal foam	Female	Good	Easy to use; no prescription required	Continual expense
Vaginal creams, jellies, tablets, and suppositories	Female	Fair to good	Easy to use; no prescription required	Continual expense; unattractive or irritating for some people
IUD (IUCD)	Female	Excellent	Requires little attention; no expense after initial insertion	Side effects, particularly increased bleeding; possible expulsion
Birth-control pills	Female	Excellent	Easy and aesthetic to use	Continual cost; side effects; requires daily attention
Sterilization	Male or female	Excellent	Permanent relief from contraceptive concerns	Possible surgical/medical/psychological complications
Abortion	Female	Excellent	Avoids unwanted pregnancies if other methods fail	Expensive; possible medical complications; psychologically or morally unacceptable to some

Source: Herant A. Katchadourian and Donald T. Lunde, *Biological Aspects of Human Sexuality* (New York: Holt, Rinehart & Winston, 1975).

little applause, except in certain circles. There are a number of "folk" devices that are employed even today. These range from regulation of intercourse according to the phase of the moon, and stepping over graves, to using cellophane sandwich-wrapping as condoms and douching with soft drinks (the carbonic acid and sugar are believed to be spermicidal and a shaken drink provides the propulsion for the agents to reach far into the vagina).

The effectiveness of most folk methods is unknown, so let us consider other means whose effectiveness or ineffectiveness has been investigated more carefully.

The Rhythm Method

It has been jokingly said that people who practice the *rhythm method* of contraception are called "parents," although the joke is funnier to some than others. Basically, rhythm involves periodic abstention. It is theorized that the time during which a woman may become pregnant can be predicted, and by refraining from intercourse during this time, pregnancy cannot occur. The couple must therefore avoid intercourse at least two days before and one day after ovulation, although, for safety, this interval is usually increased. Interestingly enough, although the method involves couples refraining from intercourse at the very time when the woman may be psychologically and emotionally most receptive, and although the system requires the careful use of thermometers, calendars, paper and pencils, certain groups claim that this is the only "natural" means of birth control. Unfortunately, the method is of doubtful reliability, partly because the menstrual periods of many woman are highly irregular.

Coitus Interruptus

In spite of the obvious inconvenience of withdrawing the penis from the vagina immediately before ejaculation, coitus interruptus is one of the most widely used contraceptive practices in Europe today. Again, there are little good data on the effectiveness of this method. It is likely of low reliability because the man must do just the opposite of what he is very highly motivated to do. The practice is risky, also, because sperm may leave the penis long before ejaculation occurs. Because of residual semen in the penis after the first ejaculation, there is higher risk of conception for those who are able, and have the time, to immediately copulate again.

The Condom

The *condom,* or rubber, is one of the most widely used contraceptive devices in the United States. Basically, it is a balloonlike sheath made of rubber, or other material, that fits tightly over the penis and traps the sperm. There are many grades of condoms, ranging from the two-for-a-quarter specials sold in rest-room vending machines to more expensive types made from animal membranes.

The best condoms are only 85 percent effective (partly because they are often incorrectly used). This means that out of 100 couples who depend on them for a year, 15 of the women may become pregnant. Disadvantages of the method include a reduced level of sensation on the part of the man and the interruption while the condom is frantically located and put on. One important reason for its popularity, however, is the protection it affords against the transmission of venereal disease.

The Diaphragm

The *diaphragm* is a dome-shaped rubber device with a flexible steel spring enclosed in the rim. It is inserted into the vagina and fitted over the cervix to prevent sperm from entering the uterus. Unlike the condom, the diaphragm must be individually fitted. It should be coated with a spermicidal cream or foam and left in place for several hours after intercourse. It is about 90 percent effective if well-fitted and used correctly. With a certain element of anticipation on the part of the woman, it need not interrupt the sexual activities to be inserted.

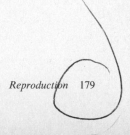

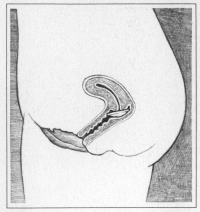

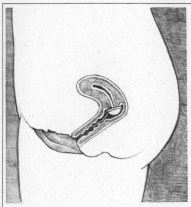

7.24 Upper: diaphragm in place. Lower: cervical cap in place. They both operate on the same principle, that is, blocking the cervix. However, the diaphragm is normally inserted each time before intercourse, while the cervical cap may be worn for weeks at a time. Also, the diaphragm is far less effective unless it is used with a spermicidal jelly, and some women find the whole process of preparation somewhat objectionable.

The Cervical Cap

The *cervical cap* functions similarly to the diaphragm. It is a small plastic or metal cap that fits tightly over the tip of the cervix. One of its advantages is that it can be left in place instead of having to be reinserted each time. In fact, it is sometimes removed only for menstruation. Its effectiveness may be somewhat higher than that of the diaphragm, since it ordinarily cannot develop leaks. However, its loss is also less likely to be noticed.

Spermicides

A number of sperm-killing foams, jellies, aerosols, suppositories, and creams are available. They are usually quickly and easily applied inside the vagina and require neither fitting nor prescription by a doctor. They are less effective than the mechanical devices, but their effectiveness increases when they are used with the diaphragm or cervical cap.

The Intrauterine Device (IUD)

The *intrauterine device,* or *IUD,* has been known as an effective contraceptive for many years. It is reported that Arab camel drivers placed fruit seeds in the uteruses of their camels to prevent pregnancies during the long, hard caravan marches. Just how the practice started is not known; there must have been better places to store seeds. Today IUDs of a variety of shapes and materials are in use. However, beyond the fact that the presence of a foreign body in the uterus somehow prevents pregnancy, no one knows exactly how the IUD works. It is theorized that the IUD may prevent implantation of the zygote, or that it may cause the egg to move too rapidly through the oviduct, even that it may alter the condition of the uterine endometrium.

The IUD has several major advantages. Once it has been installed in the uterus, it requires no further attention. The expense of the device is minimal, and its effectiveness is about 97 percent.

One problem with IUDs is that in some women the uterus rejects them. As a result, the device may be expelled from the uterus, sometimes without being noticed. Some IUDs have a string that hangs into the vagina, so there is a means to check the device. For some reason IUDs are most likely to be rejected by the uterus in women who have never been pregnant. There may also be side effects such as bleeding and pain just after insertion, and if these persist, the IUD must be removed.

The Birth-Control Pill

The *birth-control pill* is an oral contraceptive. It is composed of the hormone estrogen, either natural or synthetic, and an "artificial hormone," similar to progesterone, called *progestin.* This combination apparently inhibits the formation of egg follicles in the ovary. A pill is taken each day for 20 or 21 days, beginning on the fifth day after the onset of menstruation. The next menstrual period begins when this series of pills is finished. Regular usage of the pill on this basis results in a menstrual period precisely every 28 days. Apart from the convenience of knowing even the day of the week on which menstrual periods will begin, in many cases this regulation may alleviate the pain accompanying menstruation. The greatest advantage of the pill is that when it is used correctly it is probably over 99 percent effective. There have also been reports that regular use may delay menopause, and increase the libido of the middle-aged user, although the bases for these effects are not known.

Birth-control pills may cause undesirable side effects, especially during the first few months of use. In many ways these side effects are similar to the symptoms experienced during pregnancy. These may include weight gain, fluid retention, nausea, headaches, depression, and irritability. However, another such effect is enlargement of the breasts. Some or all of these symptoms may diminish after the first few months of use.

There may also be more serious problems associated with birth-control pills. You have probably heard reports that the pill may be associated with cancer. Early results of a study sponsored by Planned Parenthood of New York were inconclusive; although such studies indicate a higher incidence of "precancerous" areas in the cervix among women who use the pill, these areas do not invariably develop into cancer and are easily and totally curable.

Another possible risk may be thromboembolism, or blood clots, including thrombophlebitis (inflammation and clotting in the veins) and pulmonary embolism (blood clots in the vessels of the lungs). According to a British report in 1968 on women over thirty-five, those using the pill had a higher chance of dying from thromboembolism (8 in 200,000) than those who were not using the pill (1 in 200,000). Women under thirty-five had half the risk in both cases. However, about 450 women in 200,000 between the ages of twenty and thirty-five die as a result of pregnancies, so from this standpoint the pill is still far safer than pregnancy.

In any case, although the pill has produced no serious medical problems for the overwhelming majority of its users, we still have no information on its long-term use and research must be rigorously continued in this area.

Sterilization

Sterilization is 100 percent effective in ensuring that pregnancy will not occur. Either the male or female can be sterilized, although it is a much simpler matter in the male.

A *vasectomy* for a male normally takes about 15 minutes and can be performed in a doctor's office. A small incision is made in the side of the scrotum, a short section is cut from the seminal duct (the vas deferens), and the incision is then closed (Figure 7.25). The operation can sometimes be reversed to restore fertility; this has been successful in 50 to 80 percent of the cases in those areas where doctors have had the greatest experience with this operation. Another method, in which a removable plastic plug is

7.25 The simple procedure of a vasectomy. The operation has been described as similar to a visit to the dentist, with a few differences. Normally, under local anesthetic, the vas deferens are exposed by a small cut in the scrotum and a section is removed from the tube. Both ends are tied in case they show a tendency to rejoin. Although surgeons have had some success in reversing their results (particularly in California where the operation is rather common), vasectomy is recommended only for those who have made up their minds.

Sperm duct

Cut and tied

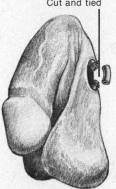

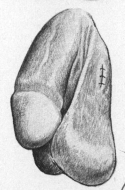

inserted in the seminal duct, shows promise of being reversible in virtually all cases.

Sperm continue to be produced after vasectomy, but instead of being ejaculated, they are reabsorbed by the body. Seminal fluid is otherwise ejaculated as before; it simply does not contain sperm. As a form of "insurance" for men who feel they may want to have children later, it is possible to have sperm frozen and kept in a sperm bank for as long as ten years. If children are desired later, the woman can be impregnated by artificial insemination.

Vasectomy, is, in itself, a simple procedure, but there are other factors that must be taken into account. For example, in the United States, as well as in some other countries, men often identify strongly with the "male role." The realization that they are no longer able to impregnate women may have a rather marked psychological effect in some individuals. For this reason the procedure is ordinarily advisable only for those men who are fairly secure in their own sexual identity. Also, in some areas it may be difficult to find a doctor who will perform a vasectomy. Although there were 38 vasectomy clinics operating in the United States in 1971, the reaction of the medical profession has, as usual, been very conservative.

Sterilization in women is usually accomplished by *tubal ligation* (Figure 7.26). The oviduct is cut and both ends are tied back. This operation is much more complicated than a vasectomy, since the abdominal wall must be opened. A method that is still in the experimental stages involves entry through the vaginal opening and, when perfected, may make the ligation a far simpler procedure.

7.26 Tubal ligation, a means of sterilization in women. Tubal ligation, or "tying the tubes," is a more serious operation than the male's vasectomy. It has traditionally been accomplished by opening the body wall, cutting the oviduct, and tying back the ends. Newer methods, however, include tying with a device that is pushed up the oviduct from the uterus. Ligation does not affect the ovaries, thus they continue to produce their hormones, so there is no change in the physiology of the woman other than totally negating the likelihood of pregnancy. Some women have reported increased sexual satisfaction with the knowledge that they cannot become pregnant, but such reactions are strictly a personal matter and cannot be predicted.

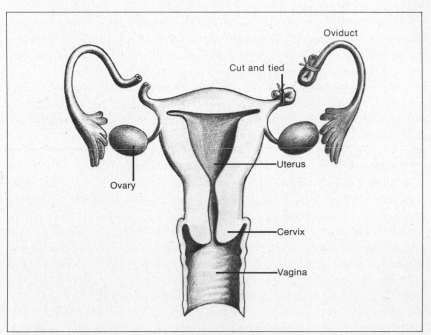

As with vasectomy, tubal ligation causes no change in hormone production, sex drive, or sexual performance.

Abortion

Abortion, the artificial termination of a pregnancy, has been one of the more common forms of birth control throughout recorded history. In most modern societies the practice is tightly regulated, but the regulations vary widely. In the United States there has been a recent relaxation of restrictions on abortion that has paved the way for new programs facilitating abortion. This policy does not necessarily mean that there are more abortions now than there were before; rather, it means the *illegal* abortions may become less common. When restrictions were in effect estimates ran as high as one illegal abortion for every two live births in the United States. Interestingly, illegal abortions are most common where laws are most restrictive. In Italy, for example, where even contraceptives have been banned, the abortion rate is estimated to be equal to the birth rate.

Illegal surgical abortions and do-it-yourself methods have resulted in innumerable deaths and permanent damage. Some of the methods employed have been as barbaric as they were hazardous. And even then, they have not always accomplished the intended result. In fact, a woman is almost always safer having a baby than attempting to abort the fetus herself without proper medical help.

When performed properly, under medical supervision, an abortion during the first three months of pregnancy is a simple procedure that entails no more risk than a tonsillectomy. The most common method is called *dilation and curettage,* or *D and C.* With this method the vagina and cervix are dilated and the lining of the uterus is scraped with a *curette,* a steel loop at the end of a handle. Another method coming into increasing use is *vacuum curettage,* in which a suction device is used to pull the embryo from the uterine wall. Another increasingly common procedure involves the injection of salt solutions into the uterine cavity.

Abortion became illegal in the United States in the nineteenth century, when it was ruled that the practice was dangerous to the prospective mother. The recent court rulings in the United States that uphold the right of a woman to undergo abortion on request have generated heated debates. One agrument has revolved around the point at which a developing embryo becomes a "person." There are those who feel that the zygote becomes a human being at the moment of conception and thus has a "right" to be born. Others argue that abortion does not constitute the destruction of a "life" until some specific later point, such as at the time of the first hearbeat, or the point at which the embryo first begins to stir or "quicken," or the point at which it is capable of surviving after birth. There are many puzzles and paradoxes in such arguments, especially since the sacredness of human life often depends on whose life we are considering. Many who oppose abortion on this ground have no qualms about supporting capital punishment or the mass destruction of

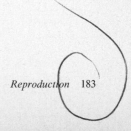

life during wars. Obviously, the question of the "morality" of abortion is not likely to be resolved on a scientific basis.

Abortion has also been opposed on the basis that it encourages sexual freedom, a possibility that is viewed with more alarm in some quarters than in others. One of the arguments heard in more than one statehouse across the land is that the woman must "pay the piper." She has had her fun, so now let her accept the consequences in the form of "compulsory pregnancy." Apart from the assumption that moral responsibility should be expected only of women, the idea that a child is considered a form of punishment is an interesting one, rarely acknowledged by its *de facto* proponents. Of course, these are only a few of the arguments that are being heard regarding abortion, but they should provide a starting point for discussion. If you are interested in hearing some other arguments, bring up the subject at your parents' next party.

As a final point of consideration, in Sweden a study was made to see what happened to children whose mothers were denied abortion. They were compared to a matched set of children who were wanted. It was found that twice as many of the unwanted ones grew up as "illegitimate," or in broken homes, or in institutions. Twice as many had records of delinquency, and twice as many were declared unfit for military service. Twice as many had required psychiatric care, and five times as many had been on social-welfare programs during their teens.

New Developments in Contraception
World population pressures, the desire of an increasing number of couples to have children by intent rather than by chance, and the wish of many to entirely avoid the rigors of parenthood have encouraged vigorous new research in birth control. The resulting ideas have been rather varied and imaginative. As with the present contraceptive methods, however, most of the new ones are for use by women.

Progestins are artificial hormones that are administered in low continuous doses. They may be administered as a "minipill" (daily, with no breaks), as an injection that lasts three months, or by the implantation of a "time capsule" beneath the skin. The capsule is made of silicone rubber and releases progestins into the blood stream at a regular rate for perhaps as long as twenty-five or thirty years.

A progestin alone does not suppress follicle development, as it does when taken with estrogen in the form of "the pill." It is not known precisely how progestins work, but they may alter the cervical mucous, preventing the sperm from entering the uterus. It has also been suggested that they may cause the egg to move through the oviduct too fast, or that they may prevent implantation or the formation of essential hormones.

Progestins are not subject to some of the risks suspected of the present birth-control pill, such as thromboembolism; however, they may cause headaches, dizziness, irregular bleeding, or the absence of menstruation. They are also reported to cause nodules in the breasts of beagles, so beagles probably shouldn't take them. They have currently been withdrawn from clinical testing, but if they are perfected, they offer hope of a simple and effective contraceptive.

Prostaglandins are hormonelike substances similar to fatty acids that are found in many kinds of animal tissues. They somehow seem to regulate hormones at their target sites in cells and organs. Prostaglandins also show promise in the treatment of a variety of ailments, from nasal congestion to hypertension. They have been used to treat both male sterility and painful menstruation, and they can induce labor at any stage of pregnancy. In controlling pregnancy, they seem to cause regression of the corpus luteum, which in turn prevents the maintenance of the pregnancy. Testing is still in the early stages, but prostaglandins show promise of providing a safe, self-administered contraceptive that could be applied only a few times a year.

The *"morning-after" pill* is a heavy dose of estrogens, to be taken by the woman for several days after copulation. It will prevent pregnancy, but the fact that it usually induces nausea for one or two days has prompted the irreverent suggestion that its principle mechanism lies in negative conditioning. These pills are now available in some American medical centers.

Stevia rebaudiana is a South American weed. The Indian women of Paraguay drink a cup of tea made from the plant each day as a contraceptive. The powdered weed has been found to reduce pregnancies in rats, and this method, as well as certain other folk remedies, are being seriously investigated in the search for the ideal contraceptive.

In the wide-ranging search for better or more effective contraceptives, many other methods are under consideration. These include long-lived spermicides, removable oviduct plugs, vaginal rings that release progestins or spermicides, IUDs that release spermicides or hormones, immunization of a woman against her husband's sperm, immunization of a man against his own sperm, a nonsteroid by-product of dynamite manufacture to be administered to men (this one stirs the imagination), and clips and plugs for the vas deferens.

All these methods, of course, must undergo rigorous testing by such agencies as the Food and Drug Administration. In the United States clinical testing on humans may be started shortly after tests on other animals are completed, but a minimum of eight to ten years of such testing is required before a drug will be released for general use. These rigorous controls have led some researchers to other countries in which the controls are less stringent. The need for birth control is increasingly urgent, but obviously no method should be extensively employed without maximum assurance of its safety.

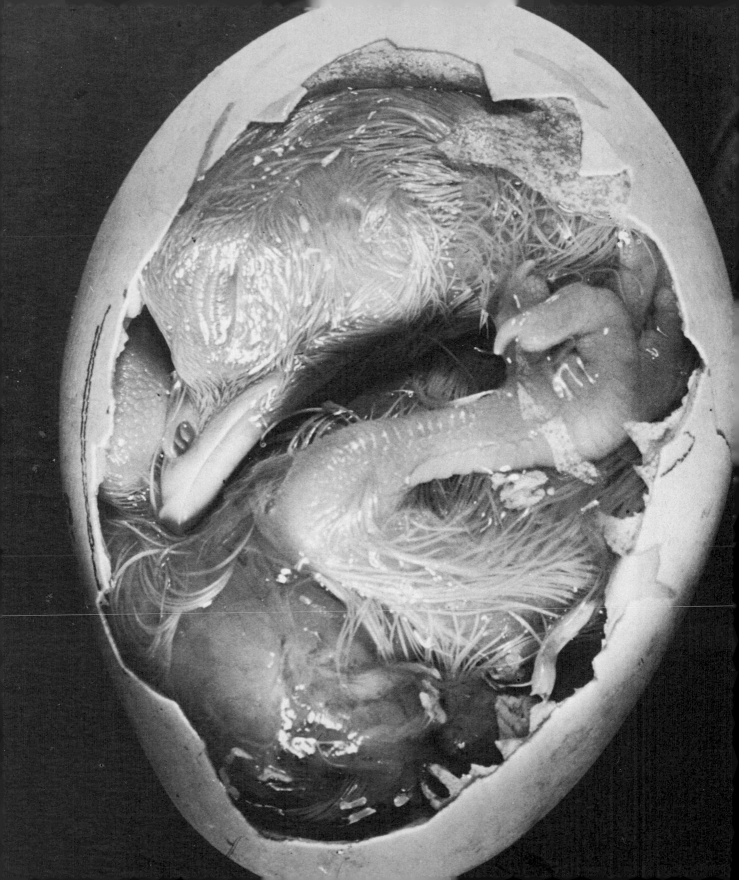

Development

The story of development concerns the events befalling that single fertilized egg after conception—how it grows and changes. Some of these events are programmed into the zygote and are highly predictable. For example, the first cell divisions occur precisely on time, and any embryologist can describe in detail the appearance of a chick embryo seventy-two hours after fertilization. In its broadest sense, the story of development might also include the learning experiences of an organism as it interacts with its environment, its aging and its death. In this chapter, however, we will concern ourselves with what happens between fertilization and birth in animals. Again, the focus will be primarily on our own species, and we will consider further certain ethical and philosophical factors associated with the production of new humans. In addition, we will discuss the motivation for having children, from a biological point of view. But let's begin by noting that animal eggs are basically of three types.

TYPES OF EGGS

Animals produce eggs of three basic types, roughly categorized according to the amount of yolk they contain. Since yolk is a food store, the embryos of those animals that derive their nutrients mainly from their mother's blood, as is the case with humans, need little yolk; perhaps just enough to last the zygote until it has implanted in the wall of the uterus. In contrast, since birds leave their mother's body at a very early developmental stage, they must carry their entire prehatching food supply with them. So, whereas a human egg is smaller than the period at the end of this sentence, you would be hard put to hide an ostrich egg with this whole book, although the size difference of both the young and the adults of these two species is not particularly great.

In the species producing a third type of egg, the new organism begins to acquire its own food at a stage some time after fertilization, but before it has arrived at what will be its final body organization. Thus, the frog egg has just enough yolk to get the developing embryo to the tadpole stage, and after that the tadpole survives on a combination of food stored in its tail and what it can find to eat. The frog zygote, then, has a yolk supply intermediate to that of mammals and birds.

EARLY DEVELOPMENT IN THE FROG

Recall that at fertilization any organism receives all the genetic information it is ever going to get. No matter what differences arise in the cells that

8.1 (pp. 189, 190)   Comparison of the development of three vertebrates, showing similarities and differences. In the early cell divisions (1, 2, 3) differences are obvious. The chick embryo, perched atop a massive and inert yolk, first undergoes incomplete cleavages. Frog and human begin similarly, but soon cell divisions lag in the larger, yolk-laden vegetal region of the frog, resulting in larger cells there. All three embryos produce a hollow blastula stage (4). Frog and human blastulae are spherical with prominent cavities. In the chick embryo, a streak of cells rises up slightly, forming only a minute cavity above the yolk.

Gastrulation (5) produces new cavities and the three germ cell layers. From these layers, future tissues and organs will be molded. The mesoderm originates differently in each. About this time the human embryo implants in the uterine wall, its meager yolk reserves depleted. Its chorion grows into the uterus, seeking nourishment. The bird too will produce membranes to assist in bringing food from the yolk. The frog grows around its food, so there is no serious supply problem.

In (6) all three produce the rudiments of a nervous system, ectoderm rising up in folds to outline the system. Subsequently discrete pockets of cells contribute to organs as systems are built. Hearts and blood vessels in the chick and human form early. In each embryo, mesodermal blocks called somites contribute to vertebrae, muscle, dermis, etc. Interestingly, each species forms pharyngeal gill pouches, a primitive vertebrate feature. In land creatures, these contribute to structures other than gills.

With continued refinement of form (10, 11) the species become easier to recognize. The tadpole will be eating long before the chick has used up its food supply, and the chick will be an adult by the time the human is born.

make up the various structures of the adult, they all derive from this single cell, with its directing set of chromosomes.

The First Cell Divisions

The first step in the process that will result in all the highly differentiated tissues and organs of the adult is the first mitotic cleavage of that primal cell, the zygote. The genetic material doubles, virtually identical chromosomes move to opposite poles, and the cell cleaves into two daughter cells. Shortly afterward, each of these daughter cells divides again.

The first two cell divisions are usually at right angles to each other, but along the same vertical axis; the third line of division is perpendicular to these (Figure 8.1). In the case of the frog this third cleavage is displaced toward one hemisphere called the *animal pole*. The reason for this displacement is that the relatively inert yolk material is concentrated in the other hemisphere, the *vegetal pole*. The cytoplasm in the area of the animal pole is less encumbered by yolk, and hence more active. The axis and polarity of these cleavages are determined by the point at which the sperm penetrated the egg at fertilization.

As the cells of the embryo continue to divide, they form a cluster, or ball of cells. These divisions are accompanied by new kinds of changes that produce a hollow cavity inside this ball, now called a *blastula*. The unequal distribution of the yolk has resulted in smaller cells of more nearly equal size having formed at the animal pole.

Gastrulation

Early in the second day after fertilization, a process called *gastrulation* begins in the frog embryo. At this stage, the embryo, logically enough, is called a *gastrula*. Gastrulation involves the migration of certain cells toward an area in the vegetal hemisphere, where they converge and begin to roll under to form an opening called the *blastopore*. The rolling under takes place along a curved line, at what will be the back, or *dorsal* side of the yolk area. The process of involution extends outward from the edges of the crescent, until finally the crescent becomes a circle. The vegetal area inside this circle is called the *yolk plug*, and this will eventually be enclosed by the expanding layer of cells around it.

At this stage three types of cells may be distinguished. The ectoderm lies on the outside, an inner *endoderm* is formed from cells of the yolky vegetal area, and the cells that had rolled under from the outside and have come to lie between the ectoderm and endoderm become the *mesoderm*. Each of these cell layers, or *germ layers,* will give rise to very specific structures. For example, the ectoderm will form the outermost layer of the skin and certain other structures including the sense organs, parts of the skeleton in various animals in the head and neck and, importantly, the nervous system. The ectoderm will also form hair and feathers. The mesoderm will form tissue associated with support, movement, transport, reproduction, and excretion (for example muscle, bone, cartilage, blood, heart, blood vessels, gonads, and kidneys). The endoderm will produce

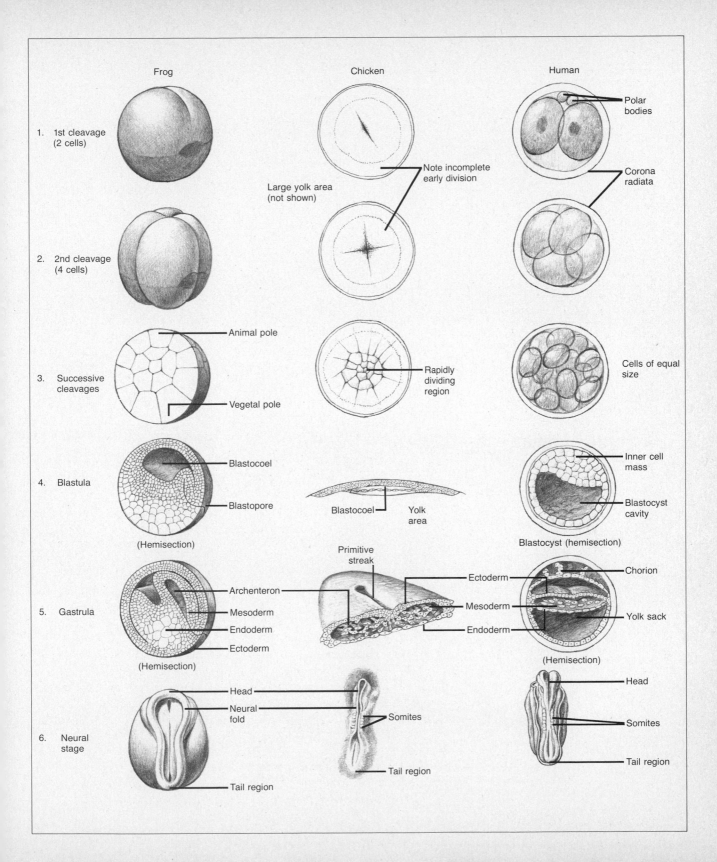

Frog   Chicken   Human

Polar bodies

1. 1st cleavage
   (2 cells)

Note incomplete early division

Corona radiata

Large yolk area
(not shown)

2. 2nd cleavage
   (4 cells)

Animal pole

Cells of equal size

3. Successive
   cleavages

Rapidly dividing region

Vegetal pole

Blastocoel

Inner cell mass

4. Blastula

Blastocoel   Yolk area

Blastocyst cavity

Blastopore

(Hemisection)   Blastocyst (hemisection)

Primitive streak

Chorion

Archenteron

Ectoderm

5. Gastrula

Mesoderm

Mesoderm

Endoderm

Yolk sack

Ectoderm

Endoderm

(Hemisection)   (Hemisection)

Head

Head

Neural fold

6. Neural
   stage

Somites

Somites

Tail region

Tail region

Tail region

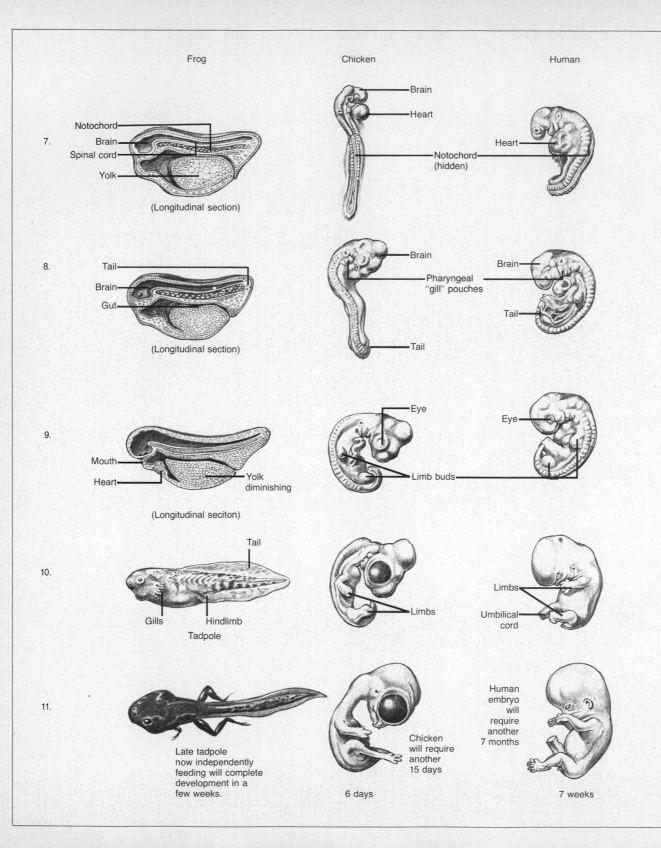

Frog   Chicken   Human

7.

Notochord
Brain
Spinal cord
Yolk

(Longitudinal section)

Brain
Heart
Notochord (hidden)
Heart

8.

Tail
Brain
Gut

(Longitudinal section)

Brain
Pharyngeal "gill" pouches
Tail

Brain
Tail

9.

Mouth
Heart
Yolk diminishing

(Longitudinal seciton)

Eye
Limb buds

Eye
Limb buds

10.

Tail
Gills
Hindlimb
Tadpole

Limbs

Limbs
Umbilical cord

11.

Late tadpole now independently feeding will complete development in a few weeks.

Chicken will require another 15 days

6 days

Human embryo will require another 7 months

7 weeks

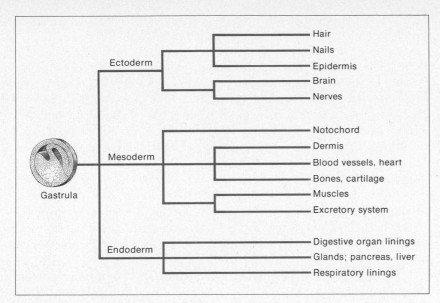

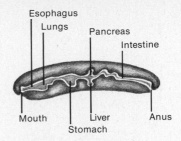

structures associated with breathing and digestion (including the lungs, liver, pancreas, and other digestive glands).

Regulation

Now, at this stage of development it is hard to see just how this peculiar ball is going to turn into a frog. The precise changes involved are difficult even for students of embryology to visualize so we won't try your patience with much detail. Some understanding of the mechanisms, though, should provide food for thought. You might be interested in knowing, for example, that even at this early stage (the second day) the fate of some cells has been sealed for hours. In other words, they are on their way to assuming their final role, and the changes that have occurred in their genetic constitution are irreversible. The genes, which have been shut down as the chromosomes direct the production of only certain kinds of *m*RNA, can never be reactivated.

Other cells in the embryo have not yet specialized, and these cells can still theoretically become part of any number of organs or systems. In fact, if some of these generalized cells are transplanted to other parts of the embryo, they will give rise to the same types of cells as their new neighbors. Had they been left in their original location, they would have formed cells of an entirely different type. As each hour goes by, more and more cells become committed, as their changes take them past the point of no return.

The Flexible Fate

We know that in some species, such as the flatworm, the fate of most cells is never sealed; or how could regeneration take place when it is cut in half? Let's consider what happens to such cells in the process of regeneration. In some kinds of salamanders, when a leg is amputated the cells in the area of the cut despecialize and revert back to the more primitive state typical of embryos. Then they begin to specialize again, some taking entirely new developmental pathways, until finally the limb is restored.

If the leg of a mouse is amputated, however, it doesn't grow back. In these higher animals the cell differentiation is more permanent, and the cells are less able to resort to an earlier condition. Epidermal cells, it is true, may migrate to grow over and protect the cut area, and some cells may change to some degree because of the wound. But general deorganization and regeneration is not possible. We might well ask at this point what determines the form a cell will take. How is development regulated?

Sequencing

For the highly complex and integrated systems of an adult vertebrate to develop, an incredible array of regulatory mechanisms must be involved. These mechanisms must see to it not only that muscle cells develop in a limb area, but that muscles of very specific sizes, shapes, and attachments are formed.

The sequence of development is also important. For example, before the nerves appear that will regulate those first timorous palpitations in the incipient heart muscles, there must be some place for the blood to go. Thus the first step is the appearance of great channels, which will eventually become major vessels, in those looseknit, undifferentiated first-blood pools. Obviously, then, the sequencing of the tissues and organs is critical, but what regulated this sequence?

The exact mechanism of sequencing and other forms of regulation are still a mystery, but we can explain them conceptually in terms of the processes we already know something about. We know, for example, that certain areas of the chromosomes can become activated while others shut down, enabling them to produce cells of various types. Certain cells may also slow down or stop their reproductive activities while others undergo a burst of reproduction, accompanied by a surge in the growth of the tissue they comprise. And, of course, cells can die. It is essential, in fact, that some cells die, such as those that lie between what will be the fingers of a developing hand. Hence the highly organized changes that take place in the embryonic body may actually be "only" the sum of a few kinds of cellular changes we already know something about.

Other changes in the cells of developing embryos probably occur as a function of time. For example, a particular segment along the length of a chromosome may have a shorter active life than another segment. After an initial period of activity, then, one set of genes might cease its activity while another continues. The question here, of course, is what determines the active life span of any set of genes. Why do some of them stop making *m*RNA so soon? Biologists are searching for the answer.

Stress

Another regulatory factor is stress. Just as karate practitioners build up certain tissues around their knuckles by pounding makiwara boards, so an

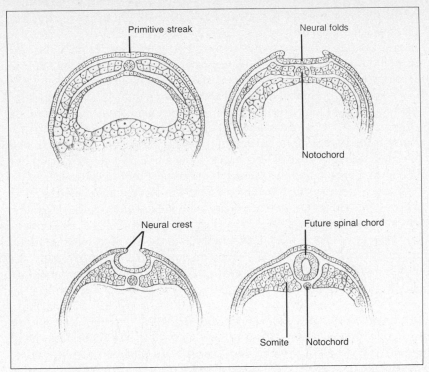

Primitive streak

Neural folds

Notochord

Neural crest

Future spinal chord

Somite   Notochord

8.4 Spinal cord induction. The rodlike notochord in this chick embryo lies just below the ectoderm. At some point it begins to cause changes in the overlying ectoderm, first causing it to thicken. Then two ridges form along its length. These grow upward and inward toward each other and join, forming a tube that will be the spinal cord. The anterior end will develop into swollen bulbs that will form the brain.

8.5 Lens induction in the vertebrate eye. Outpocketings of the highly lobed brain (called optic vesicles) come to lie under the head ectoderm, inducing that ectoderm to thicken and roll, forming the lens of the eye. The underlying brain tissue will form elongated cells that will become the light-sensitive retina of the eye. Thus part of the eye is directly confluent with the brain and both the lens and the retina are ectodermal in origin.

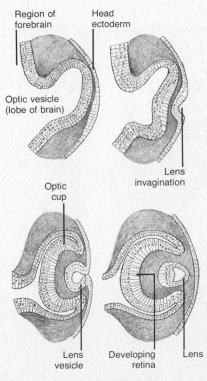

Region of forebrain

Head ectoderm

Optic vesicle (lobe of brain)

Lens invagination

Optic cup

Lens vesicle

Developing retina

Lens

embryo builds up tissue along its lines of stress. The bending and twisting of the developing embryo as it grows produces stress in certain areas. A type of mobile, undifferentiated tissue called *mesenchyme* then moves into these stressed areas, settles there, and begins reproducing and specializing as supportive tissue.

Induction

Regulation may also be accomplished by one type of tissue acting upon another. For example, after the mesodermal layer is formed, the tissue begins to thicken just under the area that will ultimately become the spinal cord. This thickened rod of mesoderm is called the *notochord*. The notochord itself doesn't form the spinal cord, but it influences the overlying ectoderm in a peculiar way. First, the ectoderm that lies over the notochord begins to thicken, then tissue starts to build up along each side of the long, thickened plaque (Figure 8.4). These two ridges, called the *neural folds*, enlarge and begin to fold toward each other until they join and fuse. The result is a hollow tube lying along the dorsal surface of the embryo. This is the early *spinal cord*. The spinal cord is then overgrown by ectodermal cells from the surrounding area.

In most vertebrates, after the spinal cord forms, the underlying notochord that has induced its formation disappears, or at least diminishes greatly. Any supportive function it might have had is taken over by vertebrae which form from mesoderm lying alongside the spinal cord, eventually enclosing it. The role of the notochord as the inducer in this case has been demonstrated by transplanting segments of the notochord to other areas of the embryo. When this is done at a certain period, before the ectoderm has differentiated too far, incipient spinal cords can be made to develop almost anywhere over the surface of the embryo.

There are a number of inducers which operate in the developing embryo. In some cases tissues that have developed from the same germ layer may interact as inducer-inducee after they have differentiated to some extent. For example, the lens of the eye is induced in ectodermal cells by an underlying area of the brain, which has also formed from the ectodermal layer. Of course, these examples cover only a few aspects of the regulatory processes that govern the complex developmental changes in the embryo. However, they give you some idea of the role of regulation and a few of the ways in which it can work.

Now let's take a brief look at development in two other kinds of animals: birds and mammals. Keep in mind that although there are important differences between frogs, birds, and humans, many of the basic patterns of development are the same.

DEVELOPING BIRDS AND THEIR MEMBRANES

Although birds' eggs are large, the early embryo is very small; in fact, the zygote is at first only a *germinal* spot. The life of a chicken, then, begins as a tiny disk on the yolk. And when cell division takes place, it is limited to that disk since cleavage of the great body of inert yolk is impossible. The blastopore in the chicken embryo does not develop as a crescent, but as a slit along what will be the body axis (Figure 8.6). As a result, the mesoderm is formed somewhat differently. The endoderm first splits away from the underside of the ectoderm, and cells from the upper ectodermal layer then roll under along the midline of the embryo to form the mesoderm.

After the spinal cord has formed, rapid growth in what will be the head and tail regions produces folds at either end of the embryonic axis. At each end of the body the ectoderm pockets in and the endoderm pockets out, bringing the two layers together. These thinned areas then rupture, forming the mouth and excretory openings, and the tube-within-a-tube structure, characteristic of vertebrates, is complete.

In the meantime, a membrane composed of endoderm plus mesoderm grows around the yolk. Blood vessels developing in the mesoderm are able then to carry food to the embryo. Another membrane, called the *allantois,* grows out of the rear area of the developing gut and forms a receptacle for nitrogenous wastes. This sac will be left behind when the chick hatches.

When the extra-embryonic membranes fuse over the embryo, the embryo is enclosed in a four-walled membrane. The inner two layers become the *amnion* and the outer two become the *chorion,* each consisting of a layer of mesoderm and a layer of endoderm.

The amnion will fill with fluid, which acts as a shock absorber to

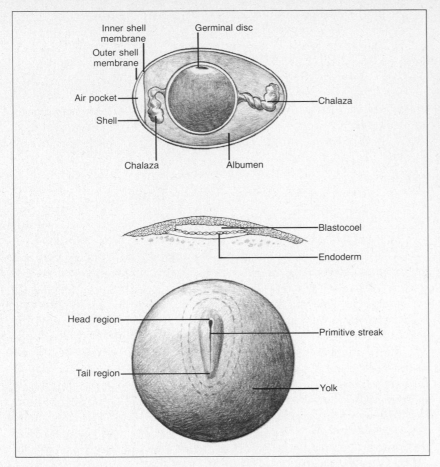

Inner shell membrane
Germinal disc
Outer shell membrane
Air pocket
Chalaza
Shell
Chalaza
Albumen

Blastocoel
Endoderm

Head region
Primitive streak
Tail region
Yolk

8.6 Top, a cutaway view of a hen's egg with a developing embryo (the germinal disc). The yolk will provide food for the embryo. Food that will be transported through embryonic vessels as membranes grow and enclose the nutrient supply. The yolk and embryo are held in place by the chalazae and cushioned by the albumen (or egg white). Later, membranes will come to lie pressed against the inside of the shell, picking up oxygen from the environment and losing CO_2 to it.

Center, the blastocoel stage of a developing chick. Compare the developmental stage with the sequences in Figure 8.1.

Lower, the embryo proper is still only a rather flattened area sitting atop the yolk. But the disc is now elongated with the outline of developing membranes growing outward from the embryo. The primitive streak is an indentation above the notochord that indicates the axis along which the spinal cord will develop. Later the head region will grow so fast as to overlap the yolk as is shown in the cross section of the head in Figure 8.1.

protect the embryo from injury. This amniotic fluid also serves as a lubricant, so that as the limb buds and other projections develop they will not fuse to the body. The mesoderm layer on the outside of the allantoic membrane will later fuse with the mesoderm on the inside of the chorion to form a three-layered *chorioallantoic membrane.* This membrane will come to surround the albumen, and thus will be just under the porous shell. Its function, as you may guess, is to pick up oxygen, which diffuses in through the porous shell, and transport it to the embryo. It also provides a surface area through which carbon dioxide from the respiring embryo can leave the egg.

As the embryo continues to develop, it eventually separates from the yolk, except for a thin stalk through which food is carried. The anterior, or head, part of the spinal cord begins to bulge and form the lobes of the brain. The mesoderm on either side of the embryo segments into blocks, or *somites,* which will soon form the vertebrae and large trunk muscles. The mesodermal membranes we have been discussing have formed from lateral extensions of these somites.

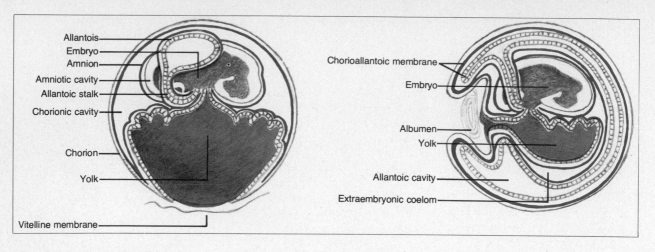

8.7 Later development of membranes in the chick embryo. The head of the darkened embryo is at the right of both figures. Note that as the yolk mass decreases, the membrane complex becomes more extensive. As extraembryonic membranes grow upward and over the embryo, finally to fuse, they form a four-walled membrane. The inner two form the protective amnion. The outer two form the chorion. The allantois first appears as a saclike extension near the rear of the embryo and is a storage site for embryonic waste. Later it will become compressed with the chorion against the eggshell and exchange between the egg and the environment will be accomplished over this chorioallantoic membrane.

A large blood vessel underneath the embryo begins to pulsate, and later it loops and fuses together to form the heart. Tiny flaps appear, which then develop into the limbs. The brain outpockets on either side to induce the great orbs that will be the eyes. Meanwhile, the endoderm also outpockets and proliferates in various places to form the highly complex glands and organs associated with digestion.

Many other barely perceptible changes occur as the embryo continues its transformation. At about the eleventh day there will be a great eruption of enzymes, and the various systems of the organism will begin to function in even greater harmony. After about three weeks the chick will begin its first feeble pecks to break free of its protective shell. When it finally emerges, it will begin to register strange and fascinating new experiences that further alter the arrangements of its constituents. Then someone will eat it.

HUMAN DEVELOPMENT

Now that we have some idea of certain general principles of development, let us consider the human animal. And, as you are doubtless aware, it is all too easy to find oneself plunged into controversy by this simple shift in species. When considering our own group there is an understandable tendency to seek "specialness" and to minimize the similarities with other, less "noble," species of the earth.

As an example of one controversy, there is an aged but growing argument over the period at which "life" begins for humans. There are those who see great importance in the zygote who would never argue that every gamete had the right to impose its genes on the world. Others would argue that a human embryo is not sacred or "one of us" until implantation occurs, or until the heart starts beating, or until the embryo begins kicking and moving (quickening), or until it draws its first independent breath. Others argue that it is *never* sacred and can be removed from the population at any point in its postnatal period without need for particular

Box 8.1 Pregnancy Tests

Although a woman cannot know for certain that she is pregnant in the first weeks through natural means, there are tests that give fairly reliable results. Missed periods are *not* sure signs, but other changes that accompany missed periods give stronger credence to this sign, such as tingling breasts, slightly increased body temperature, and morning sickness. Since implantation occurs at about the time of the next menstruation, the process may cause some bleeding. Pregnancy can cause signs that seem to indicate nonpregnancy, so these general signs are sometimes deceptive. If a woman is at least 4 weeks pregnant, a hormone will appear in her urine. When this hormone and its products are injected into a female mouse or rabbit, they will rupture the follicles of the animal and blood clots in the ovaries will appear. This is called the *rabbit test* (probably because that sounds better than the mouse test). There are other tests, such as the frog test, which give results much more quickly.

Box 8.2 Prenatal Influences

There has been a lot of discussion through the years over what influences the mother's behavior or environment may have on a developing embryo or fetus (an embryo of over two months). It was once believed, for instance, that a pregnant woman scared by a zebra might bear a striped child. A few years ago, other evidence was horrifyingly revealed that indicated a real relationship between the experience of the mother and the offspring. The tranquilizer Thalidomide was found to be responsible for babies without limbs, with grossly deformed limbs or mental retardation. Some pregnant users of the drug had their fetuses aborted, but others bore their deformed children and the shocked response around the world was to result in tightened drug laws. It has since been found that many types of sleeping aids or tranquilizers can result in abnormalities in the newborn. Other factors that can induce congenital abnormalities in children are radiation, the contraction of German measles, polio, the flu or other diseases, heavy smoking, and possibly the use of hallucinatory drugs. These are all factors that can alter the normal development of the embryo during the first trimester. Surprisingly, alcohol is most dangerous during the third trimester. For some reason the fetus in this period concentrates alcohol in its brain and liver and the result may produce abnormalities in behavior and intelligence.

concern. Yet others argue that the sacredness of human life is somehow related to political or religious beliefs. It seems to be true that a life that, as a result of its experiences, has come to agree with us, is more important than one that has reached other conclusions. A prominent Southeast Asian leader recently noted that Americans cannot be moved by the deaths of Asians. We are chagrined when trainloads of bombs, intended to be rained on people whom we have never seen, explode prematurely and kill Americans. There is no way to resolve such arguments on a scientific basis. We must establish our values and arbitrarily draw lines. Since some individuals find information useful in forming their opinions in such matters, we have an added incentive in considering the development of the human embryo.

Descriptions in bus-station novels notwithstanding, fertilization occurs with the mother-to-be totally unaware of the event. If there are sperm cells thrashing around in the genital tract at any time within forty-eight hours before ovulation to about twelve hours after, the odds are very good that pregnancy will occur. As soon as the egg is touched by the head of a sperm, it undergoes violent pulsating movements which unite the twenty-three chromosomes of the sperm with its own genetic complement. From this single cell, about $\frac{1}{175}$ of an inch in diameter, a baby weighing several pounds and composed of trillions of cells will be delivered about 266 days later.

For convenience, we will divide the 266 days, or nine months, into three periods of three months each. We can consider these *trimesters* separately, since each is characterized by different sorts of events.

The First Trimester
In the first trimester the embryo begins the delicate structural differentiations that will lead to its final form. It is therefore particularly susceptible during this period to any number of factors that might influence its development. In fact the embryo often fails to survive this stage.

The first cell divisions result in cells that all look about alike and have roughly the same potentials. In other words, at this stage the cells are, theoretically anyway, interchangeable. Seventy-two hours after fertilization, the embryo will consist of sixteen such cells. (So, how many divisions will have taken place?) Each cell will divide before it reaches the size of the cell that has produced it; hence the cells will become progressively smaller with each division. By the end of the first month the embryo will have reached a length of only $\frac{1}{8}$ inch, but it will consist of millions of cells.

The First Month
The first cleavages actually take place in the oviduct as the zygote moves toward the uterus, pushed along by the beating cilia that line the oviduct and by contractions in the oviduct walls. When the zygote reaches the uterus, it does not implant for about three days, although the uterine wall is already swollen and ready to receive it. In fact, by the time it implants it has

already reached the hollow-ball blastula stage. One side of the blastula is thin-walled, and this portion, called the *trophoblast,* will give rise to the membranes that will come to surround the embryo. The thicker-walled side, called the *inner cell mass,* will become the embryo proper.

The trophoblast secretes a *proteolytic* (digestive) enzyme that breaks down some of the blood vessels in the uterine wall beneath it. The embryo thus comes to lie in a pool of the mother's blood. The bleeding caused at this time by implantation is sometimes mistaken for menstruation. After about a week the embryo will have eaten its way deep into the uterine wall, where it will lie steeped in a bloody glycogen-laden fluid—a fluid that will sustain it during these early days. In about twelve days the injured uterine wall will have repaired itself, and the embryo, buried deep within the flesh of the uterine wall, will silently begin to change.

After implantation the trophoblast extends fingerlike projections called *chorionic villi* into the uterine wall. At the same time the mother's tissues undergo certain changes in the areas around the villi. Thus chorionic tissues of the embryo, together with these maternal tissues, form the *placenta,* which serves as the organ of transport between the embryo and mother. Through the placenta will pass food and oxygen from the mother and waste and carbon dioxide from the fetus.

At the end of a month the embryo is still smaller than a pea (Figure 8.8), but momentous changes have occurred within it. The anterior end of the spinal cord has developed bulges and lobes as it begins to form the brain. The mesoderm has divided into thirty-two pairs of dense blocks of somites. These will later give rise to much of the skin and the voluntary muscles. Three pairs of arches have also appeared in the neck region, through which arterial blood is pumped. In the human embryo these will later disappear, but in fish they become gills. The opening that will be the mouth has broken through, although no food will enter here for some time yet. Even large visionless eyes have begun to appear.

Also during this first month roaming cells in the yolk sac and chorion change in appearance and become blood cells. These, by day 24, will be pumped by a tubular incipient heart through unfinished blood vessels and will begin picking up oxygen that has diffused across the placenta from the mother's blood. Near the end of the first month primitive kidneys form, but these will function in removing nitrogenous waste from the blood for only a few days. They will be replaced by a more specialized kidney later. The first cells of the endocrine glands are now apparent. And the tiny limb buds have appeared that will eventually form the arms and legs.

The Second Month

In the second month the features of the embryo become more recognizable. Bone begins to form throughout the body, primarily in the jaw and shoulder areas. The head and brain are developing at a much faster rate than the rest of the body, so that at this point the ears appear and open, lidless eyes stare blankly into the amniotic fluid. The circulatory system is

8.8 The development of the human embryo showing relative sizes at different ages. Note that at 14 days very few features are clearly distinguishable. However, its body axis is established and the major organs have begun to form, unlikely as it seems. The embryo at the third week is strangely vulnerable to a variety of dangers from drugs to radiation. You probably can't make out the developing eyes or the tubelike heart at the fourth week. At 7 and one-half weeks and almost an inch in length, the embryo still weighs less than an aspirin tablet and has a short tail. At the end of eight weeks it will be called a fetus and by nine weeks it will take on the general appearance of a human. The head will be oversized through the embryonic period and on into childhood. By eleven weeks fingers and toes have formed as well as such refinements as the ridges of the ear. By fifteen weeks the fetus is moving frequently and has taken on the facial features of its species.

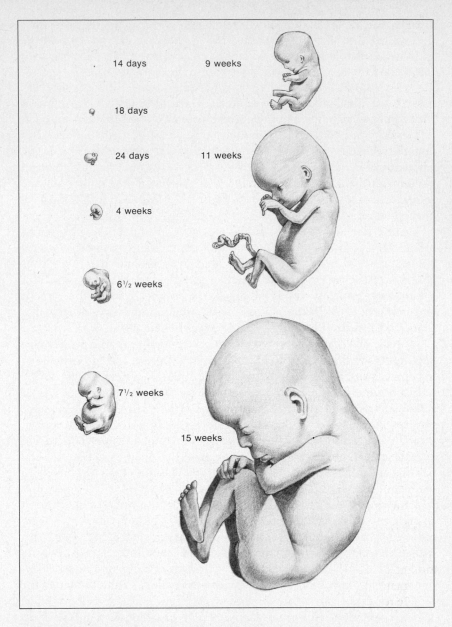

14 days

18 days

24 days

4 weeks

6½ weeks

7½ weeks

9 weeks

11 weeks

15 weeks

developing and blood is pumped through the umbilical cord out to the chorion, where it receives life-sustaining nutrients and deposits the poisons it has removed from the developing embryo. The nitrogenous wastes and carbon dioxide filter into the mother's bloodstream, where they will be circulated to her own kidneys and lungs for removal. At about day 46 the primordial reproductive organs begin to form, either as testes or ovaries, and it is now, for the first time, that the sex of the embryo becomes apparent. Near the end of the second month fingers and toes begin to appear on the flattened paddles which have formed from the limb buds. By this time the embryo is about two inches long and is more or less human in appearance; it is now called a *fetus.*

The Third Month
Growth and differentiation continue during the third month, but now the fetus begins to move. It breathes the amniotic fluid in and out of bulblike lungs and swallowing motions become distinct. At this point individual differences can be distinguished in the behavior of fetuses. The clearest differences are in their facial expressions. Some frown a lot; others smile or grimace. It would be interesting to correlate this early behavior with the personality traits that develop after birth.

At the end of the first trimester there are pronounced changes in the reproductive and excretory organs of the embryo, and its urine begins to appear in the amniotic fluid—to be filtered into the mother's blood where it is removed by her kidneys. Bone continues to form throughout the body. Each bone arises independently, usually in an area that will be near the middle of its mass; it then grows outward to meet other bones at what will be the joints. Although all the organ systems have been formed by this time, if the fetus is removed from its mother it will die.

The Second Trimester
In the second trimester the fetus grows rapidly, and by the end of the sixth month it may be about a foot long, although it will weigh only about a pound and a half. Whereas the predominant growth of the fetus during the first trimester was in the head and brain areas, during the second trimester the body grows at a much faster relative rate than the brain and begins to catch up in size with the head.

The Fourth Month
In the fourth month the back becomes less curved, and the abdominal muscles grow to enclose the viscera. The muscles become much stronger and the fetus is able to hold up its head. Fingerprints are beginning to appear as an adaptation for government employment. If the fetus is stimulated during this period, it will react at first with very general bodily movements, but later in the fourth month the fetus will respond with specific reflexes.

During the first trimester blood cells were formed mainly in the liver and spleen. In the second trimester the bone marrow begins to form blood cells. The fetal heart is now pumping strongly, and the heartbeats can be heard with a stethoscope.

The Fifth Month
The fetus is by this time behaving more vigorously. It is able to move freely within its sea of amniotic fluid and the delighted mother can feel it kicking and thrashing about. Interestingly, the fetus must sleep now, so there are

8.9 A human fetus not long before it will find itself in a different kind of world. It is now suspended in its protective amniotic fluid. The amnion and chorion are embryonic (not maternal) tissues and will be expelled as afterbirth. Oxygen and nutrients diffuse from the uterine lining into the chorion of the embryo. Nitrogenous waste materials and carbon dioxide diffuse from the fetus's blood into the mother's system from which it is expelled. The maternal and embryonic life-support tissues together comprise the placenta. Substances move mainly between the placenta and the fetus via blood vessels that converge into the umbilicus, the lifeline of the fetus.

periods when it is inactive. It is capable of reacting to more types of stimuli as time passes. For example, by the fifth month the eyes are sensitive to light, although there is still no sensitivity to sound. Other organs seem to be complete, but remain nonfunctional. For example, the lungs are developed, but they cannot exchange oxygen. The digestive organs are present, but they cannot digest food. Even the skin is not prepared to cope with the temperature changes in the outside world. In fact, at the end of the fifth month the skin is covered by a protective cheesy paste consisting of wax and sweatlike secretions mixed with loosened skin cells (*vernix caseosa*). The fetus is still incapable in nearly all instances of surviving alone.

The skeleton has been developing rapidly during the second trimester, with some bones arising anew from undifferentiated embryonic cells, and others forming through the gradual replacement of cartilage cells by bone cells. Now the mother must supply large amounts of calcium and other bone constituents to provide building materials for the fetal skeleton.

The Sixth Month

By the sixth month the fetus is kicking and turning so constantly that the mother often must time her own sleep periods to coincide with her baby's. The distracting effect has been described as similar to being continually tapped on the shoulder, but not exactly. The fetus moves with such vigor that its movements are not only felt from the inside, but can be seen clearly from the outside. To add to the mother's distraction, the fetus may even have periods of hiccups. By this stage it is so large and demanding that it places a tremendous drain on the mother's reserves.

At the end of the second trimester the fetus has the unmistakable appearance of a human baby (or a very old person, since its skin is loose and wrinkled at this stage). In the event of a premature birth around the end of this trimester, the fetus may be able to survive.

The Third Trimester

During the third trimester the fetus grows until it is no longer floating free in its amniotic pool. It now fills the abdominal area of the mother. The fetus is crowded so tightly into the greatly enlarged uterus that its movement is restricted. In these last three months the mother's abdomen becomes greatly distended and heavy, and her posture and gait may be noticeably altered in response to the shift in her center of gravity. The mass of tissue and amniotic fluid that accompanies the fetus ordinarily weighs almost twice as much as the fetus itself. Toward the end of this period, milk begins to form in the mother's mammary glands, which in the previous trimester may have undergone a sudden surge of growth.

At this time, the mother is at a great disadvantage in several ways in terms of her physical well-being. About 85 percent of the calcium she eats goes to the fetal skeleton, and about the same percentage of her iron intake goes to the fetal blood cells. Of the protein she eats, much of the nitrogen goes to the brain and other nerve tissues of the fetus.

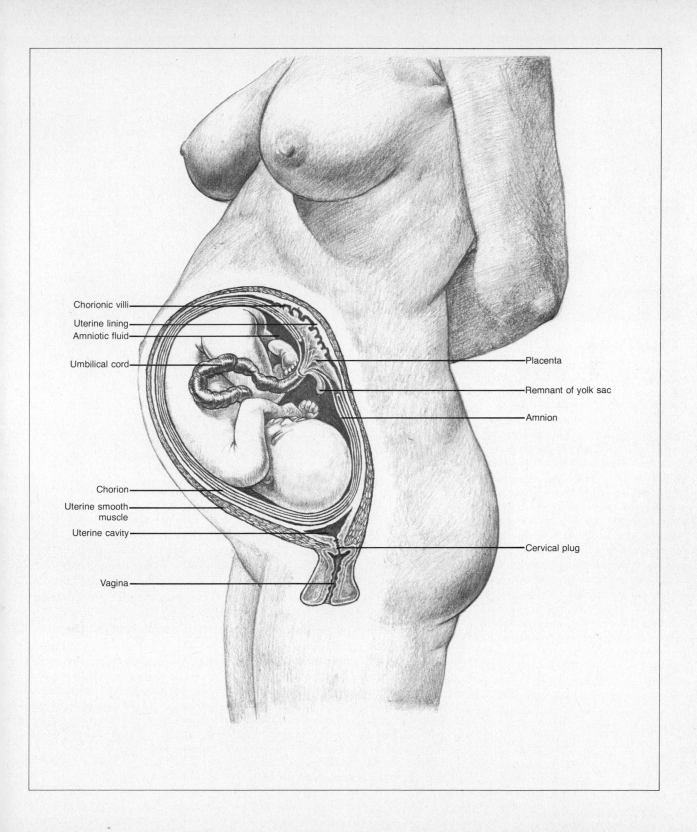

Chorionic villi

Uterine lining

Amniotic fluid

Umbilical cord

Placenta

Remnant of yolk sac

Amnion

Chorion

Uterine smooth
muscle

Uterine cavity

Cervical plug

Vagina

Human Development

In these pages we will trace the development of the human from the first pulsating moment of fertilization far up in the fallopian tube, through the grotesque and contorted stages of embryonic development, to the moment of birth. The consideration of the actual development of a human and its strong similarity to the growth of other animals should provide us with another way of looking at ourselves. Would these photographs have had an impact on Thoreau and his transcendental philosophy? Would introspective French literature have yielded before the argument that we are but another animal? Could we continue to look for answers that stem from a bulbous head with lidless eyes? Or does the knowledge of our development suggest that we should seek answers to the Great Questions elsewhere, from our physical surroundings or from some nonhuman source? In other words, does the information here suggest that we are unable to provide answers?

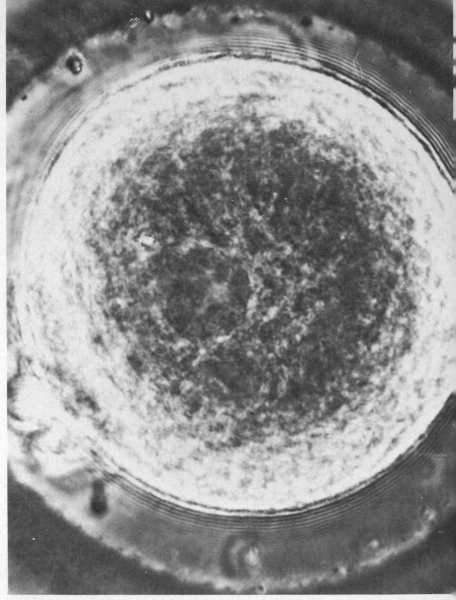

☐ This rather remarkable photograph captures the moment of fertilization of a human egg. Shown faintly in the darkened center are the chromosome-laden male and female nuclei approaching each other. At lower right are two polar bodies formed during meiosis. One might wonder at the fate of humans. How much of our behavior, and even our values, are determined at this moment? It seems strange, but we really don't know just what those chromosomes carry. We know the limits of height are more or less determined genetically. How about the limits of empathy? In any case, about thirty-six hours after this event, the zygote will undergo its first cleavage, all the while wending its way slowly toward the uterus, which will be its home.

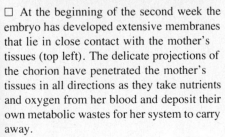

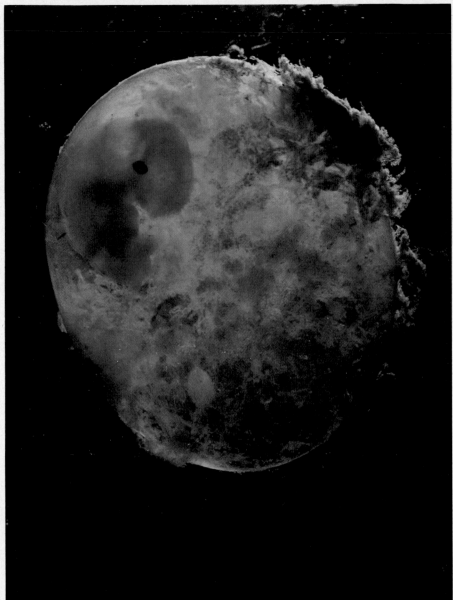

☐ At the beginning of the second week the embryo has developed extensive membranes that lie in close contact with the mother's tissues (top left). The delicate projections of the chorion have penetrated the mother's tissues in all directions as they take nutrients and oxygen from her blood and deposit their own metabolic wastes for her system to carry away.

☐ Well into the third week (bottom left) the chorionic membrane has continued to suffuse the mother's endometrium. The radiating chorion, shown here, carries blood vessels that lie closely intertwined with those of the mother, but the two do not join. The balloonlike structure is the yolk sac. The human embryo need carry only enough food to last through its first few weeks since food will later be derived from the mother's blood.

☐ The embryo at the fourth week (right). It lies protected in its amniotic sac. The dark eye is prominent and the enormous brain lies tucked against the embryonic heart. The embryo by this time has already developed primordial cells that will form its own gametes. By now the embryo is about five millimeters (0.2 inch) from head to rump and its tubular heart has begun its first timorous beats.

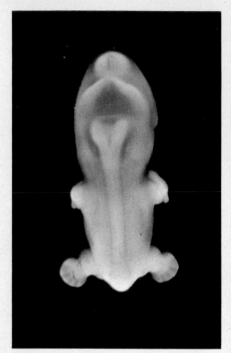

□ The human embryo at forty-two days (above), about half an inch (or sixteen millimeters) long, with the surrounding membranes removed. This is a dorsal view showing the enormous head (which helps direct the growth of the rest of the body) and the spinal cord extending to the rump. Notice the paddlelike appendages. Fingers and toes are already apparent as the tissue between them dies as a result of a mysterious chromosomal timing mechanism.

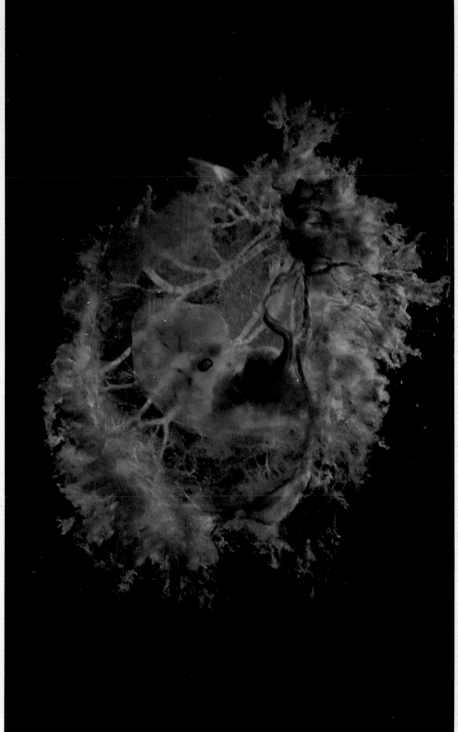

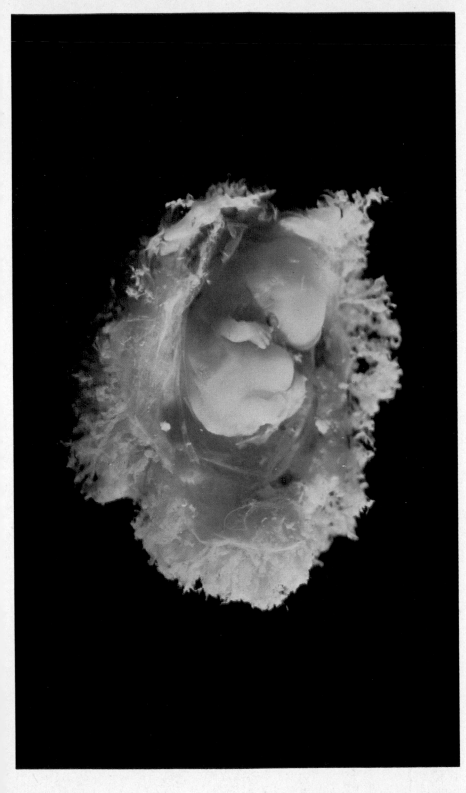

☐ At about six weeks (opposite page, right) the extensive vascular system is clearly visible leading from the projections of the embryonic chorion to the embryo itself. The organ below the eye is the now-looped heart. The embryo is still so fluid-filled that its tissues are virtually transparent.

☐ At six weeks (left), with the amnion removed, the fingers are apparent and the bulbous brain still dominates the embryo. Notice the "tail" tucked under the abdomen, apparently a vestige of our ancestry, but one which is destined to disappear. The tiny pit above the arm will become the ear.

At about this time the embryo is extremely vulnerable to all sorts of chemical agents. An increasing number of drugs have been found to produce congenital abnormalities, for example, in the growth of limbs. Even X-rays, at this time, may endanger the development of the embryo. Certain diseases are also particularly dangerous. For example, if a mother contracts German measles during the fourth to twelfth week of her pregnancy the result may be deformation in her child's eyes, ears, heart, and brain.

☐ At about seven weeks (right) the embryo, afloat in its amniotic fluid, is clearly anchored to its placenta by the twisted umbilicus through which great blood vessels pass. The abdomen is swollen due to the rapid growth of the liver, the main blood-forming organ at this time.

☐ The eight-week embryo (opposite page, left) seen in front view. The organs are now more or less complete after a startlingly rapid period of growth and development. From here on, development will primarily consist of refinements of existing structures. The skeletal system is among the last to form and bones are now evident in the arms and legs.

☐ At nine weeks (opposite page, right) lids have begun to grow down over the eyes and the outer ear begins to form. Because the plates of the skull have not fused, the head is rather flexible. During this third month the fetus may begin to move, wave its arms and legs, and may even suck its thumb. The fetus is now beginning to fill its amniotic space and will soon be forced to assume the typical fetal posture.

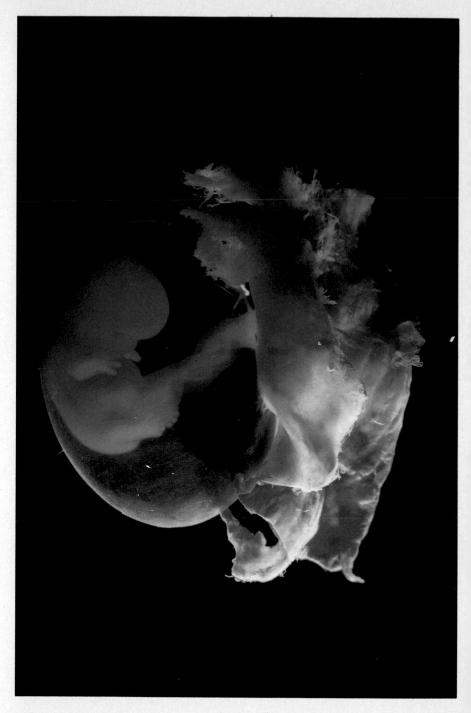

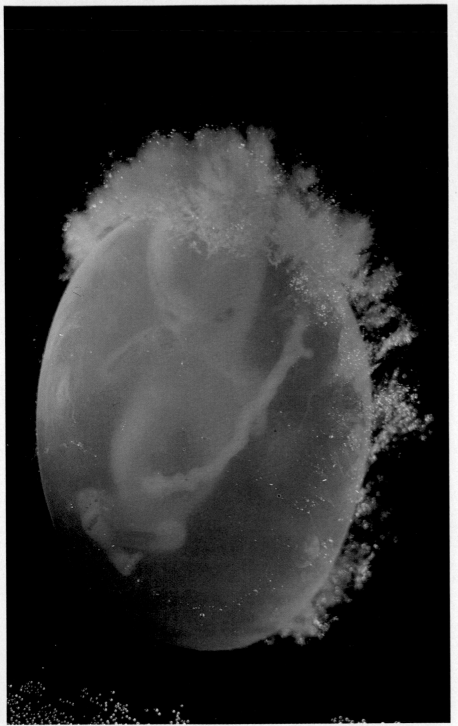

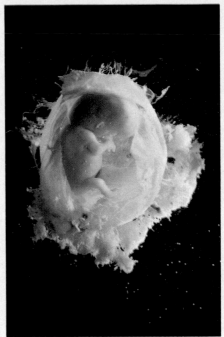

☐ At ten weeks (right) the skeleton is well along in its development. The long bones begin developing independently, growing from areas near their ends. They will join, forming joints later. In fact, the joints may not be firmly abutted by the time the baby is born. The head is still disproportionately large and will remain so, to a decreasing degree, through childhood. Notice the coiled umbilicus lying near the cuboid ankle bones showing as dark spots. The wrist bones have not yet begun to form and the jaw structure is weak indeed.

☐ At fourteen weeks (opposite page, left) the fetus is obviously human. Ribs and blood vessels are visible through the translucent skin. The vigorous movements of the fetus can now be felt by the mother. The delicate skin is actually covered with a cheesy protective coating. Refinements such as fingerprints and fingernails are now present although the fetus is only fist-sized. The uterus has grown along with the fetus but the available space is taken up by the newcomer and its attendant structures.

☐ By the end of five months (opposite page, right) the fetus is covered with fine, downy hair and its head may have started to grow its own crop. It has already started the lifetime process of discarding old cells and replacing them with new ones. The heart is beating now at a rate of 120 to 160 times per minute.

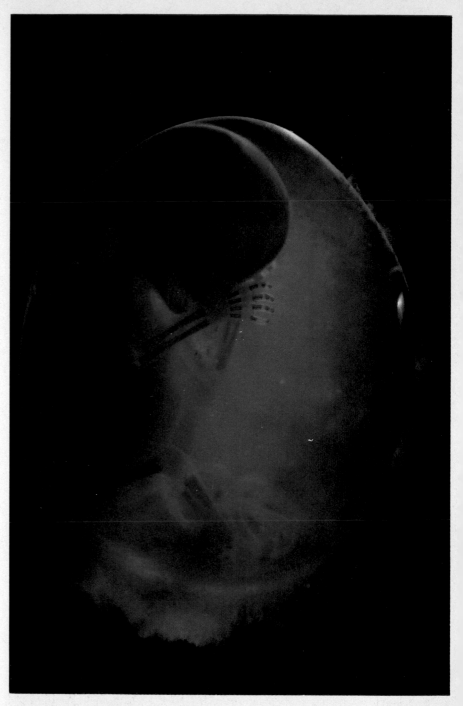

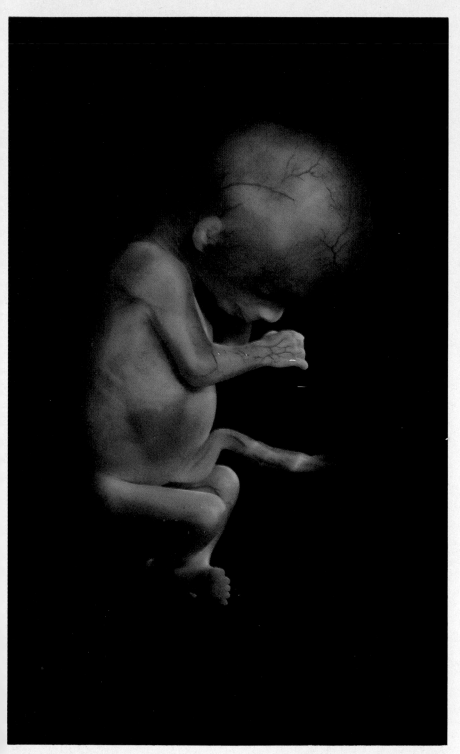
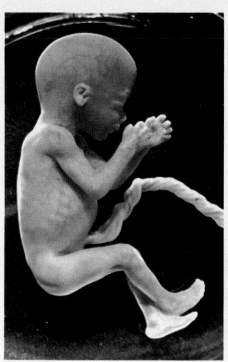

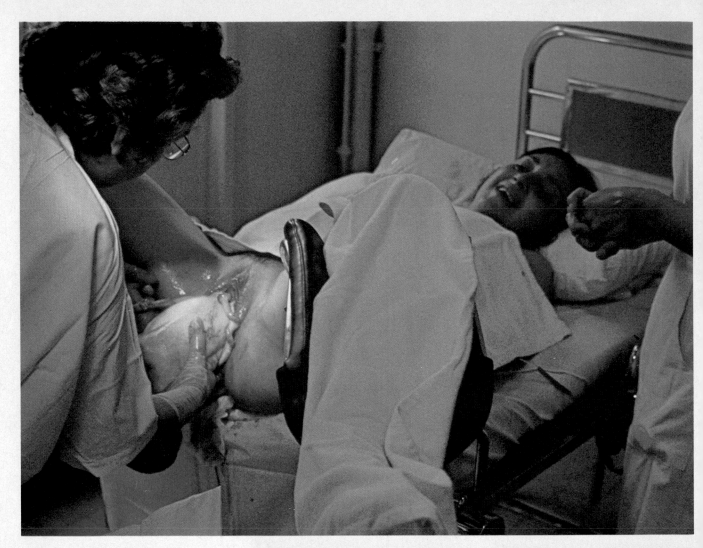

□ Delivery, a much too clinical term for what the mother is experiencing both physically and emotionally. It is a culmination of long months of changes. The infant is still not completely developed, but it is able to survive apart from the mother. Although it is virtually helpless, it will be protected by its parents and will be cared for in the long years ahead as it slowly gains independence.

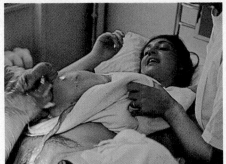

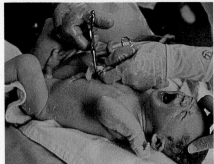

Some interesting questions arise here. If a woman is unable to afford expensive protein-rich foods during the third trimester, what is the probability of a lowered I.Q. in her offspring? On the average the poorer people in this country show lower I.Q. scores. Are they poor because their I.Q.'s are low, or are I.Q.'s low because they are poor? Is there a self-perpetuating nature about either of these alternatives?

In the third trimester, the fetus is large. It requires increasingly greater amounts of food, and each day it produces more poisonous wastes for the mother's body to carry away. Her heart must work harder to provide food and oxygen for two bodies. She must breathe, now, for two individuals. Her blood pressure and heart rate rise. The fetus and the tissues maintaining it form a large mass that crowds the internal organs of the mother. In fact, the crowding of the fetus against the mother's diaphragm may make breathing difficult for her in these months. Several weeks before delivery, however, the fetus will change its position, dropping lower in the pelvis (called *"lightening"*) and thus relieve the pressure against the mother's lungs.

There are important changes occurring in the fetus in these last three months, and some of these are not very well understood. The effects of these changes, however, are reflected in the survival rate of babies delivered by Caesarian section (an incision through the mother's side). In the seventh month, only 10 percent survive; in the eighth month, 70 percent; and in the ninth, 95 percent survive.

Interestingly, there is another change in the relationship of the fetus and mother at this time. Whereas measles and certain other infectious diseases would have affected the embryo during the first trimester of pregnancy, at this stage the mother's antibodies confer an immunity to the fetus, a protection that may last through the first few weeks of infancy.

At some point about 255 to 265 days from the time of conception the life-sustaining placenta begins to break down. Certain parts shrink, the tissue structure begins changing, and the capillaries begin to disintegrate. The result is a less hospitable environment for the fetus, and premature births at this time are not unusual. At about this time the fetus slows its growth, and drops into position with its head toward the bottom of the uterus. Meanwhile, the internal organs undergo the final changes that will enable the newborn to survive in an entirely different kind of world. Its home has been warm, rather constant in its qualities, protected, and confining. It is not likely to encounter anything quite so secure again.

Birth

The signal that there will soon be a new member of the earth's most dominant species is the onset of *labor,* a series of uterine contractions that usually begin at about half-hour intervals and gradually increase in frequency. Meanwhile, the sphincter muscle around the cervix dilates, and as the periodic contractions become stronger, the baby's head pushes through the extended cervical canal to the opening of the vagina. The infant is finally about to emerge into its new environment, one that, in time, may give it the chance to propel its own genes into the gene pool of the species.

Once the baby's head emerges, the pattern of uterine contractions changes. The contractions become milder and more frequent. After the head gradually emerges through the vaginal opening, the smaller shoulders and the body appear. Then with a rush the baby slips into a new world. As soon as the baby has emerged, the umbilicus by which it is attached to the placenta is tied off and cut. The placenta is expelled by further contractions as the *afterbirth*. The mother recovers surprisingly rapidly. In other species, which deliver their young unaided, the mother immediately chews through the umbilicus and eats the afterbirth so that it will not advertise to predators the presence of a helpless newborn. Fortunately, the behavior never became popular in our own species.

The cutting of the umbilicus stops the only source of oxygen the infant has known. There is a resulting rapid buildup of carbon dioxide in the blood, which affects a breathing center in the brain. An impulse is fired to the diaphragm, and the baby gasps its first breath. Its exhaling cry signals that it is breathing on its own.

In American hospitals the newborn is then given the first series of the many tests it will encounter during its lifetime. This one is called the *Apgar test series,* in which muscle tone, breathing, reflexes, and heart rate are evaluated. The obstetrician then checks for skin lesions and evidence of hernias. If the infant is a boy, it is checked to see whether the testes have properly descended into the scrotum. A footprint is then recorded as a means of identification, since the new individual, despite the protestations of proud parents, does not yet have many other distinctive features that would be apparent to the casual observer. And there have been more than a few cases of accidental baby-switching.

Miscarriage

Of course, not all pregnancies are terminated by delivery of a healthy offspring. The environment in which the fetus develops is optimal in terms of security and comfort; however, the route between conception and birth is fraught with risk. Part of the risk stems from the very complexity of living things. There is a phenomenal array of physical and chemical interactions going on within the body of a developing organism. If, at any point along the line, a component fails in its function or its timing, the embryo may be lost. If this happens very early, the embryo may be broken down—digested—and reabsorbed by the mother's body, perhaps without her ever knowing she was pregnant. If the embryo or fetus should fail at some period later in its development, the body of the mother often mysteriously is able to recognize that something has gone "wrong," and the fetus is aborted as a "miscarriage." Actually, about half of all miscarriages involve fetuses that are structurally or physically abnormal.

Miscarriages may be caused by a variety of factors. For example, an embryo may not implant correctly if it invades the uterine wall too soon in its development. Alcoholism or heavy smoking by the mother may cause premature birth of otherwise healthy babies. A weak cervix may not be able to support a fetus and thus the offspring may be lost. In fact, so many things *can* go wrong that the wonder is that healthy babies are ever delivered—or so it seems when one reviews the long list of hazards which threaten embryos.

WHY HAVE CHILDREN?

It may be useful at this point to consider why people have babies at all, since childbearing certainly weakens and occasionally kills the mothers, and since there is a risk of having abnormal offspring who will never achieve independence. As you are undoubtedly aware from your own experience or from that of friends, people have children for a number of reasons—reasons of various levels of merit. A couple may feel that it might help to save an otherwise shaky or unwise marriage, or a man may feel it will focus a clinging wife's attention on something else. In some cases it may be a response to questions such as, "When are you going to start your family?" (this, often, to a blissfully happy and devoted couple who thought they had), or "When are you going to give us some grandchildren?" or even, "Why don't you have any children?" In our society it is generally more permissible to ask a couple this last question than to ask why they have so many children. In early America, and in many agricultural societies today, children were a "free" labor force as well as a form of social security. Even today there are rumors of Americans having children for tax or welfare advantages, but this is harder to understand considering the expense of children. Many people have children probably because they simply had never considered any alternative. It seems that, often, cultural traditions persist long past the circumstances which gave rise to them. We are taught from Dick and Jane books onward that "life" involves growing up within a family, enduring adolescence, getting married, having children, and working hard to be a productive and accumulating member of the community all the while. It is unquestionably hard to break such cultural patterns, difficult both psychologically and sociologically, since aberrant behavior often goes unrewarded, or even punished to some degree, in our society. Benevolent smiles and well-wishes are usually reserved for those who go along.

There are also other reasons for having children, as you are well aware. Many people seem to have an innate affection for children, simply because they are children. Others seem to reserve their love for those children who are members of their own race or socioeconomic group—although such people may think of any child as "cute" (even those recovering from bombings, which these people have helped to finance).

It is often said that children contribute to feelings of fulfillment for the parents. The "fulfilling" nature of children is an interesting one. Let's consider it briefly from a biological point of view. First, we must remember that we are all the result of millions of generations of successful reproducers. If our forebears had not reproduced we would not be here. In fact, we are likely to be the offspring of the *most* hard-driving reproducers of past generations. Those individuals of any generation who reproduced with less than maximum effort left fewer genes in the following generation; those more devoted to reproduction left a greater percentage. Hence, each generation would tend to consist increasingly of the offspring of the most successful reproducers, as the proportion of the genes of the less successful ones progressively diminished in the population through time.

Now, if the motivation to reproduce (or to do the things that result in reproduction) is an innate response to the influence of genes, we, as the offspring of the most successful reproducers, can be expected to feel rather strongly about having children ourselves and caring for them intensely. Such a feeling may take a variety of forms, from a peculiar longing to see one's "own" children and to interact with them, to a general positive response or protective feeling toward any baby or child.

Since we can expect variation between individuals for any genetically determined character, we can expect to find people among us who are less fascinated by babies, or who have less longing to see their own. What will happen to the genes of these people in the next generation? (Hint: if your parents didn't have any children you probably won't either.) Instead, the individuals in the *next* generation will be composed primarily of the offspring of what sort of people in *this* generation?

It might be mentioned that reproductive ability doesn't say much about other qualities an individual may possess. There is no evidence that Leonardo da Vinci left his peculiar genes or combinations of genes to subsequent generations. Jonathan Livingston Seagull may have been a remarkable flier and philosopher, but the next gull population would certainly consist of the offspring of humdrum, 9-to-5, hardworking, child-rearing gulls. Jonathan's genes would have died with him.

Of course there are innumerable social influences that affect our attitudes toward children, or anything else for that matter, and there is no intent to say here that any such complex creature as man is blindly obeying the callings of his genes. As we learn more about the concepts of modern biology, however, it is sometimes fun, and possibly important, to speculate on the influences of our ancestry. Which traits are influenced genetically and which are simply acquired? Have we made the conscious decision to love these individuals who are so helpless, demanding, inexperienced, inarticulate, small, dirty, and broke? Are we *taught,* somehow, to love them? Or are we responding, as the descendants of the most successful reproducers, to the inner, compelling call of our heritage?

The question is not a simple one as you can see, although, if you broach the subject with many people you will get a string of responses off the tops of their heads. Part of the problem is that we tend to consider social customs as being acquired culturally or artificially through conscious decisions made in some sort of a historical vacuum, especially if these customs appear in writing in the form of codes, laws, or religious dogma. Perhaps, however, the "vacuum" is a myth. Any genetically influenced behavior will be the result of untold generations of development. As we become aware of the presence of that behavior in our society we might begin to consider it as normal or proper since it obviously works, and, in time, it might even come to be written into our traditions, religions, or laws. Then, at some point we might note that there is some enforcement of the behavior, that people who, perhaps, do not wish to adopt the behavior are constrained, in some sense, to do so. As the environment changes, there may come a time when the old codes, which have worked so well and which have come to be embraced in our traditions, are no longer necessary. Then it is an easy step for those interested in the roots of our customs to believe the behavior exists as a dogmatic cultural phenomenon, and to change it we need only to reeducate society, change our institutions, and/or rewrite our laws. We can see the problems inherent in this position in light of biological principle. Possibly our attitudes toward children, mates, society, war, competition, or territory could be more readily modified to fit our new kind of world if we could begin to realize that we are the products of evolution. Love of children, jealousy of mate, defense of territory, and aggression against competitors brought our species to whatever success we enjoy today. If we have changed our world so that these qualities are unimportant (or threaten us), we might deal more successfully with these relic tendencies by recognizing their origins. Such recognition does not imply that they cannot be overcome. After all, man is a brainy and determined creature. We have been able to overcome gravity, but we did not do it by denying its existence. Instead, we recognized it for what it was and dealt with it accordingly.

Homeostasis: Controlling the Internal Environment

About 200 years ago Dr. Charles Blagden, then secretary of the Royal Society of London, proved that he was one of the most persuasive people on earth. He talked some friends into joining him, a small dog, and a steak in a room in which the temperature had been raised to 126°C (260°F). Then he managed to persuade his friends to stay in there for 45 minutes. (The dog and the steak had little choice.) At the end of this time the men and the dog emerged unharmed, but the steak was cooked! In addition to demonstrating his polemic powers, Blagden also showed that the bodies of animals are able to compensate for extreme physiological conditions; that some living things control their internal temperatures in the face of very extreme external conditions. It has since been found that many animals not only closely regulate their internal temperatures, but a host of other physical and chemical states as well. This regulation makes sense since the delicate processes of life would be rather easily disrupted by wild fluctuations in the properties of cytoplasm.

HOMEOSTASIS

Homeostasis refers to the tendency of living things to maintain a constant internal environment (from the Greek *homoios,* same; and *stasis,* standing). Obviously no living thing strictly maintains a constant internal environment. Internal environments change, as we noted earlier with reference to the turbulent and changing activities within cells, but some constancy is essential. The external environment is forever tending to disrupt systems, to reduce living things to their most elemental components. And living things are continually struggling to keep their molecules organized. It is important that they do stay organized in some way and to some degree if life is to be maintained, and their organization is the result of many interactions that are only possible under certain conditions—conditions maintained through homeostasis. So homeostasis is the tendency to keep any changes within certain limits. But how does it work? What is the underlying principle?

Often, the steady state is maintained by *feedback* mechanisms. Feedback occurs when the product of any action influences that action. For example, when an increase in a process results in a slowdown of that process we see *negative feedback.* In such a system, a decrease in the action triggers a mechanism that increases the action. For example, the governor

9.1 A typical biological feedback system. Control sensors in the body detect when alterations in the internal milieu are needed and send messages to the proper effector, perhaps a gland. The effector acts to alter the internal environment, which, when changed, signals the control sensor, and that particular message to the effector is cut off until the sensor detects a need for it again.

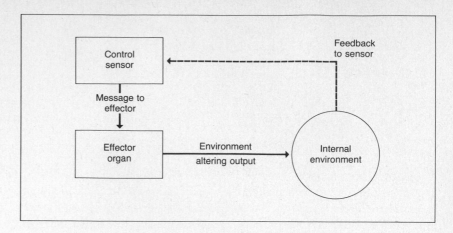

on an automobile engine (a device that does not permit the car to exceed a certain speed) works on the negative feedback principle. As the motor runs faster, valves are closed, which causes the car to slow down. As it slows, the valves are reopened. Thus the speed is kept within certain limits. *Positive feedback* works on the opposite principle: the result of the action increases that action. If accelerating a motor tended to open the carburetor, the engine would run ever-faster—until it exploded. Which feedback system, then, would you guess is more important in homeostasis?

In the human body, homeostatic mechanisms can change from one type of feedback system to the other—often with disastrous results. For example, if a man's temperature begins to rise from the normal 37°C, the rise will trigger corrective devices such as sweating, and the opening of peripheral blood vessels (producing a heat-dissipating "flush"). But at some point (usually at about 42°C), the negative feedback system breaks down and a positive feedback begins. The high temperature begins to increase metabolic activity, which . . . kills the man. Positive feedback, then, is the basis of the famed "vicious circle."

To better understand the principles of feedback control, see Figure 9.2. Note that in these simple mechanical structures, a storage tank can simplify the homeostatic processes. Look at these figures carefully to gain a better idea of the principles of homeostasis. The body, as you may have guessed, is loaded with storage tanks. For example, you don't have to constantly nibble and fast to keep your blood sugars at the proper level. When blood sugars are low, the liver simply breaks down some of its stores of glycogen and releases glucose into the blood. As blood sugars rise, the liver releases less.

TEMPERATURE REGULATION AS AN EXAMPLE OF HOMEOSTASIS

Living things must be able to adapt to temperature changes. Not only is temperature important in the immediate sense, as an animal tries to avoid becoming too warm or too cold, but evolutionarily it obviously has had great influence in the development of living systems. For example, an increase in temperature of only a few degrees causes great leaps in the rate

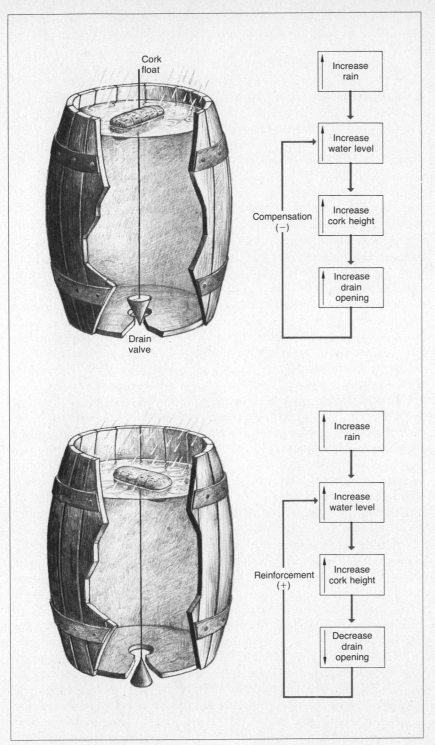

9.2 Two types of feedback systems. In the upper figure a rather constant water level is maintained as long as there is enough rain. As the amount of water increases, more is allowed to run out; as it decreases, the drain is plugged. This, then, is a typical negative feedback system. In the lower figure, as the water level rises, less is allowed to escape so the water rises out of control. As it falls, even more is released until the barrel is empty. This is a positive feedback system. Which do you suppose more closely approximates biological feedback systems?

9.3 The pupfish is a denizen of hot, briny desert springs. In natural habitats they are able to live under a temperature range of 38°F (3.3°C) to 108°F (42.2°C). Somehow, it is able to survive under such stressful conditions by a highly specialized and mysterious physiological system.

of chemical reactions. The rule is, the rate of chemical reactions doubles for every 10°C increase in temperature. At least this is true for nonliving material. Living cytoplasm refuses to respond to many such laws and so temperature may have less effect, as was demonstrated by Dr. Blagden. But the cell has had to develop mechanisms to circumvent such laws, possibly at some evolutionary or metabolic expense. (How could such expenses be incurred?)

Even living things, however, have their limits. At about −1° to −2°C, the water in cells freezes. The remaining cell constituents are left so concentrated that they are unable to function properly and life may cease (at least until the cells are warmed up in some cases). The upper limit is apparently set by the temperature at which the hydrogen bonds that hold proteins in their tertiary structures begin to break, thus unwinding (or denaturing) the protein. So animals must come to live in places that are not much colder than freezing or not warmer than about 40°C. (But certain algae are able to live in hot springs whose water may range up to 80°C.)

Cold-Blooded Animals

A cold-blooded animal is one that is not a bird or mammal. That means most animals. Cold-blooded animals (or poikilotherms, from the Greek *poikilo,* changeable) are those that do not physiologically regulate their body temperatures to any great extent. Of course, metabolic heat is produced, but there are no efficient means of conserving it. If a poikilotherm wishes to warm up, it must move to a warm place.

Saltwater fish have no great problems because the oceans are well-buffered, rarely changing over a few degrees in temperature. However, freshwater fish living in the shallows are more at the mercy of the elements. As they move to places they deem to be the right temperature for their physiological states, they may find themselves in danger of becoming landlocked and then overheated or desiccated if they are unable to escape back to deeper water. Freshwater fish that live in deeper water suffer no such threat, but because it's so cold down there, the pace of their lives must remain slow.

Among land dwellers it's a different story. Amphibians, for example, can cross inhospitable terrain to reach areas of more agreeable temperature, such as a pond, and many of them have developed the habit of *estivation,* or summer hibernation, to escape heat and drought. Toads in certain arid parts of Australia, in fact, have been known to remain buried for as long as two years while waiting for a decent rain. The toads have low metabolic rates and so do not require much food and oxygen at such times.

Reptiles cope a bit better with the extremes of land temperatures. They do not employ drastically different metabolic techniques from the amphibians, however. Instead they rely more strongly on behavioral methods. Lizards, even desert lizards, tend to let their temperatures drop with the cool of night, and so they are sluggish in the morning, their metabolic rates having slowed accordingly. If they are going to catch any insects, however, they've got to warm up, and they quickly do so by absorbing heat from the sun. They can absorb the heat of the sun even when the air temperature is near freezing. When they become too warm, they seek shade or turn toward the sun to reduce the area exposed to it. On the hottest days they may disappear underground until the cool of the afternoon. Terrestrial reptiles, then, are strongly dependent upon behavioral techniques of thermoregulation (Figure 9.4).

Warm-Blooded Animals

These are the birds and mammals. They are usually called *homeotherms (homeo,* same*)* instead of warm-blooded animals, because cold-blooded animals sometimes have warm blood. Both birds and mammals are descended from the reptiles, mammals appearing at about the time the dinosaurs made their entrance, about 150 million years ago (the birds evolving a little later). While the great dinosaurs were thundering around the earth terrorizing plants and animals alike, the mammals were existing as tiny mouselike forms that probably only terrorized insects (Figure 9.5). Then, for some unknown reason, the dinosaurs suddenly disappeared. It was once believed the great cold-blooded beasts simply couldn't cope with the earth's temperature changes. There is now strong evidence, however, that suggests this is not true; that the dinosaurs were, in fact, homeothermic. If this is so, their disappearance is yet a greater mystery.

After the disappearance of the dinosaurs, the numbers of birds and mammals burgeoned. They were well-insulated creatures, fur and feathers buffering them from the environment. And they developed other specializations as well, such as the more efficient four-chambered heart that could pump the fuel and oxygen necessary to stoke the cells' metabolic furnaces.

The modern birds and mammals, of course, come in a wide array of shapes and forms, but they have all developed incredibly complex means of regulating their temperatures. Heat is initially provided in the energy-producing metabolic reactions of the cells. Since heat travels from the inside out (the opposite in sun-basking lizards), bodies are warmer toward the inside. If you touch something that is 37°C (your body's temperature), it will feel warm because your skin is cooler than that. Although there is some variation from place to place in your body, there is very little variation from time to time. The temperature is kept under very tight control. If you measure someone's temperature with a rectal thermometer, you will not only get his undivided attention, but you will find that his temperature will not vary more than a few tenths of a degree from one time to the next, unless he gets sick. (It's lowest about 2:00 A.M. and highest about 2:00 P.M.)

The system is regulated by a thermostat, the hypothalamus (an ancient part of the brain). Certain receptor cells in this structure monitor the temperature of the brain's blood (not the skin's blood). If the blood is too warm the hypothalamus initiates a chain of events to bring the temperature down. Humans and most other large animals begin to sweat. Dogs pass air over their large tongues by panting, and cats lick themselves so that the evaporating liquid cools their bodies. In humans, peripheral blood vessels dilate to bring heat-bearing blood to the exchange area of the skin.

9.4 A desert-dwelling lizard. This animal's answer to heat is more apparent. Note the long appendages that not only dissipate heat quickly, but enable the lizard to move quickly to find cooler areas. Even desert lizards cannot stand the direct desert sun for long and so they hunt in the shade, being most active at dusk and dawn.

Then, as the blood's temperature goes down, the surface vessels contract to reduce heat loss across the skin's surface. The hypothalamus signals the pituitary to direct the thyroid glands, located in the throat area (Figure 11.1), to release a chemical that increases metabolic rate. If the temperature of the blood in the brain should continue to drop, the hypothalamus will direct the adrenal glands to produce adrenalin, thereby further increasing the metabolic rate. Finally, at about 20°C, a person will begin shivering, the effort causing the surface areas to work, producing more heat and using more energy. Have you noticed you eat more in the wintertime? Why do you suppose that is?

Occasionally the thermostat seems to become reset and *fever* results. The precise mechanism remains obscure, but infection seems to be a resetting device. When a fever is about to begin, the body temperature may still be normal, but the body behaves as if it is cold. Blood vessels in the skin contract ("You look pale; do you feel well?") and shivering may begin. Then the rate of metabolism may increase. All these things can act together to bring on higher temperature. Finally the temperature stabilizes at its new, higher setting. When the fever breaks, the opposite responses occur. The skin becomes flushed and sweating occurs until enough heat is lost to restore the body to normal temperature.

Just as your body's temperature is kept within narrow limits, so are other physiological factors. Actually, *every* activity involving growth and maintenance is controlled by complex regulatory mechanisms, but one of the clearest stories involves the excretion of water. You have probably heard that the body to which you are attached is composed mostly of water, and physiologists have glibly quoted rather precise numbers to tell you just what the percentage is. But how can they be so sure? It is true that they usually don't account for bodies that have just completed marathons, or those that have been drunk for three days, but still, they are confident because they can count on the delicate and intolerant mechanisms of the kidneys. Let's see how the kidneys work.

THE EXCRETORY SYSTEM

In addition to ridding themselves of solid wastes, animals have the problem of getting rid of metabolic wastes, or "cell garbage." In the amoeba and some other protozoans metabolic wastes simply diffuse across the cell membrane and out into the surrounding water. However, protozoa that live in fresh water have another problem as well. Since protoplasm has a higher concentration of particles in solution than pond water does, water constantly enters the body by osmosis, creating the danger that the body will swell and rupture. Some protozoa solve this problem with a *contractile vacuole,* a small vesicle that continually pumps water out from the cell interior. So in the case of freshwater protozoa the problems of metabolic waste and internal water regulation are solved by different mechanisms. In other animals this may not be the case, and the two problems may be closely interrelated.

In earthworms, for example, each segment has a pair of coiled tubes

9.5 The tiny primitive mammals scurrying about under the leaves at the lower right and the low-flying *Archaeopteryx* were unlikely candidates to inherit the earth, but they may have survived the ruling reptiles because the dinosaurs had a "cold-blooded" circulatory system that did not enable them to withstand the temperatures of the ice ages. Such surmising is based on the fact that most modern reptiles are not able to regulate their internal temperatures very effectively, unlike birds and mammals. Some researchers now disagree, arguing that the great reptiles were actually warm-blooded and that the explanation for their extinction must be sought elsewhere.

9.6 Nephridium of an earthworm. Each segment boasts a pair of nephridia. The tubelike nephridium passes through the body septum. The beating cilia set the fluid of the body in motion so that it flows past blood vessels lying tight against the nephridium that shares that segment. Metabolic wastes from the blood thus enter the tube and are excreted, while water and other essential products are absorbed from the body fluids back into the capillaries.

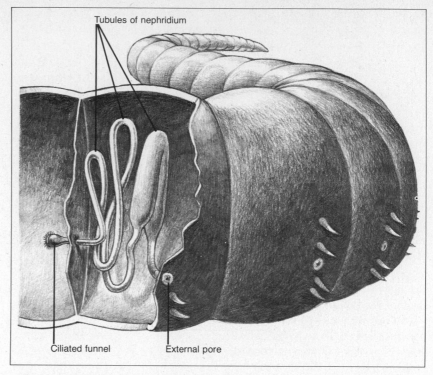

Tubules of nephridium

Ciliated funnel

External pore

called *nephridia,* which connect the body cavity to the outside (Figure 9.6). The end of the tube that opens into the body cavity is rimmed with beating cilia that circulate the fluids within the body cavity. Capillaries lie tightly coiled around the outside of the nephridia, and metabolic wastes from the blood can thus enter the tube that leads to the outside. Water, glucose, and other useful substances are reabsorbed from the body fluids back into the capillaries as the cilia-driven fluid passes out. The amount of water leaving the animal depends on how much water has moved into the body cavity. Worm urine, in case you've ever wondered, is very dilute and is excreted at a rate of about 60 percent of the worm's body weight each day.

Solutions to the Water Problem

The problems of ridding the body of metabolic wastes and maintaining proper internal water levels are solved in vertebrates in a number of ways. Freshwater fish, like freshwater protozoa, live in a *hypotonic* medium; that is, the concentration of particles in their body fluids is higher than that in the surrounding water. Hence water tends to move into their bodies by osmosis. One means by which they get rid of excess water is through a greater number of tiny clumps of blood vessels called *glomeruli* in their kidneys. We'll consider the structure of the glomerulus below, but in general this means that there is a larger surface through which fluids can leave the blood to be collected in the kidney. The result is that the metabolic waste in freshwater fish is highly dilute.

Box 9.1 Special Solutions to the Water Problem

Some insects are able to survive for years enclosed in jars of dry pepper. They do this by utilizing metabolic water and excreting almost dry uric acid.

A camel can tolerate dry conditions by adopting a number of tactics. At night it drops its body temperature several degrees so that bodily processes slow down. In the heat of the day it doesn't begin to sweat until its body temperature reaches about 105°F. In addition, the camel can lose twice as much of its body water (40 percent) without ill effect than can most other mammals. Peculiarly, the thick coat of a camel acts as insulation to keep the heat out. Also when the camel does drink, it may hold prodigious amounts of water.

Some seabirds and turtles have salt removal glands. The huge tears that may appear in the eyes of sea turtles have nothing to do with the turtle's realization that we are poisoning the oceans. They are merely the result of specialized salt removing glands. These glands, however, remove only sodium chloride from the blood.

The tiny kangaroo rat doesn't drink at all and it lives on dry plant material. It survives because it doesn't sweat and is active only in the cool of night. Its feces are dry and its urine is highly concentrated. Most of its water loss is through the lungs. Fats produce more metabolic water than other foods, so it prefers fatty foods. On the other hand, soybeans are high in protein and thus produce a lot of nitrogenous waste for which a lot of water is needed—so if a kangaroo rat is fed only soybeans it will die of thirst.

Saltwater animals, interestingly enough, have the same problem as desert animals: conserving water. Seawater is a *hypertonic medium,* that is, it has a higher concentration of solutes than the marine animal's body fluids; hence they tend to lose water by osmosis. As a result they have developed smaller glomeruli, so that less water filters out through the kidney. This mechanism may be augmented in some species by special glands that excrete the salt taken in with the seawater the animals drink. In other species the gills may function in eliminating nitrogenous waste as the blood passes through them. Most sharks and rays have a different strategy. They retain some urea, or waste, in their blood, and thus raise the osmotic pressure of their body fluid to more nearly that of seawater.

All terrestrial mammals, including man, also face the problem of how to flush nitrogenous wastes out of their bodies without losing too much water. A large part of human food is protein (for those who can still afford it), and protein is, paradoxically, at the source of man's excretory problems. In digestion, as we know, proteins are hydrolyzed into amino acids, which are transported to the cells where they will be utilized. Some of the amino acids are used as building blocks for proteins and other molecules, with no great change in their structure being required. Most amino acids, however, are broken down further by the liver for storage as an energy source.

Now, the first step in deriving energy from an amino acid is to get rid of the nitrogens, since no nitrogen enters into either the anaerobic or aerobic energy cycles. The nitrogens that have been stripped from the proteins usually end up with three hydrogens—and NH_3 is ammonia, a deadly poison. So, the problem involves how to dispose of the ammonia. It can't pass out through the lungs like carbon dioxide, so it must be excreted by the kidneys. However, ammonia is so poisonous that it must be highly diluted to be handled by the body, and thus its excretion would entail the loss of too much water. So carbon dioxide is added to the ammonia to form *urea,* which the body can handle in higher concentrations without adverse effects. The moderately dilute urea is then passed out of the body as *urine.*

Since freshwater fish have water to spare, they excrete dilute ammonia directly. However, some other organisms, including many egg-layers such as insects and birds, convert the ammonia to an insoluble *uric acid,* which is then excreted as a paste, with most of the precious water extracted and conserved by the body. The chemistry of the waste products of metabolism as they generally occur in fish (ammonia), birds (uric acid), and mammals (urea) are shown below.

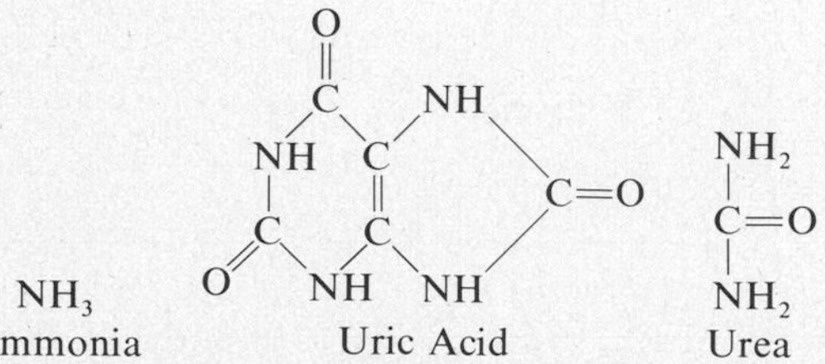

NH_3
Ammonia

Uric Acid

Urea

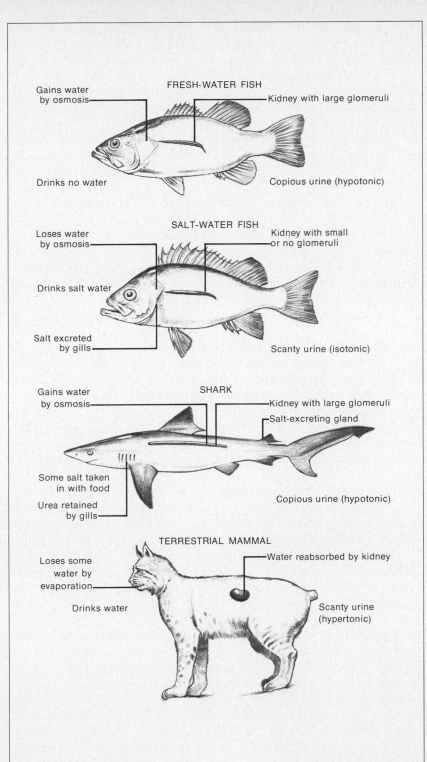

FRESH-WATER FISH

Gains water by osmosis

Kidney with large glomeruli

Drinks no water

Copious urine (hypotonic)

SALT-WATER FISH

Loses water by osmosis

Kidney with small or no glomeruli

Drinks salt water

Salt excreted by gills

Scanty urine (isotonic)

SHARK

Gains water by osmosis

Kidney with large glomeruli

Salt-excreting gland

Some salt taken in with food

Urea retained by gills

Copious urine (hypotonic)

TERRESTRIAL MAMMAL

Loses some water by evaporation

Water reabsorbed by kidney

Drinks water

Scanty urine (hypertonic)

9.7 Regulation of water and salt balances in a representative group of animals. Water enters the freshwater fish by osmosis. Great amounts of fluid cross its glomeruli daily, and it excretes large amounts of water in its urine. The saltwater fish loses water by osmosis, so it saves the remainder by filtering little fluid through its kidneys and urinating little. The shark gains water by osmosis and passes large amounts through its kidneys, urinating copiously. Land mammals drink water and lose some by evaporation. They save water by reabsorbing much of what passes through the kidney and urinate only a little hypertonic urine. Look at the cases individually and compare their problems and solutions. Why do the fish solve their problems three different ways? What can you say about their cytoplasm?

9.8 The human kidney. Note the position of the kidneys in the body. The lower part actually hangs below the protective rib cage. This is what makes kidney punches so dangerous. Also note the vast maze of blood vessels penetrating the kidney. Blood is brought into the kidney under considerable force, and many of its constituents are filtered through the glomeruli. From there they enter a tubular system from which many (including some water) pass back into the bloodstream. Those that are not recollected are passed into increasingly larger tubes until they enter the urinary bladder as urine. Here the urine is held until you stand in line for a movie. Halfway to the box office it demands to be released back into the environment. Because of its intensive filtering, urine is largely germ-free and has been used to wash wounds under field conditions when the available water was known to be impure.

Many evolutionists believe that vertebrates evolved in freshwater habitats (contrary to popular assumption), so the transition to land must have involved the development of new ways to handle the waste problem. The freshwater animal had the problem of eliminating the excess water that was continuously entering its body, and one solution was probably to allow great quantities of water to pass out through its extensive glomeruli. On land, however, the problem changed. There wasn't much water, so what there was had to be retained. In large part, the problem was apparently solved by a decrease in the size of the glomeruli.

The Human Kidney

Humans, along with other mammals, have two kidneys (although we can live with only one—nature is often redundant). The kidneys lie in the dorsal area (at the back) and extend slightly below the protective rib cage (this is why blows to the kidney area are so dangerous). Each kidney contains about a million *nephric units,* consisting of a glomerulus, a Bowman's capsule, and the connecting tubules. The structure of these units is shown, along with the kidney itself, in Figure 9.8.

The exit from the glomerulus is smaller than the entrance, so that the glomerular blood is placed under high pressure as it enters from the renal artery. As the blood, under intense pressure, wends its way through the circuitous and tortuous route of the glomerulus, much of its fluids move out through the thin-walled vessels and into the cuplike *Bowman's capsule.* Solid particles, such as blood cells and certain proteins, are too large to pass through the glomerular walls.

The filtrate received by the Bowman's capsule contains a high concentration of waste products, but also some good stuff, including water, salts, and nutrients. So as the filtrate begins to move through the tubules on its way to the urinary bladder, the blood vessels continuing on from the glomerulus head off the filtrate at the tubules and the *loop of Henle.* Here they wind around the loop and retrieve many of the usable products, both through diffusion and by active transport, allowing the urea and some water to continue on their way.

The remaining fluid in the tubules proceeds to a larger collecting tubule, and then to the *ureter,* which leads from each kidney to the *urinary bladder.* From the urinary bladder a single opening, controlled by sphincter muscles, allows the urine to pass into the *urethra,* which then leads to the outside.

Have you ever noticed that after drinking a lot of beer or other brews you have to urinate quite frequently? You may have also noticed that the quantity going out seems to be more than you thought had come in. Furthermore, you may have noticed that after a particularly eventful evening you are thirsty the next day. The reason for this is that reabsorption of water at the tubules is controlled by an *antidiuretic hormone* (ADH), which is secreted by the posterior lobe of the pituitary. Alcohol suppresses the secretion of this hormone, so that less water is

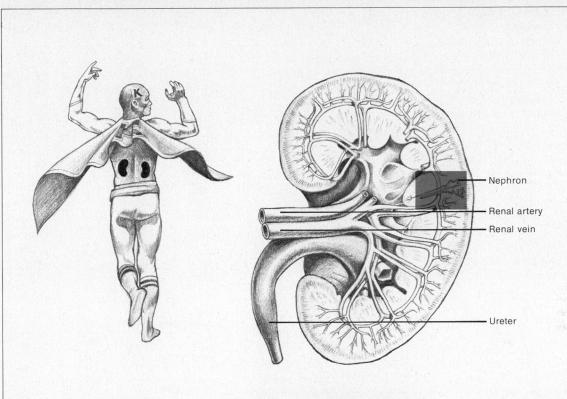

Nephron

Renal artery

Renal vein

Ureter

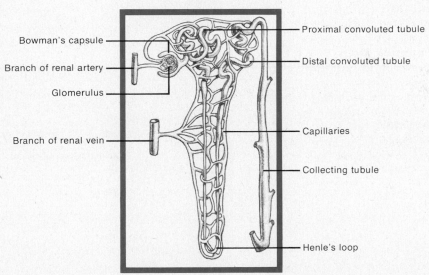

Bowman's capsule

Branch of renal artery

Glomerulus

Branch of renal vein

Proximal convoluted tubule

Distal convoluted tubule

Capillaries

Collecting tubule

Henle's loop

Box 9.2 Pressure Gradients in the Kidney Tubule

Let's look at how hypertonic urine is formed in the human nephric unit. First, body fluids filter through the capillary wall into Bowman's capsule and from there through the proximal convoluted tubule (here straightened), the long loop of Henle, the distal convoluted tubule, and then into the collecting duct. As the urine passes through the ascending loop of Henle (on the right), cells lining the loop pump sodium out of the urine into the tissue surrounding the loop. These sodium ions then diffuse over to the descending loop (on the left) where they enter the loop. From there they flow to the ascending loop where they are pumped out again. This constant circulation means that the loop of Henle and the adjacent collecting duct are always bathed in a salt solution and it is this solution that sets up the osmotic gradient that withdraws water and solutes from the urine. The ascending loop is apparently impermeable to water since no water passes out with the sodium. The collecting duct, however, is permeable to water and water freely flows out in response to the high salt concentrations in the surrounding tissue. This water is picked up by the blood vessels that come from the glomerular region and the hypertonic urine passes to the bladder.

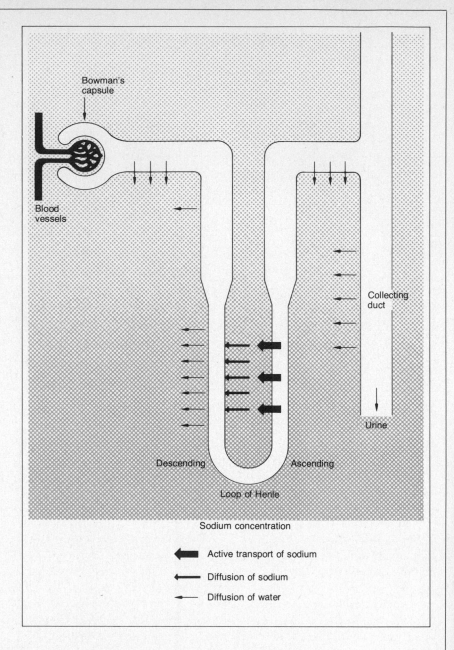

Bowman's capsule

Blood vessels

Collecting duct

Urine

Descending

Ascending

Loop of Henle

Sodium concentration

→ Active transport of sodium

→ Diffusion of sodium

→ Diffusion of water

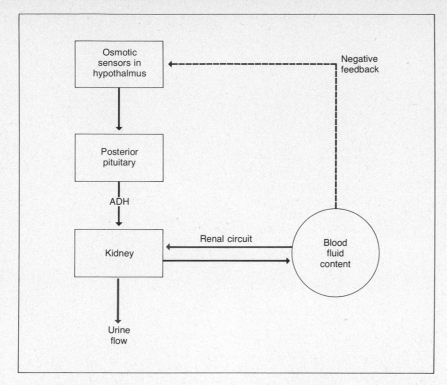

9.9 An example of a negative feedback system operating in the human body. When sensors in the hypothalamus detect that the blood is too "thick," it signals the posterior pituitary, which releases the hormone ADH. ADH causes more water to be reabsorbed through the loops of the nephric unit thus causing less urine to form. The result is an increase in water content in the blood. As the more dilute blood reaches the hypothalamus, the system is shut down and less ADH and more urine is produced until the blood is dehydrated to the point that the hypothalamus reactivates the posterior pituitary.

reabsorbed; the urine thus becomes hypotonic, or more dilute than the body fluids, and the body actually becomes somewhat dehydrated. Relevance strikes again! On the other hand, ADH secretion also decreases when you drink a lot of water since the osmotic concentration of the blood decreases (drinking a lot of water can temporarily "thin" the blood). This also results in the formation of a urine that is hypotonic to the blood.

In case you are planning to be shipwrecked sometime, you may also be interested in another bit of practical physiology. A person on a life raft cannot get the water he needs by eating fresh fish, contrary to popular rumor. The reason is that human urine cannot exceed a salt concentration higher than about 2.2 percent. Fish fluids are less concentrated than this, it is true; but fish are high in protein, so to get rid of all the nitrogen atoms, you would have to excrete a lot more water than you would be getting from the fish. You would do better eating fish than drinking seawater, of course, since the salt concentration in seawater is about 3.5 percent. Some shipwreck survivors claim that the spinal fluid of fish is a good supply of water, but all things considered, you would be better off to take some water with you.

Systems and Their Control

Have you ever wondered why, on your way to school, you have never encountered a 7-ton amoeba in the parking lot? There are probably a lot of good reasons. For one thing, an amoeba is a *protozoan*, a one-celled animal, and as you know, the volume of an object increases at a proportionately greater rate than its surface area. Thus a 7-ton amoeba would have an unsatisfactory surface-to-volume ratio. In other words, its regulatory surface membrane would be too small to accommodate its mass. Second, the unsupported weight would create such great internal pressures that the animal would probably rupture, causing a temporary parking inconvenience. Third, oxygen and carbon dioxide can move only across wet membranes, and the amoeba would have no way of keeping its membrane wet. Moreover, even if oxygen were able to move in the amoeba's body, once it crossed the membrane the only way it could reach the inner cytoplasm would be by diffusion, aided by some circulation resulting from the animal's sluggish movements. Thus the movement of carbon dioxide and oxygen through the body would be very slow and inefficient. Since amoebae have these problems, then, they must remain small and live in water.

There are 7-ton animals that live on dry land, although fortunately we rarely encounter them. These animals have had to solve all the attendant problems of large size and terrestrial life. For example, they have developed *multicellularity*. This permits groups of cells to act together to perform special tasks, such as supporting weight, circulating fluids, and gaining oxygen. These groups of cells are called *tissues*. Groups of tissues that have the same functions are called *organs*. The tissues and organs associated with a particular life process are loosely termed as *systems*. Thus we can talk about supportive, contractile, and integumentary systems, respiratory, circulatory, and digestive systems, or the hormonal and nervous systems. In this chapter and the next we will consider the various systems of multicellular organisms. We don't have the space, or for that matter, the energy, to consider the systems of all the phyla, so we will give attention to only a few groups whose body organizations illustrate the evolutionary descent of the organ systems.

SUPPORTIVE SYSTEMS

We have noted that the great stems and trunks of plants acquire their structural strength from the combined strength of many tiny cell walls.

10.1 A group of sponges, part of a colony (left). At right is a longitudinal section through a simple sponge. Note the various types of cells that make up the sponge tissue. The delicate choanocytes, or collar cells, on the interior have flagella that move nutrient-laden water through the body cavity. Among the most interesting cells are the mysterious, wandering mesenchyme cells that, although rather undifferentiated themselves, can work together to form very specific structures such as the three-pronged spicule at lower left. In sponges spicules of rather specific shapes give some rigidity to the sponge and hence act as a kind of skeleton.

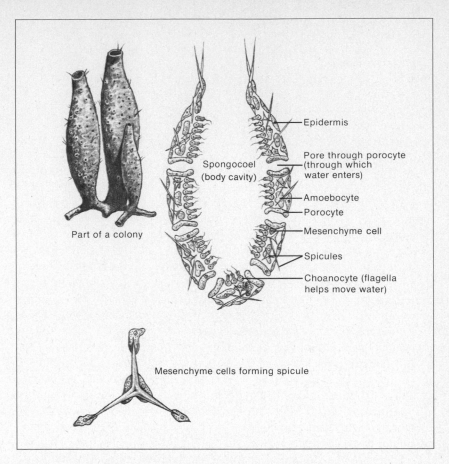

Part of a colony

Spongocoel (body cavity)

Epidermis

Pore through porocyte (through which water enters)

Amoebocyte

Porocyte

Mesenchyme cell

Spicules

Choanocyte (flagella helps move water)

Mesenchyme cells forming spicule

10.2 A molting crayfish. At rather regular intervals throughout the year, crayfish molt. Their growth is restricted by a hardened and protective exoskeleton, so growth can only be accomplished in the brief interludes between the development of new skeletons. The crayfish that has recently lost its exoskeleton may compensate behaviorally by becoming more furtive and hiding under rocks until it can venture forth in its new coat of armor.

Animal cells, unlike plant cells however, do not have cell walls, and animals' bodies are instead supported by various *skeletal* structures consisting of material that has been secreted by cells and then hardened to some degree. Such materials include bone and cartilage in animals that have internal skeletons and *chitin,* the major component of the outer skeletons of insects. In certain soft-bodied animals such as hydra, flatworms, and earthworms, a certain rigidity of the body results from the compression of cell fluids, by muscular contraction forming a type of hydrostatic skeleton.

Types of Skeletal Systems

One of the most primitive skeletal systems is that of the sponge (Figure 10.1). In these animals the body is largely supported by the water in which they live, so they have developed only weak skeletons. In fact, sponge skeletons consist only of tiny *spicules* scattered through the tissue. These spicules are formed from unspecialized mesenchyme cells, but only particular kinds of spicules develop. The 10,000 species of sponges are classified according to the type of spicules that comprise their skeletons.

The various classes have either calcium carbonate spicules, six-rayed glassy spicules, or fibrous or glassy spicules without six rays.

Some kinds of animals have skeletons outside their bodies, called *exoskeletons*. These include such species as lobsters, snails, insects, and oysters. Such skeletons serve well in protecting internal parts (as do our skull and rib cage), but animals with exoskeletons may have a problem with growth. Snails and clams simply secrete extensions to their shells as they grow, so that their shells grow with them. Lobsters, crabs, and crayfish, however, have to discard their old shells and grow new, larger ones. They do this by first withdrawing the valuable minerals from their shells, which causes the shell to soften. The shell then splits down the back. The soft, vulnerable animal crawls out, swells by absorbing quantities of water, grows rapidly for a time, and then secretes a new shell into which it redeposits the minerals.

Vertebrate animals, including man, have an *endoskeleton,* a skeleton inside their bodies. In the most primitive vertebrates, such as sharks and rays, the skeleton consists entirely of cartilage. In most other vertebrates, however, the cartilage that forms in the developmental stages is replaced by bone. Also, some parts of the skeleton form directly as bone.

Types of Connective Tissue

Cartilage and bone, as well as tendons, ligaments, and fibrous connective tissue, are all referred to as *connective tissue.* Generally, cells comprise only a small part of connective tissue; most of the mass actually consists of substances secreted by the cells. These substances are nonliving and are called *matrix.* The specific nature of each type of connective tissue is largely dependent on the qualities of the matrix (Figure 10.3).

The matrix of bone contains calcium salts, which make the bone hard, and a protein called *collagen,* which gives the bone some degree of flexibility. Because of its high density, bone tends to be heavy, but this weight is offset by the fact that most larger bones are hollow. The tissue inside these hollow cavities is called *marrow. Yellow marrow* consists mostly of fat cells, while *red marrow* is the site of the formation of red blood cells and some of the body's white blood cells. Running throughout the bony matrix are *Haversian canals,* small channels encompassing the blood vessels and nerves that serve the bone cells. The bone cells and the Haversian canals are connected by little canals called the *canaliculi,* through which the bone cells receive food and oxygen and eliminate waste.

The matrix of bone also contains cells that have the peculiar ability to withdraw calcium salts from the matrix so that the shape of a bone can change in response to the stresses placed on it. We see such changes in certain African cultures in which the practice of binding the woman's head in a tight metal ring results in a change in head shape.

In *fibrous connective tissue* the matrix consists of a heavy, interlaced mat of microscopic fibers, which are secreted by the connective tissue cells. Connective tissue is found throughout the body as a binding material,

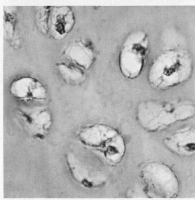

10.3   At top, a Haversian system of a bone. Through the large opening in the center run the nerves and blood vessels that serve the bone. The bone-forming cells lie in concentric rings around the canal and communicate with each other by the tiny canals, or *canaliculi,* radiating from them. Below, cartilage cells. For some reason these usually occur in pairs and they secrete the fibrous mat that is the matrix. Some bone contains cartilage cells, giving bones elasticity and resilience, but these may be replaced by bone cells with age, thus old people have a high risk of breaking brittle bones by a simple fall.

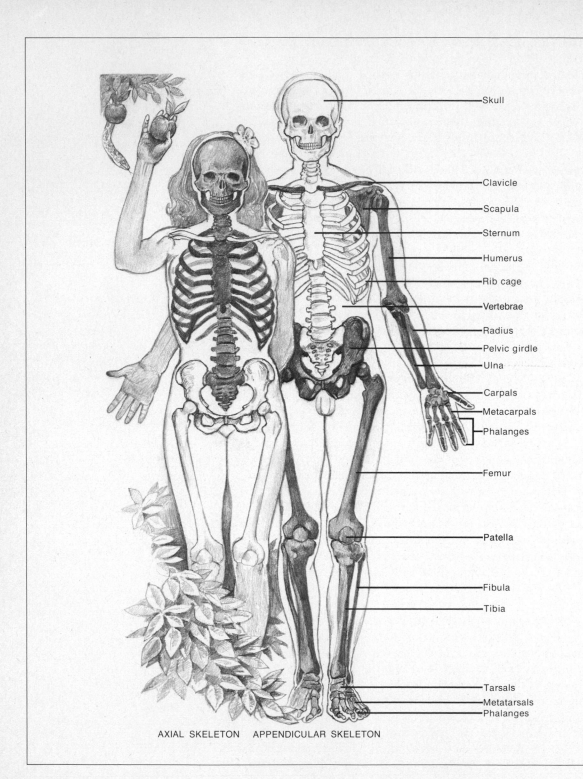

Skull

Clavicle

Scapula

Sternum

Humerus

Rib cage

Vertebrae

Radius

Pelvic girdle

Ulna

Carpals

Metacarpals

Phalanges

Femur

Patella

Fibula

Tibia

Tarsals

Metatarsals

Phalanges

AXIAL SKELETON   APPENDICULAR SKELETON

holding skin to muscles, holding glands in position, and binding bones together.

Tendons and *ligaments* are specialized types of fibrous connective tissue. Ligaments are fairly elastic and serve to bind one bone to another. Tendons are inelastic but flexible cords that connect muscle to muscle or muscle to bone. (The fibrous connective tissue just below the skin, when it is tanned, becomes leather.)

The Human Skeleton

The adult human skeleton consists of about 200 bones (the number changes as some bones fuse with age). In humans, as well as in other vertebrates, these bones make up two basic skeletal parts called the axial skeleton and the appendicular skeleton (Figures 10.4, 10.5).

The *axial skeleton* is that part of the skeleton that runs approximately along the axis of the body—that is, in the direction of the backbone. Thus it includes the skull, the spinal column, the *sternum* (breastbone), and the ribs. The skull consists of the fused bones of the braincase, or *cranium*, and the bones of the face. The backbone is made up of thirty-three separate *vertebrae*, which differ somewhat according to their position along the length of the spine (Figure 10.6).

The ribs are attached dorsally (toward the back of the body) to the vertebral column, and ventrally (toward the belly surface) to the sternum or to cartilage attached to the sternum, except for the last pair, called

10.4 The axial skeleton (dark at left) and the appendicular skeleton (dark at right). The axial skeleton is essentially comprised of the skull, vertebral column, and rib cage. In its protective function it houses the most critical parts of the human body (the central nervous system and the viscera). You may, however, notice that certain critical parts lie outside this skeletal shield. The appendicular skeleton is primarily composed of the long bones (those of the pectoral and pelvic girdles). Other animals, of course, may differ substantially in their skeletal makeup. An astute observer can tell a great deal about an unfamiliar animal's habits simply by looking at the skeleton. Would shortening the upper leg and lengthening the lower leg make humans faster or slower runners? What would it do to climbing ability?

10.5 The axial (dark) and appendicular (light) skeleton of the frog. Note the area of the skull that is devoted to protecting the tiny brain. Also note the elongated transverse processes of the vertebrae and the absence of ribs. Hopefully, with careful attention, you can discern other differences between the skeleton of a frog and a human. But you should also note the basic similarities in the body plan. What special features does the frog skeleton have that adapt the animal for its particular way of life?

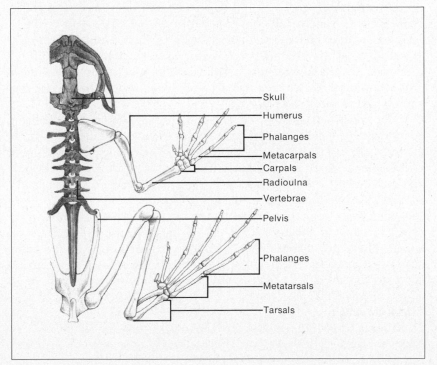

Skull
Humerus
Phalanges
Metacarpals
Carpals
Radioulna
Vertebrae
Pelvis
Phalanges
Metatarsals
Tarsals

10.6   A lumbar (lower back) vertebra showing the centrum that articulates with vertebra on either side, the neural arch through which the spinal cord passes, and the various projections and processes that provide surfaces for the attachment of muscles and ribs.

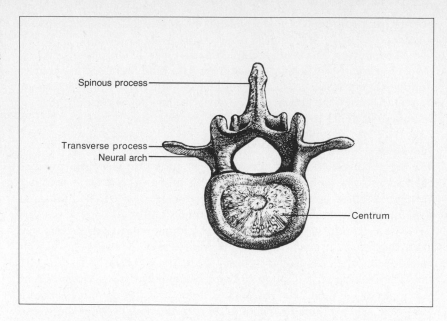

Spinous process

Transverse process

Neural arch

Centrum

10.7   The pectoral (A) and pelvic girdles (B). In humans, the bones of the pectoral girdle are loosely joined and can be flexed. In leaping animals, such as cats, the collarbone is greatly reduced so as not to interfere with the forward movement of the forelegs.

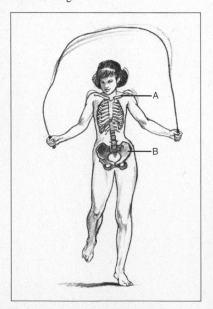

floating ribs, which are not attached ventrally. The rib cage is flexible and slightly moveable so that it expands during breathing and can absorb blows in the protection of the heart, lungs, and other vital organs.

The appendicular skeleton is comprised of the arms and legs and two girdle complexes (Figure 10.7). The pectoral girdle consists of two collar bones (*clavicles*) and two shoulder blades (*scapulas*). The pelvic girdle is composed of three bones fused to form the hips. You may find it surprising to learn that fish have rather primitive hip and shoulder girdles. On the other hand, fossil evidence indicates that the arms and legs of land animals evolved from the fins of early fish, so perhaps, all things considered, the girdles in fish are not so surprising.

The upper parts of both the arms and legs are comprised of single bones that are joined to two bones of the lower limbs (Figure 10.8). Both the arms and legs terminate in a complex of several bones. The arms end with the wrists (*carpals*) and the legs with the ankles (*tarsals*), which slide past each other to permit a rather wide range of movement for the hands and feet. The hands and feet are made up of bones called the *metacarpals* and *metatarsals,* respectively. The hands and feet each terminate in five digits, the fingers and toes. One of the identifying characteristics of humans is a highly moveable and specialized hand existing on the same body with a much less talented foot. Interestingly enough, both the human arms and legs are rather primitive structures as vertebrate limbs go.

Types of Joints
Some bones move freely, while others are only slightly moveable, and yet others are solidly fused together. The freedom of movement of any bone depends on its relationship to other parts of the body. The most moveable bones are those which form points of *articulation*, or joints, with other

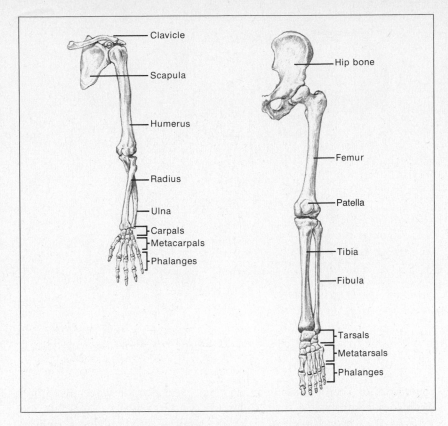

10.8 The human arm (left) and leg (right). Note that they are very similar in their basic design. Major differences include the ability to freely rotate the lower arm, but not the lower leg. This is one reason knee injuries can result from severe twisting. The shoulder and hip joints are similar as are the ankles and the wrists. The knee joint is protected by a kneecap, but the elbow, less prone to being hit, has no such protection.

Labels (arm): Clavicle, Scapula, Humerus, Radius, Ulna, Carpals, Metacarpals, Phalanges

Labels (leg): Hip bone, Femur, Patella, Tibia, Fibula, Tarsals, Metatarsals, Phalanges

bones (Figure 10.9). The *ball-and-socket joints* that connect the humerus to the scapula at the shoulder and the femur to the pelvis at the hip permit great freedom of movement. The hinge points at the knee and the elbow also provide free movement, but only in very restricted directions, back and forth. A joint such as that at the pubic symphysis permits only slight flexibility, as in the separation of the hips when giving birth. The bones of the skull, which arise separately and later fuse along *suture lines,* are completely immobile—thus the difficulty of flexing one's head.

Now let us consider how all these bones are able to move with such precision that a cigarette can be rolled with one hand, and such strength that, in the next instant, tall fences can be leaped with a single bound.

CONTRACTILE SYSTEMS
Types of Muscle

Muscles are made up of *contractile* fibers and may be divided into three types, according to their structure and the nerve stimulus that activates them. *Smooth muscle* is found in the walls of the digestive tract, and around some blood vessels and certain internal organs. *Cardiac muscle* comprises the walls of the heart. Both these types of muscles have long been termed "involuntary," since they function without conscious control. Thus you don't have to lie awake nights keeping your heart beating or

10.9 Representative joints in the human body. Some joints are sutured and do not flex at all. Others almost never move, while yet others move in very limited directions. Ball and socket joints give the greatest movement aside from sliding groups of bones such as those found in the wrist.

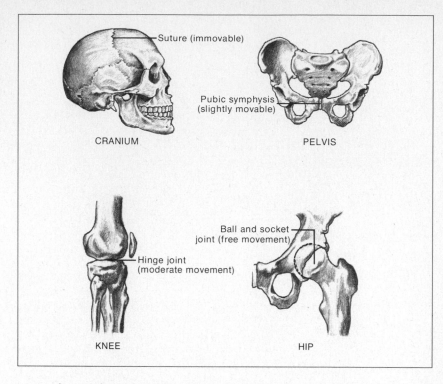

Suture (immovable)

CRANIUM

Pubic symphysis (slightly movable)

PELVIS

Hinge joint (moderate movement)

Ball and socket joint (free movement)

KNEE

HIP

10.10 Three types of muscle tissue. At left is striated, or skeletal muscle. The striations are actually due to the banding as shown in Figure 10.14. At center is smooth muscle, unstriated and spindle-shaped (although the shape is not apparent here). The nuclei are elongate and prominent. At right is cardiac muscle. Note the striations (compare with Figure 10.14). The dark, irregular band to the left of center marks the division of two muscle units.

remember to keep breathing in and out. There is evidence that some degree of control over such "involuntary" responses can, in fact, be learned. We'll discuss some of the research in this area in the next chapter. Meanwhile, what we are really interested in here is a third kind of muscle, the voluntary muscles, *skeletal muscles*. These are the muscles that enable you to move your body.

Both the skeletal muscles and the cardiac muscles have several nuclei in each fiber, or cell, whereas smooth-muscle fibers have only one nucleus each (Figure 10.10). However, in the skeletal muscles these nuclei lie just under the membrane of the fiber, instead of being scattered throughout the cell. In addition, the cells of skeletal muscles may be very long, perhaps

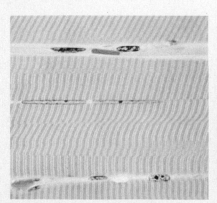

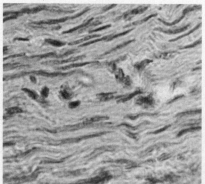

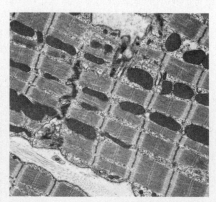

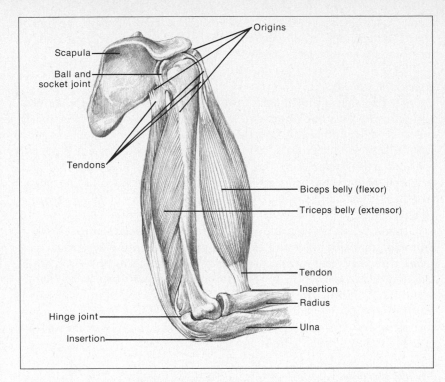

Scapula

Ball and socket joint

Tendons

Origins

Tendons

Biceps belly (flexor)

Triceps belly (extensor)

Tendon

Insertion

Radius

Hinge joint

Ulna

Insertion

10.11 The attachments of the upper arm muscles. The biceps brachii lies on the forward part of the humerus. On the back is its antagonistic muscle, the triceps brachii. Shortening of the biceps has what effect on the lower arm? Note that the origins are attached to bones that will move less than the insertion when the muscle is shortened. Belly simply refers to the body of the muscle.

even extending the entire length of the muscle. The microscope reveals that skeletal and cardiac muscles have alternate light and dark bands, called *striations,* about which we will have more to say later. The three muscle types are compared in Table 10.1 (p. 247).

Skeletal Muscles

Typically, a skeletal muscle is a bundle of millions of individual muscle fibers bound together by connective tissue. (It is the connective tissue that makes some meat so tough.) The muscle is surrounded by a very tough sheet of whitish connective tissue called the *fascia* (those people who study it are not called fascists!). The slippery fascia permits the muscles to move past each other with a minimum of friction.

Usually, the ends of a muscle are attached to two different bones by tendons. When the muscle is contracted, it becomes shorter and thicker, drawing the bones closer together. Ordinarily one of the bones moves less than the other or is completely stationary. The bone that moves less provides the area of attachment for the end of the muscle called the *origin;* the moveable bone provides the surface for the muscle's *insertion.*

Now let's consider an example. If you ask a fellow to show you his muscle, if he is a gentleman he will display his *biceps brachii* (Figure 10.11). Note that the origin of this muscle is at the joint of the humerus and scapula (*biceps* means "two-headed," so note also the double origin). The muscle inserts on the radius, the arm bone that terminates behind the thumb. As the forearm is rotated, the radius describes an arc, hence its

10.12 An old photograph of a pole vaulter, taken in 1884 by Thomas Eakins. Note that although great heights aren't reached, such an activity utilizes almost all the body's great muscle masses.

10.13 Dorsal and ventral views of the superficial muscles in the human. The darker areas are muscle bellies. The lighter areas are connective tissue, mostly tendons. Note the difference in the size of various muscles in the two sexes (male, left; female, right). You should be able to infer the action of any muscle by noting its position and attachments. For example, the deltoid is largely responsible for raising the arm toward the ball. What do you suppose the sartorius does? Why do you suppose a sprain of the lower back sometimes takes so long to heal? (Keep in mind the white connective tissue has fewer blood vessels.) In which areas of the body are the bones covered with little or no muscles? (Note the back of the pelvis and the sternum.) Why do you suppose the gluteus maximus is so much larger in humans than most other mammals? (Keep in mind our erect posture and type of leg movement.)

name. As the biceps is *flexed,* the lower arm is drawn toward the upper arm. The maximum circumference of the biceps can then be measured by those crass souls interested in such things.

Antagonistic Muscles

On the other side of the humerus from the biceps brachii there is another large muscle. This one originates at the juncture of three bones and is called the *triceps brachii.* It inserts on the elbow, or *olecranon process,* the part of the ulna that extends past the elbow joint. As the biceps is flexed, consider what happens to the triceps. Obviously, it must relax and lengthen. Then, when the triceps is flexed, the arm straightens as the biceps lengthens. Thus muscles usually work in opposing pairs in an *antagonistic* system.

In some cases antagonistic muscles must contract simultaneously. For example, simultaneous contractions of opposing leg muscles permit us to stand. The *flexor* and *extensor muscles* of the leg pull against each other and lock the knee in place. The tension, of course, must be highly regulated. Otherwise, we would be springing out of chairs and collapsing all over sidewalks. Walking, by the way, is not a simple matter; it requires a surprising degree of coordination between opposing muscles as the tension changes and weight shifts from one leg to the other.

Types of Movement

Whereas the contraction of the biceps functions in flexing the arm, other muscles have different sorts of actions. *Adductors* and *abductors* move parts of the body toward or away from the central axis of the body, as when the arm is pressed against the body or swung outward. *Levators* and *depressors* raise or lower parts of the body, as in shrugging or lowering the shoulders. *Pronators* rotate body parts downward, as when you turn your palm downward to look at your watch, and *supinators* rotate those parts upward, as when accepting M&Ms. *Sphincters* and *dilators* are rings of muscle that open and close body openings, such as the muscles of the lips. However, most muscles of this kind are involuntary, like those at the juncture of the stomach and small intestine or between the urinary bladder and the urethra.

The superficial muscles of the human body are shown in Figure 10.13. You need not memorize these, but look them over carefully to get a good working idea of how you're put together.

Muscle Contraction

The striated muscle fiber, or cell, has several nuclei, as we have noted, and is surrounded by a rather tough cell membrane. Upon closer examination, we see that each fiber is made up of small threads of protein called *myofibrils.* It is the cross banding or striations in the myofibrils that gives the fiber its banded appearance. Muscle tissue is highly active, so you shouldn't be surprised to find that it contains a rather large complement of

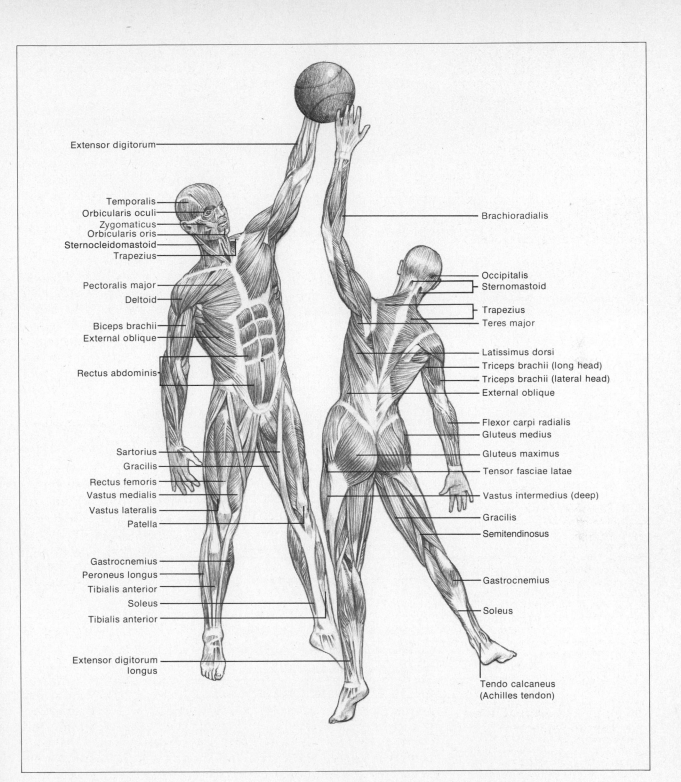

Extensor digitorum

Temporalis
Orbicularis oculi
Zygomaticus
Orbicularis oris
Sternocleidomastoid
Trapezius

Pectoralis major
Deltoid

Biceps brachii
External oblique

Rectus abdominis

Sartorius
Gracilis
Rectus femoris
Vastus medialis
Vastus lateralis
Patella

Gastrocnemius
Peroneus longus
Tibialis anterior
Soleus
Tibialis anterior

Extensor digitorum
longus

Brachioradialis

Occipitalis
Sternomastoid

Trapezius
Teres major

Latissimus dorsi
Triceps brachii (long head)
Triceps brachii (lateral head)
External oblique

Flexor carpi radialis
Gluteus medius
Gluteus maximus
Tensor fasciae latae

Vastus intermedius (deep)

Gracilis
Semitendinosus

Gastrocnemius

Soleus

Tendo calcaneus
(Achilles tendon)

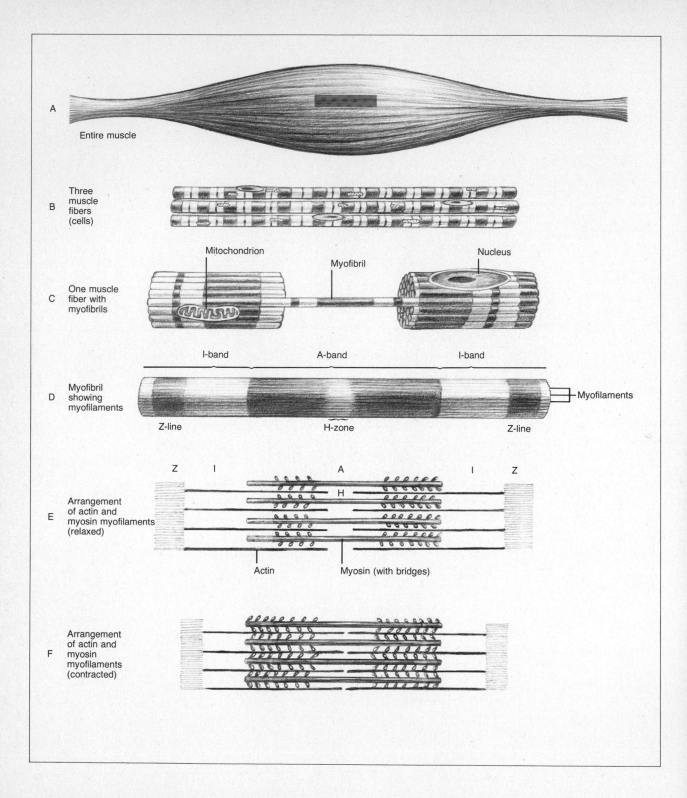

A   Entire muscle

B   Three muscle fibers (cells)

C   One muscle fiber with myofibrils

Mitochondrion

Myofibril

Nucleus

D   Myofibril showing myofilaments

I-band   A-band   I-band

Myofilaments

Z-line   H-zone   Z-line

E   Arrangement of actin and myosin myofilaments (relaxed)

Z   I   A   H   I   Z

Actin   Myosin (with bridges)

F   Arrangement of actin and myosin myofilaments (contracted)

Table 10.1 Comparison of Muscle Types

	Skeletal	Smooth	Cardiac
Location	Attached to skeleton	Walls of viscera, around arteries	Wall of heart
Shape of fiber	Elongated, cylindrical	Elongated, spindle-shaped	Elongated, cylindrical, branched
Number of nuclei	Many	One	Many
Position of nuclei	Peripheral	Central	Central
Cross striations	Present	Absent	Present
Speed of contraction	Rapid	Slow	Intermediate
Ability to remain contracted	Least	Most	Intermediate
Type of control	Voluntary	Involuntary	Involuntary

mitochondria, which, as you recall, are energy-producing organelles. The mitochondria lie snugly against the contractile units, the myofibrils.

As shown in Figure 10.14, the tiny myofibrils are made up of still smaller units, called *myofilaments*. There are two kinds of myofilaments; the thicker ones are about 100 Ångstroms in diameter and the thinner ones are about half that size. The banded appearance of the myofibril is due to the arrangement of the two kinds of myofilament in each thread.

The myofilaments consist primarily of two proteins, called *actin* and *myosin*. It is believed that the thinner filaments are composed of actin and the thicker ones are composed of myosin, since the dark bands formed by the thicker filaments disappear when the myosin is extracted from the tissue. The thick myosin filaments are found only in this darker A band; the I band contains only the thin actin filaments, although these may extend into the A band.

During contraction the A band remains constant in length while the I band shortens, and the length of the H zone within the A band also decreases. It has been suggested that the filaments themselves do not actually contract, but instead slide past each other. In other words, during contraction the actin filaments extend further into the A band, so that the H zone and the I band become narrower as the ends of the myosin filaments approach the Z line. Got that?

It is not known how the filaments are able to slide along each other or to maintain a state of contraction. It may be that the bridges along the myosin fibers (believed to be able to attach to the actin fibers) break and rejoin somewhere else along the length of the actin. Apparently the energy provided by ATP is utilized in the formation of these bridges. In fact it is theorized that one ATP → ADP conversion takes place for each bridge formed.

Contractile Systems in Other Animals

Because the skeleton and muscle systems are so strongly interrelated, we have considered the two systems in humans together. However, we should

10.14 How skeletal muscles are believed to work. The sequence shows increasingly enlarged areas of a muscle. (A). Illustrated here is the "sliding filiment" hypothesis of how such voluntary muscles work at the subcellular level. A and B give you some perspective on precisely where the Z lines are in a muscle. D-F indicate just what happens when a muscle contracts. Notice that the length of the A band remains constant. The fibers of the H zone give rise to tiny "bridges," which the I fibers can move along in a progressive fashion. The converging I bands, then, decrease the length of the H zone as the muscle shortens. The I bands are believed to be composed of actin, the H zone of myosin.

10.15 Radial symmetry, a primitive condition, is common in some types of animals, such as this sea nettle, a jellyfish, or a sea anemone. Why do you suppose bilateral symmetry (having right and left sides) ever arose? If an anemone, for example, should move head-first on the bottom, instead of creeping along on its foot, could you expect further specializations for a horizontal existence? If it came to move along more on one surface than another, that surface would be ventral. With ventrality would bilaterality arise?

also learn something about the contractile systems of other organisms before we move on.

Plants are capable of movement, as almost everyone knows. The movement, however, is usually restricted to *tropic* responses, such as a leaning toward light, or the opening and closing of flowers or stomates. Such activities are usually responses to differences in growth rate or fluid pressure within parts of the plant. It should also be pointed out that in some plant species the organism, or its gametes, can travel about quite freely under its own power.

Coelenterates are animals that show *radial symmetry* and are organized basically as hollow sacs. Animals in this group include the jellyfish, hydra, and sea anemone (Figure 10.15). Coelenterates never develop mesoderm and therefore cannot have true muscles. However they have, in their double-walled bodies, a system of contractile fibers. The fibers in the outer wall (formed from ectoderm) run lengthwise and those in the inner wall (formed from endoderm) run circularly. Since these animals are unrestricted by a rigid skeleton, they are able to perform some rather remarkable contortions.

The basic system of internal, longitudinally arranged, contractile fibers and external, circular fibers is also common in certain phyla of worms that have mesoderm, and hence, true musculature. Included here are roundworms (phylum *Nematoda*), (Figure 10.16), flatworms (phylum *Platyhelminthes*), and segmented worms (phylum *Annelida*). In addition to longitudinal and circular fibers, these more "advanced" phyla may also have additional muscle arrangements, such as diagonal or dorsoventral fibers.

In arthropods, the joint-legged group that includes the insects, the

10.16 Cross section of a roundworm, *Ascaris,* a parasite of humans. Note that the nerve cords lie both dorsally and ventrally. (What is the case in insects and vertebrates?). Also note that the ectoderm is syncytial. In other words the nuclei do not lie in separate cells. What kind of movement is possible with only longitudinal muscle fibers? Why does the worm have such a heavy cuticle? Keep in mind that it lives in intestines.

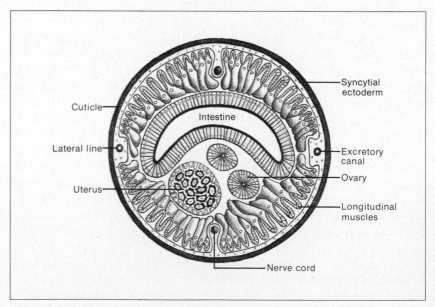

Cuticle

Lateral line

Uterus

Intestine

Syncytial ectoderm

Excretory canal

Ovary

Longitudinal muscles

Nerve cord

exoskeleton demands a different sort of muscle system. In this case the muscles, of course, lie within the skeleton. The joints are also unusual, in that they are simply thin, flexible places in the exoskeleton. Muscles may extend across a joint so that their contraction causes the joint to bend. Or a muscle may lie entirely within a single segment of an appendage, with its origin on the inner wall of the exoskeleton and its insertion on an *apodeme*—a long, thin, tough part of the exoskeleton that extends from one section of the body into the next (Figure 10.17).

The arrangement of muscles within an exoskeleton imparts remarkable strength to arthropods. You have undoubtedly noticed tiny ants carrying loads of food many times their weight or dragging along the corpses of huge insects such as crickets.

This muscle-skeleton arrangement is also responsible for the much-loved whine of the mosquito. The mechanism of the incredibly rapid contractions that move the wing once puzzled physiologists, but it has been found that only a single muscular contraction can set up a reverberation in the insect's exoskeleton. As the thoracic section vibrates, it causes the wings to move, thus a single muscle contraction may produce several wingbeats.

Most of the energy for muscle contraction in active animals involves the use of oxygen. So let's consider some of the ways living things move oxygen from the environment into their bodies.

EXTERNAL RESPIRATION

External respiration involves getting oxygen from the environment. Internal respiration, as you recall, involves intracellular metabolism and thus includes the use to which oxygen is put.

During the day the tissues of green plants may be steeped in oxygen, which the plants have manufactured themselves. But at night when photosynthesis ceases, they may have to acquire oxygen from the environment. In the case of many plants, opened stomates allow oxygen to enter the body of the plant, where it can then diffuse across the membranes of thin-walled interior cells.

There are also other ways plants may gain oxygen. For example, in very thin blades of aquatic plants, oxygen may diffuse across the outer cell walls to permeate the plant's tissue. Many types of smaller aquatic animals acquire their oxygen by a similar method. Planarian flatworms have no organs for taking in oxygen, but their thin, flattened shape provides them with a large surface area in relation to their mass. Diffusion of oxygen from the water into their bodies thus gives them all the oxygen they need for their slow-paced lives.

Larger aquatic animals, even though they lead no more active or exciting lives than the flatworm, have proportionately less surface area exposed to oxygen-laden water; hence they need some more specialized system of obtaining oxygen. Consider the chiton, a mollusk that is

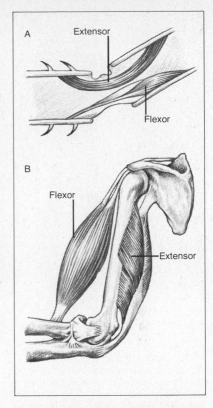

10.17 A comparison of the musculature of an insect and man. Note, particularly, the relationship of muscles and joints. Insects are much stronger than humans for their weight, so why haven't we evolved internal musculature? What are its disadvantages? Does shortening the insect's upper muscle here have the same effect as flexing the human bicep?

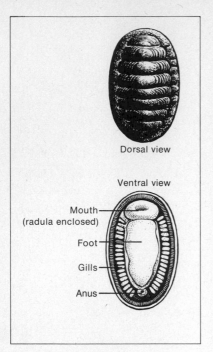

Dorsal view

Ventral view

Mouth
(radula enclosed)

Foot

Gills

Anus

10.18 The external structure of the chiton, dorsal and ventral views. The animal attaches to the rocks by suction produced by muscles of the foot. It feeds by rasping algae off rocks with a horny toothed structure called a radula. Between the foot and the upper body is a concavity in which lie 11 to 26 leaf-shaped gills. From the gills extend tiny cilia that, by their beating, keep oxygen-laden water flowing past them.

common to both the Atlantic and Pacific coasts (Figure 10.18). The Atlantic species may be less than an inch long; the Pacific species run as long as 10 inches. You have probably noticed chitons clinging tenaciously to rocks along the shoreline and may have commented on their apparent lack of personality. Gills project from a concavity around their bodies, and each gill is covered with tiny cilia whose beating circulates oxygen-laden water past the gill. The gills, as would be expected, are very thin-walled and full of tiny blood vessels, so that oxygen can enter the bloodstream easily and be carried quickly to the rest of the body.

Although fish are much more complex creatures than chitons, their means of acquiring oxygen are basically the same. The gills in fish are essentially protrusions that come into contact with the water (Figure 10.19). The cell membranes at this interface are very thin and readily permit the passage of gases across them. The gills themselves are laced with tiny, thin-walled capillaries that carry carbon dioxide out of the bloodstream and transport oxygen from the water to the rest of the body. Remember that there are normally higher carbon dioxide levels within the body, and higher oxygen levels outside it, and it is these differences that determine the direction of flow of the dissolved gases. Thus the higher level of carbon dioxide within the body causes it to move outward, and the lower level of oxygen in the body causes oxygen to move inward from the surrounding water.

Just as the chiton brings oxygenated water into contact with its gills by

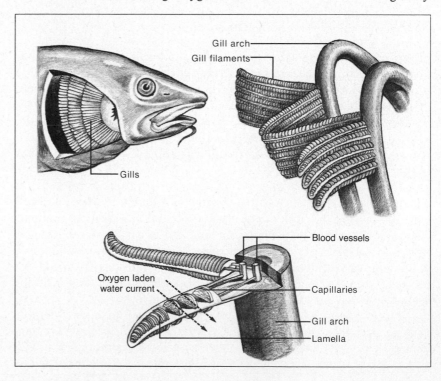

Gill arch

Gill filaments

Gills

Blood vessels

Oxygen laden water current

Capillaries

Gill arch

Lamella

10.19 The typical bony fish gill. The lamellar projections serve to increase the surface area of the gill. Note the blood vessels within each lamella. Blood passes from the gill arch into the lamellae of the gill filaments. Here gases are exchanged. The oxygenated blood then returns to the gill filaments. Here gases are exchanged. The oxygenated blood then returns to the gill arch and back to the body.

10.20 Sand shark. It is generally assumed that since sharks lack a water-pumping system they must constantly stay in motion to keep oxygen-laden water flowing over their gills. However, divers have recently reported that some species of shark sleep, lying dormant on the bottom. If this is true, the physiology of the dreaded beast may be much more complex than we had imagined. It may be that their constant motion is more directly related to their incredible appetites. A list of things found in the stomachs of sharks over the years would probably astound, amuse, and perhaps dismay even the most jaded biologist. On an expedition in which I participated, we found a Polaroid negative in the stomach of a 12-foot tiger shark. We had taken the picture and thrown the negative overboard 200 miles and two days previously. Were we being shadowed? It is most disconcerting to find a photograph of one's self in the stomach of a shark.

the movement of cilia, the fish washes water past its gills by taking in water through the mouth and pumping it out through the gill covering. Some fish, such as sharks, lack the muscles for pumping water and must swim constantly to keep water flowing over their gills. If they stop swimming they will suffocate.

Insects generally solve the oxygen problem in a different way. They breathe through a *tracheal system* of tiny tubes permeating their body tissue. These tubes open to the surface at tiny holes called *spiracles*. As shown in Figure 10.21, oxygen enters the tracheal system directly from the air and, aided by the insect's bodily movement, moves through the tracheae deep into the body tissue, where it enters directly into the cellular environment. Thus a circulatory system is not very important for external respiration in insects. (But insects do have circulatory systems. Why is this true of insects?)

The insect's system of getting oxygen probably places an inherent limitation on its size, since a tracheal system could be efficient only for organisms below a certain size of body mass. Thus an evolutionary direction taken eons ago may be responsible for the fact that you don't have to contend with 800-pound mosquitoes or 600-pound crickets.

Among those hairy milk-givers called mammals, the group that includes humans, external respiration takes place by means of special pouches called *lungs*. Oxygen-laden air is brought into the lungs by *ventilation,* or breathing. The surface of the lungs is moist, and it is this thin film of water that allows oxygen from the air to dissolve into a solution so that it can diffuse across the thin cell membranes and enter the blood-

10.21 The respiratory system of an insect. Note the spiracles in the animal's side through which air enters, eventually moving into a highly branched tracheal system or into large air sacs. The insect's movement, of course, would hasten the flow of air through these channels. It has been suggested that because of the limitations of such a system, insects have never been able to become very large.

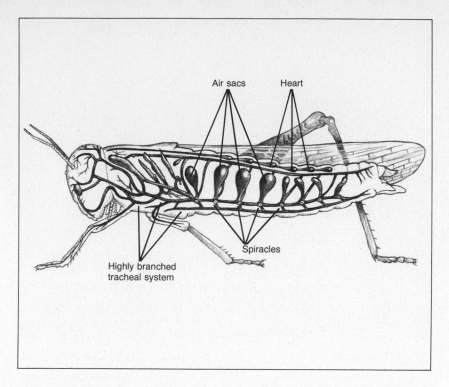

stream. The blood transports the oxygen to the tissues where it diffuses into the cells. In the cells the oxygen picks up spent hydrogen atoms from the electron transport chain and is thus converted into water.

Meanwhile, the carbon dioxide that has been produced in metabolic reactions diffuses from the cells into the capillaries, which carry it to the lungs. In the tiny, thin-walled sacs of the lungs it then diffuses across the cell membranes into the lung cavity and from there it is exhaled out into the environment. So, the study of external respiration is simply the consideration of how oxygen is brought in to pick up used hydrogens, and how carbon dioxide, which is of no direct use to animals, is discarded. Figure 10.22 describes the human respiratory system.

CIRCULATORY SYSTEMS

We will now consider a few selected species to see how oxygen and carbon dioxide, as well as food and metabolic waste products, get from one place to another within living things.

You will recall that in the vascular system of higher plants, water enters the root hairs from the soil and is carried upward through the xylem. Food is transported throughout the plant from the leaves by the phloem. The problem of circulation in algae that live in water is solved by the fact that almost the entire plant is flattened, so that every cell can exchange materials with the surrounding water or with nearby cells that are in

10.22 The human respiratory system (opposite page). The contraction of the diaphragm creates a partial vacuum in the chest and causes oxygen-filled air to rush in to fill the void. The air is warmed and moistened as it travels through the upper air passages. The air passes through increasingly smaller passages until it reaches the very thin-walled saclike alveoli, which allows oxygen to pass into the bloodstream and carbon dioxide to pass into the lungs from where it is expelled.

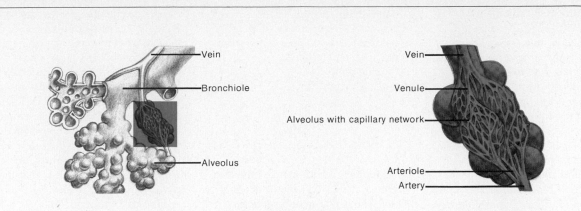

Vein

Bronchiole

Alveolus

Vein

Venule

Alveolus with capillary network

Arteriole

Artery

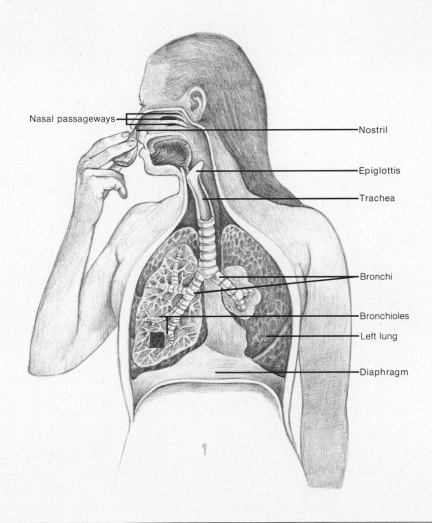

Nasal passageways

Nostril

Epiglottis

Trachea

Bronchi

Bronchioles

Left lung

Diaphragm

Box 10.1   Classification

Living things are divided, for convenience, into smaller groups. One such division separates plants and animals into two *kingdoms.* Thus, an organism in the animal kingdom could be expected to have more in common with another organism in its own kingdom than one from the plant kingdom. But among animals, certain ones share a number of traits in common, and these may be placed together into a *phylum.* Thus, all animals that have notochords at some period in their development are placed in one phylum, chordata. As finer and finer divisions are made among plants and animals, one finally arrives at a group of organisms that have so many things in common that they normally interbreed. These organisms are said to be of the same *species.* The taxonomic breakdown is:

Kingdom
  Phylum
    Class
      Order
        Family
          Genus
            Species

Members of two genera (plural of genus), then, won't share enough things in common to be able to interbreed, but they are all within the same family. And members of one family have more things in common with other members of the order than with families from another order.

There is a continuing debate over the precise definition of a species. Basically the antagonists can be labeled as "lumpers" or "splitters." The "lumpers" are those who demand rigorous criteria in order to separate a variable group into species. In other words, they might put the large grackles of Texas and the small grackles of Puerto Rico into one species, although they obviously cannot interbreed in nature. The assumption is that they would interbreed if they were to encounter each other. The "splitters" say no, they probably wouldn't, and besides it is less confusing to identify them as separate species.

contact with water. Thus there is no need for a circulatory system, and this is why you don't see "veins" in the seaweed you find washed up on the beach (Figure 10.23).

Types of Circulatory Systems

Size may be an important factor in the development of circulatory systems. In one-celled organisms and small multicellular species there may be no need for a circulatory system, since materials can easily diffuse from one part of the body to another and a large part of the total membrane area of the body is exposed to the outside environment.

Coelenterates have an unusual kind of circulatory system. Some of the cells that line the body cavity (of the inner layer of the two-layered animal) are *amoeboid*. That is, they change shape as they crawl and squirm their way through the body in amoeba fashion. When they reach the inner layer of the pouchlike animal, they may surround and engulf food particles just as an amoeba does. After engulfing food particles at the surface of the body cavity, they withdraw into the body with the nutrients and distribute them to other cells.

Flatworms also have saclike guts, with a single opening that serves both for food intake and elimination of waste. Since the body is flattened, no cell is very far from the gut, where digestion takes place, and thus nutrients can easily diffuse to the areas where they will be utilized. More complex animals, however, have specialized circulatory systems with some kind of device that pumps fluid throughout the body.

Circulatory systems may be classified according to whether they are *open* or *closed*, the open systems being more primitive. In mollusks, such as clams and snails, and in arthropods, such as insects and crayfish, a tubular heart pumps blood through vessels for a way, but then it empties into open cavities, or *sinuses*. From there it seeps or percolates through tissues until it begins to collect again in a large sinus around the heart. When the heart relaxes, blood flows in through tiny, one-way openings called *ostia* in the heart wall. When the heart contracts the ostia close, and blood is pumped out again through the vessels. Open circulatory systems, then, are those in which the blood is not continuously enclosed in vessels.

Among the most primitive of the closed systems are those of the annelids, or segmented worms. In the earthworm, for example, the heart consists of a series of five tubular rings that force blood backward through one large ventral vessel and forward through another large dorsal vessel (Figure 10.24).

Vertebrates also have a closed circulatory system, as you probably know. Actually, the term is somewhat relative, since blood may leave vessels and flow through *sinusoids*, the small open cavities of the liver and certain bones. Man is not at all unusual among vertebrates in having a large muscular heart that pumps blood into large muscular *arteries*, which in turn branch into increasingly smaller *arterioles*, and finally into *capillaries*. The walls of the capillaries are so thin that oxygen, carbon dioxide,

10.23 Seaweed may be microscopic or many meters in length, but no matter what its size, it has no use for a circulatory system. In the *Sargassum*, or brown algae (above), the "leaves" are exceedingly thin-walled, exposing a large surface area to the oxygen-laden environment. Oxygen and CO_2, then, travel directly between individual cells and the environment. When cells are not in contact with water, exchange must take place by the slow process of diffusion, perhaps aided by active transport.

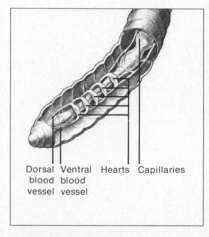

Dorsal Ventral Hearts Capillaries
blood   blood
vessel vessel

10.24 The closed circulatory system of the earthworm. The tubular rings that make up the hearts force blood backward through the dorsal vessel. Note the branches of the major vessels that lead into other parts of the body. The blood is continuously enclosed in vessels, thus the system is "closed."

Box 10.2 Blood

Among invertebrates the circulatory fluid, or blood, is quite diverse. It may be a colorless fluid or it may be (copper-containing), or even red (iron-containing). In vertebrates, the *erythrocytes,* or red blood cells, float in a colorless fluid called *plasma.* Other things floating in the plasma include proteins and various ions, as well as food and waste material. In addition to erythrocytes, the plasma may carry other kinds of cells such as leucocytes or *white blood cells,* which function mainly in combating invasions or infections, and cellular fragments called *platelets,* which function in clotting.

The red blood cells of vertebrates contain a protein-iron compound called *hemoglobin,* which is what makes blood red. Hemoglobin is able to combine reversibly with oxygen. The result is that human blood is able to carry about 60 times more oxygen than could be dissolved in plasma. In body parts where the oxygen concentration is high (such as in the lungs), hemoglobin combines readily with it. Where oxygen is low (such as in the tissues), hemoglobin tends to give up its oxygen.

Hemoglobin is comprised of four subunits, each of which has a protein chain (globin) and *heme* (a complex carbon ring structure that contains iron). Each of these subunits can hold one oxygen molecule.

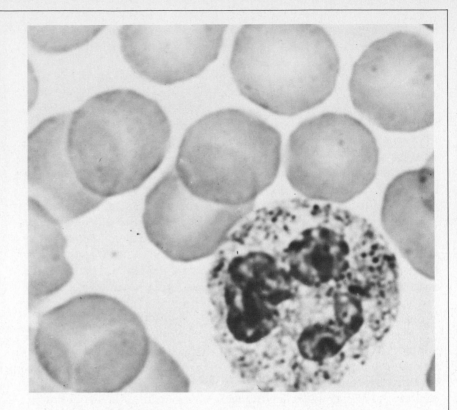

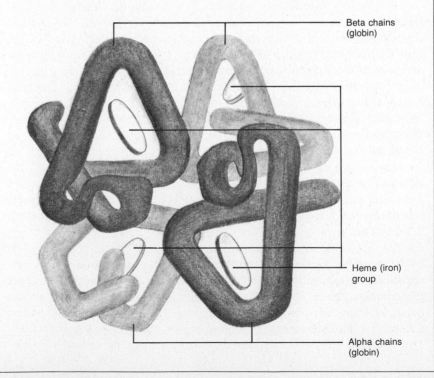

Beta chains (globin)

Heme (iron) group

Alpha chains (globin)

nutrients, and metabolic wastes are able to pass through them, to or from the tissues.

From the capillary beds the waste-laden blood flows into larger vessels called *venules,* and then into increasingly larger *veins,* which return the blood to the heart. The problem the body faces, of course, is how to get rid of the wastes it has picked up, primarily carbon dioxide and various nitrogen compounds.

Dissolved carbon dioxide gases are in equilibrium in the blood with carbonic acid and ionic bicarbonate. That is,

$$CO_2 + H_2O \leftrightarrow H_2CO_3 \leftrightarrow H^+ + HCO_3^-$$

| Dissolved | Carbonic | Ionic |
| carbon dioxide | acid | bicarbonate |

The problem here is to make sure that the waste produced by the cells is in the form of carbon dioxide when it reaches the lungs, since it is this form that passes through membranes most easily. The waste, in fact, is likely to reach the lungs as carbon dioxide because of the presence of an enzyme in the blood, called carbonic anhydrase, which moves the reactions to the left.

The nitrogenous waste products come from the body's breakdown of proteins, which leaves the unused components of the large molecules behind. Among these bits of waste matter are excess nitrogens, which can be extremely poisonous to the system. So how are the nitrogenous waste products gotten rid of? It may be in poor taste to tell you the end of an exciting story first, but you may as well know: the nitrogenous waste products are excreted in the kidneys.

The Heart

The muscular pump called the *heart* varies widely in invertebrates according to the life style of the organism, but in the vertebrates there is a rather regular progression from simple to complex.

In fish the heart is two-chambered, one chamber being a thin-walled *atrium* that accepts blood after it has been circulated through the body. When the atrium contracts, blood is forced into the *ventricle.* The atrium doesn't have to be very strong, because it doesn't take much work to pump blood only to the next chamber. The ventricle, however, must exert tremendous pressure, since it has to force the blood through the large artery called the *aorta* into the tiny mesh of the gill capillaries, where gases are exchanged, then through larger vessels into the capillary beds of the rest of the body, from where it reenters the veins. The fish ventricle, then, must be able to pump blood through two capillary beds. Not surprisingly, the fish ventricle is thick-walled and muscular.

The amphibian heart is a bit more complex in that it has two atria. One receives oxygenated blood from the lungs, and the other receives deoxygenated blood from the rest of the body. The two are mixed in the single ventricle, so the system is not particularly efficient; but it is efficient enough for the "cold-blooded" and not particularly active frog.

10.25 Various kinds of vertebrate hearts, ranging from the two-chambered to the four-chambered. The simple to complex arrangement is assumed to trace the evolutionary development of the "warmblooded" heart (that of birds and mammals). Note the increased efficiency in separating oxygenated and deoxygenated blood. These schemes are highly generalized and exceptions do exist. For example, among reptiles the crocodile heart is more strongly divided than that of snakes. In birds and mammals the developing tubular heart forms a loop, resulting in the atria coming to lie headward. In our illustration, however, the heart appears as if the looping had not occurred.

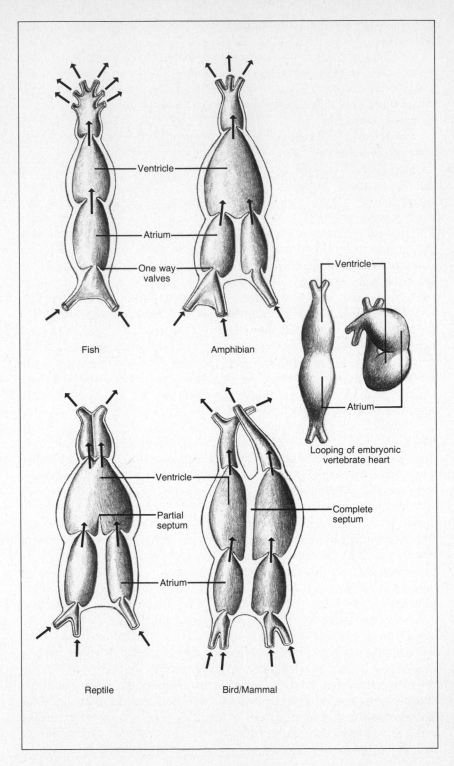

Reptiles go one step better and have two atria and two ventricles, although in most species the ventricles are not completely divided (exceptions include crocodiles and alligators). Because of the incomplete ventricular septum, there is some mixing of blood from the two sides of the heart. Still, this system is more efficient than the frog's, since one side of the heart pumps blood through the lungs and the other side pumps it through the rest of the body.

In birds and mammals the right and left sides of the heart are completely separated. The result is much greater efficiency. Birds and mammals are "warm-blooded" animals. That is, in most cases they maintain a rather constant internal body temperature. Just as it takes energy to maintain the temperature of your apartment, energy must also be expended in order for a warm-blooded animal to maintain its temperature. Thus, the metabolic rate of birds and mammals is higher than that of other animals.

The development of efficient circulatory systems and warm-bloodedness has rather interesting evolutionary implications. You may have noticed that on a cold day you can catch sluggish lizards rather easily as their metabolism drops with the temperature and they become inactive. On the other hand, it is no easier to run down a rabbit on a cold day than on a warm day. The rabbit's metabolism is relatively independent of the outside temperature. As warm-bloodedness developed in some lines of prehistoric animals, then, they would have had a distinct advantage over their cold-blooded cousins. This advantage is cited as one of the major reasons mammals eventually replaced the dinosaurs.

But let's see what the separation of the two halves of the heart has to do with efficiency. When waste-loaded blood enters the right atrium from the large veins (the *superior vena cava* and the *inferior vena cava*), it is then pumped by contraction of the atrium through a valve to the right ventricle. The right ventricle sends the blood, which is now low in oxygen, through the *pulmonary arteries* to the lungs. Here the blood picks up oxygen and then returns to the left atrium via the *pulmonary veins*. The left atrium pumps the blood into the left ventricle, which then contracts to send the blood on its long journey throughout the body. The left ventricle, then, must be very thick-walled and muscular; hence, the left side of the heart is larger. For this reason you may have the notion that your heart is on the left side. It isn't. It's right in the middle of your thoracic cavity and is about the size of your fist. Thus when you pledge allegiance to the flag, your hat is actually over your left lung.

The heart is supplied by nerves from special centers in the brain. These centers respond to a variety of stimuli. They increase or decrease heart rate according to whether arteries are being stretched (which decreases the heart rate) or veins are being stretched (which increases it) and according to the relative levels of carbon dioxide and oxygen in the blood, as we will discuss later. The system is not a simple one. For example, the diameter of specific arteries is controlled by the nervous system, so that

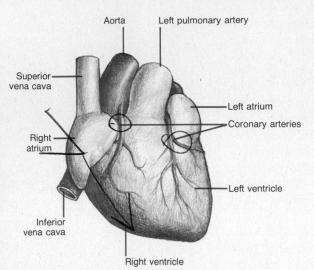

Aorta

Left pulmonary artery

Superior
vena cava

Left atrium

Coronary arteries

Right
atrium

Left ventricle

Inferior
vena cava

Right ventricle

10.26   The flow of blood through the human heart. Notice the coronary arteries that service the heart itself. Blood from the right atrium (deoxygenated and deep red) is pumped to the right ventricle and from there through pulmonary arteries to the lungs. It returns oxygenated and bright red to the left atrium and from there to the left ventricle to be pumped to the body via the aorta. Notice that the two atria contract simultaneously, as do the ventricles.

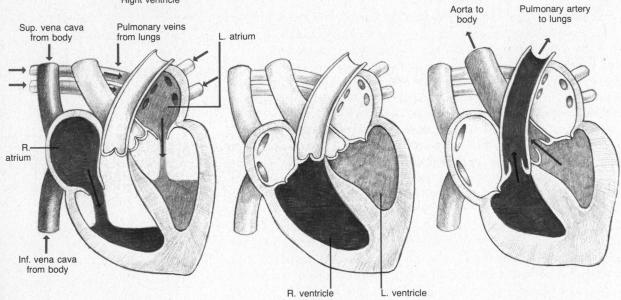

Sup. vena cava
from body

Pulmonary veins
from lungs

L. atrium

R.
atrium

Inf. vena cava
from body

R. ventricle

L. ventricle

Aorta to
body

Pulmonary artery
to lungs

Atrial contraction

Ventricles filled

Ventricle contraction

Box 10.3 Heart Attack

We have all heard the old story about the fainthearted guard dog who, upon being told the first time, "Attack," had one. Attack, in fact, is a word in common usage among almost all age groups in our country, as well it should be since heart attacks kill about 700,000 Americans alone each year. (Other countries may suffer more or less, perhaps depending upon their diet, health care, and stress levels.)

But what *is* a heart attack? It turns out to be any number of things. Often, however, it is associated with a moving blood clot that comes to block vessels servicing the heart muscle. Deprived of blood, heart tissue may die. Of course an individual can survive such an attack, but his heart is left weaker and more susceptible to the next attack. A phenomenon called *sudden death* may be due to chaotic and uncoordinated contractions of the ventricles. The heart may effectively stop pumping and after a few spasms may stop entirely.

Such malfunction is often behind the deaths of persons who mysteriously fall dead in their tracks. Many such people could have been saved if they were helped in time and, in Seattle, Washington, average citizens are being trained in cardio-pulmonary resuscitation (CPR) to help restore circulation in emergencies. The main thing is to get blood to the brain, where sensitive tissues die quickly without oxygen. In Seattle, one fifth of the population is now trained (at a cost of $1.25 each) and, so far, passersby have performed about one third of the city's resuscitations, and with a higher rate of recovery than those by professionals, since the passersby usually reach victims sooner.

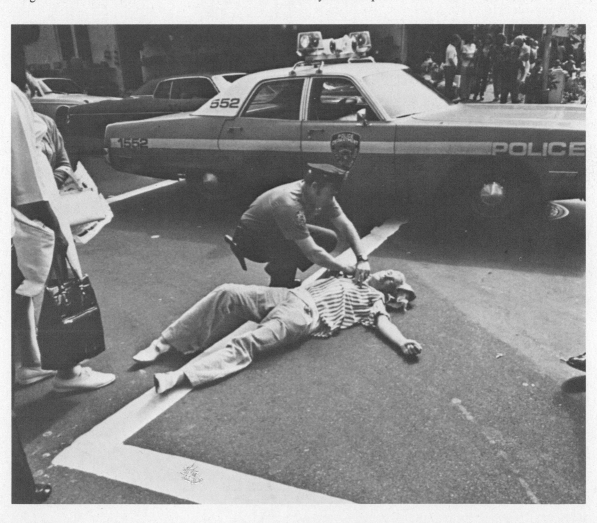

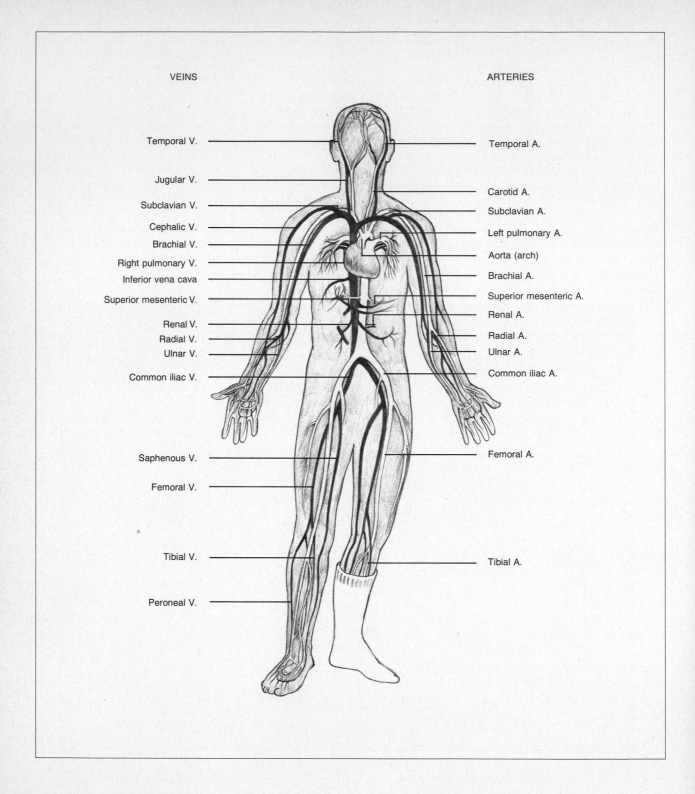

VEINS

ARTERIES

Temporal V.

Temporal A.

Jugular V.

Carotid A.

Subclavian V.

Subclavian A.

Cephalic V.

Left pulmonary A.

Brachial V.

Aorta (arch)

Right pulmonary V.

Brachial A.

Inferior vena cava

Superior mesenteric V.

Superior mesenteric A.

Renal A.

Renal V.

Radial V.

Radial A.

Ulnar V.

Ulnar A.

Common iliac V.

Common iliac A.

Saphenous V.

Femoral A.

Femoral V.

Tibial V.

Tibial A.

Peroneal V.

after a meal more blood is shunted to the stomach and intestines to aid in digestion. But the sight of a shadowy figure in an alleyway as you leave a restaurant can cause blood to be shunted back to the muscles.

Figure 10.27 shows the major blood vessels of the human body. Where are the pulmonary vessels? What is the large network below the heart on the right side of the body? You might now wonder why the good guys survive so many shoulder wounds in so many bad Westerns. According to the figure, where is a better place to be shot? Why aren't the veins on the back of your hand in the same place as those of the person sitting next to you? (You might refer to circulatory system development in Chapter 8.)

The Lymphatic System

Before we leave the topic of circulation we should take note of another, often-neglected, circulatory system, the *lymphatic system* (Figure 10.28). The lymphatic system circulates *lymph* through the body outside blood vessels and returns this fluid to the blood, where it is then called *plasma*. Blood plasma, minus the dissolved proteins, filters out into the tissues through tiny capillaries, bathes the cells, moves between them and finally winds up in small channels in the tissue called *lymph capillaries*. The lymph fluid may then move into larger channels and finally reach any of various lymph nodes, where it is filtered, thus removing any cellular debris or bacteria. Frogs and some other vertebrates have "lymph hearts" that move the lymph along, but mammals rely chiefly on muscular movements to circulate the fluid.

White blood cells called *lymphocytes* are found in great numbers in the lymph nodes. If bacteria have entered the system, they will probably eventually find themselves at one of these nodes, where they will be destroyed by the lymphocytes. The lymph flows into increasingly larger channels, or ducts, until finally, it empties into large veins near the heart.

DIGESTIVE SYSTEMS

The tired old notion of our bodies being internal-combustion engines, with our food the fuel that runs them, can still be squeezed for elements of truth. The idea is that work must be done to minimize body entropy, and the energy for this work must be released from the food in which it is bound. That energy is released through first breaking the food down into its constituents (digestion), transporting the product to the cells (circulation), and enabling the cells—which are actually the engines, if the analogy is to be valid at all—to burn the food for energy (metabolism). Of course, not all ingested material is "burned." Some of it is used as building blocks for the repair and maintenance of the machinery.

Digestive Arrangements

The basic scheme for digestion is the same in all animals, except for the

10.27 The major blood vessels of the human body. The arterial system (light) and venous system (dark) are connected by capillaries. Generally the arteries are thicker walled because they are under intense pressure by blood being pumped from the heart. The pressure is dissipated in the immense system of capillaries, some so tiny that red blood cells must move in jerky movements, single file, between the cells. The jerkiness is due to the blood's being pushed by ventricular contractions and only being allowed to back up until stopped by a system of one way valves, until it is thrust forward again. Obviously, blood leaving the capillaries of the lower extremities does not have the force to move back up the legs. This blood is forced along largely by muscular contractions. This is why, after being in one position for a time, you may feel the desire to stretch, which contracts muscles and squeezes the blood back to the heart.

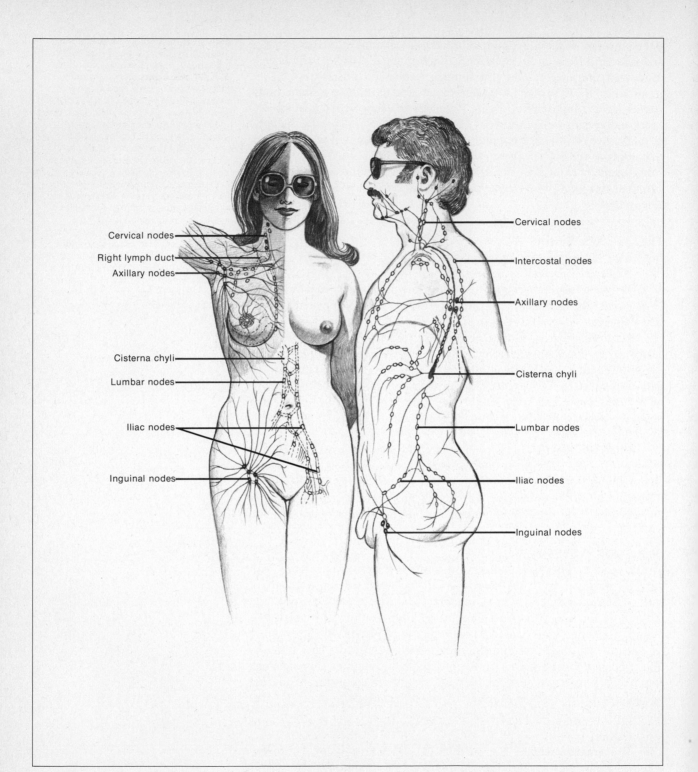

Cervical nodes

Right lymph duct

Axillary nodes

Cisterna chyli

Lumbar nodes

Iliac nodes

Inguinal nodes

Cervical nodes

Intercostal nodes

Axillary nodes

Cisterna chyli

Lumbar nodes

Iliac nodes

Inguinal nodes

details. However, these details vary widely, as you can see in Figure 10.29. An amoeba, for example, simply flows around a food particle and engulfs it, so that the particle is held in a *food vacuole*. It then pumps acidic digestive enzymes into the vacuole until the food is digested. Indigestible particles are brought to the surface of the animal and squeezed out through an opening that appears in the membrane.

The hydra and the flatworm both have pouchlike digestive systems. Whereas the hydra stings its prey with special cells on its tentacles and then hauls the prey into its digestive cavity, the flatworm is able to extrude its *pharynx,* or "throat," to capture food. In both these animals digestion begins in the *gastric cavity,* or "stomach," but before the process is completed, the particles of partly digested food are engulfed by cells of the body wall and digestion is completed within these cells. So these animals digest food by both extracellular and intracellular processes. (Which of these two processes typifies human digestion?)

In most more complex animals the digestive system is a tube rather than a pouch, with one opening through which food enters and another through which it exits—a much more civilized arrangement.

Although both humans and earthworms have tubelike digestive systems, you will notice in the earthworm several structures that are absent in man, such as the crop and gizzard. The *crop* is where food is stored, and the muscular *gizzard* is where food is ground against small stones that have been ingested. So there is now no reason for you to lie awake night after night wondering how an earthworm chews. Birds, which also lack teeth, have a similar digestive arrangement.

As another fascinating aside, earthworms have amoeboid cells in their bodies that ingest particles that move through the digestive tract as the worm literally eats its way through the soil. These cells then migrate through the body and come to rest just under the skin, giving the earthworm a color similar to the soil in which it lives, causing the early bird to come up short.

In vertebrates, such as the salamander, food enters a well-defined mouth, is swallowed by voluntary action, and then moves down the smooth-walled *esophagus* by involuntary muscle contractions. The food then enters the stomach, where digestion begins. It then passes on to the intestine, where it is further broken down by the action of secretions from the pancreas and liver. When the food is finally digested, it moves through the intestinal wall into blood vessels that lie in the wall. The nutrients are then carried by the blood to the tissues where they will be used. The indigestible particles move on through the intestine, to be eliminated through the *anus* or the *cloaca.* (The *cloaca* is a common opening for the intestine, kidneys, and reproductive organs found in amphibians, reptiles, and birds—a primitive plumbing arrangement that has been "improved" upon in mammals.)

The most important chemical process in digestion is hydrolysis. Hydrolysis, as we saw in Chapter 3, is the breaking down of substrate

10.28 The human lymphatic system, sometimes referred to as the "other" circulatory system. Lymph nodes are scattered throughout the body, but concentrated in specific areas. These nodes harbor large numbers of one kind of white blood cell and they tend to trap bacteria and other foreign matter, which these cells then attack. Lymph nodes tend to swell and become sore if they are involved in fighting an infection near them, thus they may signal infections that might otherwise go unnoticed. The lymph nodes are often affected by cancer of the breast and must be removed with the breast and its underlying muscles. Mastectomies of this sort are currently being reexamined to see whether they are merited or effective in most cases.

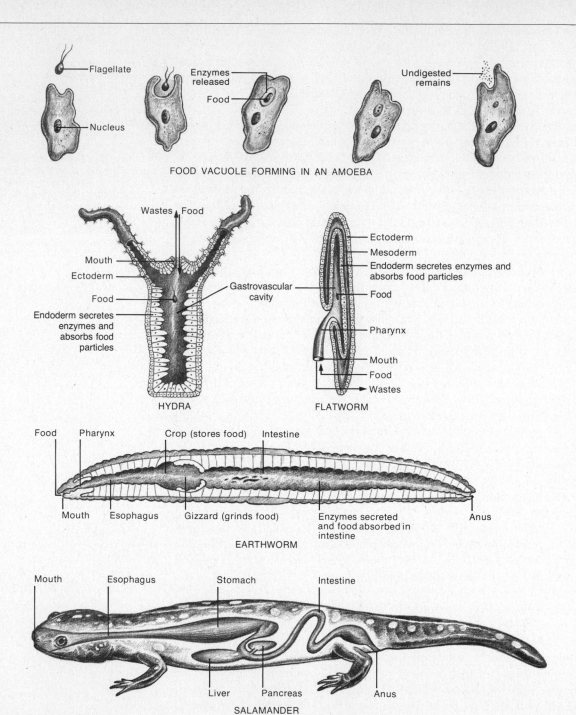

Flagellate

Enzymes released

Food

Undigested remains

Nucleus

FOOD VACUOLE FORMING IN AN AMOEBA

Wastes   Food

Mouth

Ectoderm

Food

Endoderm secretes enzymes and absorbs food particles

Gastrovascular cavity

HYDRA

Ectoderm

Mesoderm

Endoderm secretes enzymes and absorbs food particles

Food

Pharynx

Mouth

Food

Wastes

FLATWORM

Food   Pharynx

Crop (stores food)   Intestine

Mouth   Esophagus   Gizzard (grinds food)   Enzymes secreted and food absorbed in intestine   Anus

EARTHWORM

Mouth   Esophagus   Stomach   Intestine

Liver   Pancreas   Anus

SALAMANDER

matter by the chemical addition of water. Large molecules of proteins, fats, and carbohydrates are thus broken down into amino acids, glucose, glycerol, and fatty acids, which are absorbed through the intestinal wall and enter the bloodstream. Our concern here, then, is the conversion of larger food particles to smaller ones.

The Human Digestive System

In the human digestive system, shown in Figure 10.30, the breakdown of food particles begins in the mouth, where large particles are physically broken into smaller ones by the action of chewing. Chewing is especially important to animals that eat green plants, since the cellulose cell walls cannot be broken down by digestive enzymes. In contrast, animal tissue *can* be entirely broken down by digestive enzymes. This is the reason you never see cats chewing their cud. In the mouth, the food is lubricated by saliva (this is not my favorite part of biology, either). Also, some digestion begins in the mouth as starches are changed to disaccharides, or double sugars. The food then moves to the stomach through the smooth-walled esophagus, with food further moistened by glands in the esophagus.

The *stomach* is a muscular sac that churns the food as it mixes it with various secretions from the stomach wall, such as mucous, hydrochloric acid, and enzymes. The food is sealed in the stomach through all this process by two sphincters or rings of muscles, one at either end of the stomach. After the mixing is completed, the lower sphincter opens, and the stomach squeezes the food into the small intestine.

The *small intestine* is a long, convoluted tube in which digestion is completed and through which the nutrients enter the bloodstream. The inner surface of the small intestine is covered with tiny, fingerlike projections called *villi*, which greatly increase the surface area of the intestinal lining. The surface area of the villi themselves is further increased by tiny projections called *microvilli*. Each villus may have about 3,000 of these microvilli. Within the villi are numerous tiny lymph vessels, as well as tiny branches of the mesenteric arteries and veins, the blood vessels that supply much of the viscera. Some of the products of digestion move into these blood vessels.

Interestingly, fat doesn't have to be entirely broken down before it enters the circulatory system. Smaller fatty particles enter the blood vessels, but larger ones may pass directly into the lymph vessels in the villi. It is interesting and almost weird that the products of fat digestion are so quickly reconstituted to fat in the intestinal mucosal cells that after a meal high in fat content these particles may appear in the bloodstream and give the blood itself a milky appearance.

The process of absorption is accomplished in part by passive *diffusion* across the membranes of the intestinal lining, but also in part by *active transport,* a process in which energy is expended to do the work.

10.29 Digestive systems in a variety of animals. The amoeba has an unspecialized system, ingesting food through any part of its surface, simply ejecting waste through its membrane. The *Hydra* and flatworm have saclike digestive tracts, food exiting over the same route it entered. The earthworm has an essentially tubular system with a food-grinding gizzard to prepare the food for digestion. The tubular system of the salamander is embellished by a number of glands and organs that alter the food as it passes. Notice that technically food always remains outside the body until its digested products pass through the intestinal membranes. Thus, "I would like to get a hamburger in me," should perhaps be restated, "I would like to get on the outside of a hamburger."

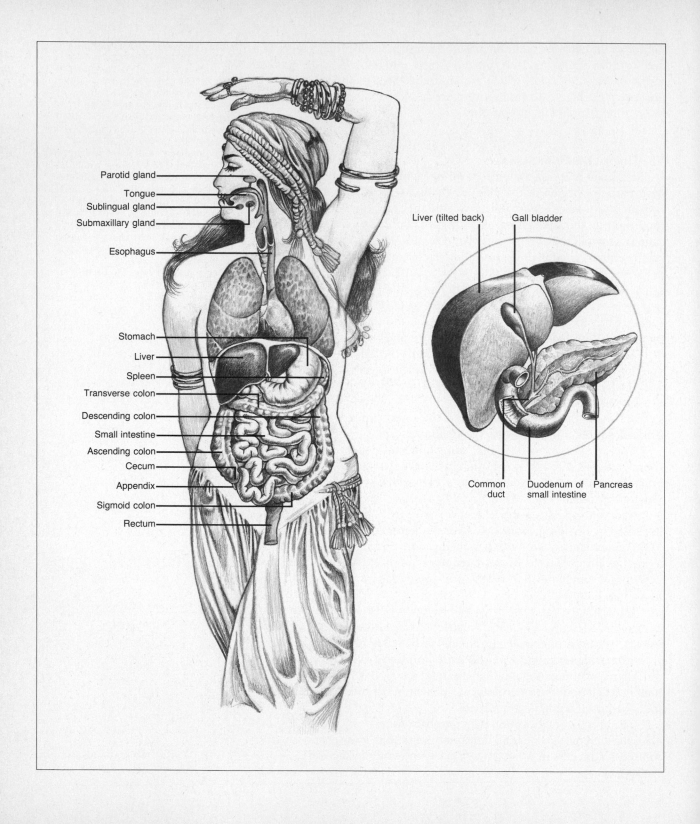

Parotid gland

Tongue

Sublingual gland

Submaxillary gland

Esophagus

Stomach

Liver

Spleen

Transverse colon

Descending colon

Small intestine

Ascending colon

Cecum

Appendix

Sigmoid colon

Rectum

Liver (tilted back)   Gall bladder

Common duct

Duodenum of small intestine

Pancreas

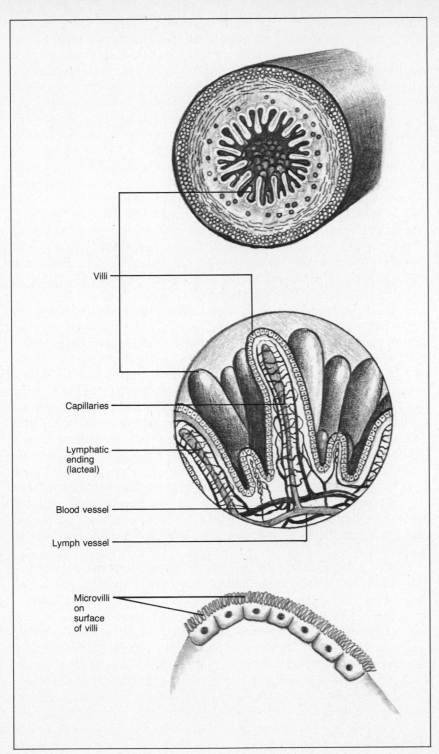

Villi

Capillaries

Lymphatic
ending
(lacteal)

Blood vessel

Lymph vessel

Microvilli
on
surface
of villi

10.30 The human digestive system (opposite page). Once food enters the esophagus, it is moved along the digestive tract largely by involuntary muscle movements. Various enzymes and other juices are added along the way to break down the food molecules so that by the time they reach the small intestine they can be absorbed into the bloodstream. Undigested or indigestible matter then moves to the large intestine where it waits for the 7:05 out. The digestive process is curiously involved with emotion somehow, so that when we are stressed the stomach may overproduce acid that, when it moves back up the esophagus a short way, causes heartburn. In addition, anxiety may reduce the efficiency of digestion or may cause sphincter muscles to remain closed, stopping the movement of food through the tract.

10.31 Cross section of a villus from the human small intestine. These tiny fingerlike projections add immense surface area to the digestive tract. They in turn are covered by about 3,000 smaller projections, the microvilli. The villi house lymph vessels. The blood vessels absorb most of the products of digestion, but the larger fat particles may pass directly into the lymphatic system. The breakdown products of fat metabolism are quickly reconstituted to fat once they enter the circulatory system.

Digestion in Humans

The events of digestion are complex and numerous, but we might briefly review the process in humans. The first ten inches of the small intestine comprises the *duodenum*. The enzymes produced by the pancreas enter the gut in this area. The pancreas secretes powerful *proteolytic* enzymes, which break down proteins. These are powerful agents, so why doesn't the pancreas digest itself? the sharp-witted student will ask. The agile-minded physiologist will quickly respond that this is prevented by the fact that the enzymes are stored in an inactive form in the pancreas and are activated by the duodenal environment. Well, then why isn't the duodenum digested? As a matter of fact, it would be if it were not "alive," with a membrane system, which actively excludes harmful substances such as digestive enzymes. This is also why, if you are unfortunate enough to be "wormy," when a worm dies, it is immediately digested. But I suppose such ultimate vengeance provides slim consolation to the host. In addition to this inherent protection, the intestine secretes a protective shield of mucus, and the high level of sodium bicarbonate from the pancreas also helps to neutralize the acids of the stomach juices.

Another substance secreted into the duodenum is *bile,* which is stored in the gall bladder after it is produced by the liver. Bile contains certain sodium salts that act as detergents (the biodegradable kind) breaking up the large fat droplets, so that the pancreatic enzymes can work on them.

Nutrient-laden blood from the intestine is carried into the *hepatic portal vein,* which leads to the liver (a *portal* vessel is one that lies between two capillary beds). The blood then filters through tiny vessels and sinuses in the liver. If it contains excess nutrients, some are removed by the liver and stored. If it contains too few nutrients, the liver responds by releasing some of its stored nutrients. Thus proper glucose and amino acid levels are maintained in the blood, despite variations in food intake. This is one more example of the amazing internal regulation, or *homeostasis,* accomplished by the body. As the body's resources of glucose drop below the amount needed to maintain a constant blood level, we have the sensation of hunger and are motivated to go out and find some glucose (in one form or another).

Food passes from the duodenum through the next two segments of the small intestine, the *jejunum* and the *ileum,* where further digestion and absorption occurs. The remaining waste products then move on into the *large intestine.* The large intestine consists of two portions, the *colon* and the *rectum.* The inner surface of the large intestine is rather smooth, and no digestion occurs here. However, the colon is a veritable hotbed of a bacterium called *Escherichia coli,* which provides some vitamins for the body in the form of its own waste product. (The presence of these bacteria in the environment is an indication of fecal contamination, an important factor in the spread of some diseases, such as infectious hepatitis.) The large intestine also extracts water from the solid waste products, or *feces,* with constipation resulting if you put off nature's call (more "relevance"!). The feces are stored in the rectum before elimination through the anus.

So we end our rather lengthy chapter on this poetic note. We have two more systems to consider, but we can save them for our next chapter. These systems, hormones and nerves, are considered separately because they share an intimate role in homeostatic mechanisms and because they set the stage for the following topic: behavior.

Hormones and Nerves

In this chapter we will look at the hormone and nervous systems together because both are intimately involved in the processes of internal regulation, and also because in some cases they act together. Regulation, as we have already seen, is a many-faceted phenomenon. Consider the regulatory function of DNA in making protein, or the regulatory function of the kidney in maintaining water balance. Regulation in its broadest sense serves to keep bodily processes within allowable limits and coordinates those processes so that the organism can function more efficiently in the part of the world it has available to it. Hormones and nerves, as we will see, are involved in regulating and coordinating an entire orchestra of bodily processes so as to minimize "dissonance," or *entropy*. In the next chapter we will deal with behavior, so it may also be good to set the stage by considering the broader role of the nervous system. Let's begin with hormones.

HORMONAL REGULATION

First, what is a hormone? (Ready with your markers?) A *hormone* is a chemical that is produced in one part of the body and carried by the blood to another part of the body, where it ultimately influences some process or activity there. Hormones are produced by *endocrine,* or *ductless, glands* that are found in various parts of the body.

Hormones perform regulatory functions in invertebrates as well as vertebrates; in fact, much important hormone research has been carried out on various arthropods, such as insects and crayfish. Hormones have been found to regulate the changes in body organization, which many insects undergo as they pass through their larval and pupal stages. A typical experiment to show the importance of hormones in such development might entail depriving part of an insect body of a hormone, noting the lack of development in that body part as the rest of the body matures normally, and then injecting the proper hormone into the inhibited part and observing the subsequent normal growth.

One of the greatest problems in arranging such an experiment is acquiring the hormones. Animal bodies actually contain very little hormone substances, since only tiny amounts in the body are necessary to produce great changes. For example, a woman produces only about a teaspoonful of estrogens in an entire lifetime. To extract only a few

11.1 The major endocrine glands in the human. Because of the immense effect of small amounts of secretions from these glands and because they often interact in complex ways, their release must be very closely regulated. The pituitary is about the size and shape of a bean and is located in the geometric center of the skull. It is the source of several hormones, some of which affect other glands, but it, in turn, may be affected by them. You should be aware of the role of each of these glands and the results of their malfunction. Hormones are formed in one part of the body and operate in another, usually after being carried there by the bloodstream. Some diseases can be diagnosed by variations in the amounts of certain hormones in the blood. In addition, some hormone levels may be changed by experience. For example, there is preliminary evidence that dominance in males causes a rise in testosterone levels (but it might be argued that the cause-effect relationship is reversed).

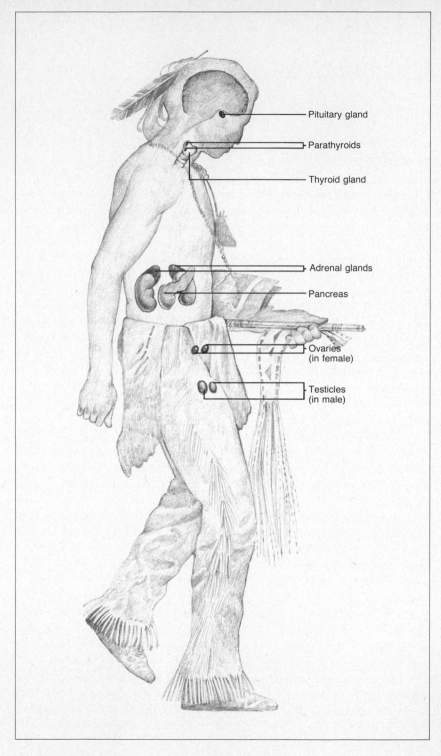

Pituitary gland

Parathyroids

Thyroid gland

Adrenal glands

Pancreas

Ovaries
(in female)

Testicles
(in male)

milligrams of another of the estrogens, estradiol, more than *two tons* of pig ovaries were required.

"Chemical messengers," as hormones are sometimes called, may function in keeping variable bodily processes within their proper limits, as does the antidiuretic hormone, or they may direct permanent changes, such as growth or sexual maturation. They may also function in emergency situations by causing the body to prepare for stress. And they may help the organism to survive better in its environment in other ways, as when amphibians or reptiles change colors to match their surroundings. Hormone action is also a part of nature's chemical warfare arsenal. Some plants, for example, manufacture substances that mimic the growth hormones of insects and thus they disrupt the normal growth of insect larvae that feed on them.

THE HUMAN ENDOCRINE SYSTEM

In the human endocrine system the *pituitary* has been called the "master gland," but this term isn't really an apt description, since many of its activities are regulated by the hypothalamus (described later). The pituitary does have a number of important functions, however, and it also exerts an effect on certain other glands. It is composed of two major parts, the *anterior* and *posterior lobes,* with a central *midlobe* area. The anterior and posterior lobes actually develop from different embryological tissue. The posterior lobe develops from nerve tissue and the anterior lobe from an outpocketing of the embryonic mouth. (What germ layers are involved?) The area between these, the midlobe, even has a function of its own.

The pituitary is about the size of a bean and is located at the geometric center of the skull, where it lies shielded by heavy layers of surrounding bones. Its small size belies its physiological complexity. In all, the pituitary secretes nine known hormones. Three of the hormones secreted by the anterior pituitary are called *tropic hormones* because they regulate the secretions of other glands—the thyroid, the adrenals, and the gonads, or sex glands.

The anterior pituitary also secretes a nontropic growth hormone called *somatotropin,* which controls bone and muscle growth. If too little somatotropin is present during childhood, the result is a *pituitary dwarf,* or *midget.* Too much somatotropin results in a *giant.* Excessive somatotropin in an adult increases the size of only certain bones that are still able to respond to the hormone—the jaw and the bones in the hands and feet. This condition is called *acromegaly.*

When the *thyroid* is stimulated by the pituitary, it produces *thyroxine,* the hormone that controls the body's metabolic rate. Deficiency of thyroxine in childhood results in a characteristic physical appearance and mental retardation called *cretinism.* It is now possible to recognize thyroid deficiency early in life, so that thyroxine can be administered artificially to promote normal development.

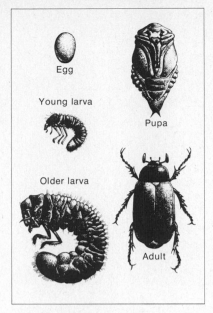

11.2 Development in a beetle. Not all insect species go through the sequence of egg, larva, pupa, adult, but whatever their pattern, each stage is brought on at a specific time by a different homone.

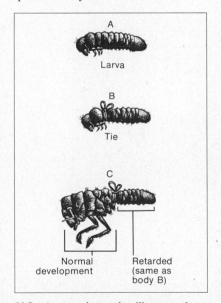

11.3 An experiment that illustrates the importance of hormones in insect development. When growth hormones produced in the head are kept from reaching other parts of the body, those parts do not mature.

11.4a Typical male and female physiques. Both bodies have reacted to different hormones, such as testosterone in the male and estrogen in the female. Also, their tissues react differently to the same hormones as target sites become selectively sensitized in the normal process of development.

The gonads, the testes, and the ovaries predominately produce *testosterone* and *estrogen,* respectively, hormones that are responsible for those secondary sex characteristics we have all come to love and admire. In males these characteristics include a deepening of the voice, broadened shoulders, hairy chests, facial hair (beards and mustaches), muscular development, sperm production, and increased production of red blood cells (which might explain the term "red-blooded he-man"). In females estrogen causes changes in the distribution of body fat, of body hair, and development of the breasts. *Progesterone,* produced by the corpus luteum, stimulates the development of the uterine lining and causes further breast development during pregnancy.

Ad- means "upon" and *renal* refers to the kidney, so guess where the *adrenal glands* are located. The adrenals have two distinct parts, the outer *cortex* and the inner *medulla.* About fifty steroids have been isolated from the adrenal cortex of various mammals, and it is believed some of these may be used in the production of different hormones, but it is not yet known just what role, if any, these steroids play. One hormone produced in the cortex is *cortisone* (manufactured when the cortex is stimulated by a secretion, called ACTH, from the anterior pituitary). *Aldosterone* is also produced in the adrenal cortex and functions in the regulation of salt and water in the body. It works by influencing the cells of the kidney's distal convoluted tubule to increase their rate of sodium reabsorption. When sodium is reabsorbed, water must follow because of the osmotic gradient that is set up. The renal cortex also produces some male sex hormones. This system is usually shut down in women, but if it should be activated, for example, by a tumor, it will cause such characteristic masculine qualities as beard growth.

The adrenal medulla secretes the hormones *norepinephrine* and *epinephrine,* which for our purposes will be referred to simply as adrenaline (which you will find out about shortly). As you will see, these hormones may produce effects similar to those caused by certain emotions. Other endocrine glands such as the thyroid, the hormones they produce, and their principal effects are listed in Table 11.1.

11.4b In 1914 these were the tallest and shortest women in Europe. The giant is seven feet four inches tall and weighs 294 pounds. She is sixteen years old. The midget is eighteen, is twenty-five inches tall, and weighs thirteen pounds. Both conditions arise from abnormal pituitary secretions.

How Hormones Work

Actually, we don't know how hormones work. Although we have known the composition of some hormones for a long time, we still aren't sure how any of them work. One hypothesis, however, states that when hormones come in contact with the membranes of "target cells," they alter certain sites on the membranes of these cells and trigger the production of a group of substances called *prostaglandins* within the membrane (see Chapter 7). Prostaglandins are involved in the production of the enzyme *adenyl cyclase,* which helps to change ATP to a form of adenosine monophosphate, cyclic AMP, in the cytoplasm. The cyclic AMP may then do any number of things. For example, it may alter enzyme activity (think of the implications) or it may stimuate the production of *m*RNA. Changes such as these could theoretically produce all the effects that are attributed to hormone action.

Feedback Systems

The regulation of hormone production is achieved, in part, through a negative feedback system. As an example, consider the interaction of the tropic hormones and their target endocrine glands. The anterior pituitary gland secretes a thyroid-stimulating hormone (TSH) that causes the thyroid gland to secrete thyroxine. As the thyroid responds and the

Table 11.1   Major Vertebrate Endocrine Glands and Their Hormones

Gland	Hormone	Principal Action
Anterior lobe of pituitary	Thyrotropic hormone (TS)	Stimulates thyroid
	Follicle-stimulating hormone (FSH)	Stimulates ovarian follicle
	Luteinizing hormone (LH)	Stimulates testes in male, corpus luteum in female
	Growth hormone	Stimulates growth of bones and muscles
	Adrenocorticotropic hormone (ACTH)	Stimulates adrenal cortex
	Prolactin	Stimulates secretion of milk, parental behavior (such as nest building in birds and fish)
	Interstitial cell-stimulating hormone	Stimulates testosterone production
Midlobe of pituitary	Melanocyte-stimulating hormone	Regulates color of skin in reptiles and amphibians
Posterior lobe of pituitary	Oxytocin	Stimulates contractions of uterine muscle, milk release
	ADH or vasopressin	Controls water reabsorption
Thyroid	Thyroxine	Controls metabolism
Parathyroid	Parathyroid hormone	Controls calcium metabolism
Testes	Testosterone	Stimulates production of sperm and secondary sex characteristics
Ovary, follicle	Estrogens	Stimulates female secondary sex characteristics
Ovary, corpus luteum	Progesterone	Stimulates growth of uterus
Adrenal medulla	Epinephrine, norepinephrine (adrenalin)	Activates sympathetic nervous system
Adrenal cortex	Cortisone, cortisonelike hormones, aldosterone	Controls carbohydrate metabolism, salt, water, and sugar
Pancreas	Glucagon	Stimulates breakdown of glycogen into glucose
	Insulin	Lowers blood sugar levels, increases formation and storage of glycogen

thyroxine level rises in the blood, the thyroxine itself has a depressing effect on the secretion of TSH, resulting in an eventual lowering of thyroxine and a release of more TSH. The levels of both hormones are thus kept within tight limits, because of the influence they exert on each other. Such feedback mechanisms are common throughout the endocrine system and furnish some of the clearest and most fascinating examples of homeostasis in living things.

THE RELATIONSHIP OF HORMONAL AND NEURAL CONTROL

Nerves may regulate bodily processes in much the same way hormones do. In some cases they may even duplicate the work of hormones. In other cases the nerves act in coordination with hormones or independently of them. Whether such regulation is hormonal or neural (pertaining to nerves), it is accomplished *chemically*.

The dividing line between regulation by nerves and hormones is not clear-cut. For example, the pituitary gland itself is partly formed from embryonic neural tissue. Also, it lies directly under the part of the brain called the *hypothalamus*. The anterior pituitary is at least partly under the control of substances released from the hypothalamus. The posterior lobe of the pituitary contains the ends of nerves that originate in the hypothalamus. Also, two hormones associated with the posterior pituitary are known to be actually manufactured in the hypothalamus and only stored in the posterior pituitary.

The adrenal glands provide other examples of close nerve-hormone interaction. *Adrenaline* is the hormone responsible for those astonishing physical feats we hear of people performing in emergencies (the story about a little old lady lifting a car off her trapped husband crops up periodically). Adrenaline is released as a result of neural stimulation of the adrenal medulla. Neural stimulation also causes the pituitary to manufacture more ACTH and thus to boost the production of cortisonelike hormones by the adrenal cortex.

PLANT HORMONES

The seeming simplicity of plants sometimes leaves us unprepared for their complexities. As a case in point, researchers have for years been trying to unravel the mysteries of plant hormonal control. Many aspects of plant growth and development are under the control of these itinerant chemicals, and so the search is not only fascinating, but may prove to be imperative as we grow more reliant on these intermediate links with our life-giving sun. Let's briefly consider three of the better known groups of plant hormones: auxins, gibberellins, and cytokinins. Keep in mind that these are only three; there are others.

Auxins

Some of the first experiments on plant hormones were conducted by one Charles Darwin and his son, Francis, and reported in an 1881 publication

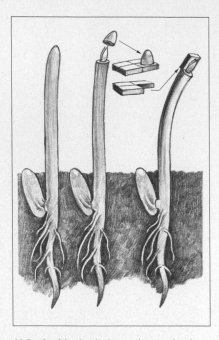

11.5 In this classical experiment, the tip was cut from an oak seedling and placed on an agar cube to absorb any juices, including hormones, from the tip. Then the block was placed off center on a decapitated seedling. As the seedling grew it bent away from the side with the agar block, demonstrating the growth-inducing properties of plant hormones.

entitled *The Power of Movement in Plants.* After noticing that plants tend to bend toward light, they covered the shoot tip of growing plants and exposed it to light from the side. Nothing happened; the plants continued to grow straight up. If the tips were capped with transparent glass, however, normal bending occurred. They stated, "We must therefore conclude that when seedlings are freely exposed to a lateral light some influence is transmitted from the upper to the lower part, causing the latter to bend."

In 1926, the Dutch physiologist, Frits W. Went, discovered something else about that "influence." He cut off the tips of emerging oat seedlings and set them on a block of gelatinous *agar* for about an hour. The decapitated plants were kept in a dark place. Then they were removed and pieces of the agar on which the tips had rested were placed along one side of the stump. Within an hour the plant began bending in the opposite direction (Figure 11.5). Thus the "influence" was believed to be chemical and was called *auxin* (from the Greek *auxein,* to increase). It was later found that auxin is generally produced in the stem tips and seeps through the cell tissue to the rest of the plant, instead of traveling in pipelines of xylem and phloem.

But how does auxin control elongation in plants? No one knows. But it is known that auxin softens the rigid cell walls of plants and allows water to swell the cells. Thus growth may be accomplished by cell enlargement rather than cell division.

The only naturally occurring auxin is indoleacetic acid, or IAA, but a synthetic form, 2, 4-D, is a powerful weedkiller (as is auxin at high concentrations) and was used by the military in Southeast Asia.

Small amounts of auxin may stimulate the growth of roots, and although it doesn't affect the growth of leaves, it plays a role in the leaf drop. With the approach of fall, the plant draws certain ions, amino acids, and sugars from its leaves and auxin helps to break down the cells that hold the leaf to the stem. Auxin also can stimulate fruit to grow without having ever been pollinated, thus producing seedless varieties.

Gibberellin

While Went was performing his agar experiments, E. Kurosawa in Japan was looking for the cause of the "foolish seedling disease" of rice. The disease caused the plants to become spindly and pale and to simply collapse. It turned out that the culprit was a parasitic fungus that contained a substance called *gibberellin.*

Gibberellins are found in most, if not all, plants and they have some rather surprising traits. For example, they cause dramatic increases in stem length by stimulating cell division and cell elongation in both leaves and stems. In my first garden, I planted lettuce seeds in a cool climate and my lettuce, to my puzzlement, grew to about five feet tall. The effect of cold weather on several plants, it turns out, is about the same as if they had been stimulated by gibberellin (Figure 11.6).

Gibberellins stimulate pollen germination and the growth of pollen tubes in a number of plant genera, and can break the dormancy of many kinds of seeds that normally can be aroused by nothing but cold or light.

Cytokinins

In 1941, a Dutch physiologist, J. van Overbeek, found that coconut milk (which is really a liquid endosperm, or "seed food") contained a peculiar growth factor, unlike anything known. The factor, whatever it was, not only accelerated the development of plant embryos, but it increased the growth rate of isolated cells in a test tube. Also, a drop of coconut milk could stimulate mature, nonmitotic cells to begin dividing again.

Coconut milk is rich in a variety of substances, however, and years of research in an effort to isolate the growth factor proved futile. In a new attack, researchers began looking for the factor in something easier to work with. It took a graduate student to find out that old herring-sperm DNA could make tobacco cells divide. Then it was found that any stale DNA would provide the factor. Apparently the factor was a breakdown product of nucleic acid. It was isolated and called kinetin, of the family of cytokinins (from cytokinesis, cell division).

Cytokinins can react with auxin to produce a variety of growth effects such as rapid cell division. It can also work alone to enhance germination once it has started. In addition, cytokinin can somehow keep leaves from turning yellow after they are removed from the tree. Apparently it keeps the DNA that functions in young, healthy leaves from turning off. In general, however, the workings of the cytokinins remain a mystery.

THE NERVOUS SYSTEM

In considering the nervous system we will begin by learning something about that mysterious structure called the brain and the great nerve to which it is connected. But first let's see how this complex system arose by tracing some nervous systems from simple to more complex forms. We could say from "lower" to "higher" forms, except that the term *higher* is admittedly prejudiced. What is usually meant by the word is any characteristic similar to those of humans, such as a large "thinking" center. The implication of this usage is that all species tend to evolve toward humanlike characteristics. Nothing could be further from the truth.

The Evolution of the Central Nervous System

Clues to the evolutionary development of nervous systems can be gleaned from a cross-species survey, a review of animal nervous systems from the simple to the more complex (Figure 11.7). One of the simplest nervous systems is that of the freshwater hydra. It consists simply of a two-dimensional net of interconnecting *neurons,* or nerve cells, spread throughout the outer body layer. The entire surface of the animal is about equally covered. There is no part that controls the rest, no nerve center that funtions in the regulation or coordination of the nerve net.

11.6 Treatment with gibberellins may cause bizarre growth in some plants. The normal cabbages on the left are shown with hormone-treated cabbage plants that have "bolted."

11.7 Examples of animal nervous systems. It is likely that the evolutionary route of vertebrate nervous systems was much the same. A, the *Hydra,* with its nerve net composed of connecting neurons; B, the planarian, or flatworm, with its longitudinal nerves connected by transverse nerves and its nerve concentration in the head; C, the earthworm, with its single ventral nerve cord and well-defined cerebral ganglia; and D, the frog, with its dorsal hollow nerve cord and the well-developed brain protected by bone. The trend is toward condensing the nervous system into a longitudinal arrangement (concomitant with the development of bilateral symmetry). A segmented body innervated by a segmented nervous system sets the stage for specialization along the nerve length. The nerve cord moves to a dorsal position in vertebrates as the brain becomes more complex and the segmentation becomes less regular.

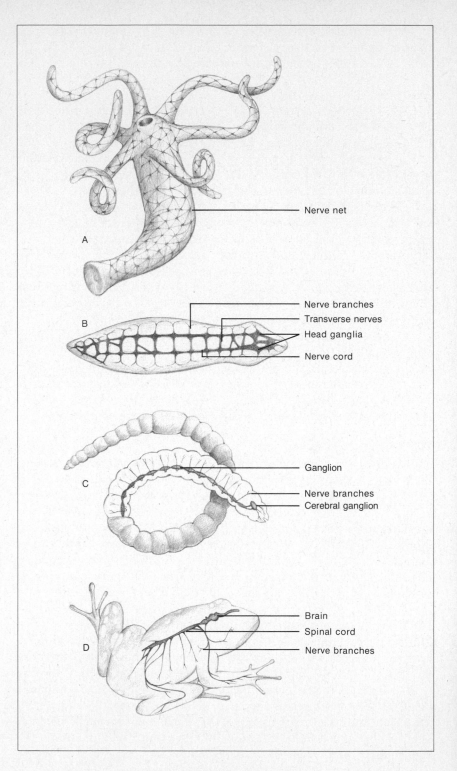

The flatworm has a somewhat more specialized nervous system. The neurons are arranged in two longitudinal *nerves* (a nerve is composed of many neurons) connected by *transverse commissures.* Your keen eye will undoubtedly also have noted the aggregation of nerves in the head region. These are the *cephalic ganglia.* A *ganglion* is a clump of neural cell bodies.

Have you ever wondered why the brain is located in the head in most animals? There have been exceptions, such as the huge, herbivorous dinosaur, the *Brontosaurus,* which had a second "brain" at the base of its tail to help direct its immense body as it browsed in prehistoric lakes. But the evolutionary reason for nerve centers in the *anterior,* or head, region may have been to permit quick analysis of the environment into which the animal would be moving. These centers would in all probability have come to be associated with the organs of perception, the specialized receptors we refer to as the *senses.* And as we know, the senses of sight, sound, smell, and hearing are commonly located in the head region. Thus, the environment into which the animal is moving may be quickly assessed. If the brain and these special receptors were located at the posterior end instead, the animal might find itself in an inhospitable environment by the time it realized its predicament.

The earthworm has only a single longitudinal nerve, but even this one nerve shows indications of a paired arrangement, in that it is two-lobed, much like two cords pressed together. In the earthworm the nerve is ventral, with the heart and digestive tract lying dorsal to it. Note the distinctness of the cerebral ganglia and the obvious nodes of the nerve cord, each with paired nerves reaching into the body. The nodes correspond with the segments of the earthworm's body, each node innervating a particular segment.

The frog represents a relatively primitive vertebrate system. In vertebrates the longitudinal nerve is dorsal, and it is hollow (remember how it develops), filled with fluid, and protected by bone. At the anterior end it swells into a large specialized organ that is more than an enlarged ganglion, although it is probably derived from such nerve clumps through evolutionary changes. The vertebrate brain shows marked specialization, since different parts of it and structures within it are intimately concerned with very specific functions. As in all vertebrates, there is the basic paired and segmented arrangement indicative of a common annelidlike ancestry. These characteristics appear in the two-lobed structure of the spinal cord in cross section and in the paired nerves that branch off from the spinal cord and brain. The branching is no longer so regular and apparent because of specialization along the spinal cord as the vertebrate body plan became more complex.

The Vertebrate Brain
Actually, there is no such thing as "the vertebrate brain," because the vertebrates include widely diverse groups, each of which is highly specialized and distinctive. In the midst of such diversity, however, it is

11.8 A comparison of the brains of a fish, a reptile, and a mammal. Notice the diminutive size of the cerebrum relative to the olfactory bulb in the fish. Also note the relative mass of the lower brain (here, the cerebellum and medulla) compared to that of the cerebrum. Reptiles have a somewhat larger, but still smooth, cerebrum and a reduced olfactory area. The brain of the cat is dominated by the convoluted cerebrum. The cerebellum, involved in coordination, is well developed in the cat, as the olfactory bulb. It is important to realize that all areas of intelligence do not necessarily increase as one moves toward cerebration. Fish, for example, can learn some things easier than reptiles can, even though reptiles are generally more flexible and adaptable behaviorally and are generally believed to be more intelligent.

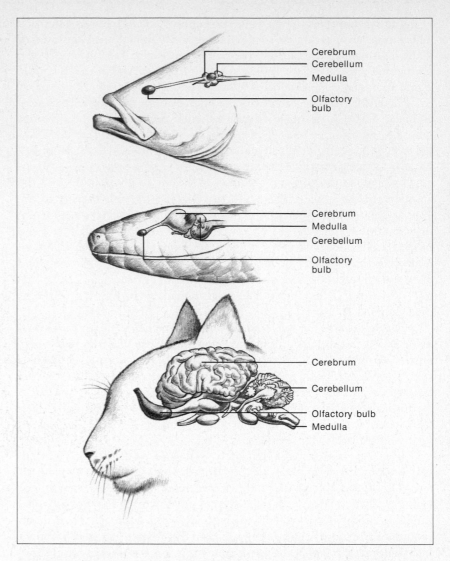

possible to detect trends that give some clues to the general pattern of brain development in animals with backbones. Generally, those traits that are most like the hypothetical or presumed ancestral forms are called *primitive;* those that are more unlike this early condition are called *advanced.*

The sequence in Figure 11.8 illustrates relative changes in parts of the brain from fish to reptile to mammal. The *medulla* is simply the specialized anterior end of the spinal cord. The *cerebellum* is associated with various lower, or unconscious forms of behavior, and the *cerebrum* is the "gray matter," the thinking part of the brain. We will discuss all these structures, but there are a few preliminary points you might find interesting. First, not only is there an increase in general brain size (in

Box 11.1 The Octopus

The octopus is a cephalopod, and like all cephalopods, is a vigorous predator. It can walk about on its "arms" or use jet propulsion by squirting water from its mantle cavity. It is soft-bodied and thus spends much of its time hidden safe in crevices of rocks. The octopus can quickly change its color or pattern to match its surroundings or blanch to a pale "white" (thus accenting its eyes) to escape predators. It may also project clouds of ink when threatened. The ink may temporarily blind an attacker and numb its smell receptors. If placed exposed into a tank it will immediately build a house out of any material available. The octopus is curious, active, and perpetually hungry, so it is an excellent animal to use in behavior tests.

The octopus has large eyes and it sees about as well as man. It is primarily a visual hunter and at the sight of a crab, it turns dark, breaks out in blotches of color, and rushes toward its prey. If the prey is behind glass the octopus will press against the glass in its excited efforts. If a tentacle should reach over the glass and contact the prey, the success doesn't seem to register with the octopus. It will continue to "pursue" the prey—almost as if the information from the tactile (touch) and visual receptors were not integrated.

The octopus can quickly learn to discriminate between horizontal and vertical visual images (possible because the neurons in its optic lobes are arranged at right angles rather than randomly as in most animals). But the octopus is unable to distinguish between cylinders A and B, which have both been cut away 30 percent. On the other hand, it can distinguish these from C, which has been cut away 20 percent. Apparently, discrimination is on the basis of what proportion of the object comes in contact with the suckers.

If an octopus learns to perform a pattern with one tentacle, there must be a time lag before that knowledge is transmitted to other tentacles. Such information has helped to formulate principles of long-term versus short-term learning.

11.9 An evolutionary tree showing the variety of organisms that have sprung from a common stock. All the groups indicated exist at present and hence are still evolving. In this scheme, general intelligence increases, in an imprecise sense, as one moves from right to left across the branches. Certainly the more cerebrated of creatures are at left, but who can say a roundworm is smarter than a flatworm? The tree is designed this way partly on grounds other than intelligence. For example, the single nerve cord of the roundworm more closely approximates the nerve cord of vertebrates than does the double nerve cord of the flatworm. Do you think this diagram makes too many assumptions?

relation to body size) as we go from fish to mammal, but there is also an increase in the size of the cerebrum in relation to other parts of the brain. In contrast, the olfactory lobe doesn't follow this pattern. Olfaction has to do with the sense of smell, so which of these animals do you suppose would rely more on a sense of smell? (Remember to consider the olfactory lobe in relation to total brain size.) Other sensory lobes could also be singled out, such as the optic lobe, which has to do with vision. Which animal do you suppose would have a larger optic lobe with respect to its brain mass, an eagle or an elephant?

Note that there is no implication here that as one moves up the evolutionary family tree, each species is smarter than the ones below it. There may be such a trend within certain groups, but there are also many exceptions. For example, the octopus, which is a mollusk like the snail, is more intelligent than many species of vertebrates. The octopus brain, in fact, is structurally similar to the mammalian brain in terms of its complexity and organization.

The evolutionary development of the mammalian cerebrum, which culminates in the human brain, is undoubtedly one of the most crucial events in the history of life on earth. Such a statement is admittedly somewhat grandiose, but its validity becomes apparent when we consider the impact of the human species on the fragile life system that thinly covers the planet. It would be interesting, therefore, to know how such a brain came to be. What spurred its development? And, as usual, no one knows. One prevailing theory, however, is that the cerebrum developed as a tool to aid in the survival of one small, seemingly insignificant group of reptiles that lived in an exceedingly dangerous world dominated by the great dinosaurs. The line of reptiles that was to give rise to mammals branched off from other reptiles 180–200 million years ago.

Because of their size, these premammals were certainly no match for the speedy, incredibly powerful, and voracious dinosaurs. Since they couldn't outrun or outfight the "ruling reptiles," a premium was placed on their mental agility. In order to survive, the premammals were forced to *outthink* the dinosaurs. The dimmer members among them would rapidly have fallen to predators and thus would have failed to leave their "dim genes" in the next generation. So, with their very existence at stake, the premammals must have rapidly increased their reasoning ability, starting the cerebrum on its way toward dominating the brain of at least one line of animals.

The rapid development of the cerebrum and increasing reliance on intelligence must have been accompanied by other changes among the early mammals. Since the cerebrum is the learning center of the brain, and since learning occurs with experience, animals that were becoming highly cerebrated would have had to develop a life pattern that gave them a chance to learn before they were exposed to the dangers of the world. Hence parental care developed. Even today the offspring of "learning animals" stay with their parents for extended periods. (How old were you

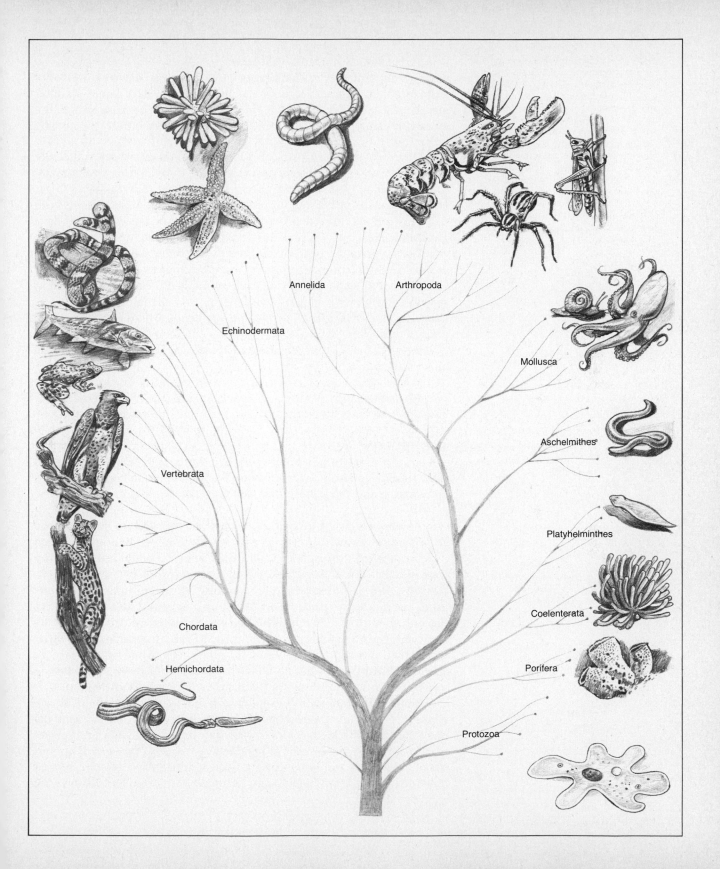

Echinodermata

Annelida

Arthropoda

Mollusca

Aschelmithes

Vertebrata

Platyhelminthes

Coelenterata

Chordata

Porifera

Hemichordata

Protozoa

11.10 Four neurons (opposite page) found in human beings that show the diversity of these cells. A and B are cells with different kinds of cell bodies, axons, and dendrites. C, a motor neuron with axons that run from the nervous system to the effector (in this case, a muscle). A sensory neuron, D, running from the receptor to the spine. Note that the sensory neuron has no true dendrites. The nodes in the myelin sheath are where one Schwann cell ends and another begins. A single nerve cell may be nine feet long such as those that run from the base of a giraffe's spine down its hind leg.

when you left home?) During this time they are cared for and protected by their parents until they can gain enough experience to cope with the world. Also, by parental association, they usually learn from their parents or other members of the group. The term *group* is appropriate because most of the more cerebrated species are social animals, although there are notable exceptions.

Of course, in a high-risk world increased cerebration is not the only course open to natural selection. As an example of another evolutionary route, consider the "strategy" of birds. The noted psychologist Oscar Heinroth is said to have commented, "Birds are so stupid because they can fly." It is certainly true that if an animal can fly, it doesn't have to have great intelligence to escape most predators. A bird doesn't normally "outfox" its predators, since it has only to give a few flaps of the wings to escape terrestrial enemies. Even to escape airborne predators, such as hawks, most birds do not rely on their cunning. Instead, they respond to such threats by employing a few very steretyped behavior patterns, such as attacking in mass, aggregating to confuse predators, taking rather specific evasive maneuvers, or giving warning cries. Birds just have never had to develop great mental capacities, so don't be misled by the discerning frown on an eagle's face.

We will take a closer look at the complex and intriguing human brain, but let's first find out something about what nerves are, how they work, and how they can function in the regulation of bodily processes.

THE NEURON

A *neuron* is simply a nerve cell—although you may reach the conclusion that "simply" is hardly the word for it. Neurons come in a variety of sizes and shapes, and there are billions of them in the human body (Figure 11.10).

A neuron is composed of a *cell body,* which contains the nucleus of the cell; *dendrites,* which receive stimuli from other neurons and conduct impulses toward the cell body; and *axons,* which conduct impulses away from the cell body. Dendrites usually branch profusely, providing increased points of contact with other neurons. Most neurons have only one axon, and it is longer than the dendrites. The axon may also be branched, but not as much as the dendrites. For our purposes we can differentiate between dendrites and axons on the basis of their interaction with other cells. Dendrites can be stimulated by other cells, while axons can't be stimulated. And axons, but not dendrites, can stimulate other cells. (You won't be surprised to learn that there are exceptions to these rules.)

Some vertebrate axons are enveloped in a fatty *myelin sheath,* which consists of the tightly rolled membranes of *Schwann cells* that surround the fiber. The myelin sheath serves to speed up the transmission of impulses along the axon. The presence of myelin gives tissue of the *central nervous system,* the brain, and spinal cord, a whitish appearance. Nonmyelinated fibers are usually gray. The white myelinated fibers lie on the outside of the

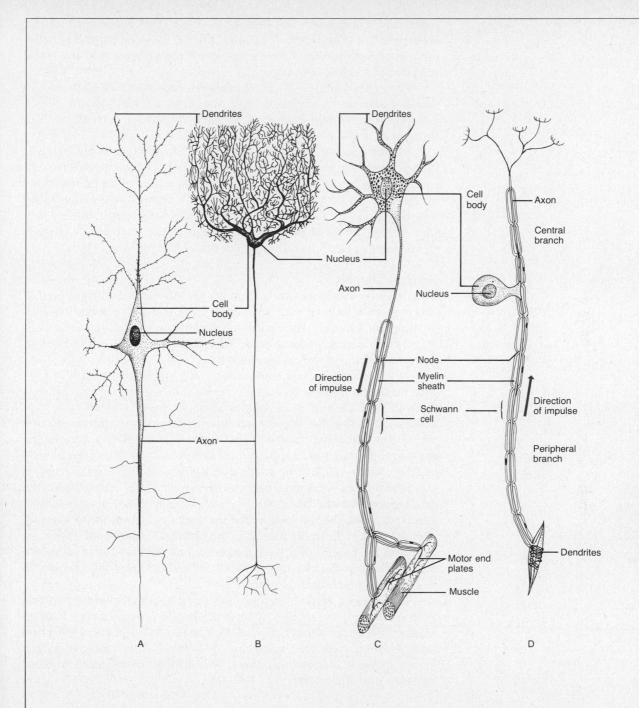

Dendrites

Dendrites

Cell
body

Axon

Central
branch

Nucleus

Cell
body

Cell
body

Axon

Nucleus

Nucleus

Node

Direction
of impulse

Myelin
sheath

Direction
of impulse

Axon

Schwann
cell

Peripheral
branch

Motor end
plates

Dendrites

Muscle

A   B   C   D

spinal cord, but retreat to the inside of the brain mass, leaving gray matter visible on the outside. Also, in most cases only myelinated fibers are capable of regeneration if they are severed; the gray areas of the spinal cord do not regenerate if damaged.

Nerves are bundles of whitish neurons, each wrapped in its myelin sheath. Some nerves are quite large, about the diameter of your finger. Although a nerve may contain many parallel neurons, each neuron can transmit an impulse independently of the others.

Another kind of cell in the vertebrate nervous system is the *glial cell.* No one is sure just what glial cells do, although they outnumber nerve cells by ten to one in the nervous system. Their branches often permeate the nerves, weaving between the neurons, and it is thought that perhaps they somehow support or nourish neurons. It is also suggested that they may store certain kinds of molecules necessary to the business of impulse transmission.

Impulse Pathways

If you step on a tack, and are a normally sensitive and intelligent person, you may wish to take your foot off it. For you to be aware of your predicament, however, the condition of your foot has to register in your brain, which is some distance away. This message travels over neurons, as an *impulse.* But let's start at the bottom.

To begin with, the tack stimulates certain *receptors* in the sole of your foot. Receptors come in a variety of types, and each type specializes in a particular sensation, such as cold, pressure, or pain. In some cases bare nerve endings function as receptors. But in any case the stimulation of a receptor results in the transfer of that impulse to an *afferent neuron,* one that carries impulses *toward* the central nervous system. The impulse may be relayed from one neuron to the next until it enters the spinal cord, is transferred to an ascending neuron, and reaches the brain, where the message is registered: "Sharp pain in foot!" If the foot has not already been lifted by a reflex action (which we'll discuss later), you may say to yourself, "I really must remove my foot from the tack," whereupon an impulse is sent from the brain, down the spinal cord, out the spinal cord along the proper *efferent nerve,* one that carries the impulse toward the *effector*—in this case the muscle that raises the leg.

The junction of two neurons—the point at which the axon of one neuron transmits the impulse to the dendrite of another neuron—is called a *synapse.* Whereas theoretically an impulse can travel in either direction along a neuron, it can cross the synapse in only one direction; so in a sense the synapse acts as a one-way valve. We will look more closely at these fascinating junctures later.

The Mechanism of the Impulse

When a nerve fiber is at rest, certain chemicals are found in greater abundance within the cell, while others are more concentrated outside the cell. Furthermore, some of these chemicals are ionic, or electrically charged. For example, the resting neuron is literally bathed in sodium ions (Na^+), but these ions are scarce inside the cell. This difference results in a net positive charge outside the cell. In the cytoplasm of the neuron, however, the situation is reversed: the cytoplasm is negatively charged. It is true that there are a lot of positively charged ions in the cell fluid, such as potassium (K^+), but these are no match for the myriad of negatively charged ions found there, especially chlorine (Cl^-) and negatively charged proteins.

The electrical potential—the difference between the two levels of electrical charge—that is set up by this peculiar distribution of charged particles is called the *resting potential*. Typically, this potential difference is about 70 millivolts, or about 5 percent as much electrical energy as found in a regular flashlight battery. The difference in the ion concentrations inside and outside the cell is maintained by considerable work on the part of the cell. As fast as positively charged Na^+ particles diffuse into the cell, the cell pumps them right back out again to maintain the resting potential. The energy for this pumping activity is provided by ATP.

Now, when a neuron is excited by either an electrical or a chemical stimulus, the cell becomes leaky to sodium ions, so that they rush in faster than they can be pumped out. The result of this shift in ions is that there is a change in the membrane potential of the cell. The change doesn't occur simultaneously along the entire length of the cell; beginning at the point of stimulation, a wave of inrushing sodium ions sweeps along the length of the cell, and as it passes, the cell fluid behind it immediately restores itself to its normal resting potential. Since the membrane is no longer separating positively and negatively charged particles, the cell at such points can be said to be *depolarized*.

This depolarization sweeps along the neuron at a regular rate of speed, depending on the neuron. The inrush of sodium ions at any point along the membrane results in a momentary net positive charge inside the cell at that point, so that the polarity is reversed. In other words, for an instant, the outside becomes slightly negatively charged and the inside slightly positively charged. This momentary positive charge inside the neuron is called the *action potential*. The impulse, then, is the passage of this action potential along the neuron. See Figure 11.11 for more explanation.

The *repolarization* of the cell begins as positively charged potassium ions rush out of the cell faster than the sodium can move in. This shift sets up a net positive charge outside again. As the membrane becomes less permeable to sodium, the "sodium pump" begins to move the sodium, which has rushed in, back out again. In the meantime, some ionic potassium begins to diffuse back to the neuron's interior. Thus the membrane potential of the cell is restored. Again, all this takes place in the briefest of periods so that most neurons can be depolarized and repolarized several times each second.

11.11 The action potential occurs as the inside of the cell gains a positive charge. The charge is from about −70 to +20 millivolts as ion concentrations inside and outside the cell change. Changes in charges occur as an impulse passes. Note, the impulse travels from right to left. The absolute refractory period designates the period at which the nerve cannot respond to another stimulus. The relative refractory period designates the period when the cell can respond only to an unusually strong stimulus.

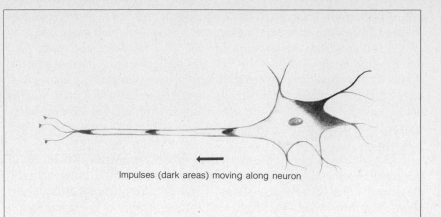

Impulses (dark areas) moving along neuron

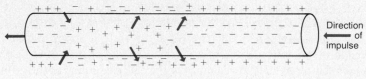

Direction of impulse

Movement of charges (ions) across segment of neuron

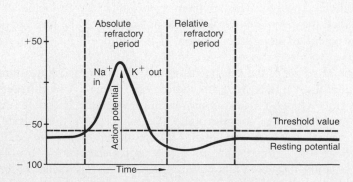

Action potential (oscilloscope tracing)

Box 11.2 Botulism

The poison produced by the bacteria *Clostridium botulinum* is the most deadly poison known. Only 60 billionths of a gram will kill a man. The spores of the organism are found as a common inhabitant of the soil around the world. Fortunately, however, specific conditions must be met in order for these spores to grow. An anaerobic and nutrient-rich medium must be provided. Thus canned food that has not been properly sterilized is an ideal home for the microorganism. If canned food has been contaminated, however, it will produce the telltale clues of foul smell and buildup of gas (which may cause cans to swell). The signs of botulism are vomiting, constipation, and paralysis of eyes and throat as breathing becomes more difficult. These begin one to three days after eating the poisoned food. The poison functions by interfering with the release of acetylcholine. This is the physiological basis of the paralysis.

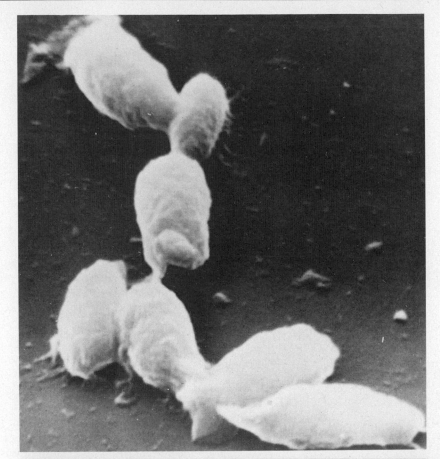

Top, *Clostridium botulinum,* the organism that produces botulism. Bottom, commercially prepared food cans swollen by these bacteria. Tasting spoiled food can spell death.

11.12 Synaptic transmission. The axon endings of one neuron lie very near the dendritic endings of the next neuron. As the impulse reaches the end of its axon it stimulates the release of a neurotransmitter, such as acetylcholine. The process requires the expenditure of energy, thus the axonal endings are peppered with mitochondria. The neurotransmitter is released through the terminal portion of the axon into the space between that axon and the dendrite of the next neuron. The neurotransmitter thus stimulates that dendrite to initiate an impulse. The neurotransmitter is quickly destroyed by an enzyme, such as acetylcholinesterase, so that the second neuron does not continue firing in the absence of a real impulse.

Two important qualities mark neural impulses. One is the fact that an impulse does not increase or decrease in strength as it moves along a neuron. Second, impulses operate on an "all-or-none" basis. This means that if a stimulus reaches the *threshold value* for a neuron, the lowest intensity at which a stimulus can be registered for that neuron, the cell fires, and it fires at only one intensity no matter what the strength of the stimulus.

If neurons operate in this all-or-none manner, how is it that we can detect various intensities of stimuli? If you put your hand on a stove, how do you now whether it is simply nice and warm or dangerously hot? Actually, we can make such discriminations in a number of ways. First, the more intense the stimulus (the hotter the stove), the higher the rate of impulses that pass along the neuron (although there are maximum limits). Also, different neurons in the same nerve may have different threshold levels. If some are easily stimulated and others require heat, then the stronger the stimulus, the more neurons will be stimulated. Hence the brain can interpret either the *frequency* of the impulses or the *number* of neurons stimulated to tell you whether you should remove your hand from the stove. Fortunately, the process doesn't take as long as the explanation.

The Synapse

Now let's consider how an impulse travels from one neuron to the next. Actually, what we are concerned with here is how one neuron activates another so that the impulse continues along a chain of nerve cells. How does the behavior of a neuron initiate a wave of depolarization along the length of another neuron? The story is an interesting one that centers on the secretion of certain remarkable chemicals.

When an impulse traveling along a neuron reaches the axon tip, it causes the buttonlike ending of the axon to secrete a chemical into the space between itself and the dendrites of the next neuron. The axon is enmeshed in the highly branched dendrites of the next neuron, so that there is a large total surface area over which the two neurons communicate. Any chemical secreted by the axon endings of one neuron, then, can be expected to immediately affect the next neuron. And this is just what happens.

In many of the neurons in our body the transmitting chemical is believed to be *acetylcholine.* The transmitting chemical is stored in small packets in the swollen tips, or "buttons," of the axon endings (Figure 11.12). When an impulse reaches the end of the axon, some of these packets move toward the membrane and rupture, thus releasing their transmitter substance into the synapse. The chemical quickly diffuses across the gap and contacts the dendrites of the next cell, where it sets up another action potential in that neuron.

After a few impulses have traveled down a neuron, causing release of the transmitter chemical into the synaptic space, it might seem that the

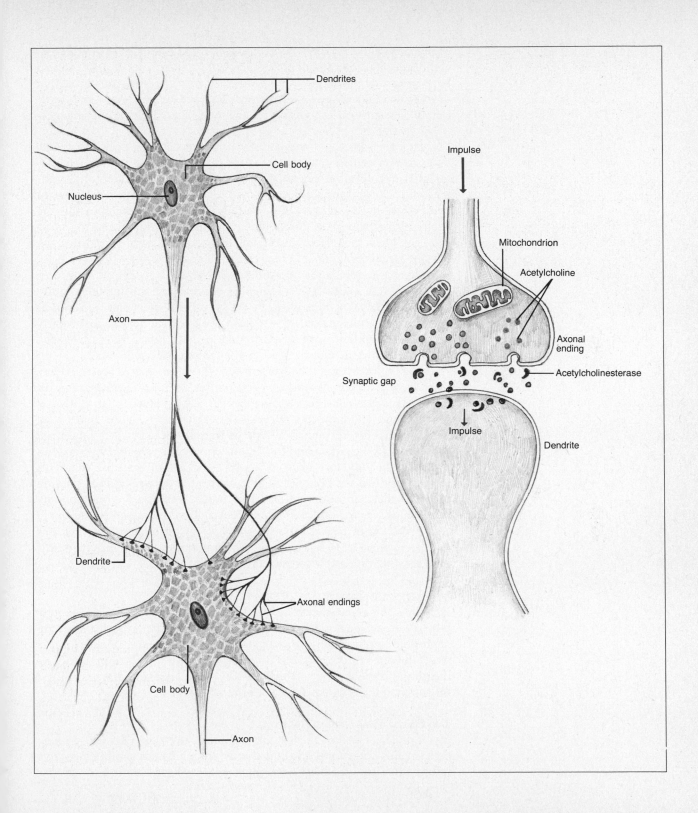

Dendrites

Cell body

Nucleus

Axon

Dendrite

Axonal endings

Cell body

Axon

Impulse

Mitochondrion

Acetylcholine

Axonal ending

Acetylcholinesterase

Synaptic gap

Impulse

Dendrite

chemical would build up in the synaptic area and cause the next neuron to continue firing in the absence of a real stimulus. But do not despair. Almost all types of dendrites secrete a substance into the synaptic space that quickly neutralizes any residual transmitter substance. When the transmitter substance is acetylcholine, the neutralizer is the enzyme *acetylcholinesterase.*

Humans sometimes accidentally come into contact with substances that inactivate these important neural enzymes, with horrible consequences. For example, certain insecticides, such as Malathion or Parathion, which are sprayed on crops across the country, have the advantage of being short-lived—unlike DDT. Within a few days after spraying the chemicals are believed to be relatively harmless. These chemicals, however, can block the effects of acetylcholinesterase. Keep in mind that the enzyme functions not only to keep chains of neurons from firing blindly in response to accumulating transmitter substance, but also to keep acetylcholine from diffusing over to nearby neurons and causing them to fire as well. What do you think might be the behavioral effect on a farmworker exposed to the insecticides? There have been several deaths, particularly among Chicano farmworkers, attributed to these chemicals. Conversely, other substances are known to block the release of acetylcholine.

In addition to excitatory transmitter substances such as acetylcholine, there are also numerous *inhibitory transmitter substances,* which are as yet largely unidentified. These substances, it seems, function at *inhibitory synapses* and counteract the effect of the excitatory synapses. Since each neuron manufactures only one type of transmitter substance, inhibitory substances are manufactured by inhibitory neurons, which converge with the axons of excitatory neurons at the dendrites of the next cell. This opposing system has been found at every central nervous system synapse that has thus far been extensively studied. Inhibitory synapses may function as a means of keeping neural excitation under control, so that a chain reaction of neural activity is not set up throughout the nervous system (Figure 11.13). It is believed that uninhibited neural activity of this kind may be responsible for the convulsions that occur in epilepsy.

THE HUMAN NERVOUS SYSTEM

Now that you know something about the way nerves work, let's see how they are organized in the human nervous system. The principles here are applicable to most mammalian systems since the neural organization is much the same throughout this diverse group. We will focus primarily on the central nervous system, which consists of the brain and spinal cord, and the fascinating peripheral nervous system.

The Spinal Cord and the Reflex Arc

If you pride yourself on the fact that you are a "thinking animal," you may be a little disappointed to realize that your spinal cord can often receive information from the body's receptors, process it, and initiate the proper

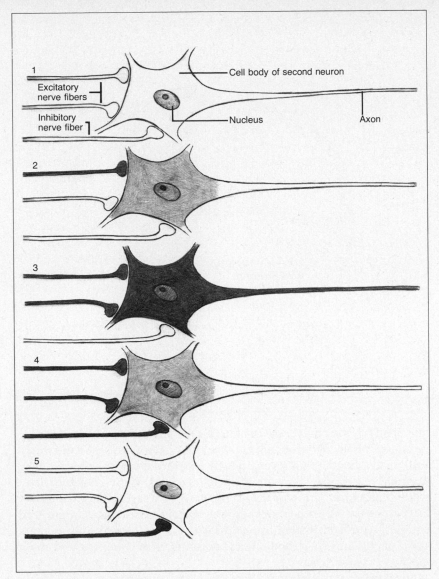

Excitatory
nerve fibers

Inhibitory
nerve fiber

Cell body of second neuron

Nucleus

Axon

11.13 How a neuron is excited or inhibited. Here, two excitatory neurons and one inhibitory neuron synapse with the dendrites of another neuron. The darker axons are firing. In (1) the neuron is at rest with no axonal action; in (2) one excitatory neuron fires, but it does not stimulate to threshold levels; in (3) both excitatory neurons fire and the second neuron is stimulated to fire; in (4) the inhibitory neuron is fired at the same time so that the threshold levels of the second neuron are raised above the effects of even both excitatory neurons; in (5) the inhibitory neuron firing alone will, of course, fail to excite the second neuron.

response before your brain even "knows" what has happened. Such a chain of events describe the *reflex arc* and illustrates something about the makeup of the spinal cord.

As illustrated in Figure 11.14, the impulse is generated in a stimulated receptor and transmitted to a sensory neuron, which enters the spinal column over the *dorsal nerve root*. The impulse is then transferred to a connector neuron, which in turn transfers it to the proper motor neuron. The motor neuron leaves the spinal cord over the *ventral nerve root* and travels outward to the effector (usually a muscle group).

Note that impulses can cross the spinal cord, so that effectors in the

11.14 A cross section of the spinal cord showing the distribution of white and grey matter and the pathway of a simple reflex arc. The arc involves three neurons (sensory, connector, and motor) but simpler arcs may have only two. In such a case the sensory neuron would connect directly to a motor neuron. It will also be noticed that the motor neuron acts by stimulating an effector (muscle fiber). Thus muscle fibers also react to transmitter substances. The sensory nerve is stimulated by a receptor that secretes its own transmitter substance. In each case the transmitter substance may be the same.

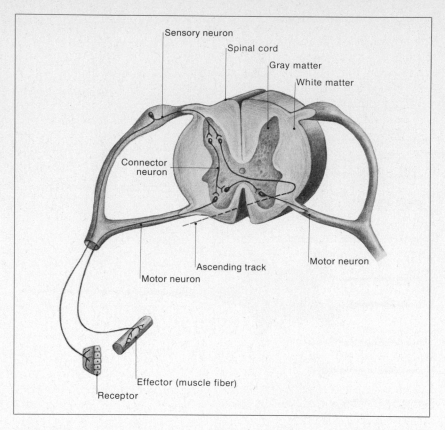

other half of the body can be stimulated. Also, the impulse can be transmitted to the dendrites of yet other neurons, which ascend the spinal cord to the brain. Thus even though the response has already been accomplished reflexively, your brain is informed of the change. Direct neural routing from the spinal cord in a reflex arc saves time because the distance the impulse has to travel from receptor to effector is shorter. And no time is spent in deliberating over the decision. You might come to think this is important if you should crawl out of your sleeping bag one night and step on a hot coal.

The reflex arc is structurally simple, but it may produce a highly coordinated response. For example, the reflex arc that takes place in the familiar knee jerk involves sensory and motor pathways that lie on the same side of the body and is thus an example of the simplest form of arc. However, the jerk of the knee entails a contraction of a *group* of muscles on the front of the leg, acting together. Each muscle must be activated separately. At the same time the motor neurons going to the back of the leg must be inhibited from firing. It wouldn't do to have both groups contracting at once. So even in this "simple" case, *several* two-neuron arcs enter into the reflex. When a reflex involves connecting neurons, it is apparent that the response can be increasingly complex and highly coordinated.

The Brain
The brain itself is an exceedingly complex structure, and there are many things we just don't understand about it. However, the research goes on. We do know some things, and you're about to hear a few.

To begin with, the human brain is divided into three parts: the forebrain, the midbrain, and the hindbrain. The *forebrain* consists of the two cerebral hemispheres and certain internal structures, such as the thalamus and hypothalamus. The *midbrain,* logically enough, is the area between the forebrain and hindbrain. The *hindbrain* consists of the medulla, the cerebellum, and the pons (Figure 11.15).

The Medulla
As a rough generality, the more subconscious, or mechanical, processes are directed by the more posterior parts of the brain. For example, the hindmost part, the *medulla,* is specialized as a control center for such visceral functions as breathing and heartbeat. As we have already seen, it also functions as a connecting area between the spinal cord and the more anterior parts of the brain.

The Cerebellum and Pons
Above the medulla and somewhat further toward the back of the head is the *cerebellum,* which is concerned with balance, equilibrium, and coordination. Do you suppose there might be differences in this part of the brain between athletes and nonathletes? Do you think this "lower" center of the brain is subject to modification through learning? The *pons,* which is the portion of the brainstem just above the medulla, connects the cerebellum and the cerebral cortex, accenting the relationship between the cerebellar part of the hindbrain and the more "conscious" centers of the forebrain.

The Thalamus and Reticular System
The *thalamus* and *hypothalamus* are located at the base of the forebrain. The thalamus is rather unpoetically called the "great relay station" of the brain. It consists of densely packed clusters of nerve cells, which presumably provide connections between the various parts of the brain—between the forebrain and the hindbrain, between different parts of the forebrain, and between parts of the sensory system and the cerebral cortex.

The thalamus contains a peculiar neural structure called the *reticular system,* an area of interconnected neurons that are almost feltlike in appearance. These neurons run throughout the thalamus and into the midbrain. The reticular system is still somewhat a mystery, but several interesting facts are known about it. For example, it bugs your brain. Every afferent and efferent pathway to and from the brain sends side branches to the reticular system as it passes through the thalamus. So all incoming and outgoing communications to the brain are tapped by this system. Also, these reticular neurons are rather unspecific. The same reticular neuron may respond to stimuli from the hand, the foot, the ear, or the eye. We

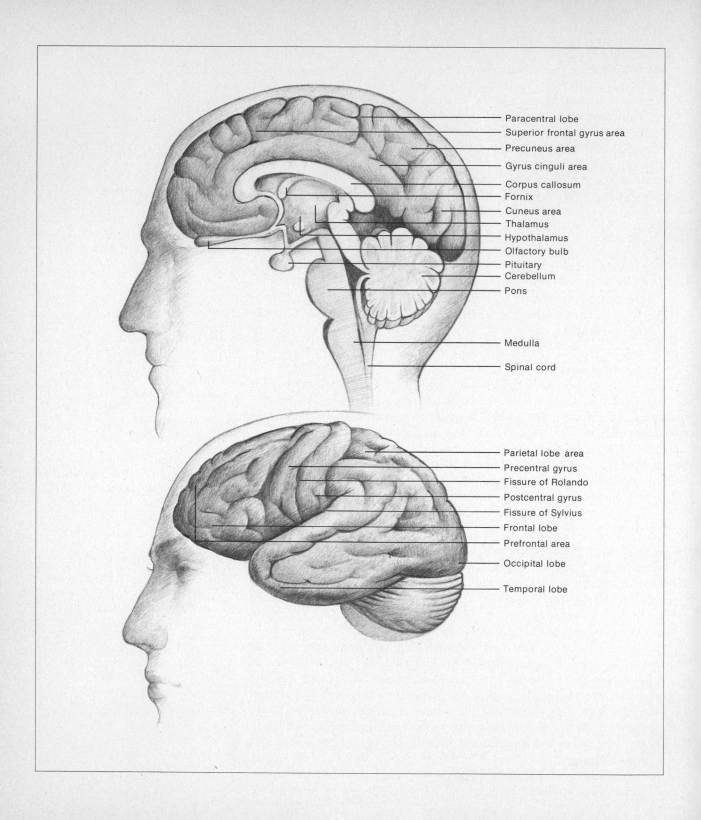

Paracentral lobe
Superior frontal gyrus area
Precuneus area
Gyrus cinguli area
Corpus callosum
Fornix
Cuneus area
Thalamus
Hypothalamus
Olfactory bulb
Pituitary
Cerebellum
Pons

Medulla

Spinal cord

Parietal lobe area
Precentral gyrus
Fissure of Rolando
Postcentral gyrus
Fissure of Sylvius
Frontal lobe
Prefrontal area
Occipital lobe
Temporal lobe

don't know the functions of this system, but it is believed that the reticular apparatus serves to activate the appropriate parts of the brain upon receiving a stimulus. In other words, it acts as an arousal system for certain brain areas. The more messages it intercepts, the more the brain is aroused. Thus the reticular system may function importantly in sleep. You may have noticed it is much easier to fall asleep when you are lying on a soft bed in a quiet room with the lights off. Under these conditions there are fewer incoming stimuli; as a result, the reticular system receives fewer messages, and the brain is lulled rather than aroused. On those nights when you suffer from the "big eye" and just can't sleep, the cause may be continued (possibly spontaneous) firing of reticular neurons (Figure 11.16).

The reticular system may also regulate which impulses are allowed to register in your brain and which will be filtered out or diminished. When you are engrossed in a television program you may not notice that someone has entered the room. But when you are engaged in even more absorbing activities, it might take a *general* stimulus on the order of an earthquake to distract you, whereas the *specific* sound of just a turning doorknob would immediately attract your attention. Such filtering and selective depression of stimuli apparently take place in the reticular system.

The Hypothalamus

The *hypothalamus* is a small body densely packed with cells, which is extremely important in regulating the internal environment as well as certain general aspects of behavior. For example, the hypothalamus helps to control heart rate, blood pressure, and body temperature. It also plays a

11.15 The human brain viewed in sagittal section (top) and in surface view (bottom). Note that there is enough regularity in the convolutions that some have been named. The midbrain area is composed of a number of different tissues enclosing an opening, shown here in lighter shades. The incredible activity of this midbrain area is not remotely suggested by the simple structures that comprise it, whether it is viewed in its entirety or microscopically. There is increasing evidence, however, that our moods, motivations, and some would say even our values, stem in large part from these tough, rubbery structures. No matter what behavioral tendencies are mediated by them, however, they may in some cases be overridden by the powerful cerebrum. The hindbrain is believed to have evolved first and subsequently to have been modified to coordinate it with the highly complex anterior areas.

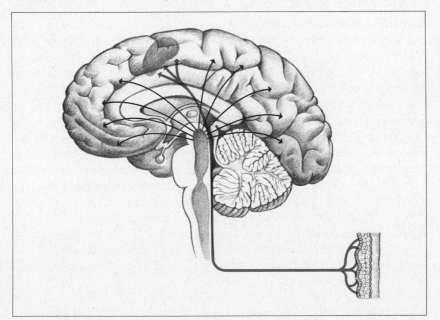

11.16 The reticular system of the human brain. The impulse originating at the lower right passes through the reticular system, a structure composed of untold millions of neurons. The smaller arrows indicate that, in this case, the entire cerebrum has been alerted, but a specific part (the shaded area) is the target of most of the impulses. It is likely, then, that this area of the cerebrum will be required to deal with whatever initiated the stimulus in the first place.

part in the regulation of the pituitary gland, as we learned earlier. And it controls such basic drives as hunger, thirst, sex, and rage. So now you know what's to blame for all your problems. Experimental stimulation of various centers in the hypothalamus by means of electrodes can cause a cat to act hungry, angry, cold, hot, benign, or horny.

The Cerebrum

For many people the word *brain* conjures up an image of two large, deeply convoluted gray lobes. What they have in mind, of course, is the outside layer of the two cerebral hemispheres, the dominant physical aspect of the human brain. The cerebrum is present in all vertebrates, but is assumes particular importance in humans. In some animals it may serve essentially as an elaborate refinement to implement behavior that could be performed to some degree without it. In more advanced animals it takes on a more functional importance.

For example, if the cerebrum of a frog is removed, the frog will show relatively little change in behavior. If it is turned upside down, it will right itself; if it is touched with an irritant, it will scratch; it will even catch a fly. Also, sexual behaviors in frogs can occur without the use of the brain—but we'll try not to extrapolate from that. Rats are more dependent on their cerebral cortex. A rat that is surgically deprived of its cerebrum can visually distinguish only light and dark, although its patterns of body movement seem unimpaired. A decorticated cat can meow and purr, swallow, and move to avoid pain, but its movements are sluggish and robotlike. A monkey whose cerebral cortex has been removed is severely paralyzed and can barely distinguish light and dark. The result in humans is total blindness and almost complete paralysis. Although such persons can breathe and swallow, they soon die.

It is apparent from this information that the cerebrum is more than just the center of "intelligence." It seems that, from an evolutionary standpoint, with increasing cerebration, more and more of the functions of the lower brain are transferred to the cerebrum. In this progession, even when the neural "center" of a certain function is not shifted to the cerebrum, a certain element of its control may be transferred there.

Generally, the degree to which neural control has been shifted to the cerebrum is reflected in the size of this structure. In other words, more "intelligent" animals have larger cerebral components in their brains. Increased dependence on the cerebrum may also be reflected by factors other than increased cerebral size. For example, the cerebrum may be more convoluted in more intelligent species. The deep convolutions of the human brain are lacking in the brain of the rat. Even the convoluted cerebrum of the highly touted and undoubtedly intelligent dolphin has fewer "layers" than the human cerebrum.

Hemispheres and Lobes

The human cerebrum consists of two hemispheres, the left and the right, each of these being divided into four lobes. At the back is the *occipital lobe,* which receives and analyzes visual information. If this lobe is injured, black "holes" appear in the part of the visual field that is registered in that area.

The *temporal lobe* is at the side of the brain. It roughly resembles the thumb on a boxing glove, and it is bounded anteriorly by the *fissure of Sylvius.* The temporal lobe shares in the processing of visual information, but its main function is auditory reception.

The *frontal lobe* is right where you would expect to find it—at the front of the cerebrum, just behind the forehead. This is the part that people hit with the heel of the palm when they suddenly remember what they forgot. One part of the frontal lobe is the center for the regulation of precise voluntary movement. Another part functions importantly in the use of language, so that damage to this part results in speech impairment.

The area at the very front of the frontal lobe is called the *prefrontal area,* if you follow that. Whereas it was once believed that this area was the seat of the intellect, it is now apparent that its principal function is sorting out information and ordering stimuli. In other words, it places information and stimuli into their proper context. The gentle touch of a mate and the sight of a hand protruding from the bathtub drain might both serve as stimuli, but the two stimuli would be sorted differently by the prefrontal area. A few years ago parts of the frontal lobe were surgically removed in efforts to bring the behavior of certain aberrant individuals more into line with the norm. Fortunately, this practice has been largely discontinued.

The *parietal lobe* lies directly behind the frontal lobe and is separated from it by the *fissure of Rolando.* This lobe contains the sensory areas for the skin receptors, and also the areas that detect bodily position. Even if you can't see your feet right now, you have some idea of where they are, thanks to receptors in the parietal lobe. Damage to the parietal lobe may cause numbness, but it can also cause a person to perceive his own body as wildly distorted. In addition, he may suffer from an inability to perceive spatial relationships in the environment.

By probing the brain with electrodes, it has been possible to determine exactly which area of the cerebrum is involved in the body's various sensory and motor activities. We can see the results of such mapping in the rather grotesque Figure 11.17. The pictures are distorted not out of any appreciation of the macabre, but to demonstrate that the area of the cerebrum devoted to each body part is dependent not on the size of that part, but on the importance it has come to have in the natural history of man. Thus we have the greatest number of *senses* in the face, hands, and genitals, but the greatest amount of control only in the face and hands (as you may already have found out).

Also note that the sensory and motor areas are not randomly scattered through the cortex, but that there is some order in their appearance. Thus the index-finger control area lies near the thumb control area, and the elbow area lies closer to the finger area than does the shoulder area.

One important fact about the cerebral hemispheres is that although they are roughly equal in potential, the two hemispheres are not identical.

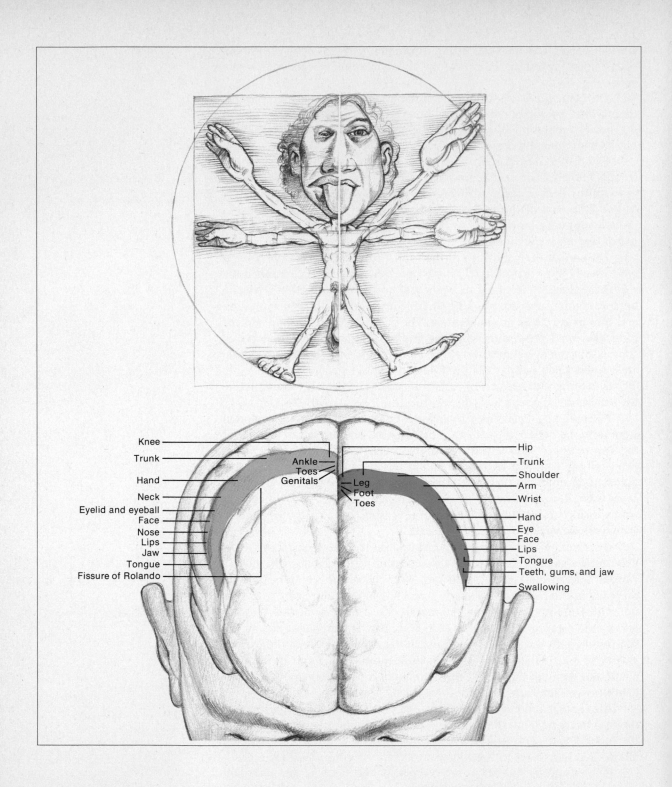

Knee
Trunk
Hand
Neck
Eyelid and eyeball
Face
Nose
Lips
Jaw
Tongue
Fissure of Rolando

Ankle
Toes
Genitals

Leg
Foot
Toes

Hip
Trunk
Shoulder
Arm
Wrist
Hand
Eye
Face
Lips
Tongue
Teeth, gums, and jaw
Swallowing

Moreover, any chemical or structural differences between them are accentuated by learning, because the control for certain learned patterns takes place primarily in only one hemisphere. One example of the difference between the hemispheres is seen in handedness. (It is interesting that less cerebrated animals such as rats and parrots also show handedness. Willy, my chinchilla, is a southpaw.) Other functions, such as speech, "I.Q.," and perception, are also more likely to be located in one hemisphere than the other. The primary speech center of righthanded people is located in the left hemisphere. In lefties, on the other hand (no pun), damage to the right temporal lobe may result in poor performance on I.Q. tests, but perceptual-test performance is relatively unaffected.

The primary route of communication between the right and left cerebral hemispheres is the *corpus callosum.* It seems that if one side of the brain learns something—for instance, by covering one eye while learning a visual task—the information will be transferred to the other hemisphere. However, if the corpus callosum has been cut, the material has to be taught anew to the other hemisphere by uncovering the other eye and repeating the training program.

It should be pointed out that a certain degree of compensation is possible between the two halves of the brain. For example, if one hand is injured, it is possible to learn to use the other hand. So, in spite of the natural handedness you were born with, retraining is possible as muscles controlling movement of the other hand develop, and neural events in the opposite side of the brain begin to permit the efficient use of that hand. It has also been demonstrated that if one hemisphere of the brain is injured, the other side can take over some of its tasks.

The Peripheral Nervous System

Pairs of thick, white nerves emerge from the brain and either side of the spinal cord and innervate every receptor and effector in the entire body. These nerves comprise the *peripheral nervous system.* There are twelve pairs of *cranial nerves,* stemming from different parts of the brain, and these primarily innervate the senses, glands, and muscles of the head. These twelve pairs of nerves are found in all reptiles, birds, and mammals (Figure 11.18). Fish and amphibia have only the first ten pairs.

The human spinal cord gives rise to thirty-one pairs of tough, white spinal nerves. Each nerve is formed from the union of a dorsal and a ventral nerve root that emerge directly from the spinal cord. The dorsal nerve root is swollen by a huge ganglion at about the level where it enters the spinal column. The ganglion houses the cell bodies of all the sensory neurons entering the spinal column. The cell bodies of the neurons comprising the ventral nerve root (the motor neurons) lie imbedded in the spinal column, so these nerves need not be disfigured by ganglia. The two great nerves fuse just outside the spinal column and travel together for a way before increasingly smaller nerves begin to branch off from them—nerves which will ultimately branch into delicate neurons to innervate every part of the body.

11.17 Function maps of the cerebral cortex showing sensory areas (left) and motor areas (right). The sensory section is taken posterior to the fissure of Rolando, the motor section anterior to it. Note that a large part of the human brain is devoted to face, hands, and genitalia. Also, the eyes and hands are given more motor than sensory space in the brain in spite of our great dependence on them for sensory information. Our lips and genitalia have a rather surprising amount of brain tissue allocated for their sensory input. Perhaps we have so much of the brain devoted to facial areas because the face is so important in the subtle processes of human communication.

11.18 Human cranial nerves. These twelve large nerves originate in the brain and most have sensory and motor functions in the head and neck. An exception is the vagus nerve, which passes down into the trunk and serves the heart (slowing it) and part of the digestive tract. Compare the olfactory bulbs in humans with those of some other animals (Figure 11.8). Whereas these nerves directly stem from the more central areas of the brain itself, similar nerves rise from the spinal cord, further back, and service other parts of the body.

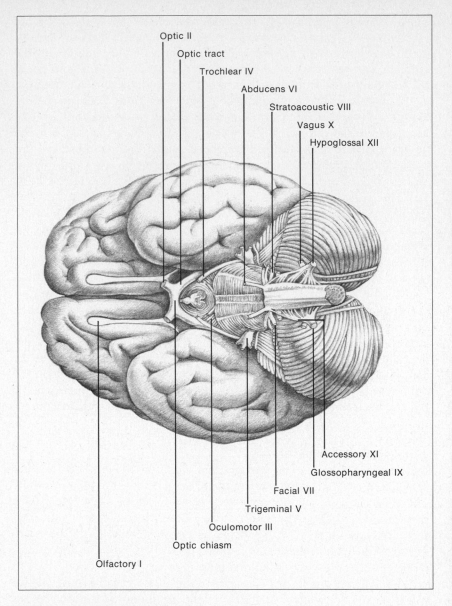

11.19 The autonomic nervous system, showing the sympathetic and parasympathetic components. The central nervous system is shown at left and next to it is a column of nervous tissue, which comprises the peripheral nervous system and through which autonomic impulses must pass. Impulses can also originate in the peripheral system as is attested to by often-active sex lives of people who have suffered injury to the spinal cord. The internal organs are innervated by both sympathetic and parasympathetic nerves, the two systems having opposite effects.

The Autonomic Nervous System

The *autonomic nervous system* is formed from a special set of peripheral nerves that serve the heart, lungs, digestive tract, and other internal organs. This system is largely under the control of the hypothalamus.

The autonomic nervous system may be divided into the *sympathetic* and *parasympathetic* nervous systems, which act antagonistically (Figure 11.19). Generally, the sympathetic system works to speed up certain body processes, with an increase in energy expenditure, and the parasympathetic system works to slow them down.

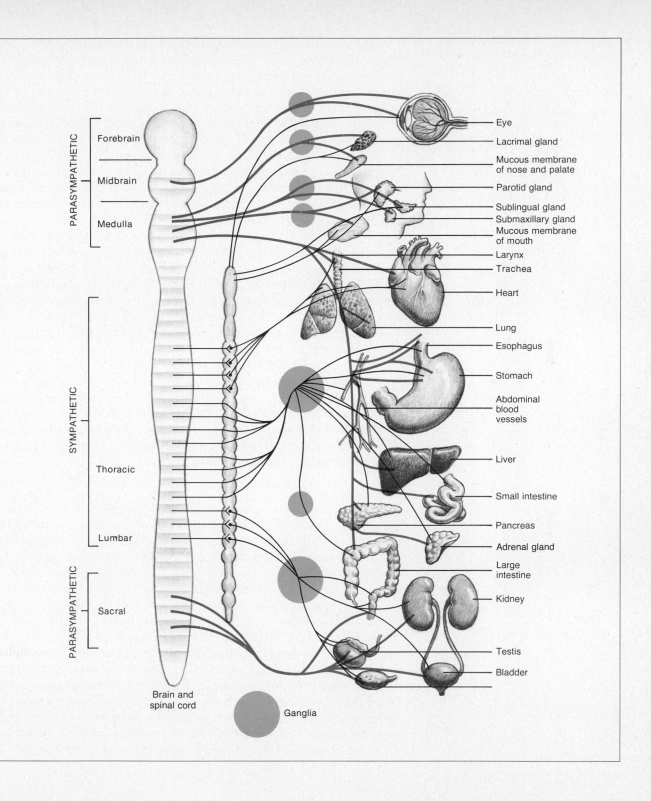

PARASYMPATHETIC

Forebrain

Midbrain

Medulla

SYMPATHETIC

Thoracic

Lumbar

PARASYMPATHETIC

Sacral

Brain and
spinal cord

Ganglia

Eye

Lacrimal gland

Mucous membrane
of nose and palate

Parotid gland

Sublingual gland

Submaxillary gland

Mucous membrane
of mouth

Larynx

Trachea

Heart

Lung

Esophagus

Stomach

Abdominal
blood
vessels

Liver

Small intestine

Pancreas

Adrenal gland

Large
intestine

Kidney

Testis

Bladder

The sympathetic nervous system is activated in what has been called "fight-or-flight" reactions. It also functions in reproductive behavior, but perhaps we should avoid further alliteration. The fight-or-flight syndrome becomes apparent in certain emergency situations. For example, if a bear rushes into the room where you are quietly reading, your body will react "sympathetically." The pupils of your eyes will dilate, the better to see the bear; peripheral blood vessels will decrease in diameter, so blood loss will be minimized in case you are scratched on your way out; heart and breathing rate will increase, bringing oxygen to your running muscles; and digestion will almost stop, since your visceral blood is needed elsewhere. After you or the bear has left, the parasympathetic system will take over and reverse all these reponses.

It is interesting that, whereas the transmitter substance in the parasympathetic axon is acetylcholine, the sympathetic system's transmitter substance is *adrenaline*. You might wonder why adrenaline couldn't simply be released from the adrenal glands in an emergency situation, to travel in the blood and elicit these same changes throughout the body — why it was necessary to develop another, but similar, emergency system. The adrenal gland, in fact, may secrete adrenaline into the bloodstream in a stress situation, but the two emergency reactions illustrate an important difference between hormone and neural regulation. The sympathetic system releases its adrenaline *directly* into the proper effector. Only small amounts are released, so although the response is immediate, it is of short duration. Greater amounts of adrenaline may be secreted by the adrenal gland, and while these take longer to reach the effector, their effect is more long lasting. As a general rule, neural regulation is more immediate and short-term than hormonal regulation.

Autonomic Learning
The autonomic nervous system is traditionally described as "involuntary," in that we do not normally exercise conscious control of our breathing rate, heartbeat, or urine formation. While it is true that the autonomic system can, and normally does, function in the absence of conscious control, there is evidence that some measure of conscious control is possible. In other words, we may be able to learn to control our autonomic reactions.

L. V. Dicara and N. E. Miller of Rockefeller University were able to teach rats to increase or decrease heart rate, blood pressure, intestinal contractions, blood vessel diameter, and even rate of urine formation. This was done by monitoring patterns and noting normal fluctuations over a period of time. As natural variations occurred, the ones in the desired direction were rewarded by electrical stimulation of the "pleasure center" of the brain and variations in the other direction were punished by unpleasant shock. If the rat was to slow its heart rate, when the heart slowed naturally, the animal would be rewarded; when it accelerated naturally, punishment would follow. Soon the heart beat more slowly.

The researchers have produced similar results in humans and have taught at least one person to control his blood pressure. The amazing feats accomplished by some practitioners of Yoga are believed to be due to autonomic learning. Mountain tribesmen of the Himalayas are able to withstand extreme cold in their bare extremities. For example, climbers have reported their Sherpa guides sleeping barefoot under frigid conditions. It has been suggested that this is possible because they have "learned" to withdraw body fluids from these extremities so that cell membranes cannot be disrupted by the formation of ice crystals. The idea of autonomic learning has not yet received the attention it deserves in the Western world, but your imagination can give you some idea of its various possibilities.

PENFIELD'S MAPPING

Wilder Penfield, of the Montreal Neurological Institute, was one of a group of researchers who produced a fascinating body of evidence by stimulating the human brain with electrodes. Most of Penfield's work was done on epileptic patients in the course of surgical attempts to correct their condition. Epileptic seizures, which vary in intensity from a brief tingling or numbness to convulsions and unconsciousness, may result from injury to the cerebral cortex. Uncontrolled neural discharges from the injured area sporadically spread throughout the brain, causing these epileptic attacks. The surgery was an attempt to identify and remove the damaged area from which the discharges were emanating without removing other essential cortical matter. The patients were conscious throughout the operation (the brain feels no pain) and could respond to questions when various parts of their brains were stimulated. By carefully positioning the electrodes and recording the patient's response, Penfield was able, in effect, to "map" the brain, to determine which parts were correlated with what kinds of activities. Such mapping has resulted in the type of information we saw in Figure 11.17, as well as some important information about the nature of memory.

One of the most startling of Penfield's findings was that the human brain actually "remembers" virtually everything it has ever experienced, and that these experiences can be elicited anew. He found that stimulation of certain areas caused patients to "relive" some of their experiences. One patient *heard* an old familiar strain of music as if it were being played in the operating room when a certain point on the brain was touched. When the electrode was removed and reapplied to the same place, the music did not pick up where it had left off, but began again from the beginning.

Another patient saw himself sitting with relatives and friends in their house in South Africa. The event had actually occurred years before. Another patient "watched" a play she had seen years before, and another heard again Christmas carols being sung in her old church in Holland.

It was fascinating to learn that it is not only the important events that

11.20 A Sherpa woman and child. These hardy mountain people may have developed physiological protection against the elements partly through autonomic learning. They are able to endure very low temperatures without incurring frostbite. Yogis may work their wonders by the same principle, learning to alter the condition of their tissues. Western physiologists have not yet explored these possibilities as fully as they might.

can be summoned from the past, but even everyday, humdrum occurrences that had been long forgotten. It seems that even extremely minute details of our lives are stored in the brain, literally millions of bits of information, which we may not be able to summon through normal attempts at remembering. The amount of detailed information that we store away unaware has also been demonstrated by hypnosis. For example, an elderly bricklayer under hypnosis described in detail the bumps on the bricks in a wall he had built when he was in his twenties. When the wall was checked, it was found that the bumps were there, just as he had said.

The events elicited by Penfield's probings are more than simple vivid recollections. The patients seemed actually to relive the experiences. Penfield himself has described the storage of such memories as being similar to recording by film or wire recorder. The recorder is able to run only forward, and it has only one speed, normal. In other words, electrode stimulation cannot elicit events in any sequence other than the one in which they happened, and when elicited, these remembered events are reexperienced at the same rate they occurred. As long as the electrode is held in place, the experience of a former day goes forward. It cannot be held still, turned back, or crossed with other periods.

As Penfield and Lamar Roberts state in their book, *Speech and Brain Mechanisms:*

The thread of time remains with us in the form of a succession of "abiding" facilitations. This thread travels through the ganglion cells and synaptic junctions. It runs through the waking hours of each man, from childhood to the grave. On the thread of time are strung, like pearls in unending succession, the "meaningful" patterns that can still recall the vanished content of a former awareness.

Penfield's work has provided many surprises and has given us new information on the vexing question of what memory really is. One theory is that memory stems from RNA changes in neurons, brought about by experience; it has also been suggested that memory involves the formation of new neural circuits developed in a learning situation. Moreover, there is some evidence that there is more than one kind of memory, and that each may involve different processes. In any case, the eventual explanation will have to account for the sometimes bizarre information provided by Penfield and his co-workers.

Now that we know something about the astonishing array of information stored in the brain, it is exciting to imagine its possible uses, if we could develop a precisely controlled and less drastic means of retrieval. On long train rides (remember trains?), or in boring company, we could reread a favorite book. There would be little need for testing pure "memory" as is so often done in classroom examinations, since each student would, in effect, be taking an open-book test. (In a hypnosis experiment one high school girl was able to "see" her notes on the blackboard during an examination after having read through them only once.) Other possibilities also come to mind. In cases of terminal disease, instead of facing death in apprehension and racked with pain, perhaps selective brain stimulation could enable a dying person to "relive" happier youthful days, to experience again long hikes with loved ones, robust optimism, and the companionship of old friends, now gone, as life dwindled away.

Behavior

12

The field of study loosely called "behavior" can't exactly be called rowdy, but it certainly hasn't been as sedate or circumspect as some other areas of science. One reason is that early in its development, researchers branched off in several different directions. As each branch developed more or less independently and as researchers in each group became vaguely aware of what the others were up to, the air filled with criticism. Some considered certain behavior patterns to be genetically determined. Others said that all behavior could be accounted for through learning. Some believed that behavior should be studied under natural conditions. Others believed it should be studied in the laboratory where the environment could be carefully controlled. And no one had much use for anyone else's ideas. It would be nice if we could say that the dissension is past history, that we are now beyond all that, but many basic disagreements remain to this day. Let's see if we can discover what all the hollering is about.

ETHOLOGY AND COMPARATIVE PSYCHOLOGY

The study of animal behavior certainly isn't new. After all, the ancients had theories to account for such events as migration and the territorial behavior of certain animals. In the 1930s, however, new dimensions were added to the discipline with the development of a unified theory of *instinct*. The idea was born in Europe, although it is hard to trace to its precise origins. The development of the instinct idea is attributed primarily to the Austrian Konrad Lorenz and his co-workers, who described their theory in 1937 and again in 1950. The Dutch zoologist Niko Tinbergen, at Oxford University, amended the scheme in 1942 and 1951, and an idea resulted that, in spite of its imperfections, demanded serious consideration. And then the arguments began.

Lorenz and Tinbergen, who shared the 1973 Nobel prize in medicine and physiology with the German biologist Karl von Frisch, were the founders of an approach to behavior called *ethology*. Ethology, a discipline cultivated in Europe, traditionally concerns itself with innate behavioral patterns. Research is normally done in the field, under conditions as natural as possible, although ethologists have also relied in part on observations of the many tame or semitame animals that can usually be found sharing their house and yard.

The strongest disagreements with the ethologists' instinct theory came from American *comparative psychologists*. Their research has traditionally

12.1 Konrad Lorenz and goslings. Lorenz, a rather aloof Austrian, has time and again proven himself to be one of the most astute observers and synthesizers in the area of animal behavior in this century. Drawing upon the work of his predecessors and coupling it with his intimate knowledge of animals, he was, with Niko Tinbergen, primary in developing the modern instinct theory. He has been strongly criticized for his assertion that humans are instinctively aggressive. Life Nature Library, *Animal Behavior.* Photograph by Nina Leen, © 1965 Time, Inc.

12.2 Niko Tinbergen, a humanistic and imaginative researcher shared the Nobel Prize with Lorenz and Karl von Frisch, an impeccable experimenter who worked primarily with insects. Tinbergen, at Oxford, worked in close association with Lorenz in the early years of ethology and summarized the concept, with his own modifications, in his landmark book *The Study of Instinct.* Tinbergen, like Lorenz and von Frisch, entered retirement by continuing to work. Tinbergen was a hyperactive child who, at school, was allowed to periodically dance on his desk to let off steam. So in "retirement" he entered a new arena, stimulating the use of ethological methods in autism.

dealt with learning processes in laboratory animals, primarily the Norway rat. Although many of the arguments continue, there has been a closing of the gap between ethologists and comparative psychologists. Probably the final integration of the two approaches depends on further data from a third area of study, *neurophysiology,* which is concerned with how the nervous system works. Such information, however, is hard won and, at present, there just aren't enough researchers working in the nebulous gray areas. But let's take a look at the ethologist's concept of instinct and see what all the argument has been about.

INSTINCT

It was once believed that, whereas human behavior is a result of learning associated with high intelligence, other animals respond only to unalterable "instincts" that are indelibly stamped into their nervous systems at birth. Since mounting evidence over the years did not bear out this sweeping generalization, the concept of instinct fell into some disrepute. In fact, many behavioral scientists declined to even use the word. The problem was compounded by the fact that while ethologists were seeking to clarify the concept as an explanatory tool, the word *instinct* found its way into the layman's vocabulary. Thus, as some were attempting to define the exact parameters of the instinct concept, others were using the term with wild abandon to explain everything from a baby's grip to swimming.

With all the confusion over the term itself, it is best to start with an explanation of what the instinct concept actually is all about. Let's first consider the components of an instinct.

Fixed Action Patterns and Orientation

It has long been known that animals are born with certain behavioral repertoires. The first time a tern chick is given a small fish, it manipulates

Box 12.1   History of the Instinct Idea

Instinct was referred to in Greek literature at least 2,500 years ago, and, for some reason, it continued in the thinking of man with almost no supporting evidence, as very few other ideas have been able to do. In the first century, the Stoic philosophers developed the idea that animal behavior takes place without reflection. This was the beginning of the movement to separate the behavior of man from that of other animals. Man was officially removed from his place on the natural scale of animals (where Aristotle had placed him at the top, just above the Indian elephant) by Albertus Magnus (Big Albert) in the 13th century. St. Thomas Aquinas, a student of Albertus, supported the separation of man and other animals based on the idea that man has a "soul," divinely placed in the embryo sometime before birth. This marked the beginning of a religious explanation of instinct. The reasoning was: There is a life after death and the part of man that continues to live is called the soul. The precise conditions, good or bad, under which the soul continues after death are determined by how the person behaved during life on earth. In order to make the "correct" decisions reason is necessary. Animals do not share in the hereafter and do not need reason. Instead, they are guided by blind instinct.

This philosophical notion of instinct was rejected by many people. For example, Erasmus Darwin, Charles's grandfather, believed that behavior has no innate quality at all, but that it is entirely learned. Charles disagreed, and, in fact, was willing to treat behavioral characters just as morphological ones, species-specific and subject to natural selection. The instinct idea was to fall into some disrepute in the last part of the 19th century, partly because of the tendency of some researchers to name an action as instinctive and to let the matter drop. Such an attitude obviously would not encourage the search for verifying data. So it came to be that in this century the idea of instinct was scoffed at, not considered valid subject matter for serious investigation.

Even by the end of the 19th century, however, some workers had not abandoned the instinct idea. C. O. Whitman, working on the behavior of pigeons at the University of Chicago, was accumulating evidence that in 1898 led him to declare, "Instincts and organs are to be studied from the common viewpoint of phyletic descent." Oskar Heinroth in 1910 described the behavioral similarities of related species and later noted that certain characters would appear in animals reared in isolation. An encompassing theory, however, was to be left to Heinroth's student, Konrad Lorenz.

A leopard frog orienting to catch a fly. The tongue flick, a fixed action pattern, is not released until the central nervous system is stimulated by the sight of an insect in the proper position (close and in the midline). The orientation is obvious as the frog shifts its position, but once the tongue flick has started, no adjustment is possible; it will continue to its completion.

12.4 In a classical experiment, Lorenz and Tinbergen showed that the behavior of a greylag goose rolling an egg back to her nest had two components: fixed and orienting. When they removed the egg while she was retrieving it, she continued the chin-tucking movements (the fixed component) but stopped making the slight side-to-side movements that kept the egg moving in a straight line (the orienting component).

the fish so that it will be swallowed head first, with the spines safely flattened. Dogs that have spent their entire lives indoors will attempt to bury a bone by making scratching motions on the carpet as if they were moving earth. The same dogs will turn around several times before lying down, although there is no grass to trample nor spiders to chase out. Such patterns, which are innate and characteristic of a given species, are called *fixed action patterns*. When fixed action patterns are coupled with orienting movements, we can speak of an *instinctive pattern*. But this probably isn't too clear just yet.

To illustrate the relationship between a fixed action pattern and its orientation, consider the fly-catching movements of a frog (Figure 12.3). For the fixed action pattern to be effective, the frog must carefully orient itself with respect to the fly's position. The performance of the fixed action pattern, the tongue flick, results in the frog's gaining a fly. It is important to realize, however, that once the fixed action pattern is initiated, it cannot be altered. If the fly should move after the pattern is initiated, the frog misses since the sequence of movements will be completed whether the fly is there or not. When a frog catches a fly, one can easily distinguish the orientation movements from the fixed action pattern.

With other instinctive patterns the orientation may be indistinguishable from the fixed action itself. For example, if a goose sees an egg lying outside her nest, she will roll it back under her chin (Figure 12.4). She keeps the egg rolling in a straight line by lateral movements of the head. In an interesting experiment, it was found that the fixed and orienting components of the retrieval pattern could be separated. The orienting movements depend on the direction taken by the rolling egg. If the egg is removed during the retrieval the goose will continue to draw her head back (the fixed component)—but in a straight line, without the orienting side-to-side movements. Thus we see that the fixed action pattern, once initiated, is independent of any environmental cue, so that once it starts it continues. The relationship between the fixed action pattern and its orientation component is similar to that between a car's engine and its steering wheel. Once the engine is started, it continues to run, but each change of direction requires a new movement of the steering wheel and thus depends on the specific environmental circumstance.

It has been found that an animal may not possess all its fixed action patterns when it is very young. Many of these patterns develop as the animal matures, as is the case with morphological characteristics. For example, wing flapping or other patterns associated with flight normally do not appear until about the time a bird is actually ready to begin flying, although there may be some incipient flapping just prior to this time. Also, the fixed action pattern and its orientation component do not necessarily appear simultaneously in the animal's behavioral repertoire. A baby mouse may scratch vigorously with its leg flailing thin air; later it will come to apply the scratch to the itch.

Now let's consider another drama, one that might put some of this together for us. After daylight a peregrine falcon begins its hunt. At first its behavior is highly variable; it may fly high or low, or bank to the left or the right. It is driven by hunger and would be equally pleased by the sight of a flitting sparrow or a scampering mouse. Suddenly it sees a flock of teal flying below. Its behavior now becomes less random as it swoops toward them in what is the next stage of "teal-hunting behavior." First the random search, and then the swoop.

Almost all falcons perform this dive in just about the same way, although there are some variations. This is a sham pass to scatter the flock, and the falcon is likely to fall through any part of it. The teal, at the sight of the falcon, have closed ranks for mutual protection. (After all, it is hard to pick a single individual out of a tightly packed group. A predator is likely to shift its attention from one individual to another, and in failing to concentrate on a single prey, it may come up empty-handed.) The sham pass works though, and the terrified teal scatter. The falcon immediately beats its way upward, and after picking out an isolated target below, it begins a second dive.

If we were to watch this same falcon perform such a dive on several occasions, we might notice that the procedure is much the same every time. And if we were to watch several different falcons, we might notice that they all perform this action in much the same way. The greatest amount of variation is during the earliest part of such dives, since, at this stage, what the falcon does will be dictated in part by what the teal does. The falcon's options become fewer, and its behavior becomes more and more stereotyped, as it descends. Finally, there is a point at which the falcon's action is no longer variable. It now initiates the last part of the attack, the part that is

no longer influenced by anything the teal does. At this instant the falcon's feet are clenched, and it may be traveling at over 150 miles an hour. Unless the teal makes a last split-second change in direction, it will be knocked from the sky, and the falcon's hunt will have been successful.

The falcon's energy will then be spent in following the teal to the ground, where, with rather unvarying and stereotyped movements, it will pluck the teal, tear away its flesh with specific movements of its head, and swallow the meat. The sequence of muscle actions in swallowing is very precise and always the same. The swallowing action itself seems to bring a measure of relief, so that the falcon does not feel like hunting again for a while. Theoretically, it is the desire to perform swallowing movements that brought the falcon out to hunt in the first place. In other words, the falcon was searching for a situation that would enable it to swallow. This probably seems a bit bizarre, but let's go on.

We can derive several points from this example. First, the state of the animal is significant. The falcon had not eaten for a time and was hungry. Hunger thus provided the impetus, or "motivation." In general, if an instinctive action pattern has not been performed for a time, there is an increasing likelihood of its appearance. Second, the earlier stages of this instinctive sequence were highly variable, but the behavior became more and more stereotyped until the final *fixed* action was accomplished. Third, the performance of this final action provided some relief. We have to assume this relief from the fact that once the falcon has performed this final action (swallowing), it does not immediately seek out conditions that will permit it to do so again. You will also note that the fixed action pattern provides the animal with a biologically important element—in this case, food.

Courting behavior in birds is also believed to be instinctive. In one experiment Daniel Lehrman of Rutgers University found that when a male blond ring dove was isolated from females, it soon began to bow and coo to a stuffed model of a female—a model that it had previously ignored. When the model was replaced by a rolled-up cloth, he began to court the cloth; and when this was removed the sex-crazed dove directed his attention to a corner of the cage, where it could at least focus its gaze. It seems that the threshold for release of the behavior pattern became increasingly lower as time went by without the sight of a live female dove. It is almost as though some specific "energy" for performing courting behavior were building up within the male ring dove. As the energy level increased, the *response threshold*, the minimum stimulus necessary to elicit a response, lowered to a point at which almost anything would stimulate the dove.

Invoking a mysterious "energy" seems less necessary to explain the hunting behavior of the falcon, since the reason the falcon hunts seems apparent: it is hungry. But then how do you explain hunting behavior in well-fed animals such as house cats? Perhaps the hunt is an instinctive action pattern in itself, so that its performance provides a measure of relief.

12.5 The phenomenally swift and violent attack of a peregrine falcon. At this point, the bird's behavior is highly stereotyped, and if the teal makes a last minute change, the falcon will miss. This last stereotyped effort, however, was preceded by a series of modifiable acts, the first and most variable being the search for prey. Each step after that was performed in increasingly inflexible ways until the final strike. A successful strike may then initiate the next instinctive sequence—the one that results in swallowing and that provides the greatest and longest lasting relief.

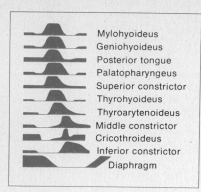

Mylohyoideus
Geniohyoideus
Posterior tongue
Palatopharyngeus
Superior constrictor
Thyrohyoideus
Thyroarytenoideus
Middle constrictor
Cricothroideus
Inferior constrictor
Diaphragm

12.6 Diagrammatic summary of the sequence, timing, and intensity of the muscular contractions involved in swallowing in the dog. A fixed pattern such as swallowing, then, may have many components. In order for the pattern to be performed correctly each component must function in a precise way. Such precision and complexity builds a certain conservatism into the overall pattern since one part cannot change drastically without affecting the other parts.

Appetitive and Consummatory Behavior

There are labels, of course, for the things we have been talking about. First of all, instinctive behavior is usually divided into two parts, the *appetitive* and *consummatory* stages. Perhaps the clearest examples of appetitive behavior are seen in feeding patterns, from which the name is probably derived. Appetitive behavior is usually variable, or nonstereotyped, and involves searching—for example, for food. Some stages of appetitive behavior may be more variable than others, as we saw in the increasingly stereotyped hunting behavior of the falcon. The specific objective of this appetitive behavior is the *performance of consummatory behavior.* Consummatory behavior is highly stereotyped and involves the performance of a fixed action pattern. Swallowing is the consummatory pattern in feeding behavior and involves a complex and highly coordinated sequence of contractions of throat muscles (Figure 12.6). These contractions are the components of a fixed action pattern. As you have learned earlier, the performance of consummatory behavior is followed by a period of rest or relief, during which the appetitive patterns cannot be so easily initiated again.

Numerous experiments have demonstrated that the performance of consummatory behavior is the actual goal of appetitive behavior. This means, in effect, that a hungry animal is not looking for something to fill its stomach, but something to swallow. In an experimental demonstration the food swallowed by a hungry dog was shunted away from the stomach by tubes leading to a collecting pan on the floor. As soon as the dog had *swallowed* the proper amount of food, it stopped eating, apparently satisfied, although its stomach was still empty. The performance of the consummatory swallowing patterns had apparently provided relief.

Releasers

Instinctive actions are released by certain very specific signals from the environment. In the case of the falcon the diving sequence was released by the sight of the teal flying below. Courtship bowing is normally released in male blond ring doves by the sight of an adult female. Environmental factors that evoke, or release, instinctive patterns are called *releasers.* The releaser itself may be only a small part of any appropriate situation. For example, fighting behavior may be released in territorial male European robins, not only by the sight of another male, but even by the sight of a tuft of red feathers at a certain height within their territories. Of course such a response usually "works" because tufts of red feathers at that height are normally associated with the breast of a competitor. The point is that the instinctive act may be triggered by only certain parts of the total environmental situation.

The exact mechanism by which releasers work isn't known, but one theory is that there are certain neural centers, called *innate releasing mechanisms*—or IRMs. When these centers are stimulated by impulses set up by the perception of a releaser, the centers trigger a chain of neuromuscular events. It is these events that comprise instinctive behavior. Many such IRMs may be involved in any complex pattern, so that specific releasers must be presented in sequence. Or, as we have learned, the performance of one consummatory act (such as a falcon's knocking a teal to the ground) may act as the releaser for the next part of the pattern (such as plucking the teal before eating it).

Action-Specific Energy and Vacuum Behavior
The energy that is believed to build up for each instinctive pattern is called *action-specific energy*. The continued buildup of this energy in the absence of a releaser may cause a progressive lowering of the threshold for release of a pattern, as we saw in the case of the male dove. Eventually, if no releaser is provided at all, the instinctive pattern may "go off" on its own. This is called *vacuum behavior*. Once, Konrad Lorenz noticed one of his tame starlings flying up from perches on the furniture and catching insects on the wing. He hadn't noticed any insects in his house, so he climbed up to see what the starling was catching. He found nothing. The well-fed starling had had no chance to release its insect-catching behavior, and so now the behavior was apparently being released in the absence of any stimulus.

The Behavioral Hierarchy
Niko Tinbergen proposed a *behavioral hierarchy* to define the relationships of the different levels of an instinctive behavioral pattern. According to this model, increasingly specific stimuli are needed to release each successive level of an instinctive pattern that lies "lower" on a hierarchical scale. As an example, he described the pattern of breeding behavior in the male stickleback fish—a common, but fascinating, denizen of European ditches. The reproductive behavior is activated by the lengthening days of spring, and at that time the visual stimulus of a suitable territory may elicit either aggression or nest building. Nest building will commence unless another male appears. In this case the sight of the competitor will elicit any of a number of aggressive responses, depending on what that male does. If the intruder bites, the defender will bite back, and so on.

According to Tinbergen's model, the energy for an instinct passes from higher centers to lower ones (from left to right in Figure 12.8). This happens only when a block, or inhibitor, for a particular action is removed. Hence, the energy of the general reproductive instinct passes to a "fighting" center on sight of another male. The sight of that male fleeing then removes a block at the next level of behavior, so that energy flows to the centers for chasing behavior.

It is thus apparent, then, that the term *instinct* carries with it very specific implications. For a behavioral pattern to qualify as instinctive, it should meet the kinds of conditions we have talked about. For example, does the pattern show an appetitive phase? Does it ever appear as vacuum behavior? Is there a period immediately following its performance when the threshold for its release is raised?

12.7 The sequence in courtship and mating in the stickleback, a fish found in European ditches and streams. The male is attracted by the sight of a female swollen with eggs. He is torn between chasing her and leading her to the nest he has built, and the result is a peculiar zig-zag dance. Once she is maneuvered into his "tunnel of love" he pokes her tail with his snout and thus induces her to lay eggs. He then promptly chases her off, returns to fertilize the eggs, and then he spends days fanning oxygen-laden water over them until they hatch.

12.8 The hierarchical organization of the reproductive instinct of the male three-spined stickleback, according to the model devised by Tinbergen in 1942. The overall reproductive instinct is composed of a number of different patterns. The pattern that is expressed is determined by what happens in the environment. These patterns are, in turn, composed of yet more specific ones, their expression also environmentally determined.

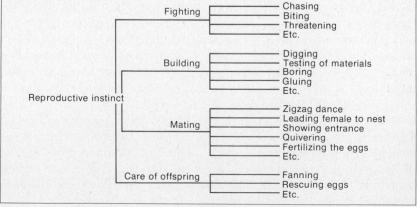

The critical student will note that there are also certain problems with the instinct model. For example, a strictly hierarchical organization makes it hard to account for the fact that different appetitive actions can lead to the same consummatory act. Also, how is it that an animal can perform a complex chain of fixed action patterns without intervening appetitive patterns? The instinct model, as we will see, cannot be rigidly applied to all seemingly innate patterns. But generally, it works. It helps to explain what we see. What we have presented is the classical ethological interpretation of instinct. The scheme has been criticized by neurophysiologists who, for example, can find no neural basis for the notion of action-specific energy. Perhaps the model's usefulness lies in behavioral prediction and as a springboard for further research.

LEARNING

The importance of learning for an animal varies widely from one species to the next. For example, there are probably very few clever tapeworms. But then why should they be clever? They live in an environment that is soft, warm, moist, and filled with food. The matter of leaving offspring is also simplified; they merely lay thousands and thousands of eggs and leave the rest to chance—to sheer blind luck. In contrast, chimpanzees live in changeable and often dangerous environments, and they must learn to cope with a variety of complex conditions. They are long-lived and highly social, and the young mature slowly enough to give them time to accumulate information. We see, then, that learning is more important to some species than others. The *way* animals learn also differs from one species to the next, but we are only beginning to understand some of the differences.

Although learning may theoretically take place in a variety of ways, we will consider only three types of learning: *habituation, classical conditioning,* and *operant conditioning.*

Habituation

Habituation is, in a sense, learning *not* to respond to a stimulus. In some cases the first time a stimulus is presented, the response is immediate and vigorous. But if the stimulus is presented over and over again, the response to it gradually lessens and may disappear altogether. Habituation is not necessarily permanent, however. If the stimulus is withheld for a time after the animal has become habituated to it, the response may reappear when the stimulus is later presented again.

There are several ways in which habituation is important in the lives of animals. For example, a bird may learn not to waste energy by taking flight at the sight of every skittering leaf. A coral-inhabiting fish may come to accept the familiar sight of neighbors living within its own territory. A strange fish, though, which might be searching for a territory, would immediately be driven away. Habituation is often ignored in discussions of learning, perhaps because it seems so simple, but it may well be one of the

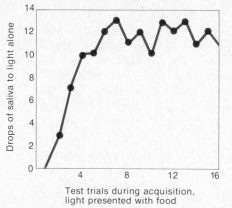

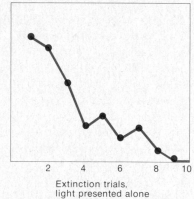

Drops of saliva to light alone

Test trials during acquisition,
light presented with food

Extinction trials,
light presented alone

12.9 The Russian biologist Ivan Pavlov and the apparatus he devised to demonstrate classical conditioning. Upon presentation of a light, meat powder would be blown into the dog's mouth, causing him to salivate. Later he came to salivate at the sight of a light alone. His salivation, then, was *conditional* on the light. Note in the first graph that the dog salivated at maximal levels after only eight trials. When the experiment was reversed and food no longer followed the light, the dog stopped salivating after only nine trials.

more important learning phenomena in nature. For example, it may function internally as a component of satiation. It may also help to explain why animals continue to react aversively to predators they see only rarely, while ignoring commonly seen but harmless species.

Classical Conditioning

Classical conditioning was first described through the well-known experiments of the Russian biologist Ivan Pavlov (Figure 12.9). In *classical conditioning* a behavior that is normally elicited by a given stimulus comes to be elicited by a substitute stimulus. Pavlov found that a dog would normally salivate at the sight of food. He then presented the dog with a signal light 5 seconds before food was dropped onto the feeding tray. After every few trials the light was presented without the food. On the basis of numerous experiments, Pavlov found that the number of drops of saliva elicited by the light alone was in direct proportion to the number of previous trials in which the light had been followed by food.

This conditioning process also worked in reverse. When the conditioned animal was presented over and over again with a light signal that was not followed by food, the salivary response to the light diminished until it finally was extinguished.

Pavlov's experiments demonstrated two important properties of classical conditioning: *generalization* and *discrimination.* It was found that if a dog had been conditioned to salivate in response to a green light, it would also respond to some degree to a blue or a red light. In other words, the dog was able to generalize regarding the qualities of lighted bulbs. However, it was found that, with careful conditioning, the dog could be taught to respond only to light with certain properties. Dogs do not see color well, so they are more likely to respond to intensity differences rather than color differences. If food was consistently presented only with a green light, and never with a red or a blue light, the dog would come to respond only to green lights, thus demonstrating clearly that it was able to discriminate differences between stimuli (discrimination) as well as the properties they had in common (generalization).

Operant Conditioning

Operant conditioning differs from classical conditioning in several important ways. Whereas in classical conditioning the reinforcers (or rewards, such as food) follow the stimulus, in *operant conditioning* reinforcers follow the behavioral response. Also, in classical conditioning the experimental animal has no control over the situation. In Pavlov's experiment all the dog could do was wait for lights to go on and food to appear. There was nothing he could do one way or the other to make it happen. In operant conditioning the animal's own behavior determines whether or not the reward appears.

In the 1930s B. F. Skinner developed an apparatus that made it possible to demonstrate operant conditioning. This device, now called a *Skinner box,* differed from earlier arrangements involving mazes and boxes from which the animal had to escape in order to reach the food.

Once inside the Skinner box, an animal had to press a small bar in order to receive a pellet of food from an automatic dispenser. When the experimental animal (usually a rat or a hamster) was first placed in the box, it ordinarily responded to hunger with random investigation of its surroundings. When it accidentally pressed the bar, lo! a food pellet was delivered. The animal did not immediately show any signs of associating the two events, bar pressing and food, but in time its searching behavior became less random. It began to press the bar more frequently. Eventually, it came to spend most of its time just sitting and pressing the bar. This sort of learning Skinner called operant conditioning.

It is possible to build "stimulus control" into operant-conditioning situations. For example, a pigeon in a Skinner box may be required to peck

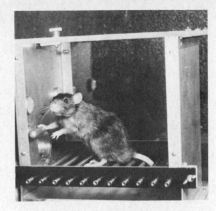

12.10 Rat in a Skinner box. Boxes are designed to promote operant conditioning. For example, a hungry rat may move randomly, searching for food, until it accidentally presses a bar that delivers a food pellet. Each delivery means a greater probability that the rat will press the bar yet again, until finally the rat learns to simply press the bar each time it wants a pellet.

a bar in order to receive food. If the apparatus is designed so that food is delivered only when the bar is pecked within a certain period after a sign lights up that says "peck," the pigeon will come to peck the bar mostly at that time. It will largely ignore the bar until the "peck" sign is lighted. If the pigeon pecks after a "don't peck" sign lights up, and that peck is not followed by food, the pigeon will come to ignore that sign through discrimination learning. It should also be pointed out that pigeons could be trained to peck in response to the "don't peck" sign. Pigeons can't really read.

The relative importance of each type of learning to animals in the wild isn't known at this point. It is likely that most adaptive, or beneficial, behavior patterns arise in nature as interactions of several types of learning.

The kind of learning of which an animal is capable is largely related to its evolutionary heritage. For example, M. E. Bitterman of Bryn Mawr demonstrated that there are strong species differences in "habit reversal" ability. In these experiments, once an animal had learned to discriminate between two choices for its reward, the reward was transferred to the opposite choice. Monkeys and rats showed the best performance, fish the worst, and turtles scored somewhere in the middle. The turtle scores were rather peculiar. They progressively improved in making spatial choices (for example, in differentiating high and low), but not in visual choices (such as in differentiating between a circle and a square). Thus it appears that habit reversal requires a specific kind of learning ability that is present in mammals, but not in fish, and exists in turtles only with respect to a certain class of problems.

HOW INSTINCT AND LEARNING CAN INTERACT

Much of the argument in the field of behavior has been over one simple question: is a certain behavior innate or learned? After years of bickering, the question itself is no longer regarded as valid. The supposition that a particular behavior stems from one source or the other neglects the myriad ways in which behavioral components interact. In a sense, it's like asking whether the area of a triangle is due to its height or the length of its base. A better question is how do innate and learned patterns interact in the development of a particular behavior?

As an example of such interaction, consider the development of flight in birds. Flight is usually considered a largely innate pattern. Obviously, a bird must be able to manage it pretty well on the first attempt, or the bird will crash to the ground as surely as would a launched mouse. It was once believed that the little fluttering hops of nestling songbirds are incipient flight movements, and that the birds are, in effect, learning to fly before they leave the nest. But then someone performed the inevitable experiments. In one set of experiments some nestlings were allowed to flutter and hop up and down, while others were reared in boxes which prevented any such movement. At the time in their lives when the young birds would

Box 12.2 Reward and Reinforcement

Animals can be trained to perform rather complex acts by a system that involves rewarding those actions most closely approximating what the experimenter wants the animal to do. As time goes by, the animal is rewarded only when he does more and more specific acts. For example, in teaching a dog to open a door he may receive food for just walking to the general area of the door in his random strolls about the room as he looks for food. Then he may be only rewarded when he touches the knob, then only when he bites the knob and, finally, only when he turns the knob thus opening the door.

Food is often used as a reward because of its rather universal appeal. By using the word "appeal" we point out a problem with the word "reward." Reward implies some knowledge of how the food works in achieving conditioning. The psychologist B. F. Skinner prefers the term "reinforcement," which is defined as any character, such as food, that alters the probability of an animal performing a particular behavior. Thus, it is an operational term implying nothing about mechanism.

The success of any class of reward, of course, is dependent upon the nature of the animal. A bowl of water may serve as a stronger reward for a Norway rat than for a kangaroo rat, since the latter derives water from food sources.

Reward, or reinforcement, may be positive or negative. If you stay in school and obey the law you will be rewarded by approval of peers and parents, not to mention by money and possibly power. Thus the classroom where you sit is crowded with people who probably all obeyed traffic laws on the way to school. On the other hand if you walk into a bar and slap the largest man there across the face with a wet fish, you may get all your teeth knocked out. Such behavior is negatively reinforced and, partly for this reason, has not become popular in our society.

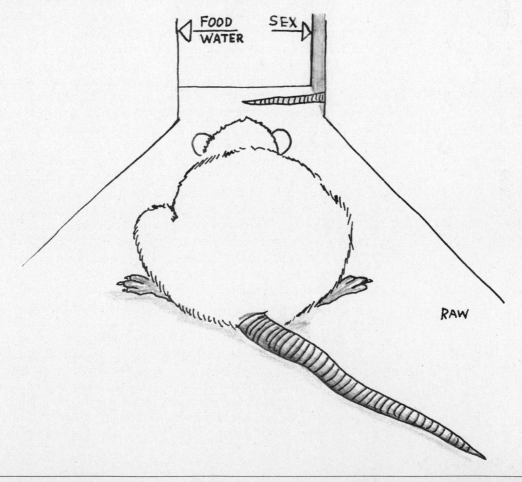

have normally begun to fly, both groups were released. Surprise! The restricted birds flew just as well as the ones that had "practiced"!

Flight behavior, then, is obviously an innate, or unlearned, pattern. It is important to realize, though, that generally young birds do not fly as well as adults. The innate pattern can be improved upon by learning through practice.

In other instances the innate and learned components of a behavior can be more clearly differentiated. The process by which a squirrel opens a hazelnut has been shown to be a complex of learned and innate behavior. If you have ever watched a squirrel open a nut you have probably been impressed with its speed and efficiency. Red squirrels, for example, first cut a groove along the growth lines on one or both of the flat sides of the nut, where the shell is thinnest. Then they insert their incisors into the groove and break the shell open.

In one study the ethologist Irenaüs Eibl-Eibesfeldt reared baby squirrels without any solid food they could practice handling. When they were finally presented with a hazelnut, they correctly performed the gnawing actions and even inserted their incisors into the grooves correctly. Thus these patterns are apparently innate. The problem was that the young squirrels gnawed all over the surface of the nut and started several grooves without really completing any of them. Eventually they were usually able to break open the shell, but only after a great expenditure of energy. In time, however, their performance improved. After the squirrels had handled a few nuts, they began to make their grooves at the thinnest part of the shell, where gnawing was easier, so that the shell then opened more easily. Most of the squirrels soon became proficient at opening hazelnuts, as their innate gnawing pattern was modified through learning to produce the complex adaptive result.

The relative importance of innate and learned components of behavior may vary widely. For example, whereas reproductive behavior in sticklebacks may be largely innate and nut opening in squirrels perhaps less so, the song in some songbirds may be largely learned.

Moreover, white-crowned sparrows, in order to learn their song properly, must hear the song of their species during a brief "sensitive period" early in life. During this period they are not even able to sing yet. If they are exposed to the song after this period, they will produce abnormal songs, lacking the finer details. Thus it may be said that learning plays an important part in the development of song in this species.

We see that some behavioral patterns are almost completely innate, such as the reproductive behavior of the stickleback fish or the sucking response in young mammals, whereas others are almost entirely learned, such as songs in certain birds or the backstroke in humans. Behaviors that entail both instinct and learning may depend more on one element than another. Such information suggests that it is better to ask how a certain behavior develops than to ask whether it is innate or learned.

Let's now take a look at two areas of animal behavior that have long been a puzzle to scientists and see if we can impart to you some of that same spirit of confusion.

RHYTHMS

Why does an apartment-dwelling cat become restless as evening draws on? How is it that some people are able to wake up at whatever time they want to, almost to the minute? Or better yet, how does a potato know what time it is? Questions about timing in living things have been asked for years, and in some cases for hundreds of years. Of course, scientists have come up with all sorts of data and theories, but the fundamental questions just haven't been answered yet.

The rhythmic nature of life should not be unexpected, in view of the cyclic nature of the world around us. The earth revolves around the sun with regularity, and it rotates at a constant rate on its tilted axis. And the moon has a predictable and recurring relationship with the earth. Our lives are segmented by daily periods of light and dark and seasonal periods of cold and warmth. Since the earth itself functions in rhythmic cycles, we should not be surprised that, through the eons of evolution, a certain rhythmicity or sensitivity to environmental rhythm has winnowed into the sensitive cytoplasm of living things.

About 300 years before Aristotle, men described plants that raise their leaves during the day and lower them at night (Figure 12.11). About 2,400 years passed, however, before anyone noticed that the leaves would continue their "sleep movements" even when the plant was placed inside and apparently deprived of its environmental information concerning the time of day. The basic question hinges here. Some organisms continue to show rhythmic behavior after they are placed under "constant" conditions. Is the rhythm a response to environmental clues of which we are not aware, or do the organisms have their own *internal* cycles? If an organism takes its timing cues from the environment, it is said to be under *exogenous* control. If it has its own innate internal cycle that operates independently of the environment, its control is *endogenous.*

Part of the difficulty in determining the mechanisms of biological rhythms is the fact that there is such a profusion of these cycles. There are rhythms in the staccato firing of neurons. Hibernating animals such as bears build up layers of fat every autumn. Human births are most likely to occur in the early morning hours than in the early evening. Cytoplasmic substrates and enzymes ebb and surge with a certain regularity.

Most of the rhythms that have been studied are those that occur in a daily cycle, or approximately every twenty-four hours. It is undoubtedly not coincidental that this is about the length of the solar day. Interestingly, when organisms with a twenty-four-hour cycle are placed under constant laboratory conditions, their rhythms may drift to a twenty-three- or a twenty-five-hour cycle. These periods of about a day are called *circadian rhythms* (from *circa,* meaning "about," and *diem,* meaning "day").

Circadian rhythms have led most researchers to agree that timing is

12.11 We have known for thousands of years that plants show "sleep movements." Their leaves rise and fall on a daily basis. It was once believed that the leaves were responding to daily changes in heat or light or temperature, but the patterns persist even when the plants are exposed to constant conditions in the laboratory.

12.12 Diagram of the dual nature of biological rhythms, according to the exogenous control theory as developed by Frank A. Brown. The geophysically dependent rhythms form the core of the timing mechanism and are responsive to such environmental variables as magnetic fields. These rhythms are not changeable individually, but are the product of eons of evolution. Superimposed on these are a set of adjustable physiological factors that time the organism more precisely to its specific environment. These factors include such cues as light, temperature, and food abundance.

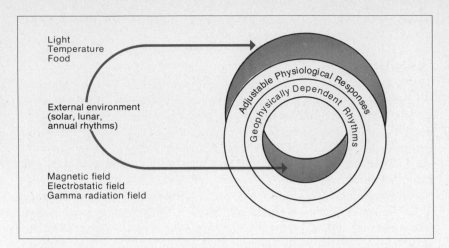

internal. It is suggested that *approximate* twenty-four-hour clocks are built into circadian organisms, but that environmental cues, such as day length, are needed to keep the clock precisely set. In other words, the rhythm is innate, but the fine tuning is environmentally dependent. The fact that some rhythms go through several cycles every twenty-four hours while others are monthly or annual does not upset the theory at all. It is suggested that these rhythms either are governed by some other internal clocks or are based on a circadian clock but have some longer or shorter cycle. A familiar example is the two hands of a clock that move at different rates within the same twenty-four-hour period, but are driven by the same mechanism. If you were to stick your finger in your ear every three minutes, the cycle of this behavior would be three minutes, but the timing would still be done with a twelve-hour clock. Another example is people who keep different hours. Some people are "day people" and others seem to be "night people," but they both use the same twelve-hour clock—and they usually marry each other.

Frank A. Brown, one of the more vigorous exponents of the exogenous control theory, conceptualizes a two-layered ring of rhythmic control (Figure 12.12). He visualizes a "core" of unchanging geophysical constants, which is overlain by a number of variable rhythms dependent on such environmental factors as light, temperature, or humidity. The geophysical influences, he believes, can penetrate all experimental situations so that the organism can receive changing environmental clues under what were thought to be constant conditions. In this scheme, only the outer conditions can be experimentally controlled. Among these subtle geophysical influences are atmospheric tides (measured as barometric highs and lows), which have secondary effects on such geophysical factors as gravity, geomagnetic field, electrostatic field, and background radiation.

Brown describes an experiment in which potatoes were kept under environmentally "constant" conditions and also shielded from the direct effects of barometric changes. The metabolism of these potatoes, it was found, varied with what was happening outside. In fact, one could look at the respiratory charts and tell whether the outside temperature was 50 degrees above or 5 degrees below zero for any day, and whether the barometric pressure was high or low. Somehow, in spite of all the experimenters were able to do, clues were filtering through to the potatoes. An alternative explanation, of course, is that the potatoes were controlling the weather.

In another study Brown had fifteen oysters sent from New Haven, Connecticut, to his laboratory in Evanston, Illinois. He and his co-workers were investigating shell-opening activities of the oysters under constant light conditions. They expected the oysters to open, as if in feeding, in synchrony with the daily high tides in Connecticut. The oysters obliged for about two weeks, but then a strange thing happened. They began to shift their cycles until they were opening at what *would have been* high tide in landlocked Evanston. Somehow, the environmental information was getting through.

ORIENTATION

There are undoubtedly people in South America who wonder where their robins go every spring. Those robins, of course, are "our" birds; they simply "vacation" in the south each winter. If you have been keeping a careful eye on the robins in your neighborhood, you may have noticed that a specific tree is often inhabited by the same bird or pair of birds each year. You may have wondered, since these birds winter thousands of miles away, how they find their way back to that very tree each spring. Such questions perplexed people for years, until in the 1950s a young German scientist named Gustav Kramer provided some answers—and, in the process, raised new questions.

Kramer found that caged migratory birds became very restless at about the time they would normally have begun migration in the wild. Furthermore, he noticed that as they fluttered around in the cage, they often launched themselves in the direction of their normal migratory route. Kramer set up experiments with caged starlings and found that their orientation was, in fact, in the proper migratory direction—except when the sky was overcast. At these times there was no clear direction to their restless movements. Kramer surmised, therefore, that they were orienting according to the position of the sun. To test this idea, he blocked their view of the sun and used mirrors to change its apparent position. He found that under these circumstances the birds oriented with respect to the position of new "sun." This seemed preposterous. How could a stupid bird navigate by the sun when we often lose our way with road maps? Obviously, more testing was in order.

In another set of experiments Kramer put identical food boxes around the cage, with food in only one of the boxes. The boxes were stationary, so that the one containing food was always at the same point of the compass. However, its orientation could be changed with respect to the

(Text continues on p. 341.)

12.13 The question of how animals actually navigate remains largely unsolved. Many of the experiments have been done with migrators and homing pigeons, and as the problem is investigated more deeply, more problems arise. Some previously discarded theories are being refurbished to see if they can yield overlooked clues, and in some areas the results are promising. For example, it seems that magnetism may play a greater role than we believed only a few years ago. Among the most fascinating findings are those that indicate that animals may use celestial clues such as the stars.

The Variety
of Life

□ These buffoon baboons are highly social animals. Here, one female in Kruger National Park grooms another. As is often the case among primates, those with young may receive a great deal of attention from other females. Grooming is an important means of bond-building among many primates. In some species, animals of higher rank are more likely to be groomed than are lower-ranking animals. Also, in some primates adult males may form all-male grooming groups from which females and adolescents are excluded.

□ The cheetah may stalk from the brush, but it usually chases its prey in the open grassland. Otherwise its great speed, up to seventy miles per hour, would be suicidal. Note the extremely long back. The back muscles behave almost like a spring; when the back is arched it straightens very quickly, sending the cat into an extended leap with each stride. The movement of its relatively short upper leg pulls the lower leg through an unusually long arc. The cheetah is unable to suffocate prey by holding the prey's muzzle in its small mouth as a lion might, but instead strangles its prey by holding it by the throat.

□ The red wolf is now exceedingly rare in its previous geographic range, the southern United States. It is a shy animal and has not been able to adapt to human expansion over its habitat as has its cousin, the coyote. A recent attempt to introduce a breeding pair to their former habitat on a southeastern island failed when the young female died.

□ The fawn is easily visible to us because of our color vision, but it is normally hunted by animals that are virtually colorblind, thus its spots and natural tendency to freeze are an effective camouflage against its predators. In addition, many young animals have very little odor. Odor is a useful tool in communication among adults, but the helpless young had best not communicate with anyone but a watchful mother.

□ Hunters and trappers have all but eliminated the Everglades cougar, a variety of the more familiar cat of the American West. The Everglades cougar is an effective predator in the swamp, but is threatened with extinction since large predators require extensive hunting areas and its hunting range is being continually drained and cleared for "development."

These are shy animals and they attempt in every way to avoid humans, but they are all too quickly finding they have no place to go. Florida, however, continues to absorb humans fleeing the north as if its resources were limitless. The effect has been so pronounced that the influx may soon be stopped by the simple lack of potable water. The Everglades may, in itself, seem an unimportant place, but it is home to many species found nowhere else, and to bring about their extinction would be shameful. Ecologically, we need all the company we can get on our immense journey.

□ The white-footed mouse, *Peromyseus leucopus*, is prey to a wide variety of animals including birds, reptiles, and mammals. The large eyes indicate the little animal is most active at night when it would be less conspicuous. The lighter underside is called countershading and is found in a wide range of animals.

□ The bearded dragon, *Amphibolurus barbatus*, of Central Australia (opposite page). Notice the primitive unspecialized teeth and the large distended throat displayed dramatically as a threat. (As a rule, threatening animals try to look larger.) Among lizards and birds the diet is indicated by the size and shape of the mouth structure. Hearing is accomplished by rather simple ears, the opening near the back of the head. Most reptiles only listen to *other* animals, however, since they, themselves, are voiceless. The eyes can move independently, an ability that would be most distracting in an animal that thought much about what it was looking at.

☐ Vultures stand huddled around the corpse of an impala on the African veldt. It is often assumed that the birds play a unique role in the web of life by eating flesh no other animal would touch. But they often squabble with jackals drawn to the kill as well as with the returning predator who made the kill in the first place. Weakened animals rarely die of disease or old age in the wild. Instead, they are taken by predators when they are no longer able to defend themselves. It has been suggested that bacteria act to render meat putrid so that they can reserve it for themselves, but a little rankness wouldn't deter this motley crew.

☐ Penguins are most unbirdlike birds, unable to fly, but hunching their necks and swimming torpedolike at speeds up to fifteen miles per hour, sometimes leaping entirely free of the water, like porpoises. The seventeen species are mostly found in frigid Antarctic areas (three live there exclusively), but one species, the Galapagos penguin, reaches the equator and lives in the tropical islands side-by-side with stilted flamingos. The Galapagos penguin is the smallest penguin species on earth, demonstrating "Bergmann's rule," which states that species tend to be smaller in warmer areas (as a means of dissipating heat by increasing surface area in proportion to its volume). All penguins require cool ocean currents, and are insulated by small, hard, overlapping feathers shielding a downy undercoat and a layer of fat below the skin. Penguins look cuddly, but their bills can cut like knives. Whereas the bones of most birds are hollow, penguins have solid bones that they use as ballast.

□ The crayfish (above) might seem to be an encumbered and ineffective beast as it stalks about stream bottoms continually stuffing morsels into its mouth. But it is a surprisingly efficient predator and somehow manages to capture small fish and tadpoles and even to seriously wound larger fish.

□ Sea anemones are coelenterates, related to *Hydra*. Their deceptively soft tentacles are loaded with thousands of stinging cells, small poisonous darts that spring forth when touched, for example, by a small fish. Other fish, however, are immune to the poison and swim freely among the tentacles, protected from predators.

☐ The purple pitcher plant (above) grows in the soggy soils of a few southeastern areas. The cuplike structure may hold water, and when a hapless insect, such as this grasshopper, enters the cup, the downward-directed barbs prevent it from leaving. Days from now, its corpse will be found floating in the deadly pool, drained of its vital juices. There is some disagreement over the importance of heterotrophy in the lives of these plants.

☐ Banyon trees form dramatic cascades as branches grow downward to enter the soil and behave as roots, bracing the great trees and forming impenetrable walls. Banyons may form a few great pillars and the open areas beneath are occasionally designated as sacred meeting places by African tribesmen.

□ The boy is paddling his dugout between the enormous leaves of giant water lillies, or Victoria plants. The platelike leaves provide great photosynthetic surfaces. Note that practically the entire surface of this pond in the Far East is covered. What do you suppose this does to oxygen levels below? (Stomates of many water plants are on the upper surfaces of the leaves.) Would the great leaves shield the sun from life below the surface? What would be the effect on the diversity of life there?

□ Lichens are actually composed of two kinds of plants interacting symbiotically (to their mutual advantage). A fungus forms the attachment for the structure, but fungi lack chlorophyll and are unable to manufacture food. However, algae take advantage of the fungi's firm grip, growing on its surface and providing food through photosynthesis. Decomposing fungi, it is believed, may be the first step in soil building.

□ Coconuts (above) are huge seeds and the coconut milk is the endosperm, or seed food. You can even find endosperm nuclei suspended in the milk of young coconuts. Coconuts are great travelers and if washed to sea, may float for long periods until they land on some new shore.

□ These mangrove trees (top right) near Singapore will finally form a dense but shallow forest in the quiet waters in which they grow. More turbulent and oxygenated waters might render the airborne, and sometimes hollow, roots unnecessary, but plants have great difficulty in establishing themselves in surging seas.

□ Spanish moss (bottom) is an epiphyte, a plant growing on another plant, but it is not parasitic in that it does not derive food from its anchoring host. It is able to obtain water and minerals from the humid coastal air.

surroundings by revolving either the inner cage containing the birds or the outer walls, which served as the background. As long as the birds could see the sun through any window, no matter how their surroundings were altered, they went directly to the correct food box. Whether the box appeared in front of the right wall or the left wall, they showed no signs of confusion. When the food was moved to a different box, they continued for a time to approach the old food box until they discovered the new compass orientation of the new food box. On overcast days, however, the birds were completely disoriented and unable to locate their food box easily.

In experimenting with artificial suns Kramer made another interesting discovery. If the artificial sun remained stationary, the birds would shift their direction with respect to it at a rate of about 15 degrees per hour, the sun's rate of movement across the sky. Apparently, the birds were assuming that the sun they saw must be moving at that rate. When the real sun was visible, the birds maintained a constant direction as it moved across the sky. In other words, they were obviously able to compensate for the sun's movement. This meant that some sort of biological clock was operating—and a very precise clock at that (Figure 12.14).

Another study of starlings was conducted by K. Hoffman, who trained birds under natural light conditions to go to one particular food box, whose position was fixed. He then put the birds on an artificial light-dark schedule that was six hours behind local time. After about two weeks of this artificial regimen, when the birds were returned to normal light-dark conditions, they immediately approached a food box that lay at a 90-degree angle from the one they had been trained to choose. Their day had been altered by one-quarter (six hours), and their orientation was shifted correspondingly by a quarter-circle. After a week or two under natural light conditions the birds returned to their original directional behavior.

Such sun-compass orientation implies, of course, that starlings know the time of day and that they also know the normal course of the sun. In other words, incredible as it seems, they apparently know where the sun is supposed to be in the sky at any given time. It seems that part of this ability is innate. In another experiment, a starling reared without ever having seen the sun was able to orient itself well, although not as well as birds that had been exposed to the sun earlier in their lives.

Sun-compass orientation exists in a variety of other animals including insects, fish, and reptiles. There is also some evidence of celestial navigation in nocturnal birds—navigation by the stars. More recent evidence indicates that some species of navigating birds may even be guided by the earth's magnetic field. It should be pointed out that this is an extremely vigorous field at present, and new data are appearing almost daily. Any synthesis at this point would be extremely tenuous. However, these examples provide some idea of the value of innate characteristics in orientation, as well as the surprising sensitivity of some species to environmental cues.

12.14 Kramer's orientation cage. The bird can be allowed to see the real sun, or the sun's apparent position can be shifted by the use of mirrors. It was found that once a bird was trained to approach a food box that lay in a particular compass direction, it would always move in the correct direction no matter how the fixtures in the cage were shifted. However, if the apparent direction of the sun were shifted a certain degree, the bird would move at an angle to the proper box, the error being at the same angle as that of the sun shift.

12.15 One of the best navigators in the world, the golden plover. In summer, the birds are dispersed rather continuously along the northern coast of Alaska and Canada, including many Canadian islands. When winter draws on in the north the western (Alaskan) group flies nonstop to Hawaii. The eastern (Canadian) group flies toward Nova Scotia, turns southward over the Atlantic, and heads to the southeast coast of South America. This group returns over a different route, mostly overland, following the coast of Central America, but flying several hundred miles over the Gulf of Mexico.

COMMUNICATION

If you should come upon a large dog eating a bone, walk right up to him and reach out as if to take the bone away. You may notice several changes in his appearance. The hair on his back may rise, his lips may curl back to expose his teeth, and he may utter peculiar guttural noises. He is communicating with you. To see what his message is, grab the bone.

It would be a remarkable understatement to say that communication is an important phenomenon in the animal world. It is accomplished in many ways and with innumerable adaptive variations. Its ultimate function, however, is to increase reproductive success in those animals able to send and receive signals. Communication may be directly involved with reproductive success as a component of mating behavior or precopulatory displays. It may also indirectly increase reproductive output by helping the offspring to avoid danger when parents give warning cries, or by simply helping the reproducing animal to live better or longer so that it may successfully mate again. Remember, the charge to all living things is "reproduce as much as possible or your genes will be lost." Communication helps animals to carry out that reproductive imperative.

Let's consider a few general methods of communication, and in so doing perhaps we can learn something about the ways in which animals influence the behavior (or the probability of behavior) of other animals. That, after all, is what communication is all about.

Visual Communication

Visual communication is particularly important among certain fish, lizards, birds, and insects—and to some primates as well. Visual messages may be communicated by a variety of means, such as through color, posture or shape, movement, or timing. As we have seen, color serves as a component of the releaser of territorial fighting in European robins. In grackles the female solicits the attention of the male by assuming a head up posture while fluttering her drooping wings. Some male butterflies are attracted to the female by her performance of a specific flight pattern. And as an example of communication by timing, fireflies are attracted to each other on the basis of their flash intervals, each species having its own frequency.

Because of the high information load possible with visual signals, subtle variations in the message can be conveyed by gradations in the display (Figure 12.16). Of course, graded displays are useful only to those species that are sensitive and intelligent enough to be able to utilize such subtleties. Another advantage of the high information load in visual signals is that the same message may be conveyed by more than one means. Such redundancy may be used either to modify the message or to emphasize it in order to reduce the chance of error in interpretation.

A visual signal may be a permanent part of the animal, as in the elaborate coloration of the male pheasant and the striking facial markings of the male baboon. These animals advertise their maleness at all times and

12.16 An example of a "graded" display. The bird, *Fringilla coelebs,* is showing, from left to right, a low, medium, and high intensity threat. The bird may change continuously from one of these postures to the next, so there are intermediate positions reflecting various levels of aggression. In other cases, an animal may show an all-or-none display. These are performed with full intensity or not at all and their advantage is that when they appear they are unmistakable. Graded displays may cause some ambiguity on the part of the observer, who may not be able to distinguish very low intensity signals, but graded displays have the advantage of carrying more information since they rather precisely reflect the level of arousal of the signaller.

are continually responded to as males by members of their own species. On appropriate occasions they may emphasize their "machismo" by behavioral signals, such as the strutting of the pheasant and the glare of the baboon. In other cases the visual signal may be of a more temporary nature, such as the reddened rump areas in female chimpanzees and baboons during estrus. As a more short-term signal, a male baboon may expose his long canine teeth as a threat (Figure 12.17).

Short-term visual signals have the advantage that they can be started or stopped immediately. If a displaying bird suddenly spots a hawk and "freezes," its position will not be given away by any lingering images in the environment. Also, in visual messages the recipient is notified of the exact location of the sender. The recipient can then respond to him in terms of his precise location, as well as his general presence and behavioral state (aggressive, romantic, or whatever).

Visual signals also have certain disadvantages. For example, if a sender can't be seen, it can't communicate. And all sorts of things can block vision, from mountains, trees, and fog to oversize hats. Visual signals are often useless at night or in dark places (except for light-producing species). Also, since visibility weakens with distance, long distances diminish the usefulness of such signals. As distance increases, the signal must become bolder and simpler; hence it can carry less information. The only way to intensify a visual signal is to increase its visibility, either by decreasing the distance or by increasing its size or contrast. It isn't possible to pump more energy into the signal itself, as with certain other forms of communication.

Sound Communication

Sound plays such an important part in communication in our own species that it may surprise you to learn that it is limited, for the most part, to arthropods (joint-legged creatures such as insects) and vertebrates. Surprised, eh? We're all familiar with the song of the cricket and the cicada, so

12.17 A male baboon "yawns" as a threat. His open-mouth display shows, to good effect, his fighting weapons. This particular individual is indeed formidable, being full-grown, but so young that his canine teeth show little wear. His hair, standing on end, makes him appear even larger. This is a temporary signal. He can close his mouth and flatten his hair, but the colorful stripes on his muzzle will remain as a permanent reminder that, whatever his mood, he is a male.

you are probably aware that insect sounds are usually produced by some sort of friction—such as by rubbing the legs together or the legs against the wings. These sounds are characterized by differences in cadence rather than in pitch or tone, as with most birds and mammals.

Vertebrates use a number of different forms of sound communication. Fish may produce sound by means of frictional devices in the head area or by manipulation of the air bladder. Land vertebrates, on the other hand, usually produce sounds by forcing air through vibrating membranes in the respiratory tract. Of course, they may communicate by sound in other ways. Rabbits thump the ground, gorillas pound their chests, and woodpeckers hammer your drainpipe early on Sunday mornings.

Most of the lower vertebrates rely on signals other than sound, but some species of salamanders can squeak and whistle, and there is one that barks! Frogs and toads advertise territoriality and choose mates at least partly by sound. Most reptiles do not communicate by sound, but territorial bull alligators can be heard roaring in more remote areas of the Southern swamps, where a few have escaped becoming status symbols as belts, shoes, and purses. Darwin described the roaring and bellowing of mating tortoises when he visited the Galapagos Islands in 1839.

Sound may vary in pitch (low or high, depending on the frequency of the sound waves), in volume, and in tonal quality. The last is demonstrated as two people hum the same note, yet their voices remain distinguishable. It is possible to graph the characteristics of sounds by use of a *sound spectrogram.* In effect, this is a translation of sound into visual markings, which makes possible a more precise analysis of the sound.

The function of the message may dictate the characteristics of the sound. For example, Figure 12.18 shows a sound spectrogram of the mobbing cries of several bird species. Such calls are brief, low-pitched *chuk* sounds, whose source is easy to locate. When a bird hears the repetitious mobbing call of a member of its own species, it is able to locate the caller quickly and join it in driving away the object of concern—usually a hawk, crow, or other marauder. In contrast, when a hawk or other predator is spotted flying overhead, the warning cries of songbirds are usually a high-pitched, extended *tweeeeee,* which is difficult to locate. The response of a bird hearing this call is quite different from its response to the mobbing call. The warning cry sends the listener heading for cover, often in a deep dive into protective foliage, from where it may also take up the plaintive, hard-to-locate cry.

Sound signals have the advantage of a potentially high information load through subtle variations in frequency, volume, timing, and tonal quality. They are distinguishable at low levels, but they can also be made to carry over great distances by pumping more energy into them—in effect, becoming louder. In addition, sounds are transitory; they don't linger in the environment after they have been emitted. Thus an animal can cut off a sound signal should its situation suddenly change—for example, with the appearance of a predator. A further advantage is that an animal doesn't ordinarily have to stop what it is doing to produce a sound. As everyone knows (except perhaps builders of student housing), sounds, unlike visual images, can go around or through many kinds of environmental objects.

Sound communication also has its disadvantages. Sounds are rather useless in noisy environments. Thus some seabirds that live on pounding, wave-beaten shorelines rely primarily on visual signalling. Also, sounds attenuate with distance. And the source of sound is sometimes hard to locate, especially underwater.

Chemical Communication

You have probably seen ants rushing along single file as they sack your cupboard. You may also have taken perturbingly slow walks with dogs that stop to urinate on every bush. In both cases the animals are communicat-

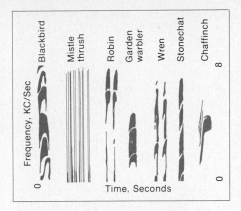

12.18 Sound spectrograms of the mobbing calls of several species of British birds while attacking an owl. The sounds have qualities that make them easy to locate. Such calls are low-pitched *chuk* sounds. Different species of birds have developed similar calls through convergent evolution. That is, their calls serve much the same purposes and were developed under relatively similar conditions, thus the qualities of the sounds came to be somewhat alike.

12.19 Sound spectrograms of five species of British birds when a hawk flies over. Such calls are high-pitched, drawn out, and difficult to locate. They, too, achieved their similarity through convergent evolution.

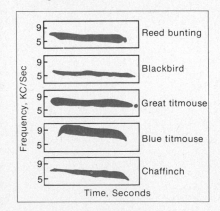

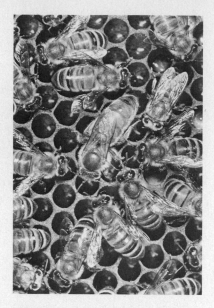

12.20 Life within a beehive is very tightly regulated by a complex system of pheromonal control. The air in the hive is filled with molecules being wafted here and there, but their message is often exceedingly critical. Here sterile female workers surround their only hope of producing their kinds of genes, their queen. As long as she lives she will produce a chemical that causes all developing females to become workers. If she should die, some of the developing females may begin the transformation to queens. The first queen to emerge will crawl about the hive inspecting the cells and killing any other potential queens. Her mate will be from among the sons produced by the reactivated ovaries of some of the workers.

ing by means of chemicals. The ants are following chemical trails laid down by their fellows, and the dog is advertising his presence.

Insects make extensive use of chemical signals, and certain chemicals elicit very rigid and stereotyped behavior. An "alarm chemical" is produced when an ant encounters some sort of threat, and as the chemical permeates the area, every ant of that species that detects the chemical reacts by rushing about in a very agitated manner. Female moths produce a "sex attractant" that attracts males from astonishing distances. The substance is called *bombykol,* and it is so potent that a male moth will be aroused by a glass rod dipped in a solution diluted to one part in a trillion!

The immediate and invariable response produced by certain chemical signals is comparable to the action produced by hormones. Whereas hormones work *within* an individual, however, these chemicals work *between* individuals and are called *pheromones* (from the Greek *pherein,* meaning "to carry," and *hormon,* meaning "to excite").

It was at first believed that pheromones are found only among the insects, species whose behavior is largely genetically programmed. But then it was found that pheromonal interactions also occur in mammals. For example, if pregnant rats of certain species smell the urine of a strange male, some component of that urine will cause them to abort their fetuses and become sexually receptive again. As we have already seen, the females of most mammal species also signal their sexual receptivity by some chemical means, with varying degrees of influence on the male, depending on the species.

Honeybees provide another interesting example of chemical communication in insects. The queen produces a "queen substance" that, when fed to developing females, causes them to become sterile workers. When the queen dies, the inhibiting substance is no longer present, and new queens begin to develop. At this time the undeveloped ovaries of some adult workers also suddenly begin to mature. Their eggs will produce drones—royal consorts, one of which will mate with the new queen on her nuptial flight.

The molecular makeup of several chemical signals in the animal world has been carefully analyzed. It was found that the backbone of the molecule usually consists of five to twenty carbon atoms. The molecular weight of the chains ranges between 80 and 300. It is reasoned that molecules with at least five carbons are needed to provide the variation and specificity for discrete messages. The maximum molecular size is more likely to be set by other factors. For example, such molecules are usually carried in the air, and they must therefore be small and light. Also, more energy is expended in the manufacture of larger molecules.

Chemical signals have the advantage of being extremely potent in very small amounts. Also, because of their persistence in the environment, the sender and receiver do not have to be precisely situated in time and space in order to communicate. Chemicals can also move around many sorts of environmental obstacles.

12.21 The golden-fronted woodpecker (left) looks a bit scruffy while molting, but is still remarkably similar to the red-bellied woodpecker (right). The range of the golden-fronted extends through dry areas from Honduras northward to central Texas where it rather abruptly stops and that of the red-bellied begins. The red-bellied's range extends to the eastern coast where it lives primarily in wooded habitats. The birds overlap in a narrow range in Texas, but in spite of strong similarities in appearance, vocalization, and behavior they do not interbreed. Whereas they do not recognize each other as sexual partners, each does see the other as a competitor and they mutually exclude each other from their territories as if they were of the same species. Obviously they use different cues in the recognition of mates and competitors.

The specificity of chemicals, however, limits their information load. Moreover, because chemicals do linger in the environment they may advertise the presence of the signaler to arriving predators as well as to the intended recipient of the message.

WHY COMMUNICATE?

Now that we have some idea of how communication can be accomplished, we might ask, why is it necessary to communicate at all? You can probably sit right there and think of a host of reasons, but let's see if we can't organize the ideas a bit.

Species Recognition

First, communication may permit one animal to know that another animal is of the same species. This may not seem too important—unless one is interested in reproduction. After all, animals of two different species cannot normally produce healthy offspring. But a lot of species are very similar, and in the absence of some precise means of identification an animal might waste a lot of time and energy in trying to mate with a member of the wrong species. Therefore, species identification must somehow be quickly established.

For example, the golden-fronted woodpecker and the red-bellied woodpecker often share the same woods around Austin, Texas (Figure 12.21). The two species look very similar to the casual observer. The most conspicuous difference is that the golden-fronted woodpecker has a small yellow band along the base of its red cap, which is absent on the red-bellied. Also, the red-bellied woodpecker has slightly more white in its tail. The habits of the species are much alike and their calls are very similar, at least to the human ear. Nevertheless, each of these birds is somehow able

12.22 A male salticid spider must approach the voracious female very carefully. She is larger and could easily overpower him. Of course a female that attacked her suitors would leave no genes in the next generation and that sort of behavior would die out. Still, he must approach her carefully. He advances a few careful steps at a time, stopping to wave specialized appendages in a very specific way. If he performs this correctly, he will cause changes in her nervous system, which inhibit her predatory behavior, at least long enough for him to climb astride her and rake a sperm-bearing arm across her genital opening until she is impregnated. He may then jolt her from her mesmerized state by a parting pinch as he scurries away. He may make it; he may not.

to recognize its own species. At least cross mating has never been observed, nor have any hybrids. No one knows exactly what cue the birds use, but to them, apparently, the signals are clear.

Interestingly enough, whereas the two species ignore each other sexually, they have come to treat each other as competitors and will eject a member of the other species from a territory as quickly as they would a member of their own species. Apparently, the sign stimuli that release sexual and territorial behavior are quite different in these species.

Population Recognition

Even when species identity has been established, it might also be to an animal's advantage to be able to recognize those individuals from its own particular breeding population. Consider a species of bird in which some subpopulations live in dark, moist lowlands and are therefore dark-colored; other subpopulations live on dry mountaintops and are light-colored, making them less conspicuous against the light-colored dry earth. If a lowland individual should mate with a highland individual, the offspring might be conspicuous in either type of terrain and might therefore quickly be taken by predators. Hence there would have evolved a tendency for members of each population to breed with their own type. Actually, there aren't many field data to show that population recognition is an important phenomenon in nature, but it seems that it would be. As support for the notion, it has been shown that population-specific signals exist in several kinds of animals.

Individual Recognition

It may also be important for an individual to be able to recognize specific individuals within its own population. If you have ever watched gulls at the beach, you may have noticed that the adults all look alike. This is because you aren't a gull; all people probably look alike to them. If you should visit a gullery, the place where they nest, you might see thousands of "identical" gulls. Color banding and recording playback experiments, however, have shown that gulls can recognize their own mates on sight and are able to filter out the calls of their mates from the raucous din of the clamoring gulls wheeling above.

In species in which the sexes come together only for copulation, the mate-recognition problem becomes one of sexual identification. Male salticid spiders approach the female only for mating (Figure 12.22). This approach is performed with much waving of special appendages to inhibit predatory behavior on the part of the larger female. Mating is brief, and the male quickly clears out. Afterward, should the female encounter him, she will recognize him only as prey. We can see, then, that individual mate recognition is important only in species that establish pairs and that it increases in importance with the likelihood of interaction with other individuals.

Individual recognition is important for two primary reasons. First, one should be able to recognize one's own mate, especially when the rearing of offspring is done by both parents. Mate recognition ensures that pairs will attend to the nest that harbors their own offspring, thus increasing their chances of reproductive output. This is an essential factor in species that show a high level of coordination while rearing offspring. For example, some birds bring food to their mates that are incubating eggs. In the African hunting dog, the male is more likely to regurgitate food to his own mate when she is nursing pups, than to any other females.

The second reason individual recognition is important is that it functions in the maintenance of hierarchies. Once animals in a group know their rank with respect to the others, there will be less fighting over food or other commodities. In any group where rank is unknown, or not yet established, the incidence of fighting is likely to be high. Rank, however, can only be maintained when each animal can recognize the others individually so that it can respond to each according to its status.

AGGRESSION

The old image of "Nature, red in tooth and claw" has recently been superseded by the popular notion that other animals get along with their own kind and humans are the only animals that regularly kill members of their own species. It might be a good idea to take a look at the animal kingdom to see what is actually going on.

First of all, we should note that *aggression* is belligerent behavior that arises as a result of *competition*. A lioness may spring toward a group of wild pigs, single out one in the chase, catch it, and quickly smother it or break its neck. However, she is not acting out of aggression. She is being about as aggressive as you are toward a hamburger. However, if one of the big boars should turn and charge her, the lioness may momentarily show a component of aggression—fear. Fear is an important component of aggression; it is simply at one end of the behavioral spectrum of aggressive behavior. At the other end is sheer, unbridled belligerence.

12.23 Problems of individual recognition appear insurmountable to a human observer at a colony of penguins, such as this Adelie penguin rookery off Anvers Island. However, to them *we* may all look alike. In any case, they can tell each other apart. After the female lays her single egg she may wander off to fish for weeks or months, leaving the male to stand on the egg and utilize his own body fat. When she returns to relieve him, she may approach many males before she recognizes the gaunt father of her egg.

Table 12.1 Dominance hierarchy in a flock of hens as determined by differences in access to food or water. The highest ranking bird may peck all others. The second highest may peck all but the higher one, and so on. A new bird introduced into such a group may have to fight frequently until its rank is established. In such a system not only the dominant bird benefits, but the subdominant as well. If the subdominant had challenged the dominant it would have gotten worked over and lost the commodity as well. So, rather than risk injury and lose time, energy, and the commodity, it might do better to seek an unclaimed commodity elsewhere.

Hen 1 pecks	2	3	4	5	6	7	8	9	10	11	12	13
Hen 2 pecks		3	4	5	6	7	8	9	10	11	12	13
Hen 3 pecks			4	5	6	7	8	9	10	11	12	13
Hen 4 pecks				5	6	7	8	9	10	11	12	13
Hen 5 pecks					6	7	8	9	10	11	12	13
Hen 6 pecks						7	8	9	10	11	12	13
Hen 7 pecks							8	9	10	11	12	13
Hen 8 pecks								9		11	12	13
Hen 9 pecks									10	11	12	13
Hen 10 pecks								8		11	12	13
Hen 11 pecks											12	13
Hen 12 pecks												13
Hen 13 pecks none												

Source: Adapted from W. C. Allee, *The Social Life of Animals* (New York: Norton, 1938).

Box 12.3 Sexual Dimorphism

You may have noticed that boys look different than girls . . . at least they used to. Even today, however, it's easy to tell a rooster from a hen. In some deepsea fishes the sexes are even more dissimilar, the male may be an appendage fused to a knob on the head of a female. On the other hand, it may be difficult to tell a male mouse from a female and the sexes of seagulls are almost identical. Sexual dimorphism—or the different appearance of the sexes—varies widely among animals.

There are interesting evolutionary implications of sexual dimorphism. How did it start? With respect to vertebrates, one idea is that as they came under increasingly severe competition for commodities such as food, the males increased in size and strength in order to compete more successfully. The females were restrained from such changes by the demands of gestation (which placed a premium on a certain body size) and the fact that they were more closely associated with the young (which made it important that they be physically and behaviorally less conspicuous). Males that were best able to gain commodities might then have been disposed to share those with females, for reproductive purposes. This meant that females might then be more likely to choose the more vigorous males as mates.

This tendency would have driven males toward increasingly garish appearances and bold, aggressive and conspicuous behavior in order to display their vigor and to intimidate other males in disputes over commodities. Exceptions exist, of course, but generally today males of most species, especially among birds and mammals, are larger, stronger, and more aggressive than females and dominant over them in social interactions. It provides some entertainment, and a lot of argument, to try to determine how this general condition could have arisen.

An interesting relationship between sexual dimorphism and domestic duties exists among some species. Consider an example from birds. The sexes of song sparrows look very much alike. The males have no conspicuous qualities that immediately serve to release reproductive behavior in females. Thus courtship in this species may be a rather extended process as pair-bonding (mating) is established. Once a pair has formed, both sexes enter into nest building, feeding, and defense of the young. The male may only mate once in a season but he helps to maximize the number of young that reach adulthood carrying his genes. He is rather inconspicuous, so whereas he doesn't turn on females very easily, he also doesn't attract predators to the nest.

The peacock, on the other hand, is raucous and garish. When he displays to a drab peahen, he must present a veritable barrage of releasers to her reproductive IRMs. In any case, he displays madly and frequently and is successful indeed. Once having seduced an awed peahen, he doesn't stay to help with the mundane chores of child rearing, but instead disappears into the sunset looking for new conquests. The peahen he has just left may be an inferior bird, not likely to be able to rear young successfully. He could have discovered this by a more careful process of mate selection. But no matter, perhaps the next conquest will be a better bird. Alas, just as he attracts females, he also attracts tigers and his flurry of sexual activity may be rather quickly ended. So, instead of carefully rearing one brood each season for several seasons, he maximizes his reproductive success through another means.

Think of other species of birds and mammals. Which evolutionary route have they taken? What about humans?

12.24 In some cases fighting may occur between species that normally never even interact. Here, a raccoon approached a hare in the darkness and in the scuffle, the hare leaped over the raccoon and administered a severe drubbing with its powerful back legs to the raccoon's head. Some evidence is beginning to appear that suggests that animals of different species that share the same area may recognize each other individually and may react to each other on an individual basis.

Fighting

Let's consider the most obvious form of aggression, fighting. You can discount the old films you have seen of leopards and pythons battling to the death. Such fights simply aren't likely to happen. What is a python likely to have that a leopard needs badly enough to risk its life for, and vice versa? Although such fighting might occur in the unlikely event that one should try to eat the other, fighting is much more likely to occur between two animals that are competing for the same commodity. Obviously, the strongest competitor for any animal is going to be one that utilizes the same habitat in the same way. And what sort of animal is most likely to do this? Right! A member of the same species. Moreover, if there is strong competition for mates, fighting is even more likely between members of the same sex within that species. And this is in fact where most fighting occurs—between members of the same sex of a species.

Of course, fighting between species sometimes occurs. The golden-fronted and red-bellied woodpeckers exclude each other from their respective territories. Lions may attack and kill African cape dogs at the site of a kill. The lions don't eat the dogs; they just exclude them.

Animals may fight in a number of ways, but they usually manage to avoid injuring each other (Figure 12.24). There are several apparent benefits to such a system. First, no one is likely to get hurt. The competitor is permitted to continue its existence, it is true, but the possibility of having to compete with him again entails less risk than does serious fighting. Also, since animals are most likely to breed with the individuals around them, the opponent may be a relative who is carrying some of the same genes;

12.25 Fighting male rattlesnakes. The triangular heads show that their poison sacs are full. Each could kill the other, yet they do not bite. The fight is more a test of will and strength as the snakes press against each other, belly to belly. Finally the weaker individual yields and his head is pushed to the ground by the stronger animal. He then retreats, leaving the area to the victor.

12.26 Male impalas fighting. Although the horns of these medium-sized antelope are formidable weapons, neither impala will attack the vulnerable flank of the other. Instead, a harmless pushing contest ensues as the tips of the ridged horns are engaged. These animals effectively employ horns and hoofs against species other than their own, but when confronted with a member of their own species, they are genetically constrained to behave in very circumscribed ways. (As an aside, should a "cheater" gene arise in the population so that opponents were gored, the gene would probably quickly spread for a time as males got rid of competitors once and for all. But suppose that in a herd of impalas many watchful eyes help to detect approaching lions faster than a few. A population of "gorers" would not only go around wounded much of the time, but would be less likely to spot predators. What would be the likely fate of such populations? Where are they today?)

hence there is a reproductive advantage in sparing him. (Can you figure out why? We'll discuss it later.) In fact, if the population doesn't tend to scatter much, the competitor might even be one's own mature offspring; it could also be a prospective mate.

When fighting occurs between potentially dangerous combatants, the fights are usually stylized and relatively harmless. For example, horned antelope may gore an attacking lion, but when they fight each other, the horns are never directed toward the exposed flank of the opponent (Figure 12.26). Such stylized fighting does, however, enable the combatants to

12.27 Hornless females of the Nilgau antelope have no inhibitions against attacking the flank of a competitor, but their butts are quite harmless. Keep in mind that "harmless" is used only in the immediate sense. The butt itself may not be dangerous, but if it establishes dominance so that the loser is deprived of commodities, the result can have far-reaching implications. An individual deprived of food, after all, is more likely to fall prey. If losing means that she is deprived of a mate, the result is genetic death even if she should live a long life. Her genes will die with her. If losing one fight meant that the animal would have absolutely no chance of breeding, or contributing her genes to the next generation, they might be expected to fight the death. A loss for this species, however, usually just means a temporary setback, so it behooves the loser to accept it gracefully and attempt to breed another time. Interestingly, horned males of the same species almost never attack in this way, nor do horned females of other species.

establish which is the stronger animal. Once dominance is established, the loser is usually permitted to retreat.

On the other hand, all-out fighting may occur between animals that are unequipped to injure each other seriously, such as hornless female antelope, or between animals that are so fast that the loser can escape before he is seriously injured, as in house cats.

We might ask ourselves, since it is useless to ask an antelope, why they don't gore each other. An incurable romantic might assume that it is because they don't want to hurt each other. In all likelihood, what the two antelope "want" has little to do with it. The fact is that they *can't* hurt each other. When the system works an antelope could no more gore an opponent than fly! In terms of the actual mechanism, it may be that the sight of an opponent's exposed flank acts as an inhibitor of butting behavior. Conversely, a facing view, under certain conditions, might serve as a release of very stereotyped fighting behavior. (One explanation of such behavior could be based on the instinct model. Reconstruct it yourself in terms of releasers, IRMs, and neuromuscular associations. The idea is discussed later.)

Can we assume, then, that animals do not fight to the death? As a general rule, they don't, but there are exceptions. Accidents may occur in normally harmless fighting, but there are also some species that normally engage in dangerous fighting. If a strange rat is placed in a cage with a group of established rats, the group may sniff at the newcomer carefully for a long time, but eventually they will attack it repeatedly until they kill it. Guinea pigs and mice often fight to the death. The males of a pride of lions may kill any strange male they find within their hunting area, and a pack of hyenas may kill any member of another pack that they can catch.

Is Aggression Instinctive?

Our discussion of aggression so far suggests that aggressive behavior in many animals is largely innate. But there are many who contend that aggressive behavior, especially among mammals, is directly attributable to learning. Others, of course, stress that aggressive behavior has both learned and innate components. The argument has extreme significance for our species. If aggression is socially disruptive in an increasingly crowded world, then we must ask how it can be controlled. As a practical example, does television violence serve to *release* aggression, and hence dissipate it, or does it serve as a behavioral model for aggression? Since there is a tendency in most academic disciplines to focus on aggression as a cultural, or learned, phenomenon, let's consider other evidence here that suggests an instinctive component.

12.28 Fighting firemouth cichlid fish. This is an extremely aggressive species, its belligerence rising sharply with the onset of the breeding season. Strangely enough, males of this species must fight before they are able to breed. If no male antagonist is present the female may be attacked and killed.

In certain species of highly aggressive cichlid fish, the males *must* fight before they are able to mate. If a male's reproductive state is appropriate and a female is available, the male will frantically seek an opponent upon which to release his fighting behavior. Finding none, he will often attack and kill the female. He is then ready to mate, but of course it is too late. The behavior can be described in terms of Tinbergen's hierarchical model. In removing the inhibitory block for mating (by fighting), he can move to the next level of reproductive behavior, but he has destroyed the releaser of that behavioral level, the female.

Among mammals, rats will learn mazes in order to be able to kill mice, and it is thus assumed that the killing is a consummatory act that reduces tension. Squirrel monkeys, after being angered by an electric shock, will learn a task in order to be able to attack a ball. The opportunity to attack serves as the reward.

Does an animal learn aggression when it is shielded from all opportunities to learn such behavior? Rats and mice have been reared in isolation, so that there was no opportunity for them to learn aggression. When a member of the same species was introduced into their cages, however, they attacked, showing all the normal threat and fighting patterns.

Human Aggression

Since aggressive behavior does appear to have innate components in some species, what about man? Is there any evidence that members of our own species are genetically driven to behave shabbily toward each other? It is difficult, of course, to isolate behavioral components in a species as complex as our own. This is partly because we have the capacity to override both learned and innate behavioral tendencies by conscious decision. For example, we rarely see couples copulating wildly on the tables of romantic little restaurants, no matter what their mood. This ability to override our "urges" has led some observers to assume that those urges—romantic, aggressive, etc.—must be culturally based, not innate. Hence to correct any disruptive aggressive tendencies in our group, we need only to change our culture and/or our institutions.

However, if the behavior in question does have an innate basis, cultural stifling of it can only be cosmetic. Do you think the people in the restaurant have forgotten their urge, even though they are intensely talking about something else? An aggressive person may have learned that smiling and speaking in low tones reaps him greater rewards, but if his aggression is innate, then we can expect him to be aggressive in small ways, or to direct it into harmless channels, or perhaps to show sporadic outbursts of violence when his guard is down. And, of course, smiles and soft voices can in themselves be the vehicle for extraordinary violence. Some of the most horribly bloodthirsty programs have been proposed in smiling, congenial, reassuring, and even kindly tones by seemingly gentle and eminently reasonable souls. The point is that if aggression is an innate tendency in our species, we can expect it to permeate our social patterns. It may be temporarily suppressed, and can perhaps be rechannelled or disguised, but it will exert a subtle and abiding influence in our daily lives.

In an interesting experiment it was shown that aggression could be provoked in people that, when released, caused some relief. First, an experimenter provoked a roomful of students to anger (that was the easy part), causing their blood pressures to rise. The students were then divided into two groups. Each group watched as the experimenter attempted to perform a series of simple tasks. One group was instructed to administer painful electrical shocks (or so they thought) to the experimenter whenever he made an error. The other group could only inform the experimenter of his errors by harmlessly flashing a light. It was found that the group that believed they were actually inflicting pain to the experimenter (who was also an actor) showed a rapid decrease in blood pressure, a sign of relief. The group that was not given an opportunity to release aggression maintained their high blood pressures for a much longer time. In other experiments blood pressure was found to drop quickly if the subjects were allowed to hurl verbal insults at the experimenter. In both cases the release of tension was short-term, as is the case with other types of instinctive patterns.

The arguments on both sides are much more complex than this brief summary might suggest. The important point, however, is to recognize that these alternative positions exist, and to attempt to structure our social programs as broadly as possible.

COOPERATION

Cooperation seems much nicer than aggression, and most of the animal stories of our youth (our pre-Jack London youth, that is) involved animals that helped each other in some way. Certainly cooperation is highly developed in some species, and its complexity and coordination in certain

12.29 Dolphins and porpoises are highly intelligent animals, often demonstrating what appears to be a remarkable degree of insight. For example, it is known that an injured porpoise, unable to swim up for air, may be carried to the surface by its comrades. Porpoises are highly sociable animals, swimming in groups and, for some reason, often joining other species such as tuna. We don't know how the porpoises and tuna interact, but many thousands of porpoises are caught and killed annually in tuna nets. In fact, tuna fishermen may pinpoint the school of tuna by watching for surfacing porpoises.

instances surpasses imagination. But again we have the problem of sifting fact from fiction—a difficult, if not impossible, task for those with a casual interest in the subject.

Cooperative behavior occurs both within species and between species. As examples of *interspecific* (between species) cooperation you will recall the relationship of the rhinoceros and the tickbird. The little bird gets free food while the rhinoceros rids itself of ticks and harbors a wary little lookout. The highest levels of cooperation, however, are most likely to be found between members of the same species.

Let's consider a few examples of *intraspecific* (within species) cooperative behavior. Porpoises are air-breathing mammals, much vaunted in the popular press for their intelligence. In fact, certain of their actions support the claim. Groups of porpoises will swim around a female in the throes of birth and will drive away any predatory sharks that might be attracted by the blood. They will also carry a wounded comrade to the surface so it can breathe. Their behavior in such cases is highly flexible, in that it is often adjusted to the subtleties of the environmental circumstances, rather than being a stereotyped behavior pattern. Such flexibility indicates that their behavior is not a blind response to innate genetic influences.

Group cooperation among mammals is probably most common in defensive and hunting behavior. For example, yaks of the Himalayas form a defensive circle around the young at the approach of danger, standing shoulder to shoulder with their massive horns directed outward. This defense is effective against all predators except humans, since the yaks often pitifully maintain this stance while they are felled one by one by rifles. Wolves, African cape dogs, jackals, and hyenas often hunt in packs

12.30 Weaver ants, *Oecophylla smargdina,* working together to repair a damaged leaf nest. Note their high degree of coordination and cooperation. When the sides of the leaves are pulled together, they will be sewn tight as other ants pass silk-spinning larvae back and forth across the gap, pressing them against the margins of the cut.

and sometimes cooperate in bringing down their prey. In addition, they may bring food to those that were unable to participate in the hunt.

Mammals such as the ones we have considered so far might be expected to show the highest levels of cooperative behavior as a result of their intelligence. It is a bit surprising, therefore, that among all animals social behavior and cooperation are most highly developed in the lowly insects (Figure 12.30). The complex and highly coordinated behavior patterns of insects are generally considered to be genetically programmed, highly stereotyped, and not much influenced by learning—except perhaps for very restricted classes of learning. But the arguments for this are beyond our present scope.

In honeybee colonies the queen lays the eggs and all other duties are performed by the workers, sterile females. Each worker has a specific job, but that job may change with time. For example, newly emerged workers prepare cells in the hive to receive eggs and food. After a day or so their "blood glands" develop, and they begin to feed larvae. Later they begin to accept nectar from field workers and pack pollen loads into cells. At about this time their wax glands develop, and they begin to build combs. Some of these "house bees" may become guards, which patrol the area around the hive. Eventually each bee becomes a field worker, or forager. She flies afield and collects nectar, pollen, or water, according to the needs of the hive. These needs are indicated by the eagerness with which the field bees' different loads are accepted by the house bees.

If a large number of bees with a particular duty are removed from the hive, the normal sequence of duties can be altered. Young bees may shorten or omit certain duties and begin to fill in where they are needed. Other bees may revert to a previous job where they are now needed again.

The watchword in a beehive is *efficiency.* In the more "feminist" species, the drones (males) exist only as sex objects. Once the queen has been inseminated, the rest of the drones are quickly killed off by the workers. They are of no further use. The females themselves live only to work. They tend the queen, rear the young, and maintain and defend the hive. When their wings are so torn and battered that they can no longer fly, they either die or are killed by their sisters. But the hive goes on.

ALTRUISM

We don't wish to shatter anyone, but, well . . . most of the Lassie stories aren't really true. Consider what would happen to the genes of any dog that was given to rushing in front of speeding trains to save baby chickens. The reproductive advantages would be considerable to chickens, but dogs with those tendencies might be selected out of the population by the action of fast trains. Their reproductive activities would therefore be cut short. In contrast, the genes of a "chicken" dog that spent his energy, not in chivalrous deeds, but in seeking out estrous females, would be expected to increase in the population. So what kinds of dogs are likely to predominate in the next generation?

If an animal is going to engage in "unselfish" deeds, its best reproductive bet lies in those deeds that advance the genes of other members of its own species, or perhaps the genes of those species that can help its own species. And as we will see, although selfless behavior does occur both within and between species, it is most likely to occur within species.

Altruistic behavior may at first seem to have obvious advantages, but the question is a prickly one, especially when we consider how such altruism might have evolved. *Altruism* may be defined in a biological sense as an activity that benefits another organism but at the individual's own expense.

It is easy to see how certain forms of altruism are maintained in any population. For example, pregnancy, in a sense, is altruistic. The prospective mother is swollen and slowed. Much of her energy goes to the maintenance of the developing fetus. At birth she is almost completely incapacitated and is in marked danger at this time. Pregnancy is clearly detrimental to her. So why do females so willingly take the risk? It may help to understand the enigma if we remember that the population at any time is composed of the offspring of individuals who have made such a sacrifice. Thus the females in the population may, to one degree or another, be genetically predisposed to make such a sacrifice themselves.

However, altruism on this basis doesn't explain why a bird may feed the young of *another* pair, or why an African hunting dog will regurgitate food to almost *any* puppy in the group. Why, also, would a bird that may have no offspring of its own give a warning cry at the approach of a hawk, alerting other birds at the risk of attracting the hawk's attention to itself?

To answer such questions we must look past the answers that first come to mind. It may seem cynical, but we must start with the premise that birds don't give a hoot about each other. A bird that issues a warning call isn't thinking, "I must save the others." At least, there is a simpler explanation of its behavior.

Keep in mind that the biologically "successful" individual is the one that maximizes its reproductive output. One way of accomplishing this is for the organism to behave in such a way as to leave as many individuals carrying its own type of genes as possible in the next generation. This would explain parental care. However—and this may not be so readily apparent—an individual can also leave its type of genes in the next generation by helping a relative's offspring to survive. An individual shares genes in common with a cousin, although fewer, of course, than with a son or a daughter. Hence there is theoretically a point at which an individual could increase its reproductive output by saving its nieces and nephews (provided there were enough of them), rather than its own offspring. From the standpoint of reproductive output, the organism would be better off leaving 100 nephews than one son.

To illustrate, suppose a gene for altruism appears in a population (notice that this sets up the mechanism for the continuance of the

12.31 Sometimes what seems to be altruistic behavior is really simply a mistake, as in the case of this brooding hen tending two kittens. Hens are not noted for remarkable intelligence and insight, and the kittens were probably substituted for her own chicks. They may have been accepted because of their size and the lowering of the threshold for her maternal behavior. The relationship probably could not have continued since hens don't give much milk and the kittens wouldn't have known what to do with the grain she scratched up for them. In addition, they wouldn't have responded to her warning clucks at the appearance of danger.

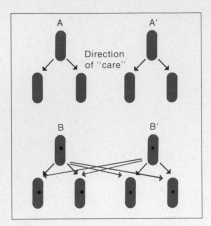

A   A'

Direction
of "care"

B   B'

12.32 In this population, A and A' are nonaltruists. They behave in such a way as to maximize their own reproductive success, but do nothing to benefit the offspring of other individuals. In another segment of the population (B and B') a gene for altruism has appeared that results in individuals benefiting the offspring of others in some way. It can be seen that, assuming the altruistic behavior is only minimally disadvantageous to the altruist, generations springing from B and B' are likely to increase in the population over those from A and A'. It should be apparent that the altruistic behavior is likely to be greatest where B and B' are most strongly related, so that B shares a maximum number of genes in common with the offspring of B' and vice versa. The idea is that B, for example, can increase its own reproductive success by caring for the offspring of a relative with whom it has some genes in common. After all, reproduction is simply a way of continuing one's own kinds of genes.

behavior). As you can see from Figure 12.32, altruistic behavior would most likely be maintained only in groups in which the individuals are related (or have some kinds of genes in common). It might be expected, then, in "nontransient" populations, those that don't move around much, and those in which there is little mixing with outside groups—in other words, where there is a high probability the proximity indicates kinship.

Keep in mind that *no conscious decision* on the part of the altruist is necessary. It simply works out that when conditions are right, those individuals that behave altruistically increase their kinds of genes in the population, including the "altruism gene." Nonrelatives would be benefited by the behavior of altruists, of course, but there is an increased likelihood of individuals nearer an altruist to be related to it.

It has been determined mathematically that the probability of altruism increasing in a population depends on how closely the altruist and "benefitee" are related. In other words, the advantage to the benefitee must increase as the kinship becomes more remote, or the behavior will disappear from the population. For instance, altruism toward brothers and sisters that results in the death of the altruist will be selected for if the net genetic gain is *more* than twice the loss; for half-brothers, four times the loss; and so on. To put it another way, an altruistic animal would gain reproductively if it sacrificed its life for more than two brothers, but not for fewer, or for more than four half-brothers, but not for fewer, and so on. Therefore we can deduce that in highly related groups, such as a small troop of baboons, a male might fight a leopard to the death in defense of the troop. By the same token, a bird will give a warning cry when the chance of attracting a predator to itself is not too great, and when the average neighbor is not too distantly related.

This model, developed by J. B. S. Haldane in 1932 and expanded by W. D. Hamilton in 1963, helps to explain the extreme altruism shown by social insects, such as honeybees. Since workers are sterile, their only hope of propagating their own genotype is to maximize the egg-laying output of the queen. In some species the queen is inseminated only once, so all the workers in a hive are sisters and have half their genes in common. (Why only half?) In such a system, then, almost *any* sacrifice is worth *any* net gain to the hive and to the queen.

The discussion of the development of altruistic behavior is admittedly somewhat esoteric. However, it seems important to consider even our most cherished and our most despised behaviors in the context of man as an evolutionary product.

We now come to one of those disconcerting changes in topics. The following material is included because you may find the information useful in your own decision-making processes.

MINDBENDING

Let's consider a few biological implications of what are loosely called "drugs." The term is not very precise, and it probably evokes images of

Box 12.4 The Sociobiology Argument

Sociobiology is an idea whose time came too early. The result was a stumbling start and a straw man for the critic's barb. It recently was framed in the form of Desmond Morris's rather lighthearted and fanciful account of how almost all behavior was the result of inborn biological urges. He said man was a naked ape and he blithely explained almost all of our behavior in unromantic evolutionary terms. As he found himself quickly driven out on a limb by people who took his words perhaps more seriously or literally than he had expected, he began to defend himself and the battle grew.

But the limb on which he found himself was weighted with others who had been driven up the same tree. The playwright, Robert Ardrey, had turned his pen to writing beautifully, if somewhat imaginatively, about man's beginnings. *African Genesis* and *The Territorial Imperative* were his vehicle to the arboreal heights. Finally, and somewhat surprisingly, these two found themselves out there with none other than Konrad Lorenz who had written about the innateness of human belligerence in *On Aggression*.

The hounds of anthropology, psychology, and sociology were relentless. Their prey, they thought, was an easy one, and after blending their voices they strutted away convinced that they had dispensed with this nonsense once and for all. The evolutionists had not relented however. They declared themselves only scratched, climbed down, cleared their throats and resumed their sermon. Their voices were weaker now, though, and fewer were listening.

Then in 1975, E. O. Wilson at Harvard published a fine and imposing text entitled *Sociobiology*. The book was hailed as a landmark in social thought but its premise was a simple one: much behavior is adaptive, being the result of natural selection. The statement seemed innocent enough and might have caused little stir were it not for the skirmishes that had preceded it.

Because of a few pages at the end of the large book that presumed to shade man along with the rest of the animals under the Darwinian umbrella, Wilson drew the ire of some of his colleagues. His detractors, originating at Harvard, were joined by thin voices from other places. The vehemence of their attack was surprising, considering the scant attention and careful phrasing Wilson had given the human condition in his book. A Sociobiology Study Group was formed, led by some of the finest minds in biology. The group's goal was apparently to confront and beat down the sociobiology argument wherever it should surface.

Because of the great prestige of the antagonists, biologists throughout the country were at first puzzled. Finally some became incensed when it seemed that the argument against sociobiology was not based on scientific evidence at all, but on some other, perhaps humanistic, perhaps political, basis. The argument seemed to be that if we admit that our behavior is innately based then we are resigning ourselves to our condition, and that we are doomed to the status quo.

Wilson, who sought to withdraw from the unseemly fray, has consistently pointed out that he did not say that humans are driven by remote control, their behavior emanating from tiny coiled molecules of DNA. He also did not resign humans to defeat by simply admitting their behavioral heritage. He said that if any of our behavioral patterns are innately based, and perhaps no longer adaptive in our new kind of world, we are more likely to be able to alter them by understanding their origins. After all, we didn't overcome gravity by simply denying its existence.

12.33 Advertisers have our number. They are well aware that most of us are very conscious of the image we project, as we behave in the way that is most carefully tailored to fit our notion of what that image should be. Thus they can sell us products that they tell us will make us appear more "male," more "female," wealthier, more sophisticated, or whatever. In a sense, they're right, because others who are subjected to the same advertising are persuaded that the use of the advertised product is chic. Sometimes the price of being chic is waking up with the peculiar suspicion that a rabbit slept in your mouth.

banal lectures during NFL commercials. Maybe "mindbenders" is a better word. In any case, what we're referring to are those substances that can produce a temporary change in neural events, attitudes, thoughts, or behavior. Since this covers a lot of things, from cigarettes and aspirin to speed and heroin, let's take them by classes and consider a few characteristics of each.

Tobacco

Tobacco is the dried leaf of the plant *Nicotiana tabacum*. When it is burned, its by-products are able to cross the thin-walled alveoli of the lungs and enter the bloodstream. The spread of the smoking habit is curious, in view of the negative reinforcement involved in the first attempts. Recent medical evidence also provides some discouragement; however, smoking was a well-established habit in Western culture long before its medical effects were known. So far cigarette smoking has been associated with higher incidences of lung cancer, emphysema, heart and circulatory ailments, and certain birth defects. Smoking may also permanently paralyze the tiny cilia that sweep the breathing passages clean, and cause the lining of the respiratory tract to thicken irregularly. Often an accompanying deep, hacking cough appears as the body attempts to rid itself of nicotines, "tars," formaldehyde, hydrogen sulfide, resins, and other goodies. To make matters worse, it seems that one invariably finds himself seated in a restaurant next to some rasping chain smoker who is trying spasmodically to clear his respiratory passages.

The word *habit* is applicable here because a smoker may actually become physiologically dependent on the products in the smoke. The smoker may suffer various withdrawal symptoms when he attempts to break the habit. In many cases even the development of serious circulatory and respiratory ailments is not a sufficient deterrent to enable a smoker to give up cigarettes. It is interesting that, in this country, millions of tax dollars are spent on research to control the effects of smoking, and on the other hand, large federal subsidies are provided to tobacco growers to ensure a continued supply.

SMOKING IS VERY GLAMOROUS

AMERICAN CANCER SOCIETY

Caffeine

Caffeine, which is one of the components found in coffee and tea, acts as a stimulant on the central nervous system. It is widely used in many cultures throughout the world. At one time in the Near East, coffee drinkers were put to death—perhaps no worse a fate to some than having to start the day without coffee.

The behavioral effects of caffeine include increased alertness and decreased fatigue and boredom. In small to moderate amounts, it may improve performance in boring or repetitive tasks but it doesn't help in more complex intellectual tasks such as reading or doing long division. It may help keep you awake so you can perform those tasks, however, since it can inhibit sleep. High doses can cause nervousness, irritability, and a "jangled" feeling.

It is possible to develop *tolerance* for the drug so that increasingly higher doses are needed to produce the same effect. Withdrawal symptoms include headaches and irritability. But the effects are minimal, so withdrawing coffee-drinkers are generally not considered dangerous. There are no known effects of the long-term use of the drug.

Marijuana and Hashish

Marijuana (also known as dope, pot, grass, or Mary Jane) is a preparation made from the Indian hemp *(Cannabis sativa).* It was cultivated throughout the United States in World War II to produce fibers for ropes when the supply of hemp from the Philippines was cut off. The wild progeny of those plants has distressed law officials no end in their effort to shield American youths from what was once called "The Killer Weed." Its active ingredient is a group of chemicals called tetrahydrocannabinols (THC). THC is highest in a preparation called *hashish* or *hash* which is made from certain types of marijuana. The effect of marijuana on most users is usually a mild euphoria, sometimes expressed as a happy or giggly mood.

The arguments that have emanated over marijuana have often been ill-conceived and irrational, as you well know, and it would be satisfying to say that the data are now in and that we have at last arrived at some valid conclusion regarding its effects. Alas! Many questions remain unanswered.

Much of the information that we do have, however, indicates that the threat from marijuana may have been overemphasized. It is known that marijuana may alter functions in the cerebral cortex and other parts of the brain, but its short-term and long-term effects seem, at present, less disastrous than those of alcohol (see below). It is probably naive to assume any drug is harmless, however. Hopefully, unbiased medical research will soon provide us with information regarding the actual physiological effects of using marijuana.

It is known that the effects of marijuana are highly variable from one individual to the next and may be partly dependent upon the setting and mood in which the drug is taken. Marijuana seems to affect short-term memory so that a person may not be able to complete a sentence because he forgot how it began. It may also lower the attention span and the ability to change quickly from one topic to another. In certain mental tasks such as information processing, marijuana may slow performance just as alcohol does. On a group of other sorts of mental tasks, marijuana has no appreciable effect. One reason for the lack of such effect may be the ability of some users to "compensate" for the effects of marijuana. Compensation refers to the ability to consciously override any effect of the drug.

With exceptionally high doses of THC the perception of the individual may change so that the world seems distorted and sensory input becomes exaggerated. The individual may not be able to direct his thought patterns cogently and he may experience paranoia (possibly brought on by the knowledge that he could go to jail).

Marijuana is generally not considered addictive, and there seems to be minimal development of "tolerance" so increasing doses are not needed. There may be a psychological "dependence" on the drug—which means that it has produced "good" effects in the past and the user may desire it again because of its rewarding effect.

Cannabis sativa has been cultivated in the Near Eastern countries since the time of the Crusades, where its THC was extracted to produce the resinous concentrate hashish. This substance was used as a drug by the *hāshshāshīn,* a Moslem terrorist group whose notorious violence was thought to be a result of addiction to the drug. More detailed accounts of the hāshshāshīn, whose name is the source of the word assassin, indicate that hashish was actually given as a reward for their murderous deeds—the "visions of glory" promised by their leader, who, by the way, was once a classmate of Omar Khayyam's.

12.34 Probably the most widely used drug in the United States is ethanol, and among young people, its most common source is probably beer (with an alcohol content of about 3 to 5 percent, as opposed to wine with about 13 percent, and whisky with about 40 to 50 percent). Among some segments of the college population, inebriation is believed to be a symbol of status, at least by the inebriated. It is often taken as a sign of *machismo*, but one which is invariably challenged by some female who swears she "can drink you under the table." The physiological results of alcohol usage are highly pronounced, causing changes not only in the nervous system, peripheral blood vessels, and liver, but also in less well-publicized but critical parts such as the pancreas.

Hashish has been widely used throughout India and Asia for centuries. Although it is highly potent and may produce marked hallucinogenic effects, there are no indications that it actually is addictive, as once supposed. It is common practice in India for nursing mothers to apply a minuscule amount of the substance to a breast to lull fretful infants as they nurse.

Alcohol

Ethyl alcohol is the active ingredient of wines, beer, and whiskies. Basically it is a central nervous system depressant. It produces its well-known stimulatory effects by depressing the nervous system's inhibitory controls. One of its major effects is an increase in the level of aggressiveness in many individuals. It produces a marked decrease in motor performance and certain mental abilities (for instance, cognitive skills). Tolerance quickly develops in users. Paradoxically, those who can "hold their liquor" may be suffering the greatest damage from the drug because various other organs may not develop the same tolerance the brain does.

Box 12.5   The Hangover

Egyptian hieroglyphics depict both priests and physicians administering to victims of hangovers. If you've had one you may prefer the former to the latter. And after all this time no one knows how to remedy the problem. (But Puerto Ricans may recommend rubbing the underarm with lemon.)

Hangover is an apparently interminable but temporary chemical imbalance in the body, caused by alcohol acting as an anesthetic on the central nervous system. Usually the results involve a dilation of blood vessels in the brain; the movement of water, and potassium and other ions from the cells outward to the intercellular spaces; depletion of magnesium from the kidneys and inflammation of the stomach lining. In addition, the sleep that follows is strangely devoid of the rapid eye movement (REM) that is characteristic of the most restful sleep.

The results are insatiable thirst, upset stomach, fatigue (many by-products of heavy physical exertion appear in the blood), grouchiness, and perhaps remorse (depending on who saw you).

The morning-after drink ("the hair of the dog that bit you") provides superficial relief while slowing down recovery, and coffee stimulates the already exhausted nervous system and prevents needed sleep.

Hangovers, however, have produced more philosophers than all the world's great books, the conventional wisdom being, "There ain't nothing, NOTHING, worth a hangover."

Alcohol is addictive and disuse may produce severe withdrawal symptoms, the most drastic of which are *delerium tremens* or DTs. Long-term use of alcohol may cause changes in the central and peripheral nervous systems, the liver, the stomach, and the intestines, but the effects vary greatly from one individual to the next. One alcoholic's mind may go, while his drinking companion may only suffer cirrhosis of the liver, a potentially fatal disease. Treatment of alcoholism is difficult and may largely involve psychotherapy such as is provided by Alcoholics Anonymous. Interestingly, a recent news article reported the relief of many parents in a large West Coast city at learning that their children were shifting from pot to booze.

Opiates

Opiates are generally what are referred to as "hard drugs." They are *narcotics* in that they relieve pain and produce sleep or stupor. Drugs in this group include *heroin, opium,* and *morphine.* Users rapidly develop marked tolerance, and in nearly all cases continued use results in addiction. Consequently, the user not only needs the drug continually, but he needs increasingly higher doses. Since such drugs are expensive, in many cases addicts resort to crime to raise the necessary money.

Injection of opiates causes a "rush" of a pleasurable sensation, or a violent negative reaction in beginning users. The rush is followed by a great sense of well-being, which is accompanied by a marked decrease in physical drive. The accompanying drowsiness produces the nodding behavior you can see in almost any New York subway. It isn't known how opiates behave in the nervous system. However, the withdrawal symptoms produced by abrupt discontinuation of the drug are usually violent. In cases of heavy addiction, withdrawal may produce such a shock to the system that death results.

The crime associated with heroin addiction has caused great consternation, especially in our larger cities. One partial solution has been the initiation of programs to maintain addicts on another kind of opiate called *methadone.* Methadone diminishes the craving for heroin, but it is also addictive. Individuals maintained on methadone through supervised medical programs, however, have been able to live relatively normal lives and to maintain satisfactory relationships with other people, including their families. Some people on methadone maintenance have held steady jobs for years while raising families and participating in community life.

Cocaine

Cocaine, or "coke," is legally classified as a narcotic, but biologically it only partly fits the description. It relieves pain, but it does not induce sleep or stupor. It acts through stimulation of the nervous system and produces a feeling of euphoria.

The use of cocaine originated with the Indians of the Andes, who chewed the leaves of the coca bush, often as they carried large loads up

12.35 Admitted to Bellevue Hospital as a result of an overdose of barbiturates. Notice the signs of general poor health in this man, such as his thinness and the condition of his skin. He is in a stupor at this point, but has been tied down because he can be expected to be violent should he ever awaken.

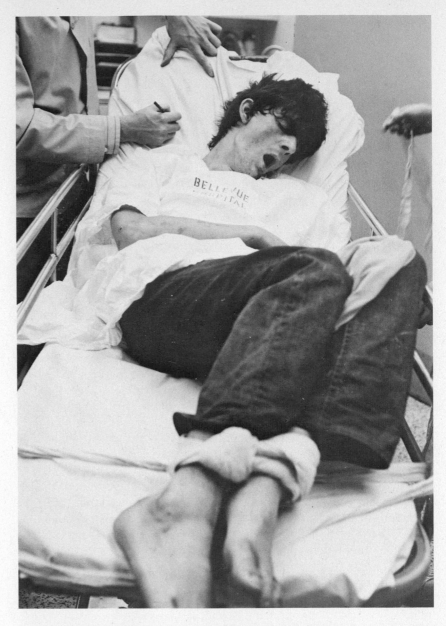

steep mountain trails. The practice in this country was to pulverize the bitter alkaline extract of the leaves and to sniff the powder. However, this habit causes destruction of the mucous membranes and may even dissolve the partition between the nostrils; so now heavy users usually inject the cocaine.

Cocaine has recently become a fashionable drug in certain segments of the American middle class, partly because it is not considered addictive. However, it can result in highly aggressive behavior and feelings of paranoia.

Amphetamines

Amphetamines, often loosely referred to as "speed" or "uppers," include a number of commercial drugs such as Benzedrine, Dexedrine, and Methedrine. Their chemical properties are similar to adrenaline, and they cause great bursts of energy that may successfully overcome any feelings of fatigue. Students and truck drivers often take low doses in order to stay awake during midnight cramming sessions and long hauls, respectively. The drugs are also used to decrease appetite in weight-watchers. Because amphetamines are effective in improving performance on rigorous physical tasks, they are sometimes taken by athletes. Low doses do not impair skills or judgment.

The greatest abuse of amphetamines occurs with injection of the drug. Injection may produce an initial rush, followed by a feeling of vigor and euphoria that may last several hours. However, this is followed by general aching, discontent, and irritability. To delay this letdown, the user may boost himself with another injection. The high may thus last for days, during which time the user usually fails to eat or sleep. The end of this period may be followed by exhaustion, severe depression, paranoia, aggressiveness, extreme irritability, and emotional overreaction.

Users develop a tolerance for amphetamines and after prolonged use may show withdrawal symptoms. These are less severe than with opiate withdrawal.

Barbiturates

Barbiturates, or "downers," are sold under a variety of trade names, including Nembutal, Seconal, Tuinal, and Amytal. They are all *sedative-hypnotics* that act on the cerebral cortex, midbrain, and brainstem areas to reduce anxiety and induce drowsiness and sleep. These effects are accompanied by loss of muscular coordination and slurring of speech—effects similar to those induced by alcohol.

Barbiturates are highly addictive and rapidly produce tolerance. Withdrawal symptoms may be as severe as with opiate or alcohol withdrawal. The heavy barbiturate user is likely to be confused, obnoxious, stubborn, and irritable. In contrast to the placid disinterest of the opiate user, barbiturate users may be particularly aggressive and violent.

Barbiturates and alcohol acting together may cause death by suppressing the breathing centers, and because both drugs cause mental confusion such accidental deaths are frequent.

Psychedelics

Psychedelics are a group of drugs that produce hallucinations and various other phenomena that mimic the experiences produced by certain mental disorders. These drugs include lysergic acid diethylamide (LSD), mescaline, peyote, psilocybin, and various commercial preparations such as Sernyl and Ditran.

LSD is probably the best known of these substances, although its use has probably diminished somewhat since its heyday in the late 1960s. LSD usually produces responses that follow a particular sequence. The initial reaction may be a sensation of weakness, dizziness, and nausea. These symptoms are followed by a distortion of time and space. The senses may become heightened and strangely intertwined—sounds can be "seen" and colors "heard." Finally, there may be changes in mood, a feeling of separation of the self from the framework of time and space, and changes in the perception of the self. The sensations experienced under the influence of psychedelics are unlike anything encountered within the normal range of experiences. The descriptions therefore can only be puzzling to nonusers.

The effects vary widely from one individual to the next, and from one "trip" to the next. The frame of mind of the user and his setting may play a large part in determining the nature of the experience. Users have often claimed new insights, but there is no evidence of psychological improvement or the ability to relate these experiences in any meaningful way even to subjective "reality." LSD, however, has been used medically to treat conditions such as alcoholism and severe paranoia. On the other hand, some users experience bad trips or "bummers," which may or may not have long-term effects. Bad trips may be terrifying experiences and can occur in experienced users for no apparent reason. There are some instances in which the mind has not recovered at all, and others in which the individual has done himself irreparable damage in the interim.

There has been a lot of discussion about possible chromosome damage due to LSD use. A nagging suspicion among users has been that this fear was generated largely by biased reports by those who oppose use of a drug that they don't use or understand. Actually, there is no unequivocal evidence that pure LSD causes chromosome aberrations. On the other hand, "street" LSD is likely to be impure, or laced with small amounts of nerve poisons such as strychnine, and the detrimental effects that have been observed could be due to such impurities. Among users of illicit LSD there is, in fact, a higher incidence of chromosome damage and a higher rate of spontaneous abortion, or miscarriage.

Communities and Competition

13

And NUH is the letter I used to spell NUTCHES
Who live in small caves known as NITCHES for NUTCHES
These NUTCHES have troubles, the biggest of which is
The fact that there are many more NUTCHES than NITCHES
Each NUTCH in a NITCH knows that some other NUTCH
Would like to move into his NITCH very much
So each NUTCH in a NITCH has to watch that small NITCH
Or NUTCHES who haven't got NITCHES will SNITCH

As any idea is sifted through the public consciousness, finally to be encompassed in a "household word," often many of the valuable nuggets of the concept are left behind. Such is the case with certain terms associated with ecology. The term *niche* has been bandied about rather freely in the past few years and has come to refer to anything from a writer's specialty, to a particular socioeconomic group, to a habitat. Even some biologists have been guilty of confusing niche and habitat. Let's see what some of these terms mean.

HABITAT AND NICHE

It has been said that if the habitat is an organism's address, the niche is its profession. The *habitat* may be defined as the place where the organism is found, and it can be described in several ways. For example, an animal may live in a desert habitat, or more specifically, in a briny desert pool. Furthermore, it may live in a certain part of that pool, its *microhabitat*. Wherever an organism lives, however, it interacts with its surroundings in myriad ways. It is affected by its environment, and may, in turn, influence this environment. The organism may also interact with other living things around it. The sum of all such interactions, along with the organism's own requirements, describe its *niche*. Actually, as you might imagine if you are beginning to gain some insight into the "scientific mind," there is quite a bit of haggling over how these terms should be precisely defined, but let's not become involved.

It has become axiomatic that two species cannot occupy the same niche indefinitely. If two species were to find themselves in such a situation, it is generally predicted that they would be unequally adapted to that niche, so that in time as one species decreased in number, the other would eventually come to replace it. In effect, this means that when you walk through the woods and see various species of small seed-eating birds, each species is probably interacting with the environment in a different way. This principle was nicely demonstrated by the ecologist Robert H. MacArthur, who showed that five species of American warblers that feed in spruce forests actually divide the trees into different feeding zones.

According to the population ecologist L. C. Birch, competition occurs where organisms utilize resources that are in short supply, or when they harm each other while seeking resources that are not in short supply. (Probably the latter cases are much rarer in nature.) Where species that share a habitat utilize its resources differently, it is assumed that they are reducing competition among themselves. For example, in an area inhabited by five species of titmice, each has a different foraging pattern—each species prefers a particular kind of tree and particular parts of trees. However, the degree of differences in their behavior changes with the season. In fact, it is found that they forage in almost identical ways during periods when food is abundant, a time when there is presumably a low level of competition.

Animals may also divide up a habitat in other ways. For example, they may utilize the same resources at different times of the day, or at different times of the year, or they may utilize different commodities within the same part of the habitat.

The real role of competition in the affairs of living things is not very well understood at present. In fact, many of our ideas are based largely on supposition or are supported by rather tenuous examples. However, never deterred, we plunge ahead.

THE LAND ENVIRONMENT

It seems safe to say that no two organisms interact with their environments in identical ways. Two tapeworms living in a raccoon's intestine, if they were asked, would probably describe their world differently, depending on their precise point of attachment and their individual perceptive and reactive tendencies. Two wildebeest living on an African plain would certainly have different experiences, and might be expected to perceive their environment quite differently. Even identical twins see the world from opposite sides of a baby carriage. The point is that it is hard to generalize about the environment of any species, So now, of course, we will generalize.

We tend to think of our earth as a ponderous place that unfailingly provides its denizens with those things necessary for life. We often fail to realize, however, that life exists only in an exceedingly thin film that veils the surface of this immense ball—a delicate shell wherein the wonderous forces of sunlight and water interact. This fragile film is responsive to a number of influences, and hence is highly variable from one place to another. Furthermore, each place is in itself unstable, so that it changes

13.1 The feeding zones of five species of the North American warblers in spruce trees. The darker areas indicate where each species spends at least half its feeding time. By exploiting different parts of the tree, the species reduce their competition for food and thus they can occupy the same habitat. Studies such as this have been done on many species since the principle was first elucidated by Gause and supported by work such as MacArthur's. It turns out that very similar species living in the same habitat tend to occupy different niches by subdividing the available resources. Since animals tend to behave so as to reduce competition, it has been suggested that competition must be an important factor in natural selection.

with time. The different kinds of "places" in which life exists over the earth at present can be categorized roughly according to their physical and biological properties. Keep in mind, however, that these are merely arbitrary divisions of the great, complex, and intergrading areas of the earth. But it makes discussion easier.

Biomes

By using some imagination we can divide the earth's land into several kinds of regions called *biomes* (Figure 13.2). Biomes are defined on the basis of the plants they support. Of course, the makeup of the plant community is dependent on other factors, such as soil conditions, weather, day length, competition, and the nature and abundance of the resident plant eaters. A moment of reflection, if all is going well, will show that the nature of any animal community is, in turn, largely determined by the nature of the plant life there, since plants are needed to trap the sun's energy and to provide shelter, building materials, and hiding places for the animals.

Let's briefly consider the six largest generally recognized types of biomes. (Lesser biomes are shown in Figure 13.2 and are in our photo essay on pages 388–397.)

Temperate deciduous forests are the type of forest that at one time covered most of the eastern United States and all of Central Europe. The dominant trees in these forests are hardwoods, such as oak, maple, beech, poplar, and hickory. These areas are subject to harsh winters, times when the trees shed their leaves, and warm summers, periods of rapid growth and rejuvenation. Before the new leaves begin to shade the forest floor in the spring, a variety of herbaceous *annual plants* may appear, plants whose life cycle consists of only one season; they bloom anew each spring, only to die in the fall. Rainfall may average 30 to 50 inches or more each year in these forests. The seasonal changes are moving and fascinating. People who live in such biomes describe a certain joy which swells within them each spring, and the secret pensiveness that overcomes them in the fall as the days darken and the forests become more silent. Perhaps we are exceeding the technical description of the biome, but these are my favorite places.

Grasslands occur in many parts of the world and are exemplified in the United States by the American prairie. They are characterized by an abundance of grasses, small bushes, and shrubs, and in some parts of the world, thickets of bamboo, which is a type of grass. The soil in such areas is usually porous. Trees may appear along streams or rivers. Rainfall averages between 10 and 40 inches each year, but may be erratic in its timing. Grasslands are present in both tropic and temperate zones. You may be surprised to learn that grasslands support more species of animals, and a greater biomass, than any other kind of terrestrial habitat.

Deserts are characteristically hot in the daytime and cold at night. The 10 inches or less of rain that falls each year usually comes in sudden downpours, so that much of it runs off, sometimes causing flash floods and

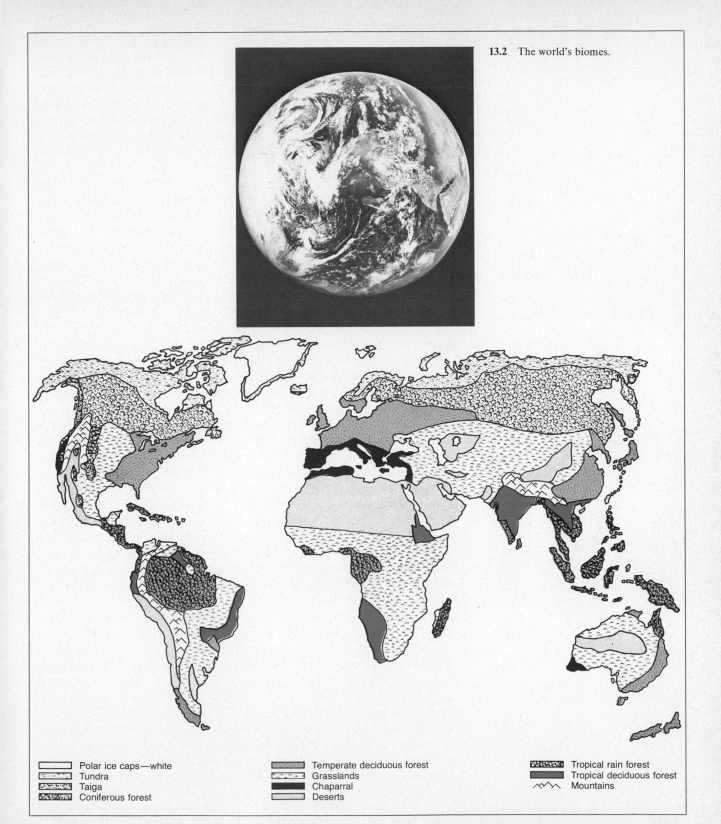

13.2 The world's biomes.

Polar ice caps—white
Tundra
Taiga
Coniferous forest

Temperate deciduous forest
Grasslands
Chaparral
Deserts

Tropical rain forest
Tropical deciduous forest
Mountains

marked erosion. Directly after a rainfall, the annual plants of the desert take advantage of the moisture and explode in an orgy of growth and seed production. Other desert plants meet the water-shortage problem in other ways, either by storing water in their tissues or by reaching far underground to tap water tables with their, of course, tap roots. Deserts harbor few kinds of animals other than reptiles and insects. The preponderance of these species is chiefly because they are cold-blooded. (Can you figure out why?)

Tropical rain forests, or jungles, are found in the Amazon and Congo Basins and in Southeast Asia. The temperature in such biomes doesn't vary much throughout the year. Instead the seasons are marked by the torrential rains that fall almost daily during the summer. In some areas there may also be pronounced winter rainy seasons. These regions support many species of plants. Trees grow throughout the year and reach tremendous heights, with their branches forming a massive canopy overhead. The jungle floor may be dark and steamy, sparsely settled by smaller plants. Jungles literally swarm with insects and birds. Animals may breed throughout the year as a result of the continual availability of food. Since there is no harsh winter season, some species are coming into production at all times. Competition is generally considered to be very keen in such areas because of the abundance of species.

Tundra regions are covered throughout most of the year by ice and snow. This biome is most prevalent in the far north, but in the United States it may be seen in the high Rocky Mountain regions. Summers usually last two to four months, just long enough to thaw a few feet of the soil above the *permafrost,* or permanently frozen soil. Thaw brings soggy ground, and many ponds and bogs appear in the depressions. The plant life consists mostly of lichens, herbs, mosses, and low-lying shrubs and grasses. Such plants must be incredibly hardy to be able to live in such a place. Their hardiness disguises their fragility. Once disturbed, these areas take very long periods to restore themselves. The marks of wagons that passed over the tundra a hundred years ago are clearly visible to this day.

Taiga refers to the northern coniferous forests that cover much of Canada and Siberia. The taiga has a longer growing season than the tundra. Annual rainfall is low, usually less than 20 inches. The forest may be dominated by a single kind of tree such as pine, spruce, or fir, and the forest floor is usually covered by a carpet of needles, with only a few shrubs or other plants pushing through. One may move silently on the muffling needles through the Canadian taiga, observing a host of mammals from porcupines to moose.

Ecosystems and Communities

Biomes can be broken down further into *ecosystems,* specific areas within the biome. Ecosystems are theoretical entities in which the component species interact as a self-sustaining unit. An ecosystem, then, is considered an independent unit (although its independence would be hard to prove).

An ecosystem might be a pond or a wooded area or a field. The organisms that live within an ecosystem comprise a *community,* and of course, they interact with each other as well as with their physical environment. Now, within a community each species has a particular role; it interacts with its environment in a certain, theoretically circumscribed way. The sum of those interactions is its *niche.* So there you are.

THE WATER ENVIRONMENT

The earth's water may be classified roughly as fresh water or salt water, although not all bodies of water fall neatly into one category or the other. For example, Lake Pontchartrain, near New Orleans, is *brackish,* or a mixture of salt and fresh water.

Fresh water has about 0.1 percent salt, whereas seawater has about 3.5 percent salt. Each has its own importance in the earth's drama. Fresh water, for example, is drinkable; seawater is not (at least for us). Also, only fresh water can be used to irrigate crops. Now let's consider fresh water and salt water separately and note the unique importance of each.

Freshwater Bodies

In general fresh water provides a less stable habitat than does seawater. Bodies of fresh water tend to evaporate to a fraction of their former size, and also to become muddied. Because of their relatively small volumes, they become polluted rather easily, and, in some cases, they can be made to age prematurely. Because of their peculiar vulnerability, human activity has, in many cases, drastically altered their character as ecosystems. In the Great Lakes, for example, the parasitic lamprey and the destructive alewife have begun to flourish, whereas these habitats will no longer support the edible lake trout and whitefish (Figure 13.3). One reason for this change is that the use of detergents and chemical fertilizers has increased the amounts of phosphate and nitrate that are washed into the water. Bacteria can break down, but only with the use of great amounts of oxygen. Algae and other plants flourish under such conditions, but as they die, they return their nitrates and phosphates to the water, causing the bacteria to use increasing amounts of oxygen. Consequently, the species that require a lot of oxygen, such as trout, are now almost nonexistent, whereas those that thrive on the algae and plankton are reproducing rapidly.

The Oceans

Oceans cover about three-fourths of the earth's surface. In fact, if the earth's surface were smoothed out, it would be covered entirely by water. The average depth of the oceans is about three miles, but there are places where the water is seven miles deep.

It has been suggested that the oceans will one day provide us with most of our food. However, we presently take only 3 to 5 percent of our food from the seas, and even if we were able to double that in the next

13.3 As a result of human activities the edible lake trout (top) and whitefish (bottom) are now almost nonexistent in the Great Lakes. On the other hand, the parasitic lamprey (attached to the trout) and the destructive alewife (center) have become common. One reason for this change is that the use of detergents and chemical fertilizers have increased the amount of phosphate and nitrate that is washed into the water. Bacteria can break these down but only with the use of great amounts of oxygen. Algae and other plants flourish under such conditions, but as they die, they return their nitrates and phosphates to the water and cause bacteria to use ever more oxygen. Finally, the lake cannot support animals that require a lot of oxygen, such as trout, but the new conditions are fine for alewives.

Box 13.1   A Dead Lake

Lake Erie was once one of the world's most beautiful lakes. Its waters were pure and abounded with life. It provided food, and lent its beauty, to man for literally thousands of years. What man, camped beside it in earlier times, spending leisurely days swimming and fishing and reveling in its quiet sunrises, could have guessed its fate? The growing human population, however, was demanding new goods. The production of those goods required great amounts of water. So factories were set up along the banks of the immense lake—factories that would use the water as a coolant, and the lake itself as a dump. Thus millions of tons of poisons of all sorts were poured directly into the water. After all, this was the cheapest means of disposal. Cheap production increased profits. But the question is, did the industrialists have the *right* to pour their garbage into a lake they didn't own? As their profits increased, the lake was changed to the extent that it became dangerous to swim in it or drink it. The industrialists, of course, when threatened with belated lawsuits in recent years, argued that their dumping of poisons into the lake resulted in cheaper goods for people, and *that,* after all, is what the people *really* wanted. You can immediately see that the question can be extended to apply to industries that vomit their poisonous fumes into the air. Whose air is it, after all?

Because of new demands being placed on the earth's fresh water, it is very likely that you will come to change your feelings toward the uses of fresh water before long. Whereas you may have, somewhere around the second grade, learned that water is something people drink and plants need, you may soon come to think of it in terms of its mining and allocation. Already, the Supreme Court has determined that only agriculturalists in one area of California may *vote* regarding water use; other citizens have no voice in the matter.

decade, our food problem would not be resolved. You may be surprised, and disturbed, to learn that most of the deep ocean waters are, like the deserts, almost completely devoid of life.

The oceans are not simply huge, still bodies of water, broken only at the surface by waves. In fact, the deeper waters are continously in motion as great silent currents hold sway beneath the surface. Actually, these currents are primarily the result of a surface phenomenon, the *trade winds*. The "trades," once so important to commercial sailors, are prevailing winds—winds that hold rather steadily from the same direction—which are caused by the difference in air temperature between the poles and the equator, coupled with the effects of the earth's rotation. These great ocean currents caused by these winds spin in wide circles, clockwise in the northern hemisphere, counterclockwise below the equator. One such ocean river is the famed Gulf Stream, which follows the Atlantic coast as far out as Bermuda and then swings eastward to northern Europe.

Where the currents are deflected upward by the mountains of the ocean floor, cold water from the ocean depths wells up to the warmer surface. This cold water carries with it the ages of accumulated sediment from the floor, sediment rich in nutrients, which forms the basis for the bloom of life at the ocean's surface. The nutrients carried by upwellings are utilized by tiny chlorophyll-bearing organisms called *phytoplankton,* which in turn serve as a food source for *zooplankton.* These minute plants and animals, barely visible to the naked eye, form the basis of the ocean's food pyramid (Figure 13.4). One cubic foot of seawater may contain over 12 million phytoplankton. These food producers, however, can live only near the ocean's surface, since the sunlight necessary for them to carry on photosynthesis cannot penetrate below about 600 feet.

Not only are phytoplankton important as primary food producers in the oceans, but they also manufacture much of the earth's oxygen. It has been estimated that they produce well over half the oxygen we breathe. Thus the importance of maintaining viable oceans is apparent. Such knowledge can only be disturbing as we continue to see the world's oceans used as a dump. The sight of the oil companies' new supertankers, which will carry dangerous, partly refined petroleum across the fragile oceans, may also cause frowns of consternation on the faces of the informed. There is slim consolation in the knowledge that even among smaller and more maneuverable oil-bearing vessels there were an estimated 5,000 collisions in one recent ten-year period. But let's not get into all that now.

Men diving to the ocean's depths in bathyscaphs have reported that they can see light at depths as great as 2,000 feet. The light that reaches these depths is pale blue, since the reds and oranges of the light spectrum have been filtered out by the water above. As a result, many of the fish that live there are reddish in color when they are viewed in the light at the surface. At their natural depths they appear dark and shadowy, since there is no red light for their pigment to reflect and their red color absorbs what blue light there is.

(*Text continues on p. 398.*)

13.4 The food pyramid of the ocean. The tiny phytoplankton at the base capture the energy of the sun. These are eaten by animals larger than themselves, which are eaten in turn by larger animals. At the top are the largest carnivores of the sea. (There are no large herbivores in the open oceans as there are on land.) What is the position of man in the food chain of the sea? Note that each level is comprised of far fewer organisms than the one below it. The reason for this will be discussed later. As food falls into increasingly short supply, the question is being asked, Should we eat from lower on the pyramid? Would it be more efficient to eat plankton than tuna?

The Water Environment

☐ Most of the earth's surface is covered with water, both salty and fresh. It thus constitutes an important part of the biosphere, but we often neglect to consider its importance because in its naturally occurring form it is essentially disruptive to much of our tissue, and our efforts to examine it are hindered in numerous ways. It puckers our skin, clogs our ears, is rarely at a comfortable temperature, and is usually hard to see through. So if we're not drinking or washing, we generally avoid it unless we are sportsmen or specialists, thus we probably imagine we know more about its role in nature than we really do. Yet our lives depend on it and hence we are usually cheered by the sight of oxygen-laden mountain brooks or long, winding rivers, such as the Yosemite's Merced River (right), one which has escaped being "straightened" by the Army Corps of Engineers.

☐ Tules, or bulrushes, which thrive on overflowed land in the American Southwest (left). Large tracts have been overgrown with these plants, such as this area at the junction of the San Joaquin and Sacramento rivers in California.

☐ A cypress swamp in winter (top). Notice the water-swollen trunks and the "knees" — roots that have risen above the waterline into the oxygen-laden air. The water in such swamps moves very slowly and hence is low in oxygen.

☐ A freshwater lake high in the Adirondack Mountains (bottom). Such lakes may be exceptionally deep and cold with a predictable turnover rate as their upper reaches are heated or cooled by the air temperature. One of the most beautiful of the world's mountain lakes is, or was, Tahoe, a name now synonymous with artifice and glitter as the struggle between developers and environmentalists continues.

☐ A mudflat (above), a saltwater area that is covered at high tide and exposed at low tide. The soil is often hard and tightly packed. This is a very difficult place to adapt to because it is subjected to repeated drying and soaking. Predators stalk these areas in search of exposed prey. Such flats may occur at estuaries, the river currents producing the prominent rills. The mudflat estuary harbors a variety of organisms if the river has not been poisoned upstream.

☐ The rocky coast of Olympic National Park, Washington. The incessant pounding of the water has carved out these striking monuments, which serve as attachment for a number of organisms, a particular type of animal dominating each vertical zone.

□ The Channel Islands off the California coast (above). This is Anacapa Island. It is rather small and, as a rule, smaller islands harbor fewer kinds of plants and animals. Whereas Anacapa lacks forests, trees abound on Santa Cruz, her larger neighbor.

□ A salt marsh estuary at Parker River in Massachusetts (far left). Salt marshes are exceedingly critical places in that the many tiny life forms living here initiate food chains that ultimately involve larger sea creatures. They are vulnerable in that they don't look important and so developers have gained easy access to them.

□ The sandy beach at Bahia Chileno, Mexico, supports few life forms, probably because the shifting sands provide very poor attachment and shallow waters tend to surge strongly.

The World's Biomes

Polar Ice Caps

□ Biomes are usually classified according to their dominant vegetation. Thus the stark absence of obvious vegetation in the polar ice caps would seemingly render these areas ineligible for the title. However, in fact, the polar seas are laden with tiny planktonic plants and animals that form the base of a complex food chain. Most foraging therefore takes place in the water, the larger organisms eating the smaller ones. Whereas Antarctic penguins may have many predators, such as leopard seals, the polar bear is so formidable as to fear nothing except perhaps the killer whale, or wealthy "sportsmen" hunting from helicopters.

Tundra

☐ The tundra is a form of northern grassland that runs in a band across North America, Europe, and Asia. Because it is visited by so

few of the earth's people, we might fail to realize that it comprises about ten percent of the earth's surface. This is the area of the famed permafrost, or permanently frozen subsoil. In winter the topsoil also freezes, thawing to a few inches in spring. In the brief growing season, about two months long, plants grow in these soggy areas only to have their roots crushed and their branches torn by driving snow as winter quickly draws on. Thus tundra plants tend to be low-lying and stunted. In some areas small trees may attempt to grow. These are the harbingers of the vast taiga forests further south. Animals in the tundra are adapted to the rigors of the habitat, as are these snowy owls, their white feathers rendering them inconspicuous as they hunt their main food source: lemmings. They also may hunt ptarmigan as they fly over herds of musk oxen and caribou in North America and perhaps reindeer in Europe. Paradoxically, many birds nest in the tundra because in the spring there are vast swarms of insects to feed on and because summer days are very long, the sun only bowing briefly to the horizon.

Taiga

☐ The taiga is comprised of vast northern
coniferous forests that are subjected to severe
winters with the ground blanketed in snow.
The carpet of fallen needles renders the soil
highly alkaline and slow to decompose, so
few annual plants or invertebrates can eke
out an existence there. The vast evergreen
areas are rent by ragged lines of deciduous
trees growing along riverbanks, making
colorful shafts through the various hues of
green in autumn. At the upper left, the tundra
blends into taiga, emphasizing the rather
arbitrary labeling of biomes. In taiga one
might expect to find porcupines, snowshoe
hares, wolverines, some songbirds, and
grouse. Their predators include wolverines
and lynxes. Wolves tend to feed on the larger
animals such as elk, moose, and mule deer.
They will not usually even attempt to attack a
confident elk that stands its ground, their
prey usually being the weak or infirm.
Wolves are highly organized animals with a
complex and fascinating social system. We
are only beginning to learn about these
magnificent animals as they rise ever-higher
on the list of endangered species.

Coniferous Forest

☐ Southern coniferous forests exist also and they are marked by less severe winters, more rainfall, and a longer growing period. On the North American continent, the taiga dips southward in a few places, reaching toward human population centers where the land is often preserved as parks. At left is a view of a coniferous forest in Banff National Park in Southwest Canada. Above, two moose graze placidly on aquatic plants in a shallow pond in Yellowstone National Park.

Temperate Deciduous Forest

☐ The growing season in the temperate deciduous forest is rather extended and accented with a moderate rainfall. In the winter the leaves drop as a mechanism to reduce water loss. The topsoil is rich and harbors countless invertebrates. Grassy fields of wildflowers may give way to shrubs near the woodlands, where the heavy vegetation harbors squirrels, raccoons, and opossums, and their predators such as the fox and bobcat. The deer are generally free of natural predators since wolves and mountain lions have been decimated. Because of harsh winters, many birds migrate to warmer areas, but a few, like the cardinal, is a winter resident. There are very few virgin stands of this forest left in the United States.

Grasslands

□ Grasslands are scattered throughout the world but are especially prominent in the North American prairies, the African veldt, the Russian steppes, and the Argentine pampas. Above, a group of pronghorn antelope grazes in West Texas and, at left, a baobab tree stands as a striking monument in the Tanzanian veldt. Grasslands are stricken by periodic drought and subjected to extreme changes in temperature. In the United States their prime sources of moisture are the Atlantic Ocean and the Gulf of Mexico. Grasslands support vast populations of small seed-eaters, such as some rodents and birds. In addition they support great numbers of large mammals including the enormous bison herds that once roamed North America and the great herds of ungulates that migrate annually across the African plain. It has been suggested that the grasslands are largely the result of man's activities, including prehistoric man who may have deliberately set fires to catch game. In any case, the biome cannot revert to forest because of the action of grazing animals and sporadic devastation by fire.

Chaparral

☐ Chaparral is restricted mainly to Southern California and certain Mediterranean areas. The plants form a dense, almost impenetrable tangle that shields a variety of small vertebrates, such as rabbits, lizards, rats, California ground squirrels (at right), and small birds, all of which are rather dull colored, as is their biome. Chaparral is subjected to long, dry summers, when the plant life turns brown and dormant, followed by brief, cool rainy seasons that cause the vegetation to spring to life. When the chaparral is green we may find mule deer wending their way through the tangled mat. At the higher elevations grows twisted, colorful trunks of the manzanita (above). It has been discovered that the plant succession following the frequent fires is partly regulated by certain plants that produce toxins, inhibiting the growth of competitors.

Desert

☐ Deserts are among the most feared and fascinating places on earth. All the great deserts, such as those of Death Valley (lower left) and Arizona (left), are located within thirty degrees of the equator. Only about five percent of North America is desert, but the Sahara is about the size of the United States. Deserts normally receive less than twenty-five centimeters (ten inches) of rain a year and the rainy period is startlingly brief. Desert plants are adapted to these conditions, however, and plants respond to brief rains by quickly flowering and producing seeds amid a riot of color. Other plants, such as the dominant saguaro cactus, have shallow roots and store water in their tissues. Some animals, including the Gila monster (America's only poisonous lizard), save water by developing thick coverings, excreting little water, and being active mainly in the cold nights. The nighttime temperatures are reached because the air is not buffered by water.

Tropical Rain Forest

☐ Tropical rain forests, such as these in New
Guinea, are deluged with rainfall throughout
the year, averaging 200 to 400 centimeters
(80 to 150 inches) annually. These are
particularly fertile places, harboring more
species of plants and animals than all the
other biomes combined. The forests are
''storied,'' the upper stories composed of the
crowns of towering giants 50 meters tall.
Beneath these is a solid canopy formed by
smaller trees with broad bases and shallow
root systems that utilize the perpetually damp
floor. Long vines may span the distance
between trees, some vines over 300 meters
long. Although the leaves are dark and the
flowers of the trees inconspicuous, certain
highly conspicuous creatures, such as the
red-cheeked parrot, accent the still darkness
of the rain forest.

Tropical Deciduous Forest

☐ In tropical deciduous forests, such as this forest in India, rainfall is rather seasonal. In the dry season, then, plants drop their leaves in order to avoid losing water through transpiration. In order to fully maximize their reproductive efforts, some species flower before putting out new leaves. Tropical deciduous forests harbor a wide variety of plant and animal life. It would be interesting to know if highly social animals, such as these monkeys, change their communication mode when the leaves drop and they can see each other more clearly. Does sound then become less important?

13.5 Viperfish, an example of a deep-sea fish. Such fish, while not noted for their personality, have enormous jaws and teeth. Thus, when they encounter another animal in their sparsely populated depths, there is a good chance that they manage to eat it. Notice the luminescent spots along its sides. Such lighting is apparently useful in signaling between members of the species. This animal, fortunately, is only about a foot long.

It was once believed that nothing could live below 1,800 feet because of the tremendous pressure of the water at such depths. At about the time Darwin wrote on the origin of species, however, a submarine cable broke in the Mediterranean and was hauled up for repairs from a depth of about 6,000 feet. To the astonishment of biologists, it was found to be covered with all sorts of living things. So notions concerning the effects of pressure had to be revised. Now it is believed that living things can exist at any depth, as long as they are able to develop pressure inside their bodies equal to the pressure outside. Fish that have been brought up alive from depths below 2,000 feet have literally exploded when they reached the surface.

The deepest reaches of the ocean must be peculiar environments indeed. We are just beginning to learn something about them, and the incoming information is fascinating. It seems that the deep ocean bottoms, untouched by the currents above them, are still and unchanging places. So far we have discovered only few kinds of animals there, and most of these are quite weird creatures (Figure 13.5). Plant life is represented only by a few kinds of fungi, species that do not need light.

The ocean bottom itself is an ooze composed of the chalky and glassy corpses of tiny plants and animals that have rained down steadily onto its surface throughout the ages. Animals that have been seen on the bottom often have long appendages to keep them from sinking into the primal ooze. Because of the utter stillness of the deepest ocean bottom, very delicate and fragile creatures can live there, but we are probably aware of only a small part of the life that exists in these abysmal regions. Some animals escape our attention by burrowing into the ooze at the bottom. Divers have seen many strange holes and burrows but no one has ever seen the animals themselves.

COASTAL AREAS

The land-ocean interface is called by romanticists and builders alike "the seashore." The romanticists would write new eulogies if they knew the full impact of the seashore in their lives. The builders might (and then again, might not) become a bit more cautious.

Life is much more abundant in coastal waters than in any other part of the ocean. The edges of the continents extend beyond the shore for anywhere from 10 to 150 miles as *continental shelves*. At these relatively shallow depths, usually less than 200 feet, sunlight easily penetrates the waters, and a wide variety of plants can grow there. The plants, of course, provide many species of animals with food, shelter, and surfaces for attachment. Coastal areas are generally composed primarily of either rock, sand, or mud. And each type of area has its own special qualities.

Rocky coasts are perhaps the most dramatic, and to coastal navigators the most awesome, with jagged, surf-bathed rock formations that only suggest the immense reefs below. These coastal areas boast a remarkable array of plant and animal life. Much of this life is distinctly zoned along the rocky surfaces. For example, a zone dominated by the periwinkle snail is clearly delineated from the barnacle zone, which is, in turn, strongly demarcated from the seaweed zone. The zones are not occupied exclusively by a single species; they are simply identified according to the most prevalent species occupying them. In each zone a number of species of plants and animals are intermixed with the predominant life form, thus enriching the life of the area.

Sandy seashores have fewer forms of life, which is perhaps a good thing for timorous bathers. Maybe you have felt some relief as you waded out into the surf and found that you didn't step on some mysterious and toothy bottom dweller. The reason they are so scarce is that wave action causes a constant shifting of the sandy bottom, depriving the less mobile species of a fixed surface for attachment. Any animal that lives in such shifting areas must be able to get around quickly, as crabs do, or be able to dig and withstand burial, like clams. Animals that live in tidal areas and are not mobile probably encounter exceedingly severe conditions at low tide, when they are subject to the parching sun and to terrestrial foragers such as shorebirds.

From an aesthetic standpoint *mud flats* are probably the least appealing part of the seashore, and this perhaps explains the general lack of interest in protecting them. They aren't especially spectacular as scenery, and they are poor places to spread a towel. Also, they tend to smell peculiar—to say the least. Mud flats do not harbor as many life forms as rocky coasts, it is true, but they support much more life than do other kinds of coastal areas.

Mud flats are submerged at high tide and exposed at low tide. Bottom dwellers, snails, insects, shrimp, crabs, fish, and birds abound in these areas. Each square yard of the mud flat can support thousands of such individuals. Many sea animals begin their lives in mud flats, only later moving out to take their place in the mysterious pageant of the open ocean. Hence the life of the ocean itself is, in a very real sense, dependent on the preservation of the unsightly mud flats. Unfortunately, mud flats and saltwater marshes have attracted the eye of commercial developers, who have not overlooked the opportunities provided by the public's lack of interest in them. Up and down the coasts our bayous, marshes, and bays are being filled in for use as industrial sites, "waterfront" housing tracts, and high-rise apartment complexes and condominiums. San Francisco Bay alone is now only two-thirds of its original size, much of it having been replaced by long stretches of paved, chromed and neon-lit tourist areas.

In order to argue effectively—or in the absence of opportunities to act, to feel properly incensed—regarding the abuse of our environment we must know something about how species interact with their environment and with each other. Is any species able to affect the well-being of another? In what ways do species interact with their environment? Does the passage of any species from the earth affect us in any real way? Often there aren't specific answers to such questions, but there is enough information available for us to make some reasonably intelligent decisions about how we should regard the life around us. Some of the fundamental relationships between living things may be easier to visualize if we compare the behavior of species on islands and continents.

ISLAND LIFE
Because islands differ from continents in important ways, we have gained some important insights from observations of how species behave in the two types of environments.

Physical Factors
Islands are, of course, much smaller than continents, but size is not the only geographic difference. To begin with, the continents are generally considered to be much older than their offshore islands. Also, islands are surrounded by water that acts as a natural barrier to the passage of living things to and from islands and thus impedes gene flow. The climate on islands may be less variable than that on continents because of the stabilizing effect of the surrounding water.

Continents generally boast a wide variety of habitat types. (Think of the range of habitat types available in the United States.) Islands, on the other hand, may only provide a few habitat types—partly as a function of their smaller size. Some islands are no more than small stretches of sand or rock, barely rising above sea level. Others, on the other hand, may be impressive bodies with tall mountains and lush valleys. So, you can see, it is difficult to generalize regarding just what sorts of habitats will be available on islands. However, in general, there are likely to be fewer habitat types available on islands than on their nearby continents.

The distance of an island from the nearest continent is another important factor. First of all, this will affect what kinds of species will reach the island, since water is a greater obstacle to some animals than to others. No one is surprised to find birds on islands. Birds are generally good travelers, so any animal reaching an island is likely to confront birds. That animal doesn't necessarily have to put up with snakes, however, since most reptiles are poor travelers. A walk through a Puerto Rican forest is enhanced by the sight of so many beautiful birds and the knowledge that no poisonous snakes are about.

Islands that lie close to a continent may be colonized by the poor travelers as well as the good ones, whereas the more distant islands would probably ordinarily be accessible only to the good travelers. Thus, fewer species normally reach the more distant islands. It should also be apparent that more species are likely to be found on older islands since the longer they have been there, the greater the probability of poorer travelers somehow managing to reach them.

We can expect that, generally, islands will have fewer species than their nearest mainland areas, since some species have particular problems in reaching islands. We can also see that there may be other reasons for the paucity of species on islands. For example, when we recall that islands may have fewer habitat types than mainlands, it is easy to see how an animal might reach an island, only to find that there was no suitable habitat. What would happen if a woodpecker landed on an island that had no trees?

If the island is low and flat, as smaller and more distant islands frequently are, the colonizer may have particular problems. As you know, the character of land changes with elevation. Hence an island that rises from sea level to lofty heights will provide more environmental variation, and hence more kinds of habitats, than a flat island with the same land area. As a result, islands with a range in elevation can be expected to support more kinds of species.

Biotic Factors

Even those island-colonizing species that are able to find suitable habitats are not likely to be as well adapted to the island habitat as they were to their mainland home. After all, the island will undoubtedly be different in some ways. A colonizing species, then, may survive rather tenuously for a short time (perhaps a few thousand years) until it becomes better adapted to the island situation.

And what about food? Any animal reaching an island is likely to confront a reduced food supply, whether it is a plant-eater or predator. The reasons for this are that probably few food species have been able to colonize the island. And those that have managed to become established may have changed during their long period of isolation so that their original predators are no longer particularly well suited for exploiting them.

Under the circumstances, then, any animal that reaches an island may have to either change its feeding habits or become extinct. As an illustration, suppose that a woodland bird species requires an acre of woodland per bird to sustain itself. When the birds reach an island, they may find that only a half-acre is available. However, if no grassland species has already colonized the island, the woodland birds may be able to enlarge their niche by feeding part of the time in the grassland. Even if they are not particularly well suited to this habitat, in the absence of grassland competitors they may be able to survive. And in time they may become well adapted to the new grassland-woodland habitat.

If a continental grassland species should appear on the island before the woodland birds become adapted, however, the latter may not be able to compete successfully with the newcomer. If it is forced from the grassland, it may become extinct. However, if the grassland species should arrive after the woodland species has adapted to grassland, the newcomer itself may be unsuccessful in colonizing. So you can see that on islands, as in romance, there appears to be an advantage in getting there first. As you can also see, competition is another reason for the paucity of species on islands.

It may seem that we are saying that a colonizing species moves into "unoccupied niches" on an island or that it adds other "niche units" to its

Box 13.2 Grass on the North Slope

One of the oil companies involved with developing the oil of Alaska's North Slope placed an ad in certain magazines some time ago. The ad was part of the oil companies' program to sell their image to the public rather than their product. In the ad they boasted that they were reseeding all the disrupted areas with a newly developed grass that grows five times faster than the grass that normally occurs there. I wrote them a letter pointing out that the faster-growing grass might easily displace the naturally occurring grasses through processes of competition. Also, I added, if the newer grass were to become dominant, would it be able to withstand the periodic aberrations in weather conditions that the natural grasses had survived for countless generations? If not, then segments of the ecosystem there, which are dependent upon grass, would be placed in a precarious position.

They sent me a brochure.

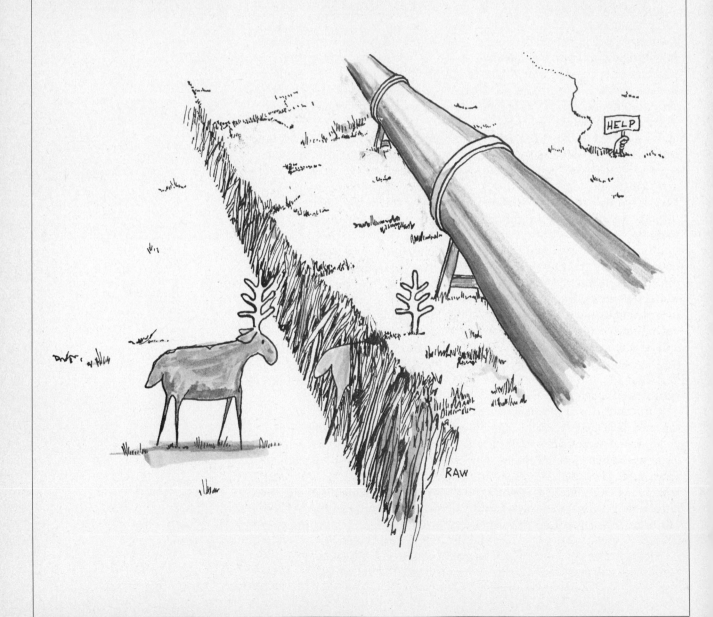

own. However, niches are not preexisting ecological slots that can be filled or vacated; a species might more accurately be said to have *broadened* its niche if, on an island, it begins to interact with its environment in more ways than it had previously on the mainland.

The evolutionist Ernst Mayr, as a result of his long studies of island species, has concluded that the most successful island colonizers among birds are generally those species that eat plants (often more abundant and accessible than other food), are social (solitary species are most easily stopped by geographic barriers), are good travelers, and are exploratory and behaviorally flexible. Flexibility is important, since any island-colonizing species is likely to be confronted by novel conditions.

Lessons From Woodpeckers

Woodpecker watchers in the United States might notice that the red-bellied woodpecker is highly successful at getting food from the branches of trees, but that this species hardly ever forages among the leaves on the forest floor. In contrast, another species of woodpecker that occupies the same forests, the red-shafted flicker, is a proficient ground forager. If the red-bellied woodpecker were to attempt to forage in the leaf litter in those areas where flickers, thrushes, and other ground specialists live, it would probably end up with less food than if it continued to search among the branches.

Now, in each generation there are undoubtedly some red-bellied woodpeckers that show a tendency toward ground foraging. However, the number of eggs laid and the number of young reared depend partly on how much food the parents can find. So those ground-foraging red-bellied woodpeckers are likely to rear fewer young than their tree-searching brothers. Thus the tendency toward ground searching is continually selected against as long as the ground specialists are around.

The red-bellied woodpecker also doesn't spend much time trying to catch insects on the wing. The woods are full of flycatchers and phoebes, species that are quite proficient at catching food in this way, and if the woodpecker were to try to compete with them, it would probably come up short again. The red-bellied woodpeckers of any generation, then, are primarily the offspring of those woodpeckers that had a tendency, possibly reinforced individually through success, to search the bark and limbs of trees for food.

On Jamaica there is only one species of woodpecker that lives there year round. This may be because not many species of woodpeckers have reached the island in sufficient numbers to become established. And those that reach it now may be excluded by a well-established woodpecker—one that now occupies an expanded niche because it didn't have to compete with similar species during its period of establishment.

We might ask why natural selection would push the Jamaican woodpecker into new adaptive zones. Why didn't the woodpecker that originally colonized Jamaica simply become better at feeding in the old way? Why branch out (no pun)? Remember that because of the problems of insect colonization the Jamaican woodpecker probably has only a few species of insects available to it. It is true that each insect species may be well represented, but populations tend to fluctuate. Many species undergo cyclical rises and falls in number, especially in areas with seasonal variations in climate. For example, on Jamaica the winter months are dry and somewhat harsh. This means that the numbers of some insect prey are likely to be lower at this time.

If the Jamaican woodpecker restricted its foraging habits to only a few bark-dwelling species, the seasonal reductions of any of these prey species would have a marked effect on the foraging success of the birds. However, if the woodpeckers expand their niche to include more flying insects, ground-dwelling insects, and plants in their regular diet, their food supply is likely to be more stable, since probably not all of these diverse species will become scarce at the same time. In other words, if a woodpecker becomes a foraging "generalist" (despecialized), then as one type of prey decreases it can switch to another, so that it can rely primarily on different foods at different times of the year. Thus it is to the woodpecker's advantage to expand its niche, and niche expansion is possible because of the absence of close competitors.

These, then, are a few of the theoretical advantages of niche expansion for the Jamaican woodpecker. But what sort of pressures could cause the woodpecker to make these changes? After all, birds aren't very bright and aren't likely to think, "I really should see about expanding my niche, since that seems to be a good move for us island colonizers." One probable impetus relates to competition. Let's see if we can put it together.

Competition

As a species invades an island, its numbers are likely to be low at first. They may remain low as the population adapts to the specific ecological situation on that island. Colonizing animals are not likely to be prepared to exploit most efficiently what the island has to offer. Their low number, however, does not mean that there is little competition between them. The level of competition would be determined by the abundance of the resources. If members of a group of island colonizers all tended to forage in the old ways, they might rapidly fall into competition for the same scarce food.

In each new generation, however, there will be those individuals that are predisposed to forage in ways a bit different from their parents. These birds would have been selected against on the species-packed mainland, but out here on the island they may be able to utilize food sources that had been previously ignored. They may even be able to find more food than their brothers who forage in the conventional way. Thus, while competition for food keeps the numbers of the strongest competitors low, the aberrant foragers may be at an advantage. By finding more food, they are able to produce more eggs and to feed more young. If the aberrant foraging pattern were to any degree genetically determined, some of the offspring of these birds might tend to forage in the new way.

Of course, as their numbers grew, there would also be increasing competition over the new foraging area, so advantages might continue to be bestowed on the more variant individuals among *their* offspring. Thus the niche of the species as a whole would continually broaden. As any other commodity, such as nest sites, fell into short supply there might be the same sort of expansion into new ecological areas. In time, then, any animal that colonizes an island may come to show a much broader range of behavior than it arrived with.

Now let us consider other possibilities that may be open to island colonizers. It probably isn't particularly important that you know about what's happening on islands as such, but by considering the possible events involved in island colonization, perhaps we can unveil certain important lessons of ecological interaction.

ADAPTIVE RADIATION

The Galapagos Islands, as you may recall, are an archipelago about 580 miles off the coast of Ecuador that is inhabited by a wide assortment of animals. The finches Darwin observed on his historic visit are perhaps not so spectacular as the large lizards that clamber from the surf-pounded rocks to graze on seaweed beneath the surging seas. However, Darwin's keen eye noticed something even more remarkable about these thirteen species of small birds. They are all 4 to 8 inches long, and both sexes are drab-colored browns and grays. There are six species of ground finches on the islands. Of the four species found on most of the island, three live on the ground and take seeds of different sizes, as evidenced by their different-sized bills (Figure 13.6). The fourth species has a long bill and eats a lot of prickly pears.

There are also six species of tree finches in the Galapagos. One of these has a parrotlike beak and eats seeds and fruit. Four have bills well suited for eating insects, each species specializing on insects of certain

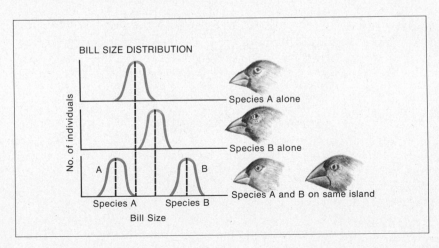

13.6 A hypothetical example of character displacement. Where species A and B do not come into contact they may have similar morphologies (here, bill sizes) and, it is presumed, they utilize their habitats in rather similar ways. Where they come into contact, however, they shift away from each other both morphologically and behaviorally, and in so doing they presumably reduce their level of competition.

types. But the strangest of all is the woodpecker finch. This little bird has a bill like a woodpecker, but it lacks the long protrusible tongue of the woodpecker with which to probe deep into crevices. So instead it pries grubs out of holes by using a long twig or cactus spine.

The thirteenth species of finch has diverged markedly from the other finches. In fact, by external appearance and behavior it should not be considered a finch at all, but a warbler. Only the examination of its internal structure unmasks it as an aberrant finch.

Speciation

So why have these little birds become so significant in the biological world? Their fame is due to Darwin's suspicion, supported by subsequent years of careful research by others, that the birds are all descended from the same stock. The prevailing idea is that long ago the islands were colonized by a species of ground-dwelling finch from the mainland. The colonizers were probably blown out to sea by some weather turbulence, and in time they became established on one or more of the islands. As their numbers increased, they could easily have spread throughout the islands. However, the rate of gene exchange among the birds of the various islands was low enough to ensure virtual isolation of each population. In time, then, as each small population was subjected to the specific and rigorous conditions of its particular island, changes would have appeared in their genetic constitution through the process of natural selection. These changes, of course, would have been reflected in the appearance and behavior of the species.

Under such conditions the population on each island would ultimately have become so different that they could no longer interbreed. The result would be *speciation,* or the formation of new species. It has been roughly calculated that each population would have to have been isolated for about 10,000 years in order to form a new species. We don't know how precise the time measurements are, but we do know that nature is in no hurry. It wouldn't matter whether the colonizers became established or not. The fact that they did was merely a fortunate interaction of characteristics between the colonizers and their new habitat.

Some of the differences between species may become more pronounced if the species occupy the same habitat. This phenomenon, called *character displacement,* may occur in response to high levels of competition between closely related species.

It is generally believed that speciation normally occurs only when populations are genetically isolated. In other words, the likelihood that the same stock will give rise to two species is greater when that stock is split into two separate populations that are prevented from interbreeding. A very few individuals could probably be exchanged between populations without interfering too much with the accumulation of genetic differences. If the exchange of genes between separate populations is too high, however, any modifications that might have appeared in either population

13.7 Darwin's finches. The darker birds on the ground and standing on the cactus are ground finches. Those in the tree are, appropriately, tree finches. Are there similarities within each group? Can you account for this? The woodpecker finch, at center, has evolved a most unusual behavioral trait, using twigs and cactus spines to prod insects from their holes. What conclusions can you draw about the diets of the different species by looking at the size and shape of their bills? The different species presumably arose from a single stock that gradually spread across the islands. The finches have a strong attachment for their home areas and are reluctant to fly across water. So once a group reached an island it was likely to be rather effectively isolated and left to follow its own evolutionary pathway. Eventually, groups differed so much that they could not interbreed. The process of a single stock giving rise to multiple species is called *adaptive radiation.*

Box 13.3 Stereotype and Opportunism

In this discussion we're talking about birds for two reasons. One, I like birds and it's my book. Two, most birds, including woodpeckers, show rather stereotyped behavior. In other words, much of what they do is apparently a result of some sort of evolutionary "programming." They certainly may alter any genetically programmed pattern as a result of learning experiences. However, their rather inflexible responses to environmental variables make them good subjects to use in the search for the ecological implications of those variables. If a lack of competitors is followed by increased variation in bill sizes that can be correlated with a wider variety of food being taken, then we have an indication of how competition may influence morphology. If, on the other hand, birds were creative and innovative creatures, in an absence of competitors they might simply change their behavior in subtle ways, each individual doing his own thing, and thus make life much more difficult for an investigator. But because of stereotype, there are many species of birds around, each specializing in its own way of taking food, building nests, etc. The changes that occur in new ecological situations do so very slowly. Recent colonizers of islands may show no change in behavior or morphology through many generations. Such stability has been demonstrated in several species that have colonized Bermuda.

Man, on the other hand, is such a flexible and opportunistic creature that not even one other member of the genus exists on the whole earth. There is no room. Man can exist on a completely carnivorous or herbivorous diet. He can live under extremely crowded conditions or isolated in the far-flung areas of the earth. Man exists in jungles and deserts and is now seeking to invade the depths of oceans and even the silence of outer space.

would be mitigated or "swamped" by the genes from the other population, thus reducing the level of differences between them.

Speciation on Continents

Of course, genetic isolation can occur on continents as well as islands. Otherwise this discussion wouldn't maintain whatever level of biological interest it has managed to muster. The Grand Canyon, for example, has divided a population of Arizona squirrels into two subgroups. The south side of the canyon is inhabited by the Abert squirrel, while the Kaibab Plateau on the north side is inhabited by the Kaibab squirrel (Figure 13.8). In the period in which they have been separated, differences between them have accumulated to the point that they are now considered a separate species.

From a biological standpoint, insular conditions may exist on continents. Each pond, for example, is isolated from other ponds to some degree. The plants and animals that inhabit ponds may or may not be good travelers, so their progeny may or may not be likely to reach other ponds. Hence if a pond remains isolated for a long period of time and doesn't dry up, its endemic species may eventually generate new life forms, perhaps not particularly different in appearance from other forms, but nonetheless unable to interbreed with them. Such genetic isolation may be found in species in lakes or mountaintops or caves or any other area surrounded by conditions markedly different from its own.

Oceans as Islands

In a sense the Atlantic and Pacific Oceans are "islands," isolated from each other by the Americas, and each has developed its own life forms. For example, the deadly sea snake ranges throughout much of the Pacific and is common along Central American coasts, but it is absent in the Atlantic (Figure 13.9). However, a sea-level canal is now being considered to replace the lock system of the Panama Canal. The new canal would allow species that have been separated for eons to suddenly come together. The sea snake, of course, would have access to the Atlantic. Imagine the surprise of bathers! The snake would also pose other problems. Ex-

13.9 Banded sea snake. These reptiles are extremely unaggressive, often being casually handled by local people on the Pacific side of Central America. However, they are in fact deadly poisonous and occasionally they do bite humans. These snakes are avoided by fish in their normal habitat, but are attacked by fish unfamiliar with them. These fish, of course, are quickly killed. If a sea level canal should replace the present Panama canal, the deadly snakes would not only be exposed to Atlantic bathers but to naive species of predatory fish.

Box 13.4 Gaining an Island Toehold

Species may arise on islands faster than on continents for two reasons: (1) the colonizers may show higher higher variability than the parent population; and (2) conditions may be stringent in the new habitat. The colonizers will, of course, not contain all the variations in the parent population, but because of their low numbers, any variations that they do carry will not be "swamped" by the genes from others, as they would in a large population. Thus differences may arise quickly in the progeny of small colonizing populations. Since colonizers are not likely to be well-adapted to their island, there may be strong selection operating on the variation so that changes in the population may take place rather rapidly.

It should be obvious, then, that the odds are against successful speciation, especially on islands. A small population attempting to colonize an area with new, and therefore somewhat inhospitable, environmental conditions is almost sure to meet with disaster. Perhaps many species of birds were blown out to the Galapagos by offshore winds, and perhaps most failed. But, finally, one didn't.

periments have shown that Atlantic fish species do not recognize the banded sea snake as dangerous, as do the Pacific species that evolved with it. As a result, the Atlantic species try to eat the snake, and of course, these attempts are fatal.

Think of the ecological chaos in terms of predation and competition that will result from the mixing of thousands of species that are now complete strangers to each other. However, there are those with political and financial clout who argue that a sea-level canal is commercially cheaper than one that utilizes a lock system.

THE WEB OF LIFE

We have already considered a number of factors that might help to explain the high extinction rate of species on islands. Another reason for this phenomenon is simplicity itself. Because islands have fewer species than nearby continents, the island ecosystem is theoretically "simpler." Ecologically, simplicity means instability, and in an unstable ecosystem extinction rates will be higher. Let's examine this point more closely, since it has important implications for our own species. Again we must deal in speculation to some degree because much of the essential data just aren't in.

Figure 13.10 shows a hypothetical and greatly simplified food web. However, these food relationships illustrate a point. Suppose the only carnivorous (meat-eating) mammal in this food web were the fox. Then, if some fox disease were to sweep the area, what would be the repercussions on the rest of the life there? We can see how the numbers of herbivores, such as rabbits, squirrels, mice, and some birds, might rise. These in turn might put new pressures on their plant food, and in so doing destroy the habitat of many other species while depleting their own food supply. The resulting initial increase in the numbers of herbivores would provide more food for owls and hawks, which could in turn be expected to increase their own numbers. The resulting abundance of such predators might then reduce the numbers of other animals, such as insect-eating birds and toads. What might then happen to the numbers of snakes? And the numbers of insects?

Now then, suppose the foxes had to share their food with lynx and wolves, as well as with hawks and owls. In this case a rampant fox disease would have markedly less effect on the ecosystem. Of course, elimination of the foxes would result in some changes in the system, since the fox has a somewhat different niche from the wolf or the hawk. But perhaps the system would continue largely as before until the fox population was restored. The point here is that a simpler system—for example, one with a single "top" predator—is inherently easier to upset.

Let's draw on a more abstract example to show the effect of the extinction of a single species in an interdependent system. Figure 13.11 shows five species that depend on each other. Dependency may be of a variety of types. For example, fleas depend on dogs for living quarters and

transportation, as well as for food. Some starlings depend on old woodpecker holes for nests. Hawks depend on mice and starlings and woodpeckers for food, but not much on fleas.

Now, suppose in this system C somehow becomes extinct. Since A, E, and D depend on C, we might reasonably expect their numbers to be reduced. B, however, depends on A, E, and D. Even a slight reduction in the numbers of all three might have such an impact on B, that it might also follow C into oblivion. This would mean that the animals on which A, E, and D were dependent would be reduced by *half.* How might this affect their fate? If A is more susceptible to the loss than E or D, its numbers might then drop quickly. What effect might this have on the two remaining species?

Now, suppose that instead of a five-species system, we have a ten-species system, as in Figure 13.12. Here the system is much more complex, although it is still far simpler than any actual biotic system would be. It is easy to see, however, that the loss of any single component might be more easily absorbed in this system than in the simpler one. Thus, the more complex a system is, the more inherent stability it has. One reason islands have fewer species than continents, then, is that they have a higher extinction rate, as components of their relatively simple and interdependent ecosystems drop out.

EXTINCTION AND US

Are there any lessons here for us? Should we care if over 150 species of animals have disappeared from the earth in the last fifty years? Should we be concerned that there are literally thousands of species whose very existence is presently endangered—largely because of our activities? Extinction, after all, is the natural end of populations. Species are born, and they mature, and they die. Some live a long time, perhaps as long as 2 million years; some die more quickly. We have hastened the extinction of many species we know about, and we have undoubtedly contributed to the demise of many more. In fact, there are undoubtedly many unobtrusive species that have lived among us during our time on earth—but that have disappeared as a result of our activities without our ever having known they existed.

It is hard to explain the rationale of many of us who are concerned about such matters. I have never seen a sei whale, yet I don't want them to become extinct. Moreover, I felt this way long before I understood anything about how they might be an important part of an ecosystem. Possibly such feelings merely reflect the cultural attitude that it is "nice" to wish other living things well; thus the attitude is rewarded. I feel nice.

There are, of course, more cogent reasons for abhorring the extermination of any species. For one thing, the kind of attitude that encourages or sanctions the destruction of other species constitutes an intrinsic threat to our own well-being. This may be similar to the sort of reasoning that causes

13.10 A hypothetical food web (opposite page). Can you tell which organisms serve as food sources for the various species? This diagram is greatly simplified—for example, there are also omnivorous birds, such as crows, and mice may sometimes eat bird food. The predatory fox may also have to share its food with lynx and wolves.

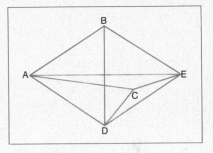

13.11 Interdependency of five theoretical species. This is a relatively simple system in which each actor has a major role. Because of their interdependency, the elimination of any could threaten the existence of all.

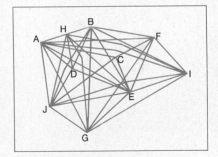

13.12 Interdependency of ten theoretical species. The system becomes vastly more complex by simply doubling the number of species. In such a case, the elimination of any one may not be crucial, since the complexity can act as a buffer. The point is, biological systems may become more unstable by becoming simplified and one way to simplify them is to reduce the numbers of their component species.

Box 13.5 The Redwoods

It stood 390 feet tall. Nothing on earth could match it. It had stood as a slender sapling in the cool coastal air, perhaps moving slightly with the light breezes, on the very day Caesar finally decided to move against Britain. But all that happened a long way from the area that would be named California. The tree lived through it all. And some of it must have been pretty strange, but not as strange, perhaps, as what went on in its last few years.

As the great ages passed, the sapling continued to grow. No one marked the time when its crown reached above all the others because it was only one of a vast forest of

such trees that ran from north of San Francisco deep into Oregon. The straight, tall, and insect-resistant trees naturally caught the eyes of lumbermen, even before the turn of this century. Then, as the population and industry of California grew, such demand was placed on the trees that some citizens tried to establish a national park to save some of the trees. But the lumbermen blocked it. In what has been called "one of the great swindles of all time," they arranged to change nearly all the redwood lands from public to private ownership.

After World War II, California experienced a population surge of unprecedented dimensions. Factories and homes were being built feverishly and lumber was needed. Redwood was ideal for lawn furniture too. The conservationists, led by the Sierra Club, pushed again for parks to be set aside. By 1960 there were two areas left that were relatively unscarred and of park caliber. While conservationists dickered among themselves over which area was the best, the lumbermen were busy in Sacramento and Washington. They also effectively muddied the waters by cynically proposing a number of other sites. The easily confused public became confused. The lumbermen were confident. One boasted that it takes five years to get a national park bill through Congress and in five years there wouldn't be anything worth fighting for. As the arguments continued, the trees kept falling.

Then, in the summer of 1965 the 390-foot giant was discovered—by the lumber companies. No announcement was made to the public but the raucous chain saws and great log movers were quickly brought in. Without fanfare, the

great tree was felled, cut into twenty-foot lengths and hauled away. In fact, every tree in the redwood stand at the junction of Bond and Redwood Creeks was brought down—right in the very heart of the Sierra Club's proposed park site. The crash of timber was drowned out only by the cheerful ring of the cash register. Trees were still falling when Ronald Reagan, then candidate for governor, spoke before a convention of the Western Wood Products Association. His words and wisdom were memorable: "A tree's a tree. How many do you need to look at? See one, you've seen 'em all."

It had been the tallest living thing in the world.

Clear-cutting the redwoods. Much of the booty will be used in the manufacture of cheap lawn furniture.

Box 13.6 Subdue the Earth

Indian farmers are incessantly waging war on rats. It has been calculated that the progeny of one pair of rats, in a year's time, can consume enough grain to feed five people. The killing is thereby justified. But who has calculated that five people can, in a year's time, consume enough grain to feed the entire year's progeny of a pair of rats? The question is patently ridiculous. But it hasn't been calculated only because rats can't calculate. Perhaps part of our problem lies here. We see ourselves as more "worthwhile" than other animals. Obviously our worthiness is subjectively derived since it is we who are doing the measuring, and it is to our own best interest to place a higher premium on ourselves and, to carry it further, on those most like us. The mechanism is self-protecting and perhaps justifiable. But can it lead to problems? Will it tend to make us, in very subtle ways, believe that we stand apart from the other species? Can such a belief, expressed or not, cause us to behave as if we are not subject to natural laws? That we are *better,* i.e., *different?*

And by the same token, do we lose something by elevating ourselves? Why is a deer's footprint in the snow "beautiful," but a man's bootprint "disfiguring?" Why is a beaver dam natural, but a human dam unnatural? Is it because we continue to see ourselves as apart from nature?

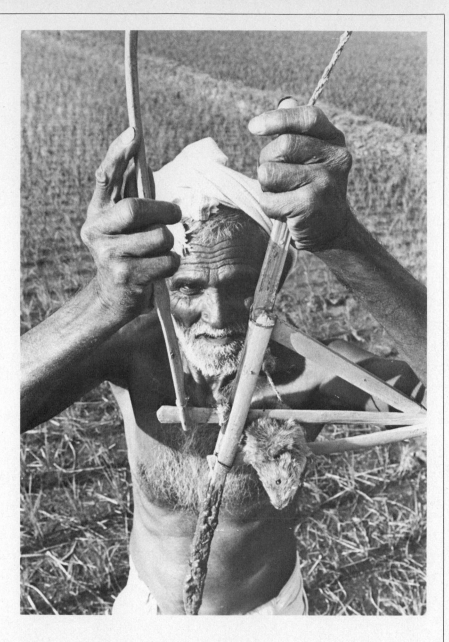

us to seek to weed out violence from our society—the fear that we, ourselves, might one day fall victim to it. The extinction of other species could also threaten us indirectly by simplifying the system of which we are a part or, more directly, by destroying parts of the ecosystem upon which we directly rely. For example, if we continue to poison the oceans because we are willing to believe only a few bottom dwellers are directly affected, we might eventually overstep some critical threshold and trigger the wholesale death of plankton, thus finding ourselves without a major source of the world's food and with our oxygen supplies dwindling.

This is all very rational, but none of this precipitated my gut response when I heard the Japanese were going to continue to hunt the whales. I simply found it sad. For some reason, I wanted those whales to continue to share the planet. If you had the same reaction, and you don't ever expect to see or eat a whale or use its oil, you might try to analyze the roots of your own response.

Populations

14

The tuntong is almost extinct! Almost no one has noticed and even fewer care. Upon hearing the news, a sensitive and intelligent person might say, "That's too bad," and then, "What the heck is a tuntong?" It is true enough that we can't be informed on the status of every species on the earth. Most of us, if fact, know of only a few of the hundreds of species on the "endangered" list. (How many do you know of offhand?) It probably isn't necessary for anyone to have such information at his fingertips, even if it were available. However, at this critical time in our history we should know certain things about how numbers of any species behave, how they change, and what causes them to flourish or dwindle.

There are a number of reasons this is important. If you have ever stood alone amid the browns, yellows, and reds of a thicketed field on a cold autumn afternoon and watched a flock of geese beating its way southward against a gray and heavy sky, the presence of those geese will be important to you. You may not be able to persuade a land developer that their breeding grounds are important on the basis of the strange stirrings that swelled within you that day, but you will have placed a new value on a remote species, one with which you have shared an important moment in your life.

The numbers of living things are also of concern to us for reasons other than aesthetic, reasons that may even appeal to the crassest profiteer. As we have seen, the more species there are, the greater the stability of the biological system they comprise. In addition, we may be dependent on other species, both directly and indirectly, for our own food, oxygen, and shelter. Finally, learning more about the population dynamics of other species may give us some insight into the status of our own. Providing information that might lead to that insight is basically the goal of this chapter.

POPULATION CHANGES

First, we should remind ourselves of the obvious: the numbers of living things change. Some of these changes are short-term or periodic and easily explainable. Grasshoppers, plentiful in summer, disappear in winter. Some birds disappear from American forests in the winter as they move to more hospitable areas. These are familiar and rather predictable phenomena. We all expect grasshoppers and birds to be back in spring—except,

14.1 The numbers of some species seem overwhelming and invulnerable. Here, wild rabbits gather at a watering hole in Australia. Their numbers swelled alarmingly after they were introduced by man to that continent, only to be drastically reduced by the dedicated efforts of ranchers with whom they competed. Such efforts included mass slaughters with clubs after the rabbits had been trapped in nets, as well as the introduction of disease. Huge natural populations have also been decimated as is made clear by a glimpse of the stuffed body of the last passenger pigeon, now in a museum. In other cases, populations may change drastically without direct human assistance. We often don't know what brings on such changes, but the study of these phenomena is one of the interests of population biology.

perhaps, for those permanently lost as a result of our use of DDT and other insecticides and our destruction of their habitats.

Apart from these short-term fluctuations, you may have noticed that the numbers of many species don't seem to vary much from year to year. For example, the number of birds around your house has probably been about the same from one summer to the next. However, if you think back, you may also notice that some species you remember as having been fairly common just aren't around anymore. Roger Tory Peterson, the noted ornithologist, has reported a decline in many species of songbirds around his home in Connecticut over the past several years. Have they simply moved to another habitat? Or are these species quietly slipping into oblivion, into a very final extinction?

And what about those sudden increases in numbers of certain animals? Where do all the locusts that suddenly appear to ravage African crops come from? Certain populations seem to blossom for some mysterious reason, and we hear of "invasions" or "plagues." Yet, in the face of an environment that allows such changes, most species seem to maintain rather stable numbers. How? Let's begin by considering how populations normally behave.

Population Growth

If we were to place a few bacteria in a suitable medium, or a few sheep on an uninhabited but ecologically hospitable island, the growth of their populations might be expected to show similar patterns, at least at first. Since we are assuming that their new environment provides them with the necessities of life, their growth will be limited primarily by the rate at which they can reproduce, rather than by environmental factors. Figure 14.2 shows the rate of increase for a theoretical population.

Note that the population size becomes greater with time because increasing numbers of individuals produce offspring. If each individual in the original group leaves two offspring, the population will remain fairly small for several generations (one bacteria divides to leave two, these leave four, which leave eight—and these are all low numbers). But once the population increases a bit, the numbers begin to skyrocket. (What would be the result after only ten generations? Twenty generations?) The reproductive capacity of any population, when it is unrestricted, is called its *biotic potential.*

Carrying Capacity

The full biotic potential of any species is not likely to be reached because of the inherent limitation imposed by the environment. *Environmental resistance* is encountered when some essential factor or factors in the environment come into short supply and thus reduce the population's rate of increase. The limit of biotic potential set by environmental resistance defines the *carrying capacity* of the environment. As a population approaches the carrying capacity of its environment, then, its numbers begin to level off (Figure 14.3).

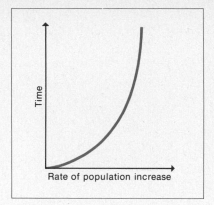

14.2 Theoretical growth. Time is plotted against population size. In such a case there is nothing operating to restrict growth, so the population is achieving its reproductive (or *biotic*) potential. The characteristic curve is produced because larger populations increase more rapidly than smaller ones.

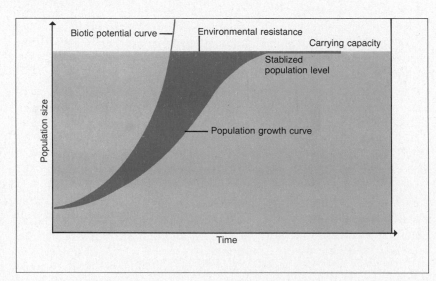

14.3 The theoretical relationships between biotic potential, environmental resistance, and carrying capacity. The biotic potential of a population is rarely, if ever, achieved due to the problems of living in a real world (the problems are collectively called environmental resistance). Some individuals don't reproduce and some fail to reproduce maximally. The characteristic growth curve is produced because a small population grows slowly, since a few individuals can generate only a small number of offspring. Once the breeding population reaches a critical size, however, it skyrockets (still well below its biotic potential) to be slowed only by the effect of its own numbers. Theoretically, it should stabilize around the carrying capacity of the environment.

In some cases a population may overshoot the carrying capacity of its environment. When this happens its growth rate may fall off abruptly only to pick up again when the numbers fall below the carrying capacity. In more stable populations the numbers may oscillate rather closely around the carrying capacity (Figure 14.4).

If a population drastically overshoots the environment's carrying capacity, the result may be a severe depletion of numbers called a *crash*

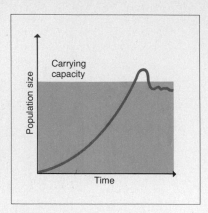

14.4 The type of population curve normally generated when a population reaches the environment's carrying capacity. Theoretically, a stable population fluctuates about the carrying capacity in a rather regular fashion.

14.5 If a population should drastically overshoot the environment's carrying capacity, or disrupt the environment and lower its carrying capacity, the population may "crash," generating what, for some reason, is called a J-shaped curve.

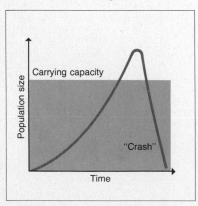

(Figure 14.5). In some species such crashes are a normal part of the population cycle. The numbers of annual plants and of many insects crash each autumn, for example, but their progeny appear each spring to reproduce madly in the few months they have available.

Some crashes are more permanent. A bacterial population may rise rapidly, but if the population lives in a petri dish on a laboratory shelf, it must eventually run out of food even as it befouls its environment with its own waste products. After it crashes, it cannot recover. As another example, if environmental resistance in the form of predators (including hunters), disease, or food and water limitations is somehow reduced for a group of animals, such as elephants, their numbers may increase to the point that they eventually devastate their habitat. Oversized herds of elephants may eat all the young plants in their habitat that, if left alone, would eventually have provided much more food—and would have continued to do so for generations afterward. The result of such devastation is a lowering of the carrying capacity of the habitat for long periods afterward. If fact, it is possible to permanently lower the habitat's carrying capacity. If stripping an area of its foliage should increase erosion, the habitat may *never* recover. Fortunately, humans are a lot smarter than elephants.

REPRODUCTION RATE

There is an argument among biologists regarding whether mechanisms are operating that maintain population numbers within certain limits—mechanisms that only occasionally break down. It seems, however, that the relative stability of so many species would indicate that some sort of stabilizing influence is operating. This influence, if it exists, could theoretically operate at two levels by affecting: (1) how many offspring are produced, and (2) the longevity or mortality of organisms in the population.

How Many Offspring Are Produced?

The reproductive potential of most species, even the slowest breeding ones, is incredibly high. Charles Darwin estimated that a single pair of elephants could leave over 19,000,000 descendants in only 750 years. The fact that the entire world was not teeming with elephants indicated to him that some elephants were not reproducing, and that the reproducers were somehow "selected" by the environment.

Whereas elephants are very slow breeders, other species are more prolific. For example, the reproductive potential, for a *single year,* of the housefly is shown in Table 14.1.

But what determines the rate at which an organism reproduces? It should be apparent that simple laws of energy would not allow an elephant to become pregnant several times a year. There is just no way that a female could find enough food to produce that many offspring. And what would happen to the helpless baby elephants already born as new brothers and sisters appeared on the scene?

It may seem puzzling, then, to learn that each animal theoretically reproduces to its biological limit. Can it be true that animals leave as many offspring as possible? You might intuitively agree when you recall that any generation is primarily made up of the offspring of the most successful reproducers. The next question we might ask regards method. How does an organism maximize its reproductive success?

Tapeworms

The tapeworm living so comfortably in the idyllic environs of your intestine is lucky to be there—although its luck is not yours. It is lucky because the life of a tapeworm is fraught with risk. The odds are against its finding a host, you. In order to reach a human host, the egg must pass out with the feces of the previous human host. Then it lies around in the sun and rain until it is eaten by a pig. Not only must it survive the elements, but it must also be found by a pig, so already the odds are against it. Once the egg has been eaten by the pig, it changes its appearance, moves through the intestinal wall of its host, and on through the bloodstream until it comes to lie in the pig's muscle. Here it forms a protective capsule around itself and waits. It waits for a human to eat the pig's flesh without cooking it too well. When this happens, the worm attaches to the intestine of its human host, grows into an adult, and begins to lay eggs.

There is obviously an overwhelming probability that not all these conditions will be met, and thus most eggs will not develop into adult tapeworms. So what is the tapeworm's answer to such a demanding life cycle? It has become capable of self-fertilization (perhaps at the risk of going blind) and it lays thousands and thousands of eggs. It's whole life is devoted to egg production. It exists only to lay eggs, eggs, EGGS! A few of these, hopefully, will wend their way, by chance, through the complex maze that is the tapeworm's life cycle.

Chimpanzees

So that's the tapeworm's response to its particular situation. But what about other animals? Are they *all* in the numbers game? Let's consider the chimpanzee, an animal that has developed a different evolutionary strategy. While the female chimpanzee is sexually receptive, she may copulate with almost any male who signals his desire. Once she has become pregnant, she may not become sexually receptive again for years. Flo, the aging but sexy female studied by Jane Goodall, did not become sexually receptive again for five years after delivering a baby. During this time, however, she attended carefully to her baby.

In the first months Flo carried her baby everywhere and diligently guarded it against danger. Later it was permitted brief forays on its own, but it was not allowed to stray far from her vigilant eye. During this time the baby would scurry back to her at any real or imagined sign of danger. As the young chimpanzee gradually became able to care for itself during those first few years, the association between mother and offspring

Table 14.1 Projected populations of the housefly *Musca domestica* for one year, a period that encompasses about seven generations. Do all these offspring survive? The numbers are based on each female laying 120 eggs per generation, each fly surviving just one generation, and half of these being females.

Generation	Numbers If All Survive
1	120
2	7,200
3	432,000
4	25,920,000
5	1,555,200,000
6	93,312,000,000
7	5,598,720,000,000

Source: Adapted from E. V. Kormondy, *Concepts of Ecology* (New Jersey: Prentice-Hall, Inc., 1969), p. 63.

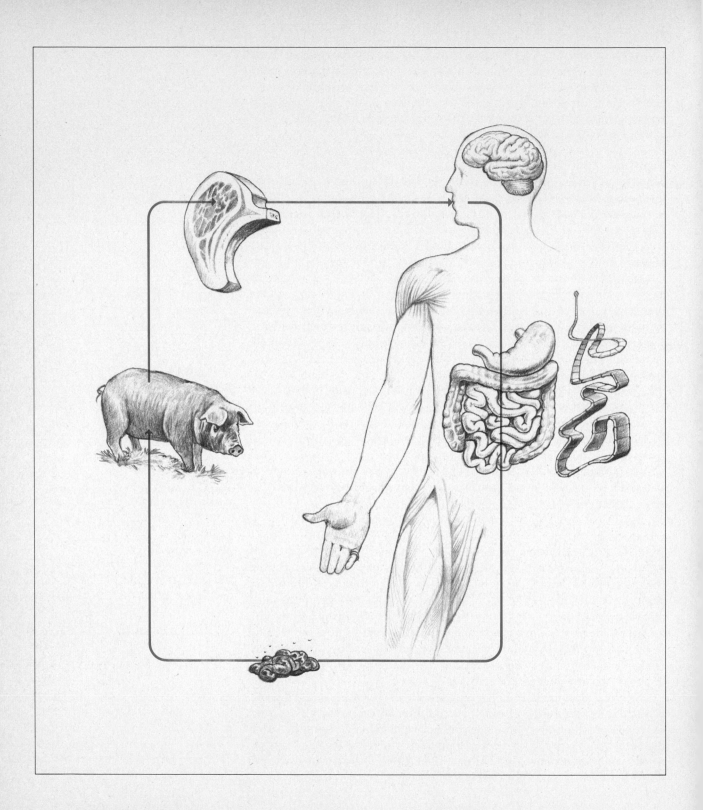

lessened, until finally Flo was free to mate again and rear another baby. Among chimpanzees, the female does not maximize her reproductive success by giving birth to large numbers of offspring. Instead, few offspring are produced and these receive very careful attention until they become independent. Why has such a strategy evolved?

Chimpanzees have historically existed in a precarious and complex world. For centuries before humans decimated the African big cats in the name of fashion the chimpanzees were continually hunted by these vigorous and intelligent predators. Their lives were complicated by other factors as well. Chimpanzees feed on a number of kinds of food, including seasonal fruits, baby baboons, and small monkeys, as well as ants, which they pick up by using tools made of broken twigs. So their diets are varied and complex. Also, they must move from place to place to exploit new food sources. Each night they must build a new nest usually of bent branches or twigs and leaves. Thus they are constantly dealing with new and sometimes threatening situations.

To cope with such an environment the chimpanzee needs not only high intelligence, but a system of extended parental care to give the young chimpanzee time to learn about its complex world. Extended parental care, however, means that the parents can successfully rear only a limited number of young. So, in this species, selection has moved the animal not toward maximum production of infants, but toward efficient care of those that are produced. In other words, chimpanzees rear as many young as possible according to the demands of their particular kind of environment. So, we see that the number of offspring attempted by members of any species is likely to be set by natural selection and that the "strategy" of natural selection will depend upon a host of historical and environmental variables.

14.6 The life cycle of a tapeworm. An infected human may suffer damage to the brain or leg muscles. However, the probability of a man failing to cook the part of a pig in which a tapeworm larva was encysted after the pig had eaten food contaminated by human feces, which had come from another human with a tapeworm, is low. So how does a tapeworm ever reproduce? It simply lays thousands and thousands of eggs and leaves the rest to chance.

14.7 Chimpanzees do not have thousands of offspring, leaving their survival to chance. A female has very few offspring in the normal course of her lifetime and, as they slowly mature, she very carefully tends to each one. The young are quite dependent for long periods as they continually acquire information about their environment. Chimpanzees may live for decades, so although they mature slowly, they still are able to survive a number of breeding seasons. (Photograph by Baron Hugo van Lawick. © National Geographic Society.)

Birds

A quick look at some data from birds provides a clearer picture of how the dictates of a particular environment have influenced the number of offspring attempted by a given species. Again, for maximum reproductive success birds lay not as many eggs as possible, but the number of eggs that will enable them to rear as many young as possible. In most cases this is the number of young the parents can feed successfully.

Ornithologist David Lack found that females of the common swift, *Apus apus,* normally lay one to three eggs. Apparently, the number has not yet stabilized or is changing because of new ecological pressures. He found that in years when food was abundant the number of young that reached feathering age was 1.9 for nests with two eggs, and 2.3 for nests with three eggs. In the nests in which Lack placed extra eggs to bring the total to four, only 1.4 young survived until feathering age. If the food supply were to remain relatively constant, one might expect the number of eggs laid by females of this population to stabilize. Where?

The number of offspring attempted by a species may also reflect the vulnerability of the young to predators. For example, certain species of gulls nest on accessible beaches, where there is a high likelihood of the young being found by prowling foxes and other nest marauders. These gulls often lay three eggs, although they can usually feed only two young successfully. However, at least one of the young will probably be eaten, so apparently the three-egg nest is an adaptation to high predation. The kittiwake, a species of gull that nests on cliffsides, normally lays only two eggs, since the young are safe from predators and both offspring are likely to reach adulthood.

There are other factors that might have made small broods reproductively advantageous even when the parents could successfully feed larger ones. Conspicuous eggs, the activity of young, and the activity of parents tending the nest can all serve to attract predators, so fewer eggs means less activity around the nest.

Time may also be a factor in determining brood size. The same quantity of food can theoretically raise a small brood quickly or a large brood slowly. The rapid development of young is important in areas where the nesting season is short—such as the polar regions. In such areas there would be greater advantage in raising a small brood quickly than in portioning the same food out among a larger brood that might not mature fast enough to escape the winter.

One other fact that bird studies have helped to substantiate is the relationship between food availability and the production of offspring. It has been shown, for example, that when food is abundant, birds lay more eggs. In some cases the increase in egg production is accomplished by multiple nestings. A pair of Jamaican woodpeckers reared three successive broods when they were fed artificially. Conversely, it appears that when food is scarce birds lay fewer eggs. The mechanism here is obvious. Food is needed by the female in order to manufacture eggs, and with less food she simply forms fewer eggs. A reduction in the number of eggs may actually

14.8 The ground-nesting gulls, such as the mew gull (top), have been forced to assume a different reproductive strategy than the cliff-nesting kittiwake (bottom). The nests of the ground-nesting species are vulnerable to marauding predators so they compensate by laying more eggs than they will be likely to rear. The ground nesters, for example, lay three eggs, whereas they can probably rear only two young, but at least one of the eggs is likely to fall to a predator. The kittiwake, on the other hand, nests on cliffs safe from predators. Since all their young are likely to survive, they lay only two eggs.

mean that more young will be reared, since each one will receive a larger proportion of whatever food there is.

There is less information on mammals, but wolves, for example, are reported to have no litters in years when food is scarce. In fact, during these periods they evidently do not mate. This may be because the hungry adults are more inclined to spend their time in foraging than in mating activities. It is possible that under extreme conditions some mammals might even be physically incapable of successful mating or of rearing young.

Humans

The reproductive strategies of any species evolve on the basis of its own particular environmental circumstances, so let's consider the factors that might have determined our own reproductive capacity. The human female has the ability to produce one child each year for over thirty years. However, does this mean that humans are physiologically and psychologically prepared to rear thirty children? Or is our reproductive ability an evolutionary adjustment to a historically rigorous life in which most, or even all, of the offspring were not likely to survive? In other words, the ability of the human female to reproduce prolifically may have evolved in response to a traditionally high death rate for children.

All this is highly hypothetical but the system might work like this. During its early years, the human is unequipped physically or mentally to deal effectively with the problems of the world. It is among the most helpless of mammalian offspring and one of the slowest to reach self-sufficiency. During this time it must be cared for by adults who strive to provide it with food, keep it warm, protect it from predators, and see to it that it doesn't harm itself. But such care can only have minimal effects on reducing the incidence of disease and parasitism in these vulnerable years. The precariousness of its first years might be reflected in a high mortality rate. Even today, in highly stressful situations, the first to die are the very young, the very old, and the sick.

Should an infant die during its first year, the mother could be expected to replace the child rather quickly by becoming pregnant again. If the child continues to survive, however, she might not reproduce again quite so soon. One reason is that in many women, suckling somehow inhibits the secretion of FSH and LH from the pituitary, so pregnancy cannot occur as long as they are nursing. In some cultures the period of nursing is extended for as long as three years, since milk production will continue that long if the child continues to suckle.

The high reproductive capability of humans might also be a response to bleak prospects of an adult surviving through the entire reproductive period, a span of approximately 30 years (from 15 to 45). There is evidence that in some earlier civilizations, the life expectancy was only about 30 years (Table 14.2). In risky environmental situations in which there is a good possibility of an individual meeting death at any age, natural selection might place a higher premium on reproductive output even at the expense of the overall well-being of the individual.

Table 14.2 Estimated Average Life Span in Human Populations

Population	Years
Neanderthal	29.4
Upper Paleolithic	32.4
Mesolithic	31.5
Neolithic Anatolia	38.2
Austrian Bronze Age	38
Classic Greece	35
Classic Rome	32
United States, 1900–1902	48
United States, 1950	70
United States, 1975	72

After E. S. Deevey, "The Probability of Death." Copyright © 1950 by Scientific American, Inc. All rights reserved.

I should add here that this business of calculating age expectancies is deceptive. For example, when the U.S. Constitution was written, the average life expectancy was 40, but the requirements for the presidency included not only great virtue but the attainment of the age of 35 as well. Did this mean that the president was expected to die in the first year of his second term? Our founding fathers may, in their wisdom, have had just such an event in mind, but it wasn't likely. They were old men themselves. The nation was full of old people. It turns out that the 40-year expectancy was due to a great number of people dying very young. *However,* if a person could make it to age 20, his or her chances of growing old were very good (Table 14.3). In fact, even in 1850, if you could make it to 20, you could probably make it to 60. And if you could make it to 60, you were likely to make it to 76. And at present *your chances are no better.* Our progress has all been at the young end. We have apparently not learned much about keeping old people alive.

Table 14.3 Life Expectancy. U.S. White Male's Probability of Achieving a Specified Age in Any Given Year of His Life

In the year:	If you made it to:			
	0 years	*20 years*	*40 years*	*60 years*
	Your life expectancy would be:			
1850	38	60	68	76
1890	42	61	67	75
1930	59	66	69	75
1950	66	69	71	76
Present	68	70	72	76

Source: Statistical Bureau of the Metropolitan Life Insurance Co. compiled from various publications of the Division of Vital Statistics, National Center for Health Statistics, and Bureau of the Census.

It has been suggested that old people do not live longer because they have been subjected to long-term, low doses of numerous substances that the human body simply cannot tolerate for extended periods. We can kill bacteria, but we can't kill lead in the air. In 1900, most adults died from disease (particularly pneumonia, flu, TB, and intestinal inflammation) and their life expectancy was 49 years. In 1950, the leading causes of death for adults were chronic and degenerative conditions such as heart disease, cancer, and brain hemorrhage, and the life expectancy was 68 years. Do you think this shift can be correlated with differences in the kind of world in which the people lived?

But what about now? As Table 14.4 shows, a child born in developed countries these days is very likely to live through its first critical years. In humans today, mortality rate decreases after the first year and is very low through the teens, then it increases gradually until about age 60, when it rises rather rapidly. This is especially true in developed countries, such as the United States, but less true in certain poorer countries.

Table 14.4   Child Mortality Rates 1–4 Years (For Latest Available Year)

Continent and country	Rate per 1,000 children (ages 1–4) per year	Continent and country	Rate per 1,000 children (ages 1–4) per year
Africa		**Europe**	
Mauritius, 1971	4.5	Austria, 1974	0.8
Réunion, 1967	5.5	Belgium, 1973	0.6
United Arab Republic,		Bulgaria, 1974	6.4[b]
1960–62[a]	37.9	Czechoslovakia, 1973	0.8
		Denmark, 1973	0.6
North and Central America		Finland, 1973	1.0[c]
Barbados, 1969	1.3	France, 1972	0.8
Canada, 1973	0.8	Germany, East, 1974	1.2[c]
Costa Rica, 1973	3.0	Germany, West, 1973	0.8
Dominican Republic,		Greece, 1974	0.8
1960–62[a]	10.8	Hungary, 1974	0.8
Guatemala, 1970	39.6[b]	Ireland, 1973	4.3[b]
Mexico, 1973	19.9[b]	Italy, 1972	0.8
Puerto Rico, 1970	1.0	Netherlands, 1974	0.7
Trinidad and Tobago,		Norway, 1974	0.7
1972	6.3[b]	Poland, 1974	0.8
United States, 1974	0.7	Portugal, 1974	2.2
		Spain, 1970	0.9
South America		Sweden, 1974	0.4
Argentina, 1970	15.7[b]	Switzerland, 1974	0.8
Chile, 1971	3.0	United Kingdom—	
Colombia, 1960–62[a]	17.4	England and Wales,	
Equador, 1960–62[a]	22.1	1973	0.7
Peru, 1960–62[a]	17.4	Yugoslavia, 1973	1.7
Venezuela, 1960–62[a]	5.9		
		Oceania	
Asia		Australia, 1973	3.9[b]
Ceylon, 1960–62[a]	8.8	Fiji Islands, 1960–62[a]	3.7
China (Taiwan),		New Zealand, 1971	0.8
1960–62[a]	7.2		
Hong Kong, 1974	4.4		
Israel, 1974	5.8[b]		
Japan, 1974	0.9		
Pakistan, 1968	17.4		
Philippines, 1960–62[a]	8.4		
Singapore, 1974	4.1		
Turkey, 1966–67[a]	14.7		

[a]Average annual; latest reliable data.
[b]Includes deaths of children under 1 year of age.
[c]Includes deaths of children under 1 year of age through 14 years.
Source: Demographic Yearbook, United Nations (1975).

The means by which the human species has increased the likelihood of survival are almost universally viewed as good and desirable. For example, we have specific medicines to combat various maladies that, in earlier days, would have proven fatal. We also have therapeutic and corrective devices to aid the sick. If someone in our midst is unable to provide for himself, that person will usually be cared for, however minimally. We have laws that state that not only must the mentally deficient be allowed to remain in the population, but that their reproductive abilities must not be tampered with. Also, our society often provides for those who, in harsher days, would have been selected against. A person doesn't have to be keen of wit and physically agile in order to cross a busy street. He simply waits for a light—a light that means it's safe to cross leisurely. The result of our social care has been a negation of many of the influences of natural selection. However, at the same time that we have ceased selection for swiftness, strength, and intelligence, we have increased the variation in our species. Thus we find among us myriad interests, talents, tendencies, and appearances. The point is that our society attempts to ensure every individual that he will live and reproduce. With our reproductive potential so high, we may be placing our species in an untenable position by our uncontrolled breeding.

To sum up then, reproductive rates are subject to natural selection, and the reproductive potential of humans has been established through the eons of our development. We have recently altered our environment so that the direction and strength of natural selection has been changed, but we are left with a reproductive capacity that better fits our earlier situation.

If this is true, or even if it isn't, we might be led to ask ourselves a number of questions. Are large families with many children "natural"? Has the relaxation of natural selection resulted in our species increasing to the extent that we are physically or psychologically unable to deal effectively with the resulting swell of our own numbers? Do crowds make you uneasy? What might be the pressures on a social system in which parents are unable to carefully rear each of their children as individuals? Perhaps the most important questions involve the future. Where do we go from here? What can we expect? How much social change will be necessary to solve the problems resulting from an overbreeding population? How much social change can we tolerate? What are the odds we can pull it off? Those morose souls who spend time with such matters have imbued many of the rest of us with a deadening pessimism—partly for reasons we will consider in the next chapter. But, looking on the brighter side, let's see how populations may be kept in check by the deaths of their members.

CONTROLLING POPULATIONS THROUGH MORTALITY

Just as the numbers that are born to any species are partly a function of environmental factors, so are the numbers that are allowed to remain in the population. What are some of these factors that tend to control populations by acting on the members themselves? What brings about death?

Abiotic Control

Just as seasonal weather changes can alter population numbers, so can irregular or unusual weather. Drought may kill many kinds of plants and animals. Many birds perish in some years because they begin their northward migration in the spring only to be caught by a late cold spell. Such physical factors, which depress population numbers, are called *abiotic controls* (*a*, meaning without; and *biotic*, pertaining to life). A drought is not alive.

It should also be kept in mind that population-depressing influences such as severe weather are *density-independent*. In a severe drought, the parching sun doesn't care how many corn plants are struggling in the field below. It kills them all. In an area saturated with DDT, most of the insects die—whether there are few or many. Their numbers mean nothing (unless they somehow tend to protect or shelter each other). Thus mortality brought about by such means is said to be independent of the density of the population.

The awesome impact of the human species on the life-supporting world environment may soon bring the reality of density-independent population control into sharp focus. For example, we are building, along with other countries such as Japan, huge tankers that are designed to carry incredible amounts of petroleum products. It is disquieting to learn that, in one recent ten-year period, there were about 5,000 collisions of smaller, more maneuverable, tankers. What do you suppose might be the effect on the oceans' biota if a single supertanker were to sink or otherwise disgorge its cargo? We might expect to see wholesale destruction of oxygen and food-producing plankton over great areas of the ocean. At the time of this writing a supertanker has run aground and has already pumped a million gallons of its oil cargo into the Straits of Magellan in order to lighten its

14.10 The beautiful beaches of Santa Barbara, California, were blackened with oil from a malfunctioning oil rig. The cleanup effort was monumental and not totally successful. In spite of a massive campaign by citizens to save oil-covered diving birds, almost all died. These are a few of the ducks washed ashore after encountering the massive slick. The pumping is continuing and new wells are planned in the geologically unstable area.

load to free itself. It's still stuck and still pumping. There have also been recent reports indicating that in certain sections of the country, industrial and automotive by-products emptied into the air are altering weather patterns. We know that the quality of air throughout the entire country has deteriorated sharply in recent years (although some people take solace in the fact that there has been a reduction of certain kinds of pollutants in some areas). We tend to sit in stoney complacency as televised reports warn our old people not to go outside, and our children to refrain from running at recess, because our air is dangerous to breathe. Should we continue to befoul and alter our environment with such recklessness, we may, indeed, come to understand more clearly just what density-independent population control really means.

Biotic Control

Biotic population control simply refers to influences on populations that are alive. For example, the organism that causes bubonic plague can reduce populations. So can a tiger. But biotic influences can operate in subtler ways as well. If a territorial bird drives a competitor into an area where there is less food, when winter comes the underfed competitor may be more likely to succumb to the rigors of the season. Thus the territory holder has indirectly brought about the reduction of the population.

Biotic controls on populations are likely to be *density-dependent*. Density-dependence means that as population density rises there will be increasing pressures on it acting to reduce that density. As the density falls,

14.11 In the early 1970s Eastern Ethiopia was plagued by a severe drought. As the drought progressed thousands succumbed in its wake. Here an Ethiopian valley is littered with rockpiles, each marking a grave. The drought and subsequent famine threatened to obliterate entire tribes. We might well ask ourselves, Is control by drought density dependent in any sense? If so, what decisions might we make about our population density?

then, the pressures on it will lessen, thus permitting the numbers to increase again. We can see how this works by considering the effects of predators.

Predation

One of the interesting facts about predation as a population regulator is that a predator normally doesn't eliminate its prey. For example, when the vigorous and intelligent dingo, a predatory relative of the dog, was introduced to Australia, it didn't kill off the primitive prey it found there. However, its hunting prowess proved so superior to that of its *competitors*, the Tasmanian devil and Tasmanian wolf, that those animals disappeared from the Australian continent. But why didn't it wipe out its prey?

Let's examine the classical theory that accounts for the density-dependent effect of predators. According to the theory (Figure 14.12), under undisturbed conditions prey numbers rise steadily, thus providing more food for predators. Then the predator numbers begin to rise. Their numbers do not rise immediately, however, since it takes time for the energy from food to be converted into reproductive efforts. When the predator numbers finally do rise, the increased predator population, of course, places increasing pressure on its prey. Then, as the prey are killed off, the predators find themselves with less food, and so their own members fall off. Because the fluctuation of predator numbers lags behind that of the prey, the prey may be well on the road to recovery before the predator population begins to rise again. In reality, the story is complicated a bit by such factors as prey-switching as the predators seek other, more available food to maintain their numbers.

Our own species is providing us with clear examples of how density-dependent regulation can fail. The great whales have been hunted

14.12 The classical predator-prey fluctuation, sometimes called the Lotka-Volterra oscillation. Notice that, as prey density falls, predator density falls also, but a little later in time. When predator density becomes low, the prey is allowed to recuperate and soon the predator density rises in response to its new food supply. This lag is given as the excuse for recent federal pleadings to allow hunters to resume shooting wolves. They argue that deer herds may never recover from the wolf predation. Look again at the curves. Does the argument make sense?

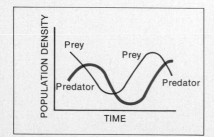

Box 14.1   The Greenpeace Effort

In 1975, at the very time the 15 nations of the International Whaling Commission were winding up their annual week-long "discussions," the Russian whaling fleet was busy in the Pacific running down and harpooning some of the last of the earth's great whales. But this time the hunters were also the hunted. The Canadian Greenpeace Foundation vessel, *Phyllis Cormack,* after 58 days of fruitless searching and woefully short of supplies, had found the Russian fleet. They intended to physically put themselves between the whales and the harpoons and hoped that the Russians would not risk an international incident by shooting anyway. The goal was twofold. First, they hoped to save a few whales. Second, they hoped to save a lot of whales—by focusing attention on the plight of the whales while the IWC was in session.

Their boat was older and slower than those of the Russian fleet, but when the Russians were spotted, they had already killed, floated, and flagged some whales, which meant that they couldn't leave the dead beasts. There were buyers at home and profit is, as always, the name of the game. The crew from Greenpeace scrambled for their small outboard Zodiacs for a closer look at the first whale they found. One of the men climbed aboard the floating carcass and cried, "My God! It's only a baby." And so it was. The infant whale was only 20 feet long—seven small steps. The crewman said the baby's skin "was warm and oily, the blood that flowed from a gaping wound hot on my hand. I . . . closed the eyelid. I felt lonely upon the ocean. . . ." But not for long. One of the Russian chase boats had turned and was bearing down on what they assumed to be poachers. A plea in Russian was to no avail—met only with vulgar retorts. As the Greenpeace crew followed, the Russians turned toward a pod of whales, gathered now, seeking to escape their tormentors. The Zodiacs gave chase, caught the whales first and rode practically on their backs to foil the harpooners. But they fired anyway, a 250-pound harpoon with an exploding warhead sinking deep into an exhausted female, too tired to dive again. The explosion shredded muscle, bone, and guts, but it didn't kill. She dived, but came up at once, thrashing in pain and screaming—literally like a child. The bull turned and rushed past the Zodiac like a great locomotive straight at the man who had fired the shot. The Russians were quite safe and were calmly reloading. As the great animal lunged near the bow, they fired again. The gallant bull made one dive under the hull and came up dead. The Canadians sat in their Zodiac and wept.

The Russians had made a profit. The IWC had heard of the effort. And millions of Americans were going about their business not really aware of how much the earth was changing.

14.13 An afternoon's fun some years ago in Holmes County, Georgia. The local people took the day off to round up foxes. Once the foxes were encircled and unable to escape, beaming parents encouraged their children to beat the foxes to death with clubs. The fox in the foreground sits exhausted and helpless awaiting his end. We might ask, If a species *enjoys* killing, will it operate according to the laws of density dependence? Or will it exert greater effort as its prey becomes rarer? Will it tend to wipe out competitor and prey alike, given the means? It has been argued that there are far more deer and ducks in the United States than ever before now that they have come under game control laws. But is this because of our increasing respect for other species, or because hunters want to increase their opportunities to kill? Many anthropologists argue that humans enjoy killing and that, given our present destructive ability, we must be restrained by a strong system of laws.

to the brink of oblivion over the past few decades as modern whaling methods have reduced personal risk while increasing profits. Although there is nothing that whales provide that cannot be gotten elsewhere, the demand for whale products has not diminished, especially in Japan. Thus, instead of man the predator relaxing his pressure and allowing the whale population to recover, whaling fleets continue to exert their depressing effect on populations of the great mammals. Then, as whales decrease in number, the price of whale products goes up, and the hunt becomes still more avid (Figure 14.14).

Parasitism

The relationship between parasite and host is, in a sense, much more delicate than that between predator and prey. After all, it is to the parasite's advantage to see that it does not kill its host or contribute to its death by weakening it. In many cases, as with tapeworms, the host is actually the habitat of the parasite, and who wants to destroy his habitat? Also, the offspring of the parasite will have greater likelihood of success if the host lives to reproduce.

In some cases the parasite and host are so well adapted to each other that they actually exist quite nicely together. In those cases where the reproduction or vigor of the host is seriously impaired by the parasite, the relationship can be considered to be a relatively new one in which the two species have not yet adapted to each other very well.

In other cases adaptation never takes place because of the disastrous effects of the parasite. For example, in 1904 the sac fungus *Endothia parasitica* was accidentally introduced into North America from China. In

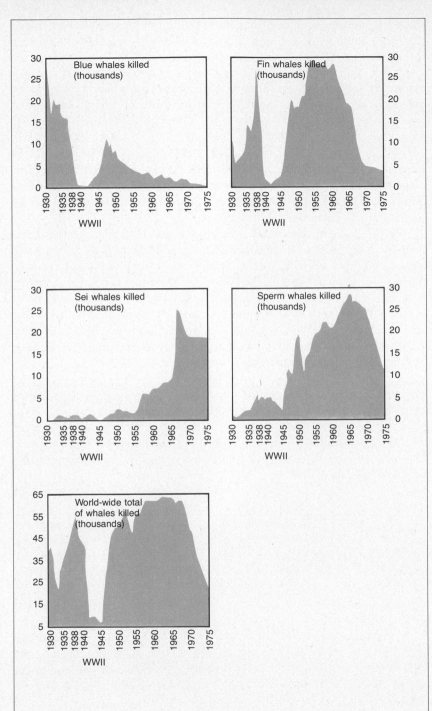

14.14 The numbers of some whales killed annually since 1930. Note that the slaughter stopped only when the civilized whale-hunting countries turned their weapons on each other. The kill began dropping in the 1960s even as boats became faster, and increasingly efficient electronic detection devices were employed. Blue whales have been legally protected since 1965, but ecologists point out that they probably will not recover because there are too few left for individuals to find mates. On a recent popular whale record (an LP you might like to hear), Roger Payne points out that all oceans are now rent with the rumblings of passing ships, making it impossible for the deep-voiced whales (like blues and finbacks) to communicate with each other across the vast ocean basins.

China the parasite was held in check by a variety of control mechanisms that do not operate in North America. Sadly, the majestic American chestnut tree proved defenseless to the fungus, and by the late 1940s the great tree that had dominated our Appalachian forests was virtually extinct. If you missed that drama, you can witness a similar tragedy as, even now, our elms are falling to the newly introduced Dutch elm disease.

Competition

As we have seen, populations may be held in check by what might be called "ecological" factors, such as competition for food. In some species, however, behavioral patterns such as territoriality or the formation of dominance hierarchies provide some members of the population with more access to commodities than others.

In *territorial* species some animals are excluded from certain areas held by other animals. If the habitat is variable, the result is that some animals will end up with "better" areas than others. If those areas hold more food, then in time of food shortage the most successful territory holders are more likely to survive. In some species of birds the male must have a good territory in order to attract females. A splendid male holding an inferior territory will not find a mate. If there aren't enough good territories to go around, then not all individuals will be able to reproduce, and thus the population number is lowered.

14.15 Animals of a wide variety of species show strong territorial behavior. The aggression is usually directed against competitors and it may range from the harmless threats and displays of prairie chickens to fights to the death among lions. Serious fighting in lions is usually directed against an intruder into a pride's territory. In territorial fighting the original territory owner almost always wins.

In species that form *hierarchies,* or "peck orders," certain animals have freer access to commodities than others. In such a system, each animal must be able to recognize others individually and to respond to them according to their rank. Hierarchies are established in various ways for different species, such as by fighting or in the play that takes place as individuals grow up together. Hierarchies reduce conflict within groups by each animal acquiescing when confronted by a dominant individual. In times of shortage of critical commodities, then, the higher-ranking animals are more likely to survive as subordinates give way before them.

In both territorial and hierarchical systems (which are not, by the way, mutually exclusive), when certain animals have freer access to commodities than other animals, harsh conditions may quickly bring about the deaths of the "have-nots," thus increasing the likelihood of survival for the "haves." This doesn't mean that the have-nots accept their role for the "good of the species." Quite the contrary. It is to their advantage to remain in the breeding population at all costs. However, a low-ranking individual may have better luck if he searches for unclaimed commodities than if he tries to challenge a dominant individual.

Disease

Finally, disease can reduce populations. We will see in this chapter how the bubonic plague drastically reduced the human population at one time. Less dramatic diseases claim thousands of lives daily. The density-dependency of disease as a population regulator is obvious. The more closely individuals are packed together, the more opportunity there is for the transmission of certain diseases. Also, the more individuals there are in a population, the greater the number of potential reservoirs in which mutant strains of disease microorganisms can develop.

In some cases disease and predation may interact to reduce populations. For example, a two-week-old caribou fawn can outsprint a full-grown timber wolf. However, caribou are subject to a hoof disease that lames them before it infests other parts of the body, and it is these lamed animals that the wolf is likely to cull out of the herd. In areas where wolves were poisoned to "protect" the migrating caribou, this hoof disease spread unchecked and in a few seasons has decimated entire caribou herds.

DEATH

Have you ever wondered why death occurs at all? Why hasn't natural selection resulted in organisms that simply live forever? If "Life" is better than "Death," then why is there no marked selection for longevity? Why are there delicate insects that don't even have mouths with which to feed themselves, but must instead spend whatever energy their frail bodies possess in finding mates before their few precious hours of life are gone? Why does man have so much trouble surviving past his "threescore years and ten"? Why death?

People with all sorts of perspectives on life have ruminated on the "meaning" of death. Ideas have been fiddled with by philosophers and theologians, as well as scores of novelists, dramatists, poets, drunks, and others of bad habits. Even aboriginal societies dealt with the meaning of death in one way or another—some very casually, not fearing it at all, treating it as if it were simply the natural extension of life. So let's join the fray and consider death from a biological point of view.

We should begin by noting two things: first, the world changes and, second, it is limited. Not only have continents moved, islands formed, ice ages come and gone, and weather patterns shifted, but man has, in a very final sense, drastically altered the earth's surface in the past few thousand years. In a changing world the organisms that live there must also change. This is what adaptation is all about, adjusting to new situations.

As a population is confronted by changing conditions certain individuals can be expected to cope with them better than others. But the changes may well be too drastic for any individual, with his inherent limits, to be able to adjust. However, as new generations appear, their myriad genetic variations may provide material on which the new environment can have its molding effect through natural selection. Many in the new generations will fall by the wayside, their kinds of genes being separated out of the population. Individuals with other kinds and combinations of genes, though, may fit the new environment well. And these organisms will live and reproduce. The point is, the arrival of new and variable individuals is important in a changing world.

Now, in any system in which the commodities are limited, such as the planet earth, as the numbers of individuals increase, there will be competition for the available commodities, such as food, shelter, land, and natural resources. So consider an organism that finishes its period of reproduction and then hangs around to compete with its offspring. Such an individual could be expected to leave fewer offspring than a parent that, having reproduced, died, so as not to interfere with its offspring's success.

David Blest has described certain moths with camouflaged markings in which the female, after laying her eggs, commits suicide. If she were not to commit suicide and stay in the area where her offspring are hatching, she might eventually be caught by a bird. That bird will then have a better idea of what the moths around there look like and will likely search for more of them—including the female's offspring. So when the female has laid her eggs, she flies around frantically, using up her energy reserves and fraying her wings beyond recognition. Then she either dies of her own efforts, or, because of her conspicuous behavior, she is captured by a bird, which is then unable to find another tattered morsel like that one.

This explanation of the moth's behavior is entirely speculative, of course, but the consideration of death as the ultimate form of altruism provides food for thought. Also, keep in mind that the moth may have no idea about the effects of her behavior. She doesn't "try" to save her young through self-sacrifice. Those moths that behave in such a way, however, leave more offspring, each of which, perhaps, carries a "suicide gene."

If new generations help to ensure success in a changing world, and a parent that remains after reproducing may be a detriment to its offspring, then the phenomenon of death can theoretically be associated with the end of the reproductive period. Why, then, isn't there selection for long lives with extended reproductive periods in all species? Since we're dealing in pure conjecture here, your own answer might be as good as anybody's. You might want to consider such matters as the limits of variability in the offspring of a single parent. Which would you expect to be more different from a parent, its children or its grandchildren? In natural selection is such difference good or bad?

When we consider death as a mechanism that helps to ensure perpetuation of a particular set of genes, we might be left with the rather irreverent idea that any organism is simply a vehicle utilized by its chromosomes to enable them to reproduce themselves. Most of us, however, prefer to feel that we have other raisons d'être than simply as carrying cases for our gonads.

We've tossed around some heavy concepts fairly loosely here, as you are aware. However, these are legitimate questions for biologists. Perhaps we are not in a position to uncover whole truths but we can begin to accumulate certain kinds of evidence. Anyhow, this sort of approach makes as much sense as dealing with the concept in subjective, artistic, abstract, or mystical terms.

HUMAN POPULATIONS

Now let's consider some points that relate specifically to the human population of the earth, something of the history of our species, and what our future looks like from here.

Early Man

Somewhere between 1 and 2 million years ago one might have noticed among the different kinds of living things on the earth a small population of primates that differed somewhat, but not much, from other animals. Strangely enough, the appearance of these animals, ambling in little bands over the vast African plains, was one of the most momentous events in the history of the planet. Who could have guessed that virtually every other kind of life, as well as great mountains, endless oceans, vast plains, immense forests, and even the air itself were to give way, change their form, before something as seemingly insignificant as that hunched figure sitting in the shade, picking at the soft parts of an insect.

We have no idea how many individuals comprised the human species in those early times. In fact all the evidence before about 1650 B.C. is entirely conjectural. We know, for example, that agriculture was almost unknown before about 8000 B.C., or possibly a bit earlier in Malaysia and the Middle East. Before that time humans must have lived by hunting and gathering their food. Considering the inefficiency of such methods, and given a land area of about 58 million square miles, it is estimated that the

earth could have supported no more than 5 million people in 8000 B.C.

It is important to realize that, of all the people who have ever lived, 90 percent were hunter-gatherers, only 6 percent subsisted by agriculture, and just a very small percentage lived as industrial men and women. But it is this latter, relatively small group, that has played by far the greatest role in bringing us to our present ecological predicament. Another way to place this group in its proper perspective is to remind ourselves that about 99 percent of man's time on earth has been spent in pursuit of the other species, hunting and gathering their flesh. Obviously, then, our spirits, bodies, and behavior must be at least partly geared for a different kind of life than we have now. And we do have a new and different kind of life now. In fact, it has fallen upon us so suddenly that within the last 1 percent of the total time we have spent on the planet, we have shifted so drastically that now only 1 percent of our species lives by hunting and gathering.

The hunter-gatherer, of course, has relatively little impact on the environment. A variety of plants and animals are available to such people, and as one falls into short supply, they will be forced to move or to utilize other materials, thus relieving the pressure on that species. Most of the food taken by hunter-gatherers, by the way, is by gathering. In studies of modern groups, the men may bring home the larger, swifter, or more dangerous protein-laden food, but the women, quietly gathering plant material, contribute the far greater part of the diet.

Later hunter-gatherers proved to be much more efficient than their forebearers, and more of a real threat to other species. Weapons were improved, bushes and trees could be cut, and men began to hunt together to bring down larger game. Language would have had to become rather sophisticated in order to maximize the results of group hunting ("Tomorrow, half of us will hunt here.") It was probably not a great step from someone saying, "You get his attention and I'll hit him with a stick," to the formation of rudimentary leadership and thence to political maneuvering. (Anthropologist Lionel Tiger has suggested that this is a reason for the overwhelmingly masculine domination of politics.)

All of these refinements in organization would have made man a more deadly hunter and, in fact, we find early signs of his awesome ability. After the ice ages ended about 10,000 years ago, about 70 percent of the North American mammals went extinct. There is a suspicious correlation in their decline and the spread of humans over their ranges. Many of their bodies have been unearthed with human weapons buried nearby. As man grew more cunning, he didn't always rely on weapons. In one place near Solutré, France, the bodies of over 100,000 horses have been found, apparently having been stampeded over a cliff. Then, as man began to use fire, he could have altered areas more to his liking; and some researchers believe that many of the world's great grasslands were created by being repeatedly burned over by early man. The point of all this is that humans began very early to make sweeping changes in their environment, apparently limited only by their primitive technology.

14.16 A scene in the day of early man. Human history is almost entirely the story of man the hunter, and only relatively small populations can be supported in this way. Agriculture has been important only in about the last one percent of our time on earth. Today the most efficient hunting by primitive groups involves groups of men working together to bring down large game, and perhaps it was the same with early man. Since nursing mothers would have been forced to stay with the children, division of labor probably occurred early. It has been suggested that male-dominated politics grew out of the hunting group in which leadership would need to be determined in order to more efficiently plan the capture of large and potentially dangerous game. Relatedness may have been high in such small groups and so altruistic behavior could have arisen as a means of perpetuating one's own kind of genes by caring for relatives. The earth was sparsely inhabited by humans in those days, bands probably rarely exceeding 200 people.

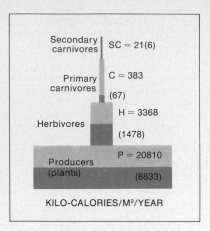

Secondary carnivores SC = 21(6)

Primary carnivores C = 383 (67)

Herbivores H = 3368 (1478)

Producers (plants) P = 20810 (8833)

KILO-CALORIES/M²/YEAR

14.17 An energy pyramid calculated for Silver Springs, Florida. The upper numbers for each tier designate the stored energy (expressed as kilo-calories) for each square meter investigated over a year. Not all the stored energy at any level is available to the level above it, so the kilo-calories that are available as food are shown in parentheses and as the shaded areas of each tier. Also, each tier is much smaller than the one below it because it is not completely efficient at recovering all the energy that *is* available. P = producers (plants); H = herbivores; C = primary carnivores (animals that eat herbivores); SC = secondary carnivores (animals that eat carnivores).

It is believed that these early human populations were separate, but interbreeding, groups, possibly divided into yet smaller bands. From studies of present-day hunters and gatherers, it has been calculated that the average size of such groups would have been about 500 men, women, and children, of whom about 200 were of breeding age. Population size was probably relatively stable and density-dependent. (Keeping in mind the food sources, can you tell why?) It is also generally believed that the number and size of the groups approached the carrying capacity of the environment.

At about 8000 B.C., however, that carrying capacity became elevated by the advent of agriculture. Men began to cease roaming and began to grow and store their food and thus they added a relatively stable element to their food supply. The lean seasons no longer marked the deaths of so many children and old people, since food could be stored. The result of the shift toward agriculture was a slow steady increase in human populations.

How could the shift to agriculture have raised the environment's carrying capacity? We have already mentioned that the ability to store such products as grain meant a more dependable food supply and a greater likelihood of surviving the harsh seasons. Something else happened as well, something which may come to be increasingly important in coming years. Agriculture permitted man to shift more toward the level of a *primary consumer* or herbivore. Let's see what this means.

The Energy Pyramid

The concept of consumer levels may be understood when we consider the relative energy efficiencies of taking different kinds of food. Let's compare the relative efficiencies of eating plant and animal food. Some animals eat plants, but it should be apparent that there is no way that all the energy stored in plant tissues can be recovered by those animals. Furthermore, animals that eat the plant eaters can't recover all the energy stored in the bodies of those herbivores. Thus much of the food stored in living things is wasted. But the wastage varies depending on how far one's food supply is removed from the level of the primary producer, the plant.

The principles have been demonstrated in what is called an *energy pyramid* (Figure 14.17). Note that of the great amount of energy stored in the tissues of the plants, less than half (8833/20810) is available to animals. The reason is twofold, as we have mentioned. First, plants have their own things to do; they aren't in the business of supporting animals. Thus, much of the food they make in photosynthesis is used to produce energy to enable them to do the things that plants do. Only a part of it is turned into "plant." Also, a herbivore uses much of the energy it retrieves from plants for its own activities. This means that only a portion of the energy it gets from plants is stored in its own tissues and thus made available to the carnivores that eat it. So, much of the energy stored by organisms at any food level is used in metabolic activities by those organisms, or is lost as heat or in other ways. It is simply not made available to the next consumer level.

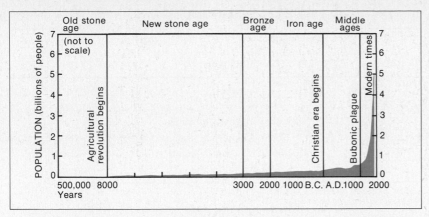

14.18 Human population growth during the last half million years. If the old stone age were drawn to scale, it would extend about eighteen feet to the left. Notice the very gradual rise as agriculture began about eight thousand years ago. Even one thousand years ago the earth's population was enormous by historical standards. But the events since the last bubonic plague and particularly in the last few hundred years could not have been predicted. Someone has rather whimsically calculated that, at our current rate of increase, the end of the next thousand years will see human bodies completely filling the known universe and expanding outward in all directions at the speed of light! That probably won't happen. Where would they sit? Of course, we can depend on natural means to bring our population under control long before then. And the thought is horrifying.

Second, animals are not 100 percent efficient at recovering energy stored in the things they eat, so much of the energy that *is* available to them is not utilized. The result of all of this is apparent when you notice in Figure 14.17 how many producers are necessary to maintain even a few herbivores and even fewer carnivores. A massive amount of food energy must be produced by plants to maintain only a few secondary carnivores.

With this in mind, then, would you say it is more efficient for humans to eat the fish that ate the fish that ate the fish that ate the algae, *or* is it more efficient to eat the algae? Should we feed grain to beef cattle or should *we* eat the grain? The diet of North Americans places *four times* the demand on agricultural resources than do the poorer diets of the less fortunate. The reason is because of our interest in eating from higher levels of the food pyramid. With such wastage involved, can you see why steaks are so expensive (politics and market manipulations aside)? Nonetheless, in spite of costs, each American eats about 200 pounds of meat each year.

Back to our earlier point, that beginning about 10,000 years ago man began to raise the carrying capacity of his environment through agriculture by insuring a more stable food supply, and by increasing his reliance on lower food levels. Man was now able to make more efficient use of plants and thus to support a larger population. All this is not to say that the agricultural revolution has been the only factor that has changed the carrying capacity of our environment. The advent of tool making also had its marked effect, and we will learn shortly of the phenomenal impact of the recent scientific-industrial revolution.

As you can see from Figure 14.18, the human population continued its slow rise through the beginning of the Christian era. It is believed that at the time of Christ, the earth's population was between 200 and 300 million and that it increased to about 500 million by about 1650, with only one period when it dropped significantly, because of the impact of the bubonic plagues.

Growth Rate

It is interesting, and tremendously significant, that not only has our population steadily increased, with few irregularities, in the last 10,000

Table 14.5 Doubling Times of the Human Population

Date	Estimated World Population	Time Required for Population to Double
8000 B.C.	5 million	1,500 years
1650 A.D.	500 million	200 years
1850 A.D.	1,000 million (1 billion)	80 years
1930 A.D.	2,000 million (2 billion)	45 years
1975 A.D.	4,000 million (4 billion)	35 years[a]
2010 A.D.	8,000 million (8 billion)[b]	?

[a]Computed doubling time around 1972.
[b]Projection of United Nations estimates.
Source: From *Population, Resources, Environment: Issues in Human Ecology,* 2d ed., by Paul R. Ehrlich and Anne H. Ehrlich, W. H. Freeman and Company. Copyright © 1972.

years, but so has our *rate* of increase. Growth rate increase can perhaps best be shown by considering "doubling times"—the time necessary for a population to double in size (Table 14.5). To go from 5 million people on earth in 8000 B.C. (the present number in only three of New York City's five boroughs) to 500 million in 1650 A.D. took between six and seven doublings over a period of 9,000 to 10,000 years. So, during that time, the human population doubled on an average of about every 1,500 years.

A glance at the right-hand column of Table 14.5 will show that, all other things being equal, in only thirty-five years we will need two cars, two schools, two roads, two houses, and two cities throughout the world for every one that presently exists. And that will only maintain our status quo as far as material goods are concerned. The problem is that not all things can remain equal over that time. Can India or Guatemala double its food? Can we double the population of the Eastern Seaboard without the inner cities falling apart? Can California stand another Los Angeles? Where will the water come from? If we double the number of our oil wells, can we double the amount of the earth's oil?

And keep in mind that if we *should* manage to withstand environmental resistance and double the earth's population in thirty-five years, it will be precious few years after that before the population would double again! And then again and again!

Those with a great faith in technology have suggested that the answer lies in increasing the carrying capacity of the earth through agricultural, technological, and medical "revolutions." They maintain that it is possible to accommodate billions more people. We are an aggressive and innovative species and there is no doubt that we *can* successfully withstand population increases, for a time. The question is, do we want to? How old will you be when the earth has 7 billion people? How old will your children be when there are 14 billion people on earth at the current rate of population increase? How hard is it to get away by yourself right now with the earth's population at only 3.75 billion? (The world's population at the end of 1976 reached 4 billion. *Never* has the earth harbored so many

Box 14.2 Stopping Population Growth—The Mathematical Dilemma

Let's consider the mathematical problems in slowing the population growth of an underdeveloped country, say, Mexico. Suppose it has a birth rate of 42 per thousand and a death rate of 7 per thousand. The growth rate is the difference between birth and death rates, or 35 per thousand (3.5 percent—doubling in only twenty years). How can it put a stop to this suicidal growth?

The only immediate solution (short of killing people already born) is to lower the birth rate so that it equals the death rate. But (and this may surprise you) the death rate is *very* low. Remember, Mexico is a rapidly growing country where there are very few old people; half the population is under 20. Obviously, then, the birth rate must also be very low, lower even than for a stable, developed country, at least for a while. Mexico would need to cut its birth rate to ¼ its current level. Can any country suddenly reduce its birth rate by a factor of four? It would seem unlikely, at least without measures they would certainly find repugnant, such as compulsory sterilization or mandatory abortion for mothers of *one or more* children.

But, you might say, perhaps the brakes can be gradually applied. The social results would therefore be less traumatic. Suppose we simply calculate the number of births this year and let that be the number allowed in any future year. Since the number of births tends to increase each year, holding the number stable would have the effect of increasing the control. Eventually, Mexico would shift from the age structure of a growing population to that of a stable one. Very neat! Only . . . in actuality, the population will be growing all the while, and it will in fact *triple* before stability is reached at birth and death rates of about 14 per thousand! Can a country now near the limit of its resources tolerate that kind of population increase?

—STEVEN VOGEL

Birth and Death Rates in Selected Nations[a]

	Birth Rate	Death Rate	Growth Rate
Africa			
Mali	50.1	25.9	2.4
Kenya	48.7	16.0	3.3
Zambia	51.5	20.3	3.1
Nigeria	49.3	22.7	2.7
S. Africa	42.9	15.5	2.7
Egypt	35.5	12.4	2.3
North America			
Mexico	41.9	7.2	3.5
U.S.	14.7	9.0	0.6
Canada	15.4	7.4	0.8
Jamaica	29.8	7.2	2.3
Dominican Republic	45.8	11.0	3.5
Costa Rica	29.5	5.0	2.5
South America			
Colombia	40.6	8.8	3.2
Brazil	37.1	8.8	2.8
Argentina	22.7	9.4	1.3
Venezuela	36.1	7.0	2.9
Asia			
Bangladesh	49.5	28.1	2.1
China	26.9	10.3	1.7
India	34.6	15.5	1.9
Japan	17.2	6.4	1.1
Indonesia	42.9	16.9	2.6
Europe			
W. Germany	9.7	12.1	−0.2
Italy	14.8	9.9	+0.5
Sweden	12.6	10.8	0.2
Spain	18.3	8.2	1.0

[a]Birth and death rates are crude, that is, unadjusted for age, season, etc. In some cases the figures are estimates.
Source: Population and Vital Statistics Report, U.N. Statistic Papers Series A, Vol. 28, No. 4.

Some crude birth and death rates. The death rates of some underdeveloped nations are low because they have so many young people. For example, half the people in Mexico are under 20 years old. To calculate the doubling times divide the growth rate into 70. It was once believed that in no human population could birth rates exceed 50 per 1,000, but some African nations have shown we were wrong.

Table 14.6   World Population

	1975	1976	Growth Rate (percent)
World Total (in millions)	3,968	4,045	1.94
Africa	401	412	2.74
North America	343	348	1.5
South America	219	224	2.3
Asia	2,256	2,306	2.2
Europe	473	476	0.6
Oceania	21	22	—
USSR	255	258	1.2

Source: Population and Vital Statistics Report, U.N. Statistic Papers Series A, Vol. 28, No. 4.

humans!) In the past 300 years the human population has increased an average of 10 million per year. But the present rate of increase is 80 million per year! How many is that each day? Each hour?

Population Changes Since the 1700s

Somewhere around the seventeenth century the world's human population suddenly began to shoot upward. Such an abrupt rise has heralded population crashes in other species, and until we have some sort of evidence that humans are immune from natural laws, we might want to carefully examine the factors behind our skyrocketing numbers.

The sharp rise in the population of Europe between about 1650 and 1750 is attributed to innovations in agriculture and the beginning of the exploitation of the Western Hemisphere. The accompanying rise in Asian populations is harder to explain. India, for example, was in turmoil following the demise of the Mogul Empire. In China, after the fall of the Ming Dynasty in 1644, the Manchu emperors brought political stability and new agricultural policies, which may account for the increase in the Chinese population.

Not much is known about Africa in this period, but its estimated population was about 100 million in 1850 when European medicine and technology were introduced. In the next hundred years, then, its population doubled. Between 1750 and 1850 the populations of Asia increased only about 50 percent, but in Europe the numbers of people swelled much more rapidly. The factors permitting such growth included newer agricultural techniques, better sanitation, and advances in medicine, such as the smallpox vaccine.

In the next hundred years the world population grew from over 1 billion in 1850 to about 2.5 billion by 1950. Since the Korean War we have added another 1.25 billion to our numbers. Between 1850 and 1950 the populations of Asia, Africa, and Europe approximately doubled, the population of Latin America increased about five times, and that of North America increased sixfold.

Box 14.3 Birth Rates and Death Rates

Birth rate may be expressed as the number of babies born per thousand people in the population per year. At the year's end the total number of births is divided by the estimated population halfway through that year. The death rate is calculated in a similar fashion. For example, if 200 babies were born to a population of 10,000 people, the birth rate would be 20 per thousand. If 180 people died in the same period, the death rate would be 18 per thousand. The rate of increase for that period, then, would be 20 per thousand per year. This is an annual growth rate of 2 percent.

The world's growth rate in 1972 was 2 percent. Thus the population will double in 35 years if the rate remains constant. The chart below describes doubling times as related to annual percent of population increases.

Annual percent increase	Doubling time (years)
0.5	139
0.8	87
1.0	70
2.0	35
3.0	23
4.0	17

The growth rate in the United States presently approaches 1 percent, but in several Latin American countries, the 3 percent figure is reached.

After 1850

Between 1850 and 1900 the death rate continued to decline in the face of advances in industry, agriculture, and medicine. The lot of the European previous to this time had been incredibly bad. The cities were garbage-filled and overrun with rats. The stench was devastating. Life in the country was not much better, with the scattered hamlets being little more than isolated slums inhabited by people of numbing ignorance. But, by the beginning of the twentieth century, the situation of these people improved as the likelihood of crop failure decreased. In addition, transportation advances meant that food could be imported to locally famished areas. Also, about this time the role of bacteria in infection became known, and so more lives were saved. Death rates in Europe dropped from about 24 to 20 per 1,000 in those fifty years.

At about the turn of the century a new and highly significant trend appeared in the world's populations. It seemed that, for some reason, industrialization was followed by a decreased birth rate. The phenomenon, which has been highly speculated upon, has never been entirely explained, but there are some good guesses at why it happened. First of all, in agricultural societies, children are a source of labor and a social security measure. They provide assistance before they leave home and they care for their aged parents later. In an industrial society, however, they are less likely to aid in production, but they maintain their status as "more mouths to feed." Also, mobility is important in industrial societies and mobility is reduced by the presence of children. Given these reasons, it is interesting that, in the early twentieth century, as birth rates dropped off in the industrial cities they did the same in rural areas. One reason was the mechanization of farming that reduced the need for labor. So, it seems that industrialization tends to depress both death rate *and* birth rate.

About the time of World War II many underdeveloped countries experienced a dramatic reduction in death rates. The cause was apparently the rapid import of public health programs, including new drugs, from the more developed countries. (Among the most important changes was the control of malaria through the widespread use of DDT.) Such importation of "death control" produced the most sweeping changes in human population dynamics in the history of the species. It is important to realize that the changes in death rate in these underdeveloped countries were not accompanied by the social and institutional changes that marked the slower reduction in death rate in the more developed countries. The changes were much more rapid and were brought about from the outside.

AGE PROFILES

In the consideration of how any population changes, it is important to consider the ages of the people in the group. Figure 14.19 shows the "age profiles" of two kinds of populations. Mauritius, a small island in the Indian Ocean, is typical of many rapidly growing countries with high birth rates and declining death rates. The United Kingdom, on the other hand, is

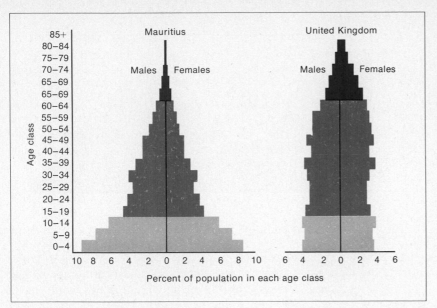

representative of older developed countries that have had low birth and death rates for many generations.

The profile of Mauritius is produced by a high birth rate, combined with a sudden recent decline in the deaths of infants and children. There hasn't been enough time for those who have benefited by imported "death control" to have moved into the older age groups. Furthermore, the childen of these young people will further inflate the youngest end of the age spectrum. However, Mauritius is an island, just as the earth is, and this kind of growth can't go on forever. A program of population control will have to be initiated, or environmental resistance in the form of famine or some other check will eventually alter the shape of the age pyramid. What would the age pyramid look like for a rapidly developing country ten years after it had suddenly instigated strict birth-control programs? Judging from the age profile of the United Kingdom, at about what age would you say that death-control programs begin to lose their effectiveness?

FERTILITY RATE

It has been pointed out by demographers, the people who study human populations, that *fertility rate* is a better predictor of the behavior of populations than birth and death rates. Fertility rate is calculated as the number of births per thousand among women of child-bearing age, established for statistical purposes as fifteen to forty-four.

The importance of considering fertility rate was apparent in the erroneous interpretation of the widely heralded drop in birth rate in the United States during the 1960s. The news media pounced on this phenomenon as indicating the end, at last, of the population explosion in this country. But, a closer look at the data showed that although couples

14.20 Fertility rates in the United States between the years 1910 to 1974. At what point were you born? How does this affect your job prospects? How does it affect your prospects if you intend to be a teacher? A manufacturer? In recent years the United States has continued its decline in fertility rate. Does this mean we will one day have a country full of old people facing a world filled with hungry young people? Or will other countries begin to follow our lead? Keep in mind that we are only about six percent of the world's population. But also remember that we use about half its resources.

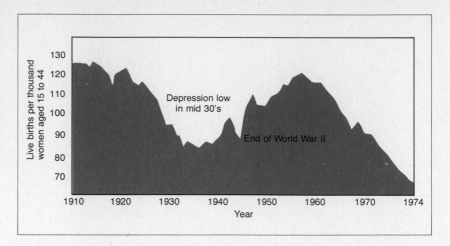

were having fewer children, a more important reason for the drop was that at that time there were simply fewer women of child-bearing age in the population. A glance at the age profile of the United States would have shown that there was a large number of girls under fifteen in the population, just waiting to move into the reproductive brackets.

What has been the course of the fertility rate in the United States? Figure 14.20 shows that it dropped between 1959 and 1968. The fertility rate declined to about 85 per thousand, but still failed to reach the low of the depression years. Then in 1970 it rose to 87.6, fell in 1971 to 82.3 and in 1972 it dropped to below 80. In early 1977 it was announced that birth rates in the United States had decreased to the point that couples were having only an average of 1.8 children, well below the 2.2 "replacement" figure. Because of our age profile though, our population can't be expected to actually decrease for some time.

But does this ultimately mean the end of the population explosion in the United States? The experts are unanimous! They don't know. Our population has historically followed erratic routes. The optimists like to point to such factors as an enlightened public, the extension of family planning services to more people, the trend toward male and female voluntary sterilization, liberalization of abortion laws, and the impact of the feminist movement. On the other hand, pessimists are quick to point out that high unemployment and a tight job market for prospective fathers has traditionally depressed population growth, and that if the economy should regain it strength, we will be on our merry way once again.

THE OTHER EFFECTS OF ZPG

We will know by the early 1980s if the current depression in American birth rate is going to be prolonged enough to change our age profile. The reason is, the great mass of women who were born during the great baby boom, from 1947 to 1957, are moving into their 30s, an age period when, if they're going to have children, they'd better get on with it. But many

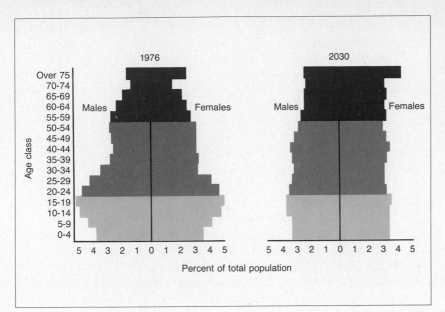

14.21 Changing profiles. If the trend in population growth continues, the population structure in the United States will change drastically in the next fifty years. Notice that the younger age groups will decline and that the older will increase. The graph seems simple enough, but what are the social implications of such changes?

women just aren't as reproductively directed as they were in 1957 when they were having 3.8 children. Should the birth depression hold then, the average age of an American citizen will rise (Figures 14.21 and 14.22). As the products of the last baby boom move along the age pyramid, it gives one the impression of a goat passing through a python, swelling each age rank as it goes.

The next question is, what would an "old America" be like? We know the effects of a "young America." Forty three million children were born between 1947 and 1957, a fifth of our present population. First they flooded the schools (in the 1950s, California opened a school a week, on the average), then the job markets. But no sooner were the schools built than they were felled by the harbingers of the "baby bust."

But what about the other end of the age spectrum? The old people weren't just quietly passing on. In fact, they were living longer than ever before. They were given less public attention and so no one noticed that three million more people have been added to the over-65 group since 1970, pushing the number to 23 million. And by the end of the century there will be 31 million people over 65. Thirty years after that the number will be 52 million. This, by the way, will happen before we reach Zero Population Growth (ZPG) which is, at our current rate, scheduled for about 2040. At ZPG all the age groups will have about the same number of people. As the age pyramid changes to an age rectangle, the youth culture of today will recede into the misty realms of ancient history.

But people tend to sicken as they grow older. So what will be the consequences on medical policy? Will they be cared for through a socialized system of medicare? Will they be left to their own devices? In

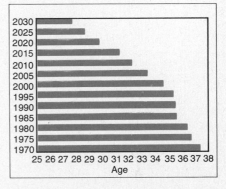

14.22 The projected median age of the population in the United States. As birth rates drop, the median age of the population is expected to rise until in about fifty years the average American, now under thirty, will be pushing forty.

1975 medical costs for the over-65 group averaged $1,360, twice the bill for those under that age.

What about pension costs? At General Motors, for example, pension payments have doubled in the last 10 years and will double again in another 10 to 15 years. In 1967 this company had ten workers on the payroll for each pensioner. Now the ratio is 4 to 1 and in the 1990s it will be about 1 to 1. Such a "dependency ratio," of course, places tremendous burdens on future employees. As another example of what the age shift can mean, in 1945 there were 35 workers for each person on social security. Today the ratio is 3.2 to 1 and by 2035 it will be less than 2 to 1. Since the system pays out more than it withholds, the present reserves will be exhausted in four years.

The result of increasing the numbers of old people, of course, is to place a greater burden on the young. There is some fear among economists that the increasing resentment of the young will result in a simple failure to support the elderly. One partial remedy, of course, is to allow oldsters to continue to work and carry their own weight—an appealing prospect to many of them. And then we might learn to stop regarding them as a different species.

The point of all this is to illustrate once again that at this delicate point in the history of our species, almost all our options contain clauses and force risks. We've heard it a million times, but maybe its true anyway, that there are no simple answers. Our task is to learn to immediately question the ramifications of the decisions we have left, to think globally, to extend our assumptions past the obvious, to arm our imaginations with hard fact. We may find that there are no obvious routes, no "right" ways, but at least we will begin to understand the risks. And some are preferable to others.

WAR

Intergroup Aggressiveness and Commodities

At this point we're going to consider a form of human interaction that has had important effects in reducing small populations but that has had less effect on the overall population of the world. The topic here is *war*. Because of its complexities we can only discuss it in a limited and rather superficial way, and perhaps there are reasons why it could better be fitted in somewhere else. However, it seems that war must be considered, at least in part, as a *population* phenomenon. It is certainly one kind of population *interaction*. In essence, what we will suggest is that war may be one expression of a long history of competition between different groups as each seeks to advance its particular types of genes over those of other groups. In other words, we will consider war as a biological phenomenon. Of course, and by *all* means, war is not simply that. We are much too complex and involuted creatures for any of our behavioral traits to be considered a result of "simply" anything. Perhaps, however, competition for commodities is an *element* of the roots of war. In any case, let's see if we can develop a few ideas along this line. Hopefully, the material will at least serve as a springboard for discussions.

14.23 Warfare is despised by almost all developed countries. Many idealistic young men have been greatly changed, sometimes in morbid ways, by the experience of war. Yet, for some reason, it is almost always viewed as a real possibility and sometimes a viable alternative. The primitive groups may engage in war much more casually, with far less moralizing and rationalization. Perhaps the primitive groups understand better what war really is. Perhaps it is a rather simple way of gaining or ensuring commodities. For them, it may also be an effective way to take revenge since tribes are so small that an attack against a group is likely to personally affect the individual who is the real target. When civilized man attempts such revenge, the result may be of the sort that leveled beautiful and nonindustrial German cities in World War II or the retaliation of Wehrmacht troops against Italian villages. The soldiers at My Lai may have felt avenged after the shooting, but, again, the women and children were not the real targets. In essence, then, the grand idea of warfare may take the form of countless personal tragedies if the belligerent countries are populous and have vast killing power. The embattled platoon may be unable to relate to problems of the distribution of commodities that brought them to the battleground in the first place. But among primitive groups the reason for fighting may be as clear to the youngest warrior as to the chief.

War's Effect on Numbers

A quick look at a world population curve will show that there has been no great depression in our numbers during the period of any war. In fact, it has been calculated that the number of *all* people killed as a result of all recorded wars is presently made up in only a few days population growth. After almost 20 years of war in Vietnam, with millions of casualties, the populations of North and South Vietnam have increased by two million each. Smaller populations, on the other hand, have been virtually decimated by warfare. We have been provided recent examples of such tragedies in Nigeria and Bangladesh.

The Evolutionary Implications of War

Let's consider our topic on an evolutionary basis. We are sure to fall into argument with some sociologists, political scientists, anthropologists, and probably some biologists. But that's all right.

In order to develop our case, we must be allowed three premises, ones that were established in previous discussions. First, we must recall that in a system of limited commodities that occasionally fall into short supply, there may be strong competition between the ones who rely on the commodities.

Second, within groups in which the members are likely to be related, there may be reproductive advantages in cooperative behavior. A helpful and cooperative individual is likely to be advancing his own kinds of genes by aiding relatives. Thus "in-group" members are likely to receive favorable treatment from each other.

But, third, those members of the species that are designated "out-group" individuals (not members of the reproductive group, perhaps of another group) are likely to be strong competitors and unrelated. So aggressiveness might be expected to be primarily directed toward such individuals.

Now, let's consider what might have been early man's situation. You will recall that the carrying capacity of the environment for hunters and gatherers is rather low at all times and that environmental resistance can be expected to increase seasonally. In the hunting and gathering groups, then, there would have been times, such as in winter, when there were severe shortages of commodities. At such times, then, an individual might be likely to share food with his immediate family (think of the reproductive advantages), but less likely to share with more distant relatives. Not only would he not be likely to share hard-won food with out-group members, but he could be expected to fall into strong competition with them for whatever food was available. Such strong competition, of course, could be expected to result in outright aggression. Even when the carrying capacity of the environment was raised by agriculture, after the population had increased to its new limits, competition for land or water could be expected to increase accordingly.

So, in such a system of limited commodities, what sort of individual

Box 14.4   Out-Group Aggression

The principle of identifying individuals on the basis of their group or subgroup status and then behaving toward them according to their affiliation is well represented throughout social animals. Hyenas live in *clans,* or groups of families, and rarely will they permit a member of another clan to enter their group or share their kills. Wolves may attack any strange wolf that enters their area. Jackals drive away other jackals that they do not recognize. Lions, amiable enough within their own pride, may kill a stranger that they encounter within their hunting area. Rats and mice are notorious killers of strangers.

Out-group hostility is not universally demonstrated among animals, as would be expected, since we're dealing with a diverse and complex kingdom. For example, chimpanzees are highly social and they do not usually greet strangers with hostility. (They also eat a variety of abundant food types and have no permanent or semipermanent nest sites. Why might this be a factor?) The general point can be made, however, that there must be a great advantage to out-group hostility in certain cases as revealed by its prevalence among social animals.

would be most likely to leave progeny carrying his genes—a friendly, cooperative individual who was willing to share with everyone, or an aggressive, competitive person who took for himself and his immediate family all he could? What do you suppose, then, would be the disposition of most of the individuals in the next generation?

The aggressive spirit, of course, would be tempered by the need to get along with in-group members. Aggressiveness could also be reduced by behavioral mechanisms, such as hierarchical systems that would effectively designate rank and thus reduce in-group strife. Among humans, aggression might also be reduced through such cultural phenomena as laws or religion.

If any of this is correct, then, it seems that there would be a tendency to identify individuals according to their group affiliation (or potential relatedness). Within groups, such as the breeding populations of 500 or so individuals that roamed the earth about 10,000 years ago, there would be subgroups, such as families, in which relatedness was higher than in the group as a whole. Aggressiveness within subgroups could be expected to be lower than between subgroups, with the highest aggression existing, potentially, between different groups. The principle of identifying out-groups as targets of aggression has not escaped political leaders, who, seeing their position jeopardized by strife among their followers, initiate war or other actions toward another, or "out" group. The result is that the followers are pulled together, at least temporarily, as their aggression is directed toward the designated group (see Box 14.4).

Commodities and Warfare
How well does the human condition fit the general evolutionary scheme described above? It matters little for purposes of our discussion whether our behavior in these areas is indelibly stamped into our nervous systems as a result of natural selection or whether our actions are the result of learned patterns, a "fault" of our culture. (If we want to discuss *changing* our condition, however, these questions become critical.) The questions we must ask here are, do we generally show out-group aggression? Are we in competition for commodities? Does our competition often result in aggression? Is war an extension of this sort of aggression?

It seems, on an intuitive basis, that the competitive nature of humans can hardly be denied. Apparently, all that is necessary to stir vague feelings of animosity toward any group is to identify it as "them." We compete at all levels. Not only are we quietly aware of whether we do any particular thing better than our friends, but our competitive nature has been directed and channeled culturally by such institutions as little league baseball, school football teams, and the system of academic grading. Needless to say, the merchants among us have very successfully capitalized on the tendency as millions of TV viewers watch athletes, with whom we somehow manage to identify, compete on the playing field each week. Then, of course, periodically there is the great international competition of the Olympiad.

14.24 Latin America is undergoing a population surge that simply cannot be maintained, as increasing numbers of young people enter the breeding population. The intensification of the search for dwindling commodities can increasingly be expected to bring populations into conflict.

And millions of people who have utterly no idea of what the game is about will devotedly follow the world championship chess tournament if one of "ours" is playing one of "theirs."

But if this competitiveness exists, does it function in gaining commodities through the more serious practice of waging warfare? At this point we might take a look at some recent events.

A Recent Lesson

In 1969 two tremendously overpopulated countries in Central America, Honduras and El Salvador, declared war. Elements in the American press played up the notion that it started over a soccer game, but the issues were a bit more deeply rooted. El Salvador, with a population of 3.3 million, had 413 people per square mile (782 per square mile of arable, or cultivatable, land). Honduras had a population of 2.5 million, with only 57 people per square mile (155 per square arable mile). The doubling time of both populations was only 21 years. Obviously El Salvador was under greater pressure. The pressure was partly relieved as about 300,000 Salvadorans moved to Honduras in search of land and jobs. They were greeted as "outsiders" by the Hondurans and friction developed between the Salvadorans and their hosts. The problem escalated into a brief but nasty war that was ended by the intervention of the Organization of American States. This international body, in an unprecedented move, acknowledged that population pressure had been the cause of the war—too many people seeking too few commodities.

A More Recent Lesson

As another, perhaps more painful, example of competition for commodities precipitating war we might consider our own policy in Southeast Asia.

14.25 "A baby was found with its head under a rock. Its head was lopsided and its eyes were masses of pus. Unfortunately, it was alive. We hoped that it would die"—*Saipan, 1944.*

Untold thousands of people have died there as a result of programs funded by American citizens. The death toll of our own young men is set at about 50,000, but this figure ignores the paralytics and amputees who were returned to us. It also says nothing of the emotionally scarred, both those who went to war and those who watched the misery each evening before beginning a large meal. One wonders, on what rational basis would the world's most powerful nation exert such a monumental effort to rain destruction on what one recent president termed a "raggedy-ass little third-rate country?" Perhaps clues can be found in the statements of some of our political leaders. Richard Nixon, in 1953, when he was vice-president, said, "If Indo-China falls, Thailand is put in an almost impossible position. The same is true of Malaya with its rubber and tin." In 1963 President Eisenhower wrote, "The loss of all Vietnam together with Laos in the west and Cambodia on the southwest . . . would have spelled the loss of valuable deposits of tin and prodigious supplies of rubber and rice."

Earlier, in 1954, Mr. Eisenhower had thought to include tungsten as an important commodity found in Indo-China. He made the statement in an interview published in the April 16, 1954, edition of *U.S. News and World Report*. The article was entitled "Why U.S. Risks War in Indo-China" and the subheading states:

One of the world's richest areas is open to the winner in Indo-China. That's behind the growing U.S. concern. Communists are fighting for the wealth of the Indies. Tin, rubber, rice, key strategic raw materials are what the war is really about. The U.S. sees it as a place to hold—at any cost.

Is it possible that, in the name of "liberty," our 19-year-olds actually died for tin, rubber, rice, petroleum, tungsten, iron, manganese, zinc, coal, and antimony—the commodities listed in another article entitled "The Big Prize—Southeast Asia"? It is interesting, and perhaps a little sad, that while people on both sides were dying those horrible deaths we've witnessed so many times, and while homes, farmland, and hospitals alike were being reduced to rubbish, American oil interests were busy mapping out and arranging oil leases in the waters off South Vietnam. Possibly economist Percy W. Bidwell, writing for the Council on Foreign Relations, summed up the American position when he stated:

Our purpose should be to encourage the expansion of low-cost production and to make sure that neither nationalistic policies nor Communist influences deny American industries access on reasonable terms to the basic materials necessary to the continued growth of the American economy.

Recently, Washington columnist Jack Anderson, a tough critic of American politics, was cautioned by a general serving under former Secretary of State Henry Kissinger not to publish a "sensitive" report on our relationship with the African country of Burundi. Anderson, not surprisingly, chose to publish the report. In it he described the normalization of American relations with Burundi, and why it came about.

In early May, 1972, the ruling Tutsi tribe of Burundi began the systematic slaughter of their ethnic rivals, the Hutus, who had attempted a coup. The scale of the slaughter was incredible. By late 1973 over 200,000 Hutus were dead, and another 100,000 had fled the country. The U.S. government showed its disapproval of the events by adopting a "minimal relations" policy with Burundi. Then suddenly, the U.S. stand warmed, almost overnight.

The thaw corresponded with the discovery of vast deposits of nickel in Burundi. Since that time, agents for U.S. corporations have swarmed into Burundi, and the country has become a much-courted friend of the U.S. government.

One top White House official wrote in a memorandum that the normalization would provide "opportunities for American corporations that are interested in exploiting the major new mineral discovery." He also wrote that resumption of U.S. aid (from our tax money) might also earn "concessions for American companies who believe there may be hydrocarbon desposits (oil) on the eastern shore of Lake Tanganyika." According to Anderson's report, then, our foreign policy can be altered in a number of ways, depending on where the *commodities* lie.

As a final example of the close interrelationship of commodities and our military posture, *Time* magazine (February 14, 1977) reported that President Carter's choice for CIA director, Admiral Stansfield Turner, had spurred brainstorming seminars on how international developments affect U.S. strategy. One topic discussed, during his tenure as president of the Naval War College, was the role of U.S. naval power at a time when the United States had begun importing increasing amounts of commodities from foreign sources. It seems increasingly naive to ask, "But what does the military have to do with trade?"

Of course, all of this sounds rather cynical, as perhaps it should. But on the other hand, it is argued that if the United States did *not* vigorously seek to insure its access to the world's goods, other countries would soon outstrip us economically and militarily. We might then be placed in the position of eventually being denied essential commodities.

Perhaps the real question should center over what should be regarded as essential. Do we really *need* the oil that lies off the coast of Vietnam? The answer is assuredly yes, if we continue to cultivate our appetite for petroleum products. Perhaps the next question is, then, will we be able to reduce the likelihood of war by controlling our appetites? This is a complex question, but one which we will approach in the next chapter.

The Proximal Causes of War
In an effort to determine the proximal causes of war, Robert C. North and others at Stanford University have correlated involvement in wars in modern Europe with population growth, rising Gross National Product (in 1972, President Nixon posed proudly for photographers below a sign showing a record GNP for the United States), and expanding military budgets. These were only correlations and, as such, were subject to debate over their interpretation. However, a later, more complex, analysis showed a causal chain that involved population growth in areas with unchanging or slowly growing resources, high technological development, increases in the demands of consumers, and a tendency to invest energy beyond normal limits. On the other hand, North found that countries in nonaggressive phases, such as Sweden, are likely to have relatively small and stable populations, high and developing technology and good access to resources, either domestic or as a result of favorable trade.

The Prospects
It's hard to guess how long we will be able to keep ourselves out of war. The prospects do not appear very good. The next conflict, for example, could

well be related to oil. We are desperately in need of prodigious amounts of the slippery stuff. The TV commericals are not wrong; we have an oil-based economy. Powerful and self-serving oil lobbies have, over the years, managed to make us utterly dependent on their product.

We can certainly assume that the price of fossil fuel will rise sharply in the next few years. Also, despite our rhetoric about self-sufficiency, the United States may become increasingly dependent on foreign oil and natural gas. And here is the essence of the problem: the likelihood of conflict increases when a commodity lies on the other side of an international boundary. Under the best of conditions those commodities can be moved peacefully through trade. It is when restrictions are invoked in hopes of using the goods as bargaining chips that the tensions increase. And the world's most powerful country needs oil from outside her boundaries. Whereas in 1973 we imported 35 percent of our oil, the National Petroleum Council predicts that we will import 60 percent by 1985. Even if we should risk drilling offshore oil wells, the problem will not be alleviated unless we can increase our refinery capacity (Figure 14.27). Of course, our efforts to become self-reliant on energy will result in pressure to relax air pollution standards and strip-mining regulation, and to develop hazardous nuclear power.

At present the Soviet Union is in a far superior position. By the year 2000 it will be the only major power who is still self-sufficient in most mineral and energy resources. It possesses an estimated 37 percent of all the oil in the world, 40 percent of the natural gas, 54 percent of the coal, and enormous uranium reserves.

Now that we are beginning to understand the meaning of "finite resources," there is some anxiety over our continued access to foreign oil. In an effort to reduce our present need for imported oil, we are currently preparing to ravage some of our most beautiful and irreplaceable Western mountains in an effort to squeeze the last drop out of their oil-bearing shale. We have already built an environmentally hazardous pipeline to funnel oil from the North Slope of Alaska to our waiting machines (and those of Japan, since we are unable to process all the oil that is being sent south—an apparently unfortunate oversight by the oilmen, which resulted in unexpected profits). To reduce our dependence on oil we are building potentially hazardous nuclear power plants (see Chapter 16). *Yet, currently, our demands for energy have recently been doubling every eight years.* We continue to respond faithfully to the imploring merchant. Each family is seeking to own more goods, newer cars, and niftier gadgets as we continue to measure our value as citizens according to the list of our possessions. Unrestrained, we good citizens will continue to propel our government to perform abominable acts to provide "things" for us.

However, restraint, by definition, implies loss of freedom. Freedom has long been important to our society. The basis of the American economy is even called "free enterprise." So here is a dilemma. If individuals continue freely in their search for more and newer goods, they

Box 14.5 Disarmament

"The basic need is evident; once again it is a change in human attitudes so that the 'in-group' against which aggression is forbidden expands to include *all* of humanity.

If this could be accomplished, security might be provided by an armed international organization, a global analogue of a police force. Many people have recognized this as a goal, but the way to reach it remains obscure. The first step might involve partial surrender of sovereignty to an international organization. It seems probable that as long as most people fail to comprehend the magnitude of the threat, that step will be impossible. At the very least we must learn to weigh the risks inherent in attempting controlled disarmament against those run by continuing the arms race. An attempt at disarmament could lead to a war, or the destruction or domination of the United States through Chinese or Soviet 'cheating.' But, if we were successful at disarming and if we achieved an international police force, the reward would be a very much safer world in which resources would be freed for raising the standard of living for all humankind. Few problems deserve more intensive study. The dynamics of disarmament appear to be even more complex than those of arms races. In spite of this, in 1970 the Arms Control and Disarmament Agency (ACDA), the only U.S. agency charged with planning in this area, had a budget of only a few million dollars (contrasted with $80 billion for 'defense'). Moreover the ACDA is heavily influenced by the State Department, a stronghold of cold war bureaucracy."

(From *Population, Resources, Environment: Issues in Human Ecology,* 2d ed., by Paul R. Ehrlich and Anne H. Ehrlich, W. H. Freeman and Company. Copyright © 1972)

Box 14.6   Values

"We have but one explicit model of the world and that is built upon economics. The present face of the land of the free is its clearest testimony, even as the Gross National Product is the proof of its success. Money is our measure, convenience is its cohort, the short term is its span, and the devil may take the hindmost is the morality.

Perhaps there is a time and place for everything; and, with wars and revolutions, with the opening and development of continents, the major purposes of exploration and settlement override all lesser concerns and one concludes in favor of the enterprises while regretting the wastages and losses which are incurred in these extreme events. But if this was once acceptable as the inevitable way, that time has passed.

The pioneers, the builders of railroads and canals, the great industrialists who built the foundations for future growth were hard-driven, single-minded men. Like soldiers and revolutionaries, they destroyed much in disdain and in ignorance, but there are fruits from their energies and we share them today. Their successors, the merchants, are a different breed, more obsequious and insidious. The shock of the assassination of a President stilled for only one day their wheedling and coercive blandishments for our money. It is their ethos, with our consent, that sustains the slumlord and the land rapist, the polluters of rivers and atmosphere. In the name of profit they preempt the seashore and sterilize the landscape, fell the great forests, fill the protective marshes, build cynically in the flood plain. It is the claim of convenience for commerce—or its illusion—that drives the expressway through neighborhoods, homes and priceless parks, a taximeter of indifferent greed. Only the merchant's creed can justify the slum as a sound investment or offer tomato stakes as the highest utility for the priceless and irreplaceable redwoods.

The economists, with a few exceptions, are the merchants' minions and together they ask with the most barefaced effrontery that we accommodate our value system to theirs. Neither love nor compassion, health nor beauty, dignity nor freedom, grace nor delight are important unless they can be priced. If there are non-price benefits or costs they are relegated to inconsequence. The economic model proceeds inexorably towards its self-fulfillment of more and more despoliation, uglification and inhibition to life, all in the name of progress—yet, paradoxically, the components which the model excludes are the most important human ambitions and accomplishments and the requirements for survival."

(From *Design With Nature*, Copyright © 1969 by Ian McHarg. Reprinted by permission of Doubleday & Co., Inc.)

will be brought into more intense competition as a result of dwindling resources and burgeoning demand. We cannot expect wars to be averted as shortages in one commodity are followed by the realization that yet another is becoming scarce. The answer must lie in some system of effective international law. All such attempts thus far have not been encouraging but perhaps even self-serving men can soon be made to see that worldwide control of commodities is in their best interest. We might also consider the creation of a world body that could effectively enforce a lowering of the destructive capabilities of nations; but here, the problems seem great also.

NEW APPROACHES TO THE OLDEST PROBLEMS

At least part of the problem in achieving worldwide solutions is that national leaders are almost always old men—even in youth-worshipping America. This is probably good since, presumably, age brings wisdom. But these leaders can be expected to be steeped in the principles that seemed to operate so well in the world of only a few decades ago, and they can be expected to be reluctant to accept the sorts of drastic measures that are now called for. This is not to say that being young automatically imbues one with new wisdom, especially if the young manage to remain ignorant. However, answers must lie in new ways of thinking. As Abraham Lincoln once said, "The dogmas of the quiet past are inadequate for the stormy present." And the present storm grows.

In most political systems, including our own, the problem is complicated by the nature of the systems themselves. From among a generally competitive populace, it seems that those who are able to rise to high office must be among the most competitive and aggressive of all. For every election winner, there are scores of losers—people who competed and lost. In other words, these kinds of political systems automatically select from among the most hard-driving competitors. It is these people, then, who make the decisions regarding the course their nations will take. What sorts of decisions can we expect from them?

Also, particularly in the United States, political campaigns are costly and those driven souls who reach high office are likely to be beholden to special-interest campaign contributors. In many cases these contributors are in the position of needing resources that cannot be furnished from within our own boundaries. Thus the likelihood of foreign policy dedicated to providing these resources increases—and so does the risk of war. It seems apparent that we must either change our system of values so that "gain" loses its status as our loftiest ideal—or we must realize that it is perhaps more important to gain freedom and a sense of peace than a faster car and a bigger house.

Resources and Energy

The unprecedented swell of human population is placing new kinds of demands on the world's resources. It seems that each new member of the species is born with a burning desire to drive a Porsche. However, in most cases throughout the world, the more immediate demand is not for the exotic symbols of status, but the "simpler" things, such as food. It has been estimated that as the world's population grows, two people join the ranks of the hungry for each one that will be adequately fed. So we might consider the human species as being composed of two more or less distinct groups, those seeking the things necessary to maintain their very existence and those who have the necessities of life and are seeking other things.

We might also divide the world's resources into two groups, those which are renewable, such as crops and timber, and those that are nonrenewable, such as oil and tin. We will explore the critical nature of this difference later, but it should be made clear that no matter what the nature of the resource, it is subject to steadily growing demands. These demands are beginning to reach such proportions that we can no longer be content with "asking ourselves questions" regarding how our resources will be used. We had *better* begin providing clear, carefully chosen answers. It may be that these answers will be unwelcome or even dreaded. Nevertheless the questions cannot be avoided. A mountain climber seeing a rockslide above him is in much the same predicament; he must make a decision or be lost. His decision to move in a certain way may save him, but he has no guarantee of that. He may be lost anyway; still, he *must* make his move.

But exactly what is the nature of our own predicament and what kinds of decisions have been made so far? Let's begin by considering how we are doing with one of our most basic necessities: food.

FOOD

You have eaten recently or you wouldn't be sitting here reading; you would be out hustling food. It may be hard for most of us with time to read books to envision the hunger and starvation that exists on our planet. But perhaps 10 to 15 people who were alive when you started this paragraph are now dead because they didn't have enough to eat.

The sight of starving people is a grim one. Once, when I was working in the West Indies, I saw Haitian children, black skinned, with swollen

15.1 Many of the difficulties centering over the world's commodities lie in their distribution. Some have more than their share, others have less. It is difficult for most of us to truly grasp the desperate plight of much of the rest of the world.

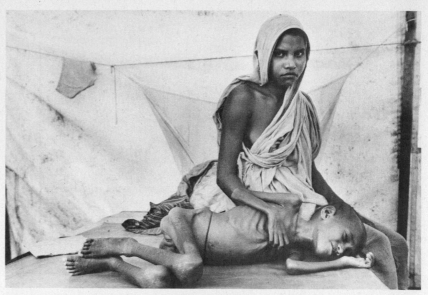

bellies, tottering about in the dust as they slowly died of *kwashiorkor*, a disease caused by protein starvation. This strange sickness, which causes the hair to become reddish and brittle, usually appears when the child is weaned. I was told by a group of missionaries at dinner one night that even if these children should begin a good diet, they would probably be mentally retarded. I found out later that this is because the brain grows rapidly in the very early years and requires high levels of protein for its normal development. According to N. S. Scrimshaw and J. E. Gordon, in their book *Malnutrition, Learning and Behavior,* if the growing brain is deprived of protein at this time *irreversible* abnormality will result. These are great human tragedies but they are not confined to small, isolated areas of our world.

Of the 60 million people who die each year, it has been estimated by French food specialists René Dumont and Bernard Rosier that 10 or 20 million of these deaths are due to starvation or malnutrition. This figure is regarded as too high by some authorities but it should be remembered that starvation (due to the lack of food) or malnutrition (due to an unbalanced diet) may *indirectly* bring about death. The coup de grace may be administered by some disease or parasite after the person has been weakened by an inadequate diet. Environmental biologist Paul Ehrlich considers deaths by starvation to include all those that would not have occurred if the person had been well fed.

The Distribution of Food
Why are people starving? Isn't there enough food to go around? In 1965, the President's Science Advisory Committee placed the world average caloric needs at 2,354 per person each day. The United Nations Food and Agriculture Organization has estimated that there were, in 1965, 2,420

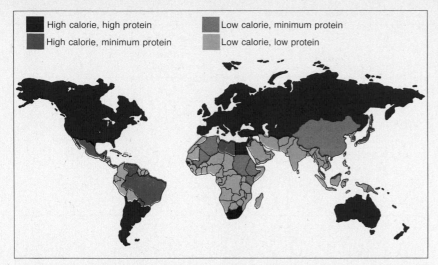

15.2 Who is fed—a comparison of protein and calorie consumption throughout the world. There are several ways this map may be viewed. How is the food divided politically? Where are the food frontiers? Who borders the hungry? How do food levels relate to weather or vegetation patterns? How are food and industry related? How does food abundance relate to birth rate? Where might political problems realistically be expected? Why would they arise?

calories per person available in the world's marketplaces. So, in terms of calories (ignoring protein, vitamins, and mineral needs), there probably is enough to go around—barely. The fact that people are starving means the problem is partly one of distribution. The rich have more than their share, the poor have less.

Another problem involves storage. At least 10 percent of the world's available food is destroyed by pests, waste, and spoilage somewhere between the marketplace and the stomach of the consumer. Look at Figure 15.2 to see how the world's food is distributed.

In the best of worlds, it would seem that there would be a net flow of food from richer areas to poorer ones in an effort to alleviate the plight of the hungry. But is this the case on *our* small planet? It is true that the developed nations send about 2.5 million tons of gross protein to the underdeveloped countries each year. The gifts are well publicized. It is less well known, however, that the developed countries *take* about 3.5 millon tons of *higher quality* protein from the poorer countries in the same period. Much of our shrimp, for example, comes from India. Protein-deficient Latin America provides us with much of the fish that we convert to dog and cat food. (I must confess to a particular ill-feeling toward the producers of those commercials of imbecilic families following a finicky cat with a cute name around with cans of various high-protein foods until they gleefully find one it will accept.) Sadly, it seems inevitable that sooner or later we are going to have to place strong restrictions on the ownership of pets. I say this reluctantly since I live with fish, a turtle, three mice, two chinchillas, a hamster, a cat, a dog, and a horse. Actually, the horse lives outside. But the problem is a serious one. For example, it is estimated that in Santa Barbara, California, with its population of about 75,000 people, there are 60,000 dogs in and about the city. Each day the 500,000 owned dogs in New York City deposit 90,000 gallons of urine and 150,000 pounds of feces on the streets. And it seems that about half of this ends up on someone's

Table 15.1 Cultivated Arable Land

| Continent | Area in Billions of Acres | | |
	Total	Potentially Arable	Cultivated
Africa	7.46	1.81	0.39
Asia	6.76	1.55	1.28
Australia and New Zealand	2.03	.38	.04
Europe	1.18	.43	.38
North America	5.21	1.15	.59
South America	4.33	1.63	.19
USSR	5.52	.88	.56
Total	32.49	7.88	3.43

Source: President's Science Advisory Committee, *The World Food Problem* (1967).

shoes! Much of the rest of the protein that is imported to developed countries is fed to poultry and livestock, since the wealthy can afford to eat from the top of the food pyramid.

Increasing the Production of Food

Because of our growing numbers, we must greatly increase the world's food production just to maintain our present level of food scarcity. So how are we doing? One effort has involved opening more land to cultivation. Table 15.1 shows the amount of land presently under cultivation as compared to potentially arable land. The term *potentially arable* is somewhat misleading, since much of this land cannot be used without irrigation. The immense problems of finding water and providing irrigation conduits mean that, in Asia for example, almost all the land that *can* be cultivated *is* being cultivated.

Then there is the problem of cost. It has been calculated that on the very conservative assumptions that one acre of land will support one person, and that it takes only $400 to develop each acre, the world would need $280 billion in the next ten years just to feed the people who will be *added* to the population!

It was once suggested that the food problem could be greatly alleviated simply by opening up the great jungles of the world to cultivation. Brazil made such an effort in setting up an agricultural colony in the Amazon Basin. The program failed dismally, however, because tropical soils are desperately poor in nutrients. You may find this surprising, since jungles seem to be so lush and rich. There are great stores in tropical foliage, to be sure, but when that foliage falls to earth and decays the nutrients are quickly washed from the soil by the torrential rains or reabsorbed into new foliage. This difficulty is compounded by the problem of clearing an area in the face of almost overnight encroachment by rapid jungle growth. And then there is the problem of trying to till this soil. In many areas when tropical soil is exposed to sunlight and oxygen it forms

Box 15.1   How Hunger Kills

What happens when someone starves to death? What does starvation do? The first thing is that the person loses weight. The living organism must metabolize and if there are no nutrients entering the gut, the victim will burn his own tissues for fuel. The body literally digests itself.

A person is considered to be starving when he has lost about 33 percent of his body weight. When he has lost 40 percent, death is almost inevitable. As starvation progresses, various organs begin to lose their efficiency. The liver, kidneys, and endocrine system may cease to function properly. Lack of carbohydrates may affect the mind so that the person becomes listless and confused, often seemingly unable to understand his plight. Kwashiorkor and vitamin-deficiency diseases can set in. Soft bones may appear in children who lack vitamin D. Lack of thiamine may cause memory lapses. Lack of niacin causes skin inflammation, diarrhea, insanity, and death.

Children are often permanently affected since improved diet cannot straighten bones or build normal brain cells, but adults can often approach starvation and then recover to a degree, as we know from the rescue of concentration camp inmates. But these people usually die sooner than their better fed contemporaries.

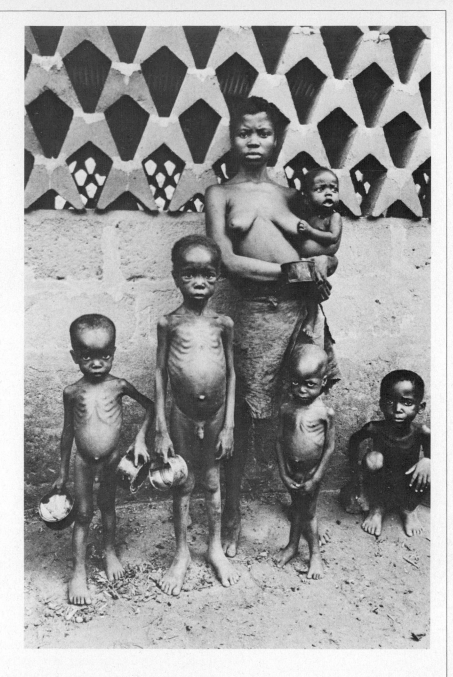

15.3 The "miracle" rice IR-8 and the usual California variety CS-M3. The taller CS-M3 tends to fall over as top-heavy plants reach fruition. But the shorter IR-8 is able to stand erect under its own weight. In this 1976 experiment the shorter variety yielded almost twice as much grain as the taller plants.

laterite, a hard rocklike substance. The beautiful and enduring temples of Angkor Wat in Cambodia are made of laterite and sandstone.

This is not to say that tropical jungles *cannot* be made to produce food for human populations, but the problem will have to be approached much more carefully than it has been thus far. We're going to have to make concentrated efforts to develop new farming techniques and to learn more about how indigenous peoples have been able to successfully cultivate such areas in the past. Whatever efforts are made to make land more productive will have to be considered very carefully. Important mistakes have been made when developers failed to consider the full impact of their schemes.

Miracle Crops

In the late 1960s there was a great fanfare heralding the impending "green revolution." New kinds of grains had been developed that yielded plants heavy with seed. For example, one strain of dwarf rice, IR-8, produced over twice as many rice grains per plant as the conventional strains. The new miracle crops do, indeed, hold great promise for the world's populations. However, they have their problems.

First of all, the new strains require special handling. For the nations most in need of food to be able to use these crops, farmers must be carefully trained in new agricultural methods. However, there has traditionally been a reluctance of people steeped in their traditions to adopt new ways. There is also a problem in underdeveloped countries of providing trained technicians to teach the farmers. In some cases the technology itself is useless. How can controlled-temperature farming be implemented in areas that have no electricity?

Miracle crops also must be heavily fertilized, and fertilizer of course costs money. Then there is the problem of the fertilizer runoff. As rains

Box 15.2   The Aswan High Dam

The Aswan High Dam, built in Egypt with Russian support, was supposed to provide hydroelectric power and to increase Egypt's food supply by controlling the unpredictable Nile River. The project meant that great art treasures were flooded as submerged land was drained for cultivation. However, only one tenth of an acre of land was made available for each person added to Egypt's population during the period of construction. One result of the dam was that the Nile no longer flooded the delta farmlands annually. These annual floods served to restore the farmland fertility with deposited silt. This no longer the case, the quality of the farmland decreased. The dam also cut off the nutrients that had been washed to the Mediterranean Sea as a result of the annual floodings. Because of this, or the change in the salinity of the sea that the dam produced, the sardine catch dropped from 18,000 tons per year to 500 tons per year. The stable lake created by the dam allowed aquatic snails to flourish. The snails serve as an intermediate host to a blood fluke that bores into humans causing the dreaded disease, schistosomiasis. The construction of the dam had important political implications at the time. These tombs had to be moved to be saved from the dam's waters. The political scene has changed now. So has the environmental one.

wash fertilizer from the fields into bodies of water, the nutrients may cause the water to *eutrophy,* or "age," prematurely. Nutrient-rich water may also be subject to great blooms of microscopic algae that can interfere with the balance of life and cause massive fish kills. The "red tides" of the oceans, when the waters strangely turn red and fish die from lack of oxygen, are the result of such blooms. An increase in agricultural production, then, may reduce the catch of fish in offshore waters.

World grain production currently amounts to about 1.6 pounds per person per day, with wheat, rice, and corn accounting for about three-fourths of the total yield. Since grains are concentrated sources of energy and are easily stored, they probably represent a good focus for continued research effort.

Ecologists such as Guy Bush at the University of Texas have warned of the consequences of relying too heavily on artificial strains of grains with low genetic variation. One of the dangers cited is that wild populations of insects or disease organisms, replete with their wide genetic variation, might through the processes of natural selection quickly accommodate themselves to any single strain and wipe it out in short order. Thus in the long run we might be better off paying the price of lower productivity for increased genetic variation in our crop species.

This point was dramatically illustrated by an unusual event in 1969. Specially developed high-yield corn plants were attacked by a new type of fungus called the southern corn blight. By 1970 the epidemic was so severe that the United States corn crop was reduced by 15 to 20 percent (a loss calculated at more than a billion dollars). The tragedy jolted many agriculturalists into a new appreciation of how food supply is a function of biological realities. The lesson was a simple one: extreme uniformity of our food crops makes them particularly vulnerable to the appearance of new kinds of pathogens or predators.

Domestic Animals as Food

One of our biggest problems is not in just providing more food, but in providing the essential kinds of food. As you recall, each time the food produced by plants is transferred to a higher level in the pyramid, most of the energy of the preceding level is lost. However, the use of animal protein from high on the pyramid is advantageous for a number of reasons. Since we are animals, the flesh of other animals provides us with the eight or ten essential amino acids in proportions that are relatively close to our needs. To get all the various proteins we need from plant tissue, we would have to consume large volumes of plant bulk and a wide variety of plants. Also, for many of us, meat tastes better than plant food.

It is important, however, which animals we choose to raise as sources of food. For example, goats are a major source of milk, but the goat is a destructive grazer and is best suited for marginal areas that have no other use. In their overgrazing, goats have contributed to ecological disasters by stripping the soil of its protective plant layer. Sheep also tend to overgraze;

Box 15.3   Clear-Cutting

One of the most controversial of the current lumbering practices is that of clear-cutting. Clear-cutting involves going into an area and chopping down all the trees there on an unselective basis. Everything goes. The area is then burned to remove debris and the land is bulldozed and replanted with a fast growing "even-age" stand of trees that can be easily harvested later. Presently over 60 percent of the U.S. annual timber production comes from clear-cutting, especially in pine, Douglas fir, and redwood national forests. You might read those last two words again.

Proponents of clear-cutting argue that some species, such as Douglas fir, cannot grow well in the shade of other trees, that the practice allows all trees to grow faster, that it is the best way for all our enormous wood and paper "needs" to be met, and, besides, it is cheaper.

Opponents argue that small, widely-spaced areas can be clear-cut, but that it is being overdone. Also, clear-cutting, they note, allows the soil to dry and increases the fire hazard. They point out that homogeneous even-age trees are more susceptible to insects, wind, and disease and that the reseeded trees are smaller and of poorer quality. In addition, it turns out that clear-cutting and selective cutting are about equally expensive. Furthermore, clear-cutting causes erosion, degrades water supplies, and depletes the soil of nutrients.

It has also been pointed out that 10 percent of the annual harvest is sold to overseas buyers, thus creating artificial shortages at home, which lumber companies can then use to get permission to cut more of our national forests. And, of course, once an area has been clear-cut, it can only be harvested in this way in future generations. As our forests fall, keep in mind that recycling a stack of newspapers 36 inches high saves one tree.

in addition, sheep are not universally popular as food items and because of their heavy wool coats they are not well suited to tropical areas. Hogs, on the other hand, can be reared in a wide range of habitats and are extremely efficient in converting plant protein to animal protein. The trouble is that their dietary habits make them direct competitors with man for the available plant food. In fact, like man, hogs are *omnivores*—that is, they eat about anything—animals as well as plants. Hence whereas they compete very directly with man, they can also subsist on man's refuse.

Cattle, because of their complex digestive arrangement, are able to survive on plants that are not food sources for humans. Since they ordinarily don't wander from their grazing areas, they don't require constant attention. Moreover, they travel well, are large enough to be used both as beasts of burden and for other work, and they provide leather as well as beef. In addition, the milk of cattle is a food staple in certain areas whereas not many people think much of pig milk. Perhaps, however, the most efficient meat in terms of its trophic cost is poultry. More flesh is produced per pound of plant tissue eaten than is produced by any of these other kinds of animals. Thus chicken is relatively inexpensive. Its dollar cost reflects its trophic cost. The less efficiently a pound of meat is produced, the more it will cost.

Speaking of cost, it may well be to our advantage to see the price of beef skyrocket. It should ideally become so expensive that very few people could afford it. The result would be better health for the general populace (beef and pork are increasingly implicated in a variety of diseases, including cancer), and it would free vast amounts of acreage that could be used to raise crops that people could eat directly. Do any of these ideas rub you the wrong way?

Fishing

Perhaps fishing should be included in the section on nonrenewable resources, because, in a very real sense, we do not replace the fish we take. Our fishing practices are more akin to mining.

There are many paradoxes associated with our fishing practices. The world fish catch more than tripled between 1938 and 1968. At that time the Food and Agriculture Organization estimated the catch at 64 million metric tons (Figure 15.4). Since then, there have been some alarming new trends. In spite of increasing worldwide efforts by the fishing industries, the fish catches are actually diminishing in many areas. However, since the oceans are a prime source of food for the world's burgeoning human population, the pressure on fish populations is *not* being relaxed. Instead, the fishing industries are diligently studying fish behavior and developing their technical arsenals as they continue their drive to meet the demands of our own hungry numbers, although fish play a relatively minor role in the world's total consumption of protein (Figure 15.5).

The management of the oceans is complicated by what biologist Garrett Hardin refers to as the "law of the commons." For the most part

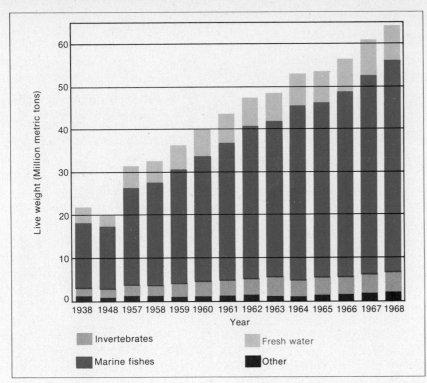

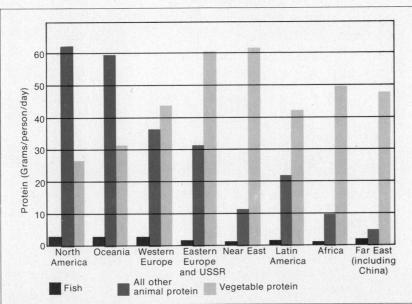

15.4 From 1938 to 1968 the world's fish catch more than tripled. By far the greatest increase was among marine fishes. Actually, humans consume only about half the catch. Most of the rest is fed to animals, either pets or those animals we expect to eat. Of course, the energy conversion factor should rule out the latter behavior by the sheer economics we saw demonstrated in Figure 14.17.

15.5 The left column shows the grams of fish eaten per person per day in various parts of the world. The middle column designates other animal protein, and vegetable protein is shown at right. Note that of the major industrial areas, North America depends more on "red meat" protein such as beef and pork than any other area, and is the only major industrial area to do so. Most of the world depends on vegetable protein. Which makes more sense in terms of the energy pyramid (Figure 14.17)? Is the ability to waste considered a symbol of status?

the decisions made by various nations are not compatible with optimal productivity. For example, the decision to add another vessel to the fleet may be marginally productive to the country, but in the long run may

Box 15.4   Triage

During World War I a "triage" system was put into effect to determine which wounded would receive priority in field hospitals. Medical attention was based on the condition of the wounded. Those who were so injured that it seemed they would probably not recover even if treated, were sedated and left to die. Time and energy were spent only on those who, with help, could probably survive.

Some scientists are now advocating such an approach on a worldwide basis. The idea is that food or other supplies should only be sent to those countries that could benefit from such assistance. The poorest countries with the highest birth rates and least likelihood of being able to survive would be left to collapse of their own weight. The suggestion is based on the belief that if an attempt is made to distribute resources to all nations the demand will be too great and no nations will be able to survive in anything like their present form.

contribute to a diminished catch because of overfishing. Yet if that nation chooses not to add another vessel, it may find itself on the short end as other nations continue to add vessels. The first nation comes up short both ways. So what do you think its decision might be? What will be the long-range effects of such decisions? Can you think of some better way of apportioning the world's resources?

Other Ideas

There are also other ideas regarding how to feed the world. Some are conceptually fascinating, but as yet unproved. For example, it is known that single-celled organisms, which are high in protein, can be grown on petroleum. If these organisms could be produced in large enough amounts and purified economically, they could conceivably provide us with high-grade protein. Other scientists are busy trying to produce yet newer grains with higher quality proteins (which means containing a better balance of amino acids). A nutritious mixture of corn and cotton seed meal enriched with vitamins A and B has been available in Central America for over ten years but unfortunately has not been accepted by the people. This is one of the great problems of food scientists—how to get people to eat what is being newly developed. Historically, the hungriest people in the world are those that recognize the fewest items as food.

And then there are a variety of other ideas being considered, such as culturing algae, ranching various rodents or antelope, converting weeds to cattle feed, extracting protein from plants or small fish, converting wood to cattle feed, and using sewage to grow edible slime. It's likely, however, that quite a few people will have trouble accepting most of these as food unless they can be made palatable through processing techniques, or unless the hunger level rises high enough.

It is interesting that of the thousands of species with which we share this planet, only a handful have been designated food items. In fact, it turns out that modern man, with his vast abilities and imagination, has virtually discovered *no* new crop plant or domesticable animal. Of course, food may be imported into one area from another. Captain Bligh was directed to take the *Bounty* to bring back a "new" plant that had long been cultivated in the South Pacific. New kinds of fish meals are still made of fish. And the miracle grains are simply variations of plants that we have used for thousands of years. Recently, a promising crop "species" was developed that is a hybrid of wheat and rye, old standbys.

One is tempted to ask whether our energies *should* be directed toward frantically trying to provide a minimal diet to teeming millions of people as their numbers grow increasingly faster. Should this be the role of our scientists and the lot of modern man? Or should we begin to put strong pressures on the swell of our numbers? Should we stop asking how we can manage to keep ever greater numbers of people barely alive, and begin to ask how we can improve the quality of life for those already living? Some of the answers being considered are grim indeed.

WATER

Although the United States is blessed with an average of 30 inches of rainfall per year, we are beginning to see the localized effects of too little water. In some areas the housing industry has been curtailed because there wasn't enough water for additional families. The residents of some areas have found a silver lining here and have used the water issue to restrict population growth in their towns, which they prefer to see stay small.

Our use of water in this country is prodigious and is growing by leaps and bounds. In 1900, for example, the United States used 40 billion gallons of water each day, an average of 530 gallons per person. In 1974, however, we used about 380 billion gallons a day, an average of 1,800 gallons per person (Figure 15.6). The average person in an underdeveloped country uses 12 gallons per day. In 1965 the Department of the Interior informed us that the water used by *only the Americans alive in 1965* for the remainder of their lifetimes would exceed all the water used by all of the people who ever lived!

Basically, we have two problems. In parts of our country, such as the Southwest, there is not enough rainfall. In other parts of the country, such as the Northeast, the Great Lakes region and Florida, there is abundant water, but it is being rendered unusable by pollutants. Other areas such as southern California, Arizona, Utah, Colorado, Texas, and Oklahoma can expect both problems.

But food and water are related to only one type of problem. Let's take a look at some of our nonrenewable resources.

NONRENEWABLE RESOURCES

The line between renewable and nonrenewable resources is a fine one. For example, water is considered cyclic in that it evaporates and falls again as rain and thus is "renewable." However, much water is tied up in various manufactured and natural products and thus is not easily retrievable. And we know that water may become so polluted that it is no longer usable. Thus, in a sense, it is at least partly nonrenewable.

For purposes of this discussion, we will consider nonrenewable resources to be those naturally occurring materials, such as minerals, whose amounts are fixed and can therefore be depleted.

In our consideration of these commodities we must remember that they do not occur uniformly throughout the earth. Hence, their use may be manipulated by nationalistic policies, thereby potentially heightening the chance of conflict over their distribution. Consider the imbalances in metals. Asia is rich in tin, tungsten, and manganese. In fact, over half the world's tin is located in Indonesia, Malaya, and Thailand. North America, on the other hand, is poor in these elements, but well-endowed with molybdenum. Gold, of course, occurs mostly in South Africa; so does platinum, but not much silver is found there. Over half the world's nickel is in New Caledonia and Cuba, and most mercury is found in Spain, Italy, and the Sino-Soviet areas. Because of the implications of such resource

15.6 Lake Mendocino, an important reservoir in Northern California, drying as a result of an extended drought. The photo was taken in March, 1977, a time when it would normally be completely full. At this point it is down thirty-five percent. Perhaps such problems would be of less magnitude if Americans realized just how much water they actually use. For example, each American uses an average of 1,800 gallons of water each day. A bath takes 30 to 40 gallons, and a 5-minute shower, 25 gallons. Less obvious use includes an incredible 40 gallons of water that goes into the production of a single egg, 75 gallons into 1 pound of flour, 150 gallons into a loaf of bread, 2,500 gallons into a pound of beef, 230 gallons into a gallon of bourbon, 280 gallons into a Sunday newspaper, 100,000 gallons into a car.

distribution, let's look at resource reserves both for the world and for the United States.

It should be apparent from Figure 15.7 that there are certain problems we can expect if we continue to "develop" (and "exploit") these resources on a "devil take the hindmost" basis. In another twenty-two years there may be only ten natural resources left in the world, and only four in the United States. (How old will you be by then?) By the year 2042 the world may have only eight of these left, and the United States three. Also, barring deliberate population-control measures or some devastating catastrophe, there may well be 15 billion people on earth by that time.

Of course, these figures are only estimates. Some will prove too small and others too large. We will undoubtedly begin to recycle more materials as we find new deposits and develop new technology. At the same time, the population will probably continue to boom for some time, each new human replete with his needs and propensity to acquire.

Geologist Preston Cloud points out that we can make the best predictions for those substances that are found associated with specific rock layers. These include the fossil fuels—coal, natural gas, and oil—and to a degree, iron and alumina. The reserves, grades, locations, and recoverability of many other critical metals are much harder to estimate. Thus the problem of developing long-term policies regarding their use is magnified. Should we continue to drain each known deposit, confident that others will be found because they *must* be found?

The approach to such problems taken by the United States is particularly important for a number of reasons. Since the United States has been technologically developed for a relatively long time, we have exploited our own resources to a greater extent than most other countries. Whereas a great wealth of undiscovered resources may exist in Russia, for

15.7 Estimated natural reserves for the United States and the world. These can only be considered very rough estimates because we really know so incredibly little about the composition and behavior of the globe we call home. Our amazing technological abilities give us the impression of having a great deal of information, but often the most basic, and therefore perhaps the most essential, data are lacking. Many of our environmental and political policies, however, are formed on the assumption that figures such as these are at least in the ballpark.

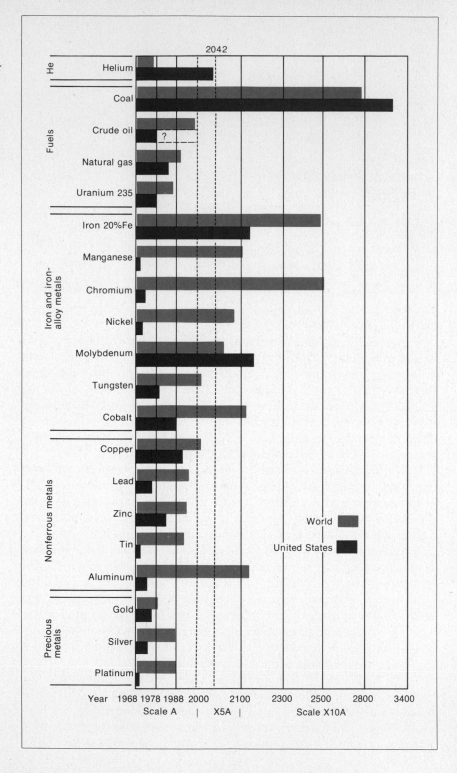

Box 15.5 Desalination

You have undoubtedly heard somewhere that desalination of salt water will solve our water problems. Unfortunately, someone is putting you on. By 1972 there were 600 desalting plants operating throughout the world, producing about 250 million gallons of water a day and projected to reach one billion gallons a day by 1976. This turns out to be 0.0007 percent of the world's daily needs and just 0.0025 percent of the water used each day in the United States alone.

The answer would seem to be, build more plants. But the problem is, the fresh water they produce is just too expensive for large-scale use (presently up to $2.15 per 1,000 gallons). In addition, there are environmental problems. What does one do with the mountains of salt that are extracted? The salt would poison any water or land areas where it was dumped.

The most talked-about energy source is nuclear. Nuclear desalination, of course, produces all the problems attendant with nuclear power. Furthermore, if nuclear plants could produce 100 million gallons of water (and, incidentally, at the same time 15,000 tons of salt) per day, we would need 4,000 of these coastal plants just to provide one third of our water needs by the year 2020. They would also render most of our beaches unusable. In addition to human desalination schemes, efforts are being made to use nature's desalination—precipitation. For example, in late 1977, the first study was conducted to determine the feasibility of towing icebergs, which tie up a large percentage of the earth's fresh water, to the Middle East.

example, we are probably much more knowledgeable regarding our own reserves. It seems that we would therefore be more acutely conscious of the limits of natural resources and would take a rather cautious approach to their removal or exploitation. For some reason, though, it doesn't work that way. Recall the power plays of the oil industries as they gained governmental approval to extract the oil from the North Slope of Alaska—to transport it by ecologically hazardous pipelines to loading facilities at the port of Valdez. Also, new offshore oil wells are being drilled in the seismically unstable Santa Barbara Channel, where the water continues to be polluted from the seepage left from the great 1968 spills. Now the oil executives are stroking their chins and wondering how much oil they can get from under the waters of the Eastern Seaboard. They have already received the permission of the government, but at the time of this writing are being temporarily restrained by a federal judge.

It seems that, in spite of increased public awareness of the importance of developing ecologically sound policies, there has been almost no reordering of priorities by our government, or the government of any other developed country. Big business goes on as usual. But it is simplistic to lay the entire blame for policy errors at any single door. Certainly, the oil companies would not so vigorously search out the earth's oil if there were no public demand for it. We need that oil. But then, the demand exists, at least in part, because the powerful "highway lobby" has such inordinate influence in Washington. Thus Americans have only woefully inadequate means of public transport and must depend heavily on automobiles, automobiles that use petroleum products. Such circularity can only be broken, it seems, by a powerful national effort to free ourselves from the sinking vortex, and such an effort is not likely to be made in the absence of a large-scale shift in values. Unfortunately, there is no sign of such a shift.

Another reason that the decisions made by the United States are especially critical has to do with our national appetite. The United States, with something over 5 percent of the world's people, currently uses about 30 percent of the world's resources! And that figure is expected to go much higher within the next ten years. (Other developed countries also consume much more than their "share." The U.S., Canada, Europe, the U.S.S.R., Australia, and Japan consumed about 90 percent of the energy and the steel produced in the world.) Considering our own limited reserves, then, it is apparent that much of what the United States uses must come from outside its borders (Table 15.2).

To see how our consumption compares to some other countries, consider the following: in 1968 our steel consumption per person (production plus imports minus exports) was about 342 times that of Indonesia, 86 times that of Pakistan, 68 times that of Ceylon, 23 times that of Colombia, 9 times that of Mexico, twice that of Switzerland and France, 1.5 times that of the United Kingdom and the Soviet Union, 1.4 times that of Japan, and 1.1 times that of Sweden. In 1968 we used over one-third of the world's energy output, over one-third of its tin, one-fourth of its phosphate, potash, and nitrogenous fertilizers, and about one-eighth of its

Table 15.2 Net U.S. Imports as a Percentage of the Country's Consumption

Resource	1965 Imports	1975 Imports
Manganese	94	99
Chromium	92	91
Tin	80	75
Nickel	73	71
Petroleum[a]	19	35

[a]Includes natural gas liquids.
Source: Statistical Abstracts of the United States, 1976, U.S. Department of Commerce, Bureau of the Census.

Box 15.6　Strip Mining

Mining companies have stopped strip mining. Now, they are engaged in "surface mining," a more circumspect and unassailable term they hope. Whatever they call it, however, they are not restricting themselves to Ohio and West Virginia as some seem to believe. In fact, about half our states are being gouged and scarred by their gargantuan machines.

Since World War II about 6,200 square miles have been desecrated by strip miners. The unreclaimed land alone comprises an area equivalent to a mile-wide swath of land stretching from New York to San Francisco. The latest attack has been on the grass-covered plains of the coal-bearing states from Montana to New Mexico.

There are four types of surface mining: (1) open pit mining, primarily for iron, copper, sand, and stone; (2) contour strip mining, performed on rolling hillsides, usually for coal; (3) area strip mining, taking 30 to 40 inches of the topmost soil, for coal and phosphate; and (4) dredging, usually in seabeds for sand and gravel.

Our impending energy crises will place renewed pressure on our natural reserves as energy-producing industries are given freer rein. The companies proudly point to a few well-groomed showcase reclamation projects in Virginia, Tennessee, and Illinois, but the truth is, many of the areas simply *can't* be reclaimed, and in most cases no attempt is made at all. When land is reclaimed properly, most of the cost is born by local, state, and federal governments. After extracting their enormous profits, mining companies in the United States rarely spend more than $300 to $400 per acre in reclamation. And the rains washing over the great scars continue to turn streams dark and poisonous.

Box 15.7   Shale Oil

Shale oil is potentially an ideal fuel. We know where it is. There is no need for costly searches with hit-or-miss exploration techniques. It is present in immense quantities (estimates range up to 3 trillion barrels). It contains little sulfur. It's right here in the United States, 70 to 80 percent of it on government-owned land. And the price of such oil would be no higher than that of oil from other sources. The problem is, the shale land is among the most beautiful that the country has to offer. It lies in the mountains of Colorado, Utah, and Wyoming. The decision has to be made, then, whether to devour the mountains to satisfy our demand for oil. The Interior Department has announced an experimental program to tap these reserves, "in the national interest."

In the extraction methods that are proposed, the shale is literally chewed up and cooked. The rock then yields a heavy oil similar to crude petroleum. Of the 8 million acres of oil-bearing shale, only six tracts of about 5,000 acres each will be involved in the initial tests. In two of the tracts, oil will probably be directly boiled out of its underground reservoir. A chamber will be blasted, then natural gas will be injected into it and set afire. The oil would then be extracted by special wells.

Another method may be the conventional "pillar-and-room" tunneling that is presently used in coal mines. But in the case of the shale the project would be gargantuan. Perhaps the most destructive method involves open-pit mining, the third alternative. This is the strip-mining that has so devastated coal-laden areas of Kentucky and West Virginia, and is presently being used to get at the copper near Butte, Montana. Here, enormous earth-moving machines scrape the earth bare of billions of tons of its natural cover to expose the shale beds 100 to 850 feet below the surface. These beds are then gouged out, forever destroying the contour of the land.

Environmentalists are concerned for a number of reasons. For one thing the waste shale occupies more space *after* it is boiled. Where will it all be put? The oil companies suggest they might fill in canyons with it, fertilize it, water it, and plant it.

cotton. All this by one-twentieth of the world's population. Thus the United States has an environmental impact far in excess of its numbers.

Not only is what we do important on the basis of our staggering appetite, but also because we are often imitated by other countries. I was once told by an emissary from a West African nation that the burning desire of his countrymen was to build another Los Angeles. Imagine that!

ENERGY

Energy is simply the capacity to do work. The problem of energy supply and utilization is, of course, a common denominator for all living organisms. Just as plants must acquire energy from the sun and animals must derive energy from plants, humans must also find energy to be able to rearrange trees and oil and iron to build houses so that they may be shielded from disruptive elements in their environment. Now, if the humans demand more or larger houses, then they need not only more resources, but more energy to rearrange the resources.

Some of the demands placed on energy resources are "legitimate" in that they relate to very basic needs such as food production or environmental shields. We may be reaching the point, however, where it will be necessary to distinguish such essential energy uses from nonessential demands. Do we really need elaborate neon-lit billboards, automatic can openers, oversized electronic amplifiers, electric toothbrushes, and over-powered six-passenger cars that are usually used to carry one person to work? But you could compose your own list of such nonessentials just as well. So, let's take a look at our energy resources and some of the demands being placed on them.

We should first make it clear that, at present, we are not running out of even the current forms of energy. The United States, for example, has vast deposits of coal. However, the utilization of these resources involves considerations other than availability. There are heavy costs to our environment both in producing energy and in consuming it, and these costs must be carefully weighed in calculating the net benefit of energy use. We will see again the enormous impact of the industrialized countries, particularly the United States, on energy resources (Table 15.3).

Table 15.3 Sources of Energy in 1975 (excluding food and the burning of wood and dung)

Source	Percentage of World Consumption	Percentage of U.S. Consumption
Petroleum	45.5	46.4
Coal	28.5	18.2
Natural gas	18.3	28.2
Hydroelectric	6.8	4.6
Nuclear	1.5	2.6

Source: Energy in Focus: Basic Data (1975), Federal Energy Administration, and F.E.A. Energy Information Analysis Office, Washington, D.C.

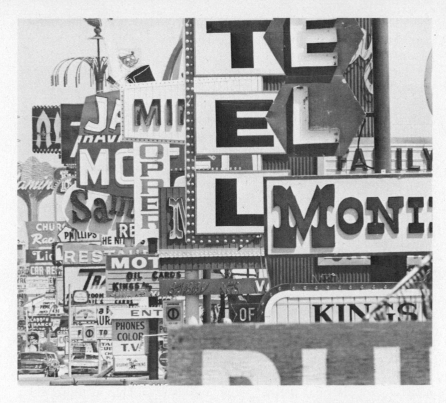

15.8 An example of the visual pollution (as seen here in Las Vegas) blighting so many areas of an essentially beautiful land. Strangely enough, almost everyone prefers to see subdued development with small, quaint signs announcing the business establishments of the area. Yet those same people allow the most garish and jangling advertisements to assault their sensibilities with impunity. The individual merchants are not entirely to blame, because if one decides to build so that his establishment blends well with the wooded environment and to put up a small, tasteful sign, his more garish neighbors will attract more attention and hence more business. Essentially, a business is forced to outshout its competitors. One answer is the development of strict zoning laws, but that requires the support of an enlightened public. And, of course, restriction implies loss of freedom—a touchy subject among Americans. What, then, is the answer?

We will also consider, in terms of dollars, our efforts in the research and development of various energy sources. In a special report to the White House in December 1973, a government task force proposed a monumental increase in such efforts. In fact, it requested $10 billion be spent over the next five years. By way of gaining perspective, this is about one-eighth of the Pentagon's budget for 1974 alone.

Water, Wind, Earth, and Sun

Moving water has been harnessed as a source of energy for many years. Waterfalls turned paddlewheels to grind corn in early America. Later, moving water was backed up behind great dams and then released to turn huge generators for *hydroelectric* energy. In a few places *tides* have even been used to produce hydroelectric energy. Since rivers and tides are likely to persist we can probably always count on hydroelectric energy to some degree. At the present, the world's *potential* production of hydroelectric power is roughly half that *now* produced by fossil fuel.

There are serious problems, however, in the use of hydroelectric power. To begin with, for a country to make full use of such a power source it must *already* be industrialized, and much of the potential hydroelectric power lies in nations that are now underdeveloped. Another problem involves dams, which are after all only temporary structures. They must be carefully monitored, and in time they must be removed or replaced. Also,

15.9 President Carter appointed America's first Secretary of Energy in 1977. He selected James Schlesinger, a brilliant, gruff, and somewhat arrogant New Yorker. Schlesinger quickly found that not much can be accomplished without the assent of Congress, and that Congress is largely composed of self-serving people with an acute sensitivity to special interests. Carter, with training in nuclear physics, had announced that he was leery of nuclear power and wished to encourage the exploitation of American coal reserves.

15.10 We require large amounts of energy to arrange the molecules around us so that they form for us a more hospitable environment. The problem is, we rely strongly on fossil fuels, priding ourselves on each new find, ignoring the fact that they produce by-products that pollute our habitat, threaten our health, and even our lives. One energy source that seems nonpolluting is hydroelectric power (opposite page, bottom left), furnished by dams. The nonpolluting title may be debatable but their enormous secondary effects are not. They alter vast geographical areas in often-critical ways. We have traditionally directed rather little research attention to nonpolluting sources, perhaps, because wind, sun, and tides are more difficult to package and sell and thus their development has not been encouraged by commercial and lobbying interests. Recently, however, we have begun turning our attention to the long neglected windmill (bottom right), to geothermal energy (top right; the "smoke" is really harmless steam), and to solar energy (top left). Wind and geothermal sources may be rather localized, more appropriate for some places than others. Solar energy is feasible in more places, but presently it is used mainly for heating air and water in individual dwellings.

they tend to "silt in," so that they can lose their control of water flow. These problems, of course, ignore a central question: do we really want to dam up all our free-flowing rivers?

How about *wind?* Wind has provided cheap and pollution-free energy throughout much of Western Europe and the American Southwest, but windmills are presently often treated as quaint relics. Part of the problem is that wind is a rather local phenomenon and is, as any sailor knows, extremely unpredictable in most places. Theoretically, however, wind might be able to provide significant amounts of energy if the proper technology could be developed for utilizing and storing its power. Although no such large-scale effort is underway, a federal research grant has recently been awarded to at least one educational institution to design and build an effective wind-powered generator.

One of the most promising sources of power is *geothermal energy.* This means using the steam that is produced when the molten masses deep below the earth's surface heat the water trapped above. A few countries, such as Italy, Iceland, and New Zealand, are already using geothermal energy effectively. However, critics say that such sources can be developed only in a few locations around the earth, and that there is no way to estimate how long the superheated water will last in these areas.

And then there is *solar energy.* It has the obvious advantage of being as common as sunlight. The amount of sunlight striking different parts of the earth varies, of course, depending on the latitude and the local weather patterns, but with our present technology it is feasible to utilize solar energy in most parts of the United States.

Because sunlight is a rather low-level energy source, a large surface is needed to collect it. In fact, a collecting device designed to meet the energy needs of a city of 750,000 people would require an area of about 16 square miles (assuming a solar-to-electric energy conversion efficiency of about 10 percent).

Sunlight is already dispersed, so that there is no need to distribute it from a central source. Therefore, the roofs of individual houses may provide surfaces large enough to run individual heating and cooling units and to heat and cool water.

Solar heating may be one of the most realistic solutions to the current energy problem, at least for space heating. However, for some reason, it is not receiving the research attention it warrants. Although $13.2 million in federal funds was allocated to solar-energy research and a total of $200 million is earmarked for such use through 1979, for purposes of comparison, this is less than one-fortieth the military budget for 1974 alone. At present, daydreaming about solar energy can only be a pacifier. We aren't going to be able to use solar energy on any important scale for a very long time and we don't show any real intention of beginning the effort.

Fossil Fuels

The supplies of petroleum, coal, and natural gas will not be quickly depleted, but they are certainly not going to hold out indefinitely. Coal, the

most abundant of these, will probably be largely consumed within the next few hundred years, but petroleum and natural gas will go much sooner. Geologist M. King Hubbert, who has formulated the most reliable estimate to date, states that these latter two resources will be depleted substantially in about 100 years. This does not mean that all the earth's fossil fuels will be gone, but the residual amounts will, for the most part, be so small that their extraction would be economically unfeasible. It is possible that mining for such products might continue if a tiny vial of petroleum should come to have the approximate value of a rare and exotic perfume. Plastics, clothing, vitamins, phonograph records, and carbonated water would, of course, be priced accordingly.

Before we reach that stage, however, we will undoubtedly have ravaged the great oil-laden shale deposits in the vulnerable mountains of Colorado, Utah, and Wyoming. Proposals are already under consideration to strip mine these mountains in what sometimes seems to be a diligent, and certainly profitable, effort to remove the last of the American oil as soon as possible.

Future shortages may temporarily be diverted by converting coal to oil and gas. It is expected that economically feasible means will be developed in the 1970s. But, then, such a program would decrease the life expectancy of coal.

So how goes the funding? Coal research, in 1973, received $167.2 million, about $400 million in 1974, with the governmental task force proposing that $2.18 billion be spent by fiscal 1979.

The Energy of the Atom

One of the most talked-about solutions to our energy problems is *nuclear power*. Presently, nuclear reactors produce less than 1 percent of the energy used in the United States. However, this percentage is doubling about every two years.

At present nuclear power is based on *fission* processes. This means that when the nuclei of uranium atoms are split, part of the mass is released as energy. Today, we are able to utilize only 1 or 2 percent of the energy produced, but the fuel is rather cheap and enormous amounts of energy can theoretically be realized.

In the future, much of our energy may come from "breeder reactors" that will permit the extraction of 40 to 70 percent of the energy of new uranium. These reactors, then, could theoretically supply our anticipated demands for electrical energy for thousands of years. But here lies an important point. Nuclear power can produce only *electrical* energy, and electricity accounts for only one-fourth of the current energy budget in the United States. This figure is expected to go up to one-half by the year 2000, but even so, nuclear power is not being considered as a replacement for fossil fuels in such areas as private transportation. Nor can it be, unless there are great changes in our concept of how we should be getting from one place to another.

Nuclear fusion may, in the long run, prove to be our best energy bet. Fusion involves the joining of heavy hydrogen nuclei under superhigh temperatures, forming helium nuclei and releasing huge quantities of energy. It is this process that produces the energy of the sun and the hydrogen bomb. We haven't yet learned to control fusion reactions for the sustained production of energy, and the technology is apparently a long way off. The great advantage of nuclear fusion over fission is that very low radioactivity would be produced and hence the system is inherently much safer and less polluting. In the discussion of radiation pollution we will consider some rather alarming reasons why we should proceed with the development of nuclear fission plants with the spirit of mating porcupines—very, very carefully.

Bioethics and Environment

It has been said that we have seen more technological advances the last ten years than in all our previous history. That may be true, but what does this information tell us? It could be cited with equal enthusiasm by a proud industrialist or a horrified environmentalist. The latter might argue that our technological abilities have outstripped our wisdom, our ability to deal with our creations. So what can we make of our obsession with technology? Perhaps Thoreau was right. Perhaps we have busied ourselves with trivialities and forgotten about the real business of living. He once refused a doormat for his cabin at Walden Pond because be figured he wouldn't have time to sweep it. He needed that time, he said, to sit or walk in the woods. Living. *Experiencing!*

In any case, most of us are completely, and probably irrevocably, caught up in the business of 20th century living. We *demand* our doormats, and our TVs, and our automobiles, and our whatever. And with our demand we create new problems. Doormats do not grow wild; they must be manufactured. Energy must be employed to rearrange molecules. If manufacturing doormats decimates the forests or produces harmful waste, then doormats cost more than the price we see marked. Because of our demands we may be forced to breathe chemical-laden air or to drink dilute solutions of asbestos—hardly noticing the small changes within us until we pay some ultimate price.

This is all a bit grim, but our alarm is understandable when, day after day, yet another synthetic substance is unveiled as a poison or carcinogen. Still our demands continue unabated as our environment becomes more artificial, complex, disruptive, and jangling.

But are we being alarmist? What sort of evidence would prompt such dismay? Let's take a look, now, at some of the pollutants with which we are surrounding ourselves. Let's see where they come from and what they do. And then let's discuss our decision-making processes and hopes for change.

POLLUTION

A *pollutant* is generally defined as a harmful product that enters the environment. Such products may be in the form of noise, heat, chemicals, or garbage. In this discussion we will consider the pollution produced by human activity, although nature may produce some of the same sub-

16.1 Air pollution has startling and pervasive powers to dissolve even buildings and monuments. The by-products produced in the burning of coal and oil, for example, may produce acid rain. Carbon dioxide and water form carbonic acid, sulfur oxides form sulfuric acids, nitric oxides form nitric acid, and so on. This means our cities are continually drenched with a most peculiar sort of rain. An Egyptian obelisk survived thousands of years in the desert, but its engravings have almost disappeared since it was placed in New York's Central Park a few years ago.

stances. For example, there is some natural seepage of oil from undersea deposits that has been entering the oceans for thousands of years. Oil companies accused of polluting oceans and beaches are fond of bringing up the possibility of natural seepage to account for the mess. And then, much of Los Angeles is lost to aerial view on days when weather dictates that the city should breathe its own effluent; but the eruption of volcanoes can also emit prodigious amount of ash and fumes into the air.

The present environment has been molded by such natural occurrences throughout the earth's history, just as the life on earth has adapted to such events. The kind of pollution that concerns us here, however, is that which, in kind or degree, is new to the earth and its creatures—that which has only recently begun to be produced by the brainiest of the earth's forms. The reason for the concern is that the new pollutants may alter critical aspects of our environment so rapidly that much of the earth's life will not have time to adapt to the changes before it is destroyed.

Of course we're not simply concerned about how to continue living or *existing* on our planet, but we might also legitimately have an interest in protecting the *quality* of our lives. The once-beautiful Lake Erie is now so polluted that many cities along its shores require boaters to have typhoid shots in case they should accidentally fall in! But, because of the factories there we now have more paper cups and more brightly painted cardboard boxes to package toothpaste in. And such goods may define life's quality to some people. I once saw an old man fishing in New York City's East River as traffic honked, screeched, and roared only a few yards behind him. I wondered if he remembered quieter days there when one could catch something besides sludge-eating bottom dwellers. I also couldn't help wondering what the term "quality of life" might mean to him.

But let's get on with the consideration of a few specific types of pollution.

Air Pollution

If you live in a large city, you might be particularly concerned about air pollution because you can *feel* it. It burns your eyes and nose and may give you terrific headaches on the worst days. It may even make your days darker in a literal sense. Air pollution, for example, cuts by 40 percent the amount of sunlight reaching Chicago. You may also find it mildly "irritating" that in many cities certain air pollutants dissolve nylon stockings, windshield wipers, and statues. You may be even more moved by the fact that air pollution can also kill crops and people.

Air pollution comes from a variety of sources, but principally from the automobile (Table 16.1). Fuel burned in stationary sources, such as in heating units for buildings, also largely contributes to foul air.

We should realize, also, that air pollution is not just the problem of big cities, nor is it restricted to industrialized countries. Meteorologists are now describing a thin veil of pollutants that hangs over the entire earth. Polluted air has even been found over the North Pole! Generally, however,

Table 16.1   Air Pollution Sources in the U.S. in 1974 (millions of tons)

Source	Carbon Monoxide	Particles	Nitrogen Oxides	Sulfur Oxides	Hydrocarbons
Fuel burned for transportation[a]	73.5	1.3	10.7	0.8	12.8
Fuel burned in stationary sources	0.9	5.9	11.0	24.3	1.7
Industrial processes[b]	12.7	11.0	0.6	6.2	3.1
Solid waste burning	2.4	0.5	0.1	0.0	0.6
Miscellaneous[c]	5.1	0.8	0.1	0.1	12.2
Totals	**94.6**	**19.5**	**22.5**	**31.4**	**30.4**

[a]Motor vehicles, aircraft, railroads, vessels, and agricultural, industrial, and construction machinery.
[b]Emissions for over 80 industrial processes/products were computed, including all major operations known to emit more than 10,000–20,000 tons per year nationally.
[c]Including forest fires, coal refuse burning, organic solvents, oil and gasoline production, and other sources.
Source: Environmental Protection Agency.

the foulest air is found in the industrialized countries where it is particularly concentrated over large cities.

How serious is the threat of air pollution in such areas? Pretty serious. During one great smog in London in 1952, 4,000 deaths were *directly* attributable to the quality of the air. Indirectly, probably many more people were affected, because air pollution works it's way in subtle and insidious ways. Since air pollution kills slowly, it is difficult to attribute human deaths to pollution levels or specific contaminants. It has been shown, however, that death rates are higher in smoggy areas. The most likely to be affected are the very old, the very young, and those with respiratory ailments.

Carbon Monoxide

Carbon monoxide is the most prevalent air pollutant, as you can see in Table 16.1. It is biologically important because of its tendency to combine with the blood's hemoglobin in the place of oxygen. The effect is to cut down the blood's supply of oxygen to the tissues, thus causing the heart to have to work harder to oxygenate the body. The increased demands on the heart place a severe strain on people with heart or respiratory ailments. Looked at another way, carbon monoxide can have the same effect as loss of blood. In fact, spending eight hours in an atmosphere with 80 parts per million of carbon monoxide has the same effect as losing over a pint of blood. It may interest you to know that in a traffic jam the air may contain nearly 400 parts per million of carbon monoxide. The people breathing this air may experience headache, loss of coordination, nausea, abdominal cramping, and even partial blindness.

Nitrogen Oxides

Nitrogen oxides are produced primarily from gasoline engines, power plants, and industry. Nature produces 10 times more oxides of nitrogen

16.2 A street scene in Washington, D.C. Each of these machines continues to burn hydrocarbons as they sit idly, emptying the poisonous waste directly into the air. Such methods of moving people are not only inefficient but dangerous as the air of our cities becomes increasingly fouled. New emission laws and requirements of vehicles with better mileage will help the problem but they can only be viewed as stopgap efforts. Obviously, we need to think through this thing one more time.

than man does, but man's products are concentrated in urban areas. Although nitrogen dioxide (NO_2) irritates lungs and withers plants, the greatest danger arises when it combines with hydrocarbons in the presence of sunlight to produce photochemical smog. This smog is particularly reactive and is one reason pantyhose don't last long in the city. But if it eats your pantyhose off, what does it do to your eyes and lungs? It dissolves automobile tires too, and may be responsible for some traffic accidents. But yet our chambers of commerce continue to cry out for more industry (Figure 16.2).

Sulfur Oxides

Sulfur oxides, produced mainly by burning coal, are thought to exert particularly severe reactions during heavy smog. Sulfur oxides combined with water can produce sulfuric acid, which damages not only lungs and plants, but can dissolve marble, iron, and steel. Acid rains in some areas have presented severe problems to art as well as health. These kinds of chemicals are very harsh elements that produce coughing, wheezing, and choking. The most serious damage is done when the by-products of sulfur oxide formation interact with ammonia and metallic salts in the air, and the resulting compounds are inhaled deeply. They are believed to figure importantly in the increased incidents of bronchitis, asthma, and emphysema, which are so frequently suffered by people in high smog areas. (Emphysema is a progressive lung disease and is considered fatal.) Globally, sulfur oxides may not present a great problem because rain washes them from the air, but locally they may be the most dangerous pollutant in the atmosphere. The United States emitted 20 million tons of sulfuric oxides into the air in 1970, and by 2000 will increase that to 95 million tons unless strict controls are created.

Hydrocarbons

Hydrocarbons, as you know, are a diverse and variable group of elements, some of which are believed to be a causative factor in respiratory cancers. The hydrocarbons emitted in automobile exhaust are rather easily trapped by the present crankcase antismog devices. However, nitrous oxides, with which the hydrocarbons would normally bond to form relatively harmless compounds, are not filtered out by these devices. Thus, partly because of improper antismog devices and partly because automobile manufacturers have increased the compression ratios of their engines, nitrous oxide pollution has soared in areas with heavy automobile traffic.

Particulate Matter

Certain industrial processes release dusts, minerals, vegetable matter, or other waste particles into the air that are large enough to be filtered. Some of these particles are organic compounds that contain chemically active groups of sulfates, nitrates, fluorides, or ammonia. Others are essentially metallic, containing cobalt, iron, lead, or copper.

Because of the difficulties in pinpointing causal relationships in subtle

Box 16.1   The Effects of Air Pollution

Because the effects of air pollution are slow to show and because such effects often contribute to an already weakened condition, other detrimental influences that set in may receive the blame. Therefore, it is hard to pin anything on air pollution. But some hard evidence does show air pollution as being harmful. For example, it is known that the amount of head colds and pneumonia is higher in areas of high air pollution. Cigarette smokers from polluted St. Louis have four times the incidence of emphysema as do smokers from Winnipeg, Canada, where the air is cleaner. Chronic bronchitis is more serious among British postmen who serve high-smog areas than those who work in the more pollution-free districts. There have generally been sharp increases in emphysema deaths in areas where air pollution has worsened.

Lung cancer deaths in America and England have been correlated with the density of smoke in the air. In Staten Island, New York, 55 men in every 100,000 have lung cancer in the smoggiest areas, but a few miles away where it is less smoggy, the incidence is only 40 in 100,000. It has been argued that the point that air pollution contributes to death can only be made statistically, but those familiar with statistics are often alarmed at the findings. In 1969, sixty members of the faculty of the Medical School of the University of California at Los Angeles released a statement urging anyone who didn't have compelling reasons to remain, to move out of the smoggier areas of Los Angeles to avoid respiratory illnesses.

processes, it has been difficult to determine the extent to which air pollution can cause deaths. This task may become easier, however, if current trends are allowed to continue. According to the U.S. Public Health Service, annual sulfur dioxide emissions are expected to increase from 20 million tons in 1960 to 35 million tons by the year 2000. The quanity of nitrogen oxides will triple to almost 30 million tons in that period, and particulate matter will increase by 50 percent to about 45 million tons. And by then, unless we make some changes, we will have four times the number of automobiles we have now. The concomitant rise in gasoline usage will not be the only problem. How long did it take you to find a parking place this morning?

Temperature Inversions

It is likely that some spectacular tragedies due to polluted air will occur in large cities before long. Such crises are most likely when *temperature inversions* occur. Normally the air temperature decreases with altitude, but sometimes a layer of warm air moves above the cooled air, thus limiting the normal upward flow of the air from below. The result, as you see in Figure 16.3, is that the polluted lower air is not allowed to escape into the upper atmosphere. This means the foul air is trapped and the people below must breathe their pollutants. When temperature inversions occur over cities, the number of deaths from respiratory ailments usually rises sharply.

Part of the problem of air pollution in cities lies with the design of the cities themselves. It is unfortunate that we have managed the growth of our cities so miserably, especially since most people in this country live in urban areas. At present 70 percent of our population lives on only 1 percent of the land, and the outlook is for even greater shifts to metropolitan areas in the next fifty years. Such trends give us a great opportunity to devise economical ways of living, since compactness can result in manageability. However, our approch to urban development has not changed noticeably in the past few years, as evidenced by the fact that we continue to develop sprawling megalopolises (such as the one that is presently emerging as a single metropolitan area extending from about Richmond to Boston). Our cities are, at present, grossly inefficient and often dangerous places to live, and it is not likely that simply increasing their sizes will help matters much.

Those who appreciate irony may glean moments of grim humor by traveling on the New York City subway system. The cultural center of the world asks those seeking its pleasures to roar and screech about in antiquated old cars that reek with a scent suspiciously urinesque. And to wait for the privilege on dirty, sodden platforms where the sounds of passing trains can impart a permanent high-pitch deafness to the frequent traveler. It is true that such transportation methods cause far less pollution per passenger mile than the myriad honking taxis lurching above. But is this the best we can do?

One might argue that, at least, an effort has been made, at least a

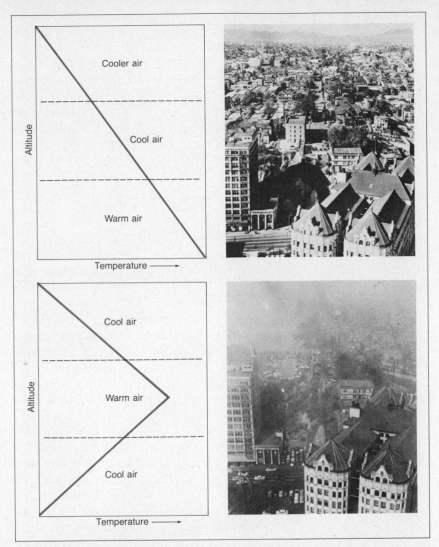

16.3 In a temperature inversion, (shown at the bottom), a warm layer of air lies over a cooler layer, trapping air pollution below it. These photographs were taken in Los Angeles in the early 1950s and demonstrate that smog is not a recent phenomenon.

lip-service to rationality. After all, Los Angeles has almost no system of mass transit other than an inadequate bus system. The citizens are almost entirely dependent on automobiles. In 1974 they voted down a one percent sales tax which would have provided funds for the beginning of a mass transit system. And when the wind is right, lettuce crops 90 miles away wilt and die.

The Prospects
The prospects for cleaning up the air on a worldwide basis are, at present, not particularly promising. Many countries seem to associate effluence with affluence (with good reason if the United States serves as a model) and are willing to put up with a fouled environment in exchange for physical

Box 16.2 Highways

Our highway system is developing according to the classic positive feedback system. More and bigger highways attract more cars, contribute to the collapse of mass-transit systems (over 200 have gone out of business since 1963) and thereby increase the need for more highways. The Highway Trust Funds (about $5 billion a year from taxes on gasoline, oil, tires, asphalt, and motor vehicles) are presently used almost entirely for new highways, and zealous politicians, particularly from oil-producing states, are ever-vigilant to see that the enormous sums are not used to break the vicious circle.

Each year one-sixth of our gross national product goes for automobiles and road building! Over four million square miles in the United States now lie paved! And although we spend over $5 billion each year to build highways alone, each mile of freeway consumes 24 acres of precious land. To date we have spent for mass-transit only 2 percent of what we have paid for highway construction and the American Association of State Highway Officials has requested that we spend $320 billion more on highways in the next 15 years (an amount about equal to our national debt).

You might also like to know that, on the average, a single mile of interstate highway costs $1,400,000.

conveniences and manufactured goods. In certain less-developed nations, the environmental movement in the United States is viewed as a preoccupation of a pampered society that can afford such luxuries.

Undoubtedly, we can make great gains toward cleaner air in the United States by the initiation of a forceful program, and the beginning of such an effort is already under way. In 1970 Congress passed a new Clean Air Act, which set strict standards for both automobiles and factories. Hydrocarbon and carbon monoxide emissions from automobiles had to be reduced 90 percent by 1976. Of course there have been powerful industry efforts to relax the standards.

Let's cite some examples. In 1968 Californians sued the automobile industry to speed up its effort at air pollution control, but for some reason the U.S. Justice Department dropped the suit. National emission standards were to be met by 1971 but in 1970 the government chose to relax those standards. In 1972 the Environmental Protection Agency gave in to industry pressure and set deadlines for 1977 (1975 in California); the national deadline must be met by 1978, unless the Clean Air Act is amended. The industry's argument is that emisson control devices reduce mileage—which is true, by about 7 percent, but 500 pounds added weight decrease mileage by 14 percent, automatic transmissions by 6 percent, and the air conditioning up to 20 percent, so the loss due to emission control is relatively small.

There is some concern that our present fuel problems, whether real or designed, may result in lowered air quality standards so that "unclean," or high-polluting, fuel can be burned in industrial operations and automobiles. Another result of a "fuel shortage" might be to relieve automobile manufacturers from the responsibility of developing low-polluting engines that, with present engine size and design, will burn more gasoline.

Other methods that might improve the quality of our air include the prompt development of good mass-transit systems and the encouragement of smaller automobile engines. This encouragement might be provided in the form of horsepower taxes—which would mean that only the wealthy could afford the luxury of pollution—or the strict regulation of engine sizes. Eventually, however, we must come to rely less on the internal-combustion engine. And there we have a marvelous example of understatement.

Water Pollution

Sewage

It has been said that the tap water of New York City contains many interesting ingredients—not the least of which is chow mein. A more precise, but no less fascinating statement is that in some localities in the United States the water one drinks has already passed through the bodies of seven or eight other good people. Under the best of conditions that water will have been filtered and treated with small amounts of deadly chlorine in an effort to poison microorganisms, so that the water will be

Box 16.3   Sewage Treatment

When our population was smaller, sewage could be safely emptied into moving rivers where bacteria and fungi (decomposers) could digest it. Those large organic molecules from excretion, fallen leaves, and other by-products of life were broken down into their elemental constituents. These constituents could then be used as building blocks by primary producers in the recycling of life's materials. When the numbers of waste-producers along such rivers became too high, the natural processes of decomposition proved too slow and sewage-treatment plants were devised. These plants are, in effect, places where the natural processes are speeded up within the confines of a restricted area.

Ideally, a modern two-stage sewage treatment plant collects the wastes discharged into our waters, gets rid of the waste, treats the water, and pumps it back into our supply. First, the largest pieces of waste are screened out. Then the water is pumped into settling tanks where finer material sinks to the bottom as sludge. The supernatant (fluid over the sludge) is pumped into tanks where air is bubbled through it to provide oxygen for bacteria and fungi, which break down the organic molecules. Sludge is again allowed to settle out and the supernatant is chlorinated to kill the bacteria and it is ready to drink again.

Unfortunately, many of our cities lack facilities for complete processing of sewage. This means that partly treated or untreated wastes are pumped into our waterways. Even those plants with adequately treated sewage present certain environmental problems. For example, the effluent from the best sewage plants is high in phosphorus, nitrogen, and certain other elements. These can act as fertilizers by enriching the water into which they are dumped. (Their effect is added to that of phosphate from detergents, which accounts for about half the phosphate overload.) The result can be clouded, smelly water covered with algae scum. Many species of algae that respond so well to inorganic fertilizers are not used as food by zooplankton and thus do little to initiate a food chain. Also, it has been determined that where pollutants raise the overall "productivity" of water, in terms of mass, it decreases the diversity of species present. Certain blue-green algae produce poisons that have even killed cattle and can produce rashes and vomiting in humans.

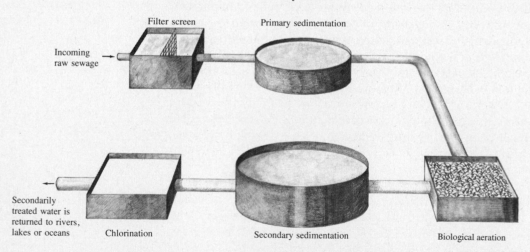

Primary Treatment

Filter screen   Primary sedimentation

Incoming
raw sewage

Secondarily
treated water is
returned to rivers,
lakes or oceans   Chlorination   Secondary sedimentation   Biological aeration

Secondary Treatment

relatively safe to drink. (Chlorine is sometimes questioned as a water additive because certain chlorine compounds can cause mutations, but the consensus is that, considering the risk of impure water, chlorination is the lesser of two evils.)

In many locales the water is not safe to drink, as evidenced by the recent outbreaks of infectious hepatitis in the United States. Infectious hepatitis is believed to be caused by a virus carried in human waste, usually through a water supply that is contaminated by sewage. There is some disturbing evidence that this virus may be resistant to chlorine, especially in the presence of high levels of organic material. Despite our national pride in indoor plumbing and walk-in bathrooms, sewage treatment for many communities in the United States is grossly inadequate, and waste that has been only partially treated is discharged into waterways. Recently the news services carried a story announcing that the New Orleans water supply may be dangerous to drink. However we have been assured that there is no cause for alarm—a committee has been appointed to study the problem!

Chemical Effluents

As we provide for our expanding population, increasing amounts of fertilizers, pesticides, radioactive particles, and strange new industrial by-products are finding their way into our water supplies. In many cases, we not only do not know the effects of the chemicals, we don't even know what the chemicals are. In the instances in which the chemicals are known, we rarely have any idea of the effects of long-term, low-dosage ingestion. The continued need for energy, food, and manufactured goods, however, ensures the entrance of a number of potentially dangerous substances into our water supply. Under these circumstances we desperately need to initiate studies on the effects of chemical contaminants as soon as possible.

The danger of our present ignorance is clear from the belated information on certain long-familiar contaminants. For example, we learned only around 1950 that nitrates, a common constituent of agricultural fertilizers and water supplies, could alter hemoglobin so as to impair its oxygen-carrying capacity. More recently it was found that certain bacteria in the digestive tract can convert nitrates to highly dangerous nitrites. Nitrate water pollution is especially dangerous in certain areas of California, Illinois, Wisconsin, and Missouri. As another example, we have known for some time that low concentrations of cadmium can be dangerous to human health, but its allowable limit in water was not set until 1962.

In many cases water pollution is so obvious that it has been the subject of wry humor, such as cartoons of a small fish begging the fisherman not to throw it back. In some cases the level of pollution is even startling. For example, areas of the Cuyahoga River in Ohio have been declared a fire hazard because of the high levels of combustible materials it carries. A recent report showed only *traces* of water in parts of the Houston ship channel. Such cases are extreme, but our country also abounds with less

16.4 A massive fish-kill of alewives as a result of pollutants being emptied into their habitat. Such kills occur with some regularity throughout the United States, sometimes on a much smaller level, perhaps as laundry unobtrusively poisons a small stream in your town. Spectacular kills such as this, however, dramatically drive home the point that we do not always act in the best interest of life on earth. If we cause such tragedies as this, how may we have changed the earth in subtle, as yet undetected, ways?

dramatic examples of pollution. Certain winding rivers passing through the beautiful Maine countryside are piled high with yellow foam from upriver paper mills. The smell emanating from such water is rather remarkable in its own right. Joggers along the beaches of southern California, running within sight of great offshore oil rigs, find their feet covered with heavy layers of tarlike substances. Backpackers in Colorado drink freely from remote mountain streams, but they must purify water as they get closer to "civilization."

Heat Pollution

A form of water pollution that is even subtler than chemicals or bacteria is *heat*. Water absorbs heat when it is used as a coolant or lubricant in

Box 16.4   A "Dead Sea" Moves on New York

The ability of man to defile his environment is being starkly demonstrated by the sea of fetid filth that is moving inexorably in Atlantic waters toward the beaches on Long Island.

For four decades, New York and its surrounding communities have been dumping their sewage sludge some 12 miles offshore where, out of sight and out of mind, it was supposed to sink and dissipate miraculously with the wash of the sea. But the heavy and continuous influx of diseased muck—five million cubic yards a year—has proved too much for the waters around it. A "dead sea" was created, which hangs like a pestilential

nightmare just beyond the city's horizon. And in 1970 that sea began to move. Its drift course was back toward the communities that, in their innocence and ignorance, had created it.

The presence of this hovering sea of ooze has been no heavily guarded secret among Atlantic seaboard officials. It has been the observable and devastating result of a rapidly conceived policy of somehow meeting the problem of an excess of human and industrial effluence. During Earth Week of 1971, in fact, it was listed as one of the top 10 ecological disasters in the world.

What has been unexpected, however, is the slow surge of this mass toward the coastal civilization that spawned it. To date the drift

defies scientific explanation. Its contamination could reach the Long Island beaches and those of the Fire Island National Seashore within three years.* The consequences of that surge, with its fecal coliform bacteria, hepatitis, encephalitis, and other diseases, have so far eluded description.

Man's inhumanity to man manifests itself in many ways. But in the drift of the "dead sea" toward the New York shoreline, he must come to understand that the recuperative powers of the seas are not inexhaustible—that the sea, too, and the men who live along its inspiring borders must pay enormous environmental prices for the abuse of its tidal blessings.

*To date this contamination has not washed ashore, but the threat still seems imminent. Although New York has been ordered by the Environmental Protection Agency to stop ocean dumping by 1981, the city continues to discharge 250 million gallons of raw sewage daily and dump sewage sludge and large materials offshore.

(Excerpted from "A 'Dead Sea' Moves on New York," *Santa Barbara News Press*, January 15, 1974, Section C, page 12. Reprinted by permission.)

Box 16.5 Spilled Oil

There has been a lot of discussion recently over the effects of oil spilled into the ocean from the operations of petroleum companies. Actually, it is estimated that the 10 million tons of oil that reaches the oceans by such means each year, is matched by natural seepage. However, an important difference exists in pollutants from the two sources. Much of the man-made contribution is partly refined and therefore its effect on living systems is much more drastic. Just how drastic isn't known. The ocean presents such severe problems to

researchers that only the most obvious and flagrant effects of oil pollution can be determined. Much of the research that has been done so far also presents some rather puzzling data. It is known that short-chain hydrocarbons damage marine larvae, and that simple ring systems are generally poisonous to living things, while complex ring systems are known to cause cancer. In spite of great efforts by local citizens, seabirds that become covered with oil almost invariably die. Even flamingos and penguins have been killed by oil. In Alaskan

waters great rafts of oil-covered corpses of birds, sea lions, hair seals, killer whales, and sea otters have been found (but in the interest of fairness to oil companies, no one wanted to say the oil had caused the deaths).

Oil contamination does not always kill. In certain cases, barnacles have grown back on oil-slicked rocks with apparently little difficulty. Some tiny isopods and stunted goose barnacles may live within the tar balls that form from oil slicks. Certain bacteria are known to digest such tar balls, but they do it so slowly, even

under ideal conditions, that the oil residue is likely to enter the food chain unchanged through another route before the bacteria have their way.

If the oil reaches the ocean floor where little oxygen exists, it is likely to persist for long periods. Inshore refined-oil spills are highly toxic and may kill worms, crustacea, mollusks, and fish for months afterward and may impart objectionable tastes to denizens of scallop and oyster beds.

manufacturing or in energy production. When it is returned to its source at this higher temperature, it may alter the biota of the water to which it is added. The temperature increase need not be great to exert profound effects. A difference of only a few degrees can markedly alter the growth rates of small aquatic organisms. Such a temperature change may also cause changes in the species composition of communities of larger organisms such as fish. Thus delicate balances, which have been established as the result of long periods of evolution, may be disrupted.

Among the greatest potential sources of heat pollution are nuclear power plants. The nuclear process generates enormous heat, and the water used as a coolant, if it is dumped offshore, could drastically alter the marine life in that area. Of course, it is in coastal waters that many of the tiny organisms at the base of the ocean's food pyramid begin their existence.

But why is the addition of only a little heat so critical? How does it exert its effect? Basically, heat pollution works in two ways, (1) by increasing the rate of oxygen consumption by aquatic organisms while, at the same time, lowering the oxygen-carrying capacity of the water (resulting in a lowering of the ability of the biological system to decompose waste), and (2) by increasing the rate of evaporation of the water, thus raising the concentration of pollutants left behind.

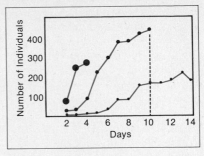

16.5 The effect of temperature variations on populations of water fleas. The largest dots represent populations at 33.6°C; the medium dots, 24.8°C; the smaller dots, 19.8°C. Note that after ten days only the populations grown at 19.8°C survived, despite the more rapid increases in the others. Small changes in the temperature of water, then, may drastically alter the composition of life in that water. On a long term basis, nuclear heating of shoreline water could drastically alter food chains in the ocean, one of our most promising sources of food for the future.

Enforcement

We have long had legislation that limits industrial water pollution—the Refuse Act of 1899. But it has only recently begun to be enforced, under the joint authority of the Environmental Protection Agency and the Army Corps of Engineers. The act was bolstered in late 1972 by the Water Pollution Control Act, which has as its goal the end of water pollution from industrial and municipal sources by 1985. So there are laws, of sorts, against polluting our waterways. But water pollution continues to become worse, partly because of lax enforcement and partly because of inadequate surveillance.

The drinking water for municipalities is monitored by the U.S. Public Health Service, but this agency is empowered only to set standards relating to the initiation or spread of disease. It has had, until very recently, little to say about the introduction of chemicals, such as pesticides, into our drinking water. The PHS is presumed to be currently revising its monitoring procedures to broaden the scope of its surveillance.

It is apparent that we need some system of firm national standards to replace the varying and flimsy set of regulations that now govern what we allow in our waters. Any such system should probably include criminal penalties for violators. Perhaps *individuals* should be held accountable for infractions instead of faceless corporations, since individuals make the decisions. One recent example of corporate punishment was the $5,000 fine imposed on an oil company responsible for a major oil spill attributed to negligence. The payment was about equivalent to a man who earns $20,000 a year being fined a nickel. It has also been suggested, perhaps too

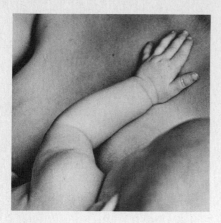

16.6 Unfit for human consumption? It has been found that the milk of some humans may contain levels of pesticides that would not be allowed in commercial cow's milk. We often eat from the top of the food chain and thereby encounter food that may, itself, have concentrated DDT in its tissues (see Figure 16.8). DDT is soluble in fat and is often stored in human fatty tissue. Sometimes the concentration there may reach such levels that some researchers have suggested that it is unwise for an overweight person to attempt to lose weight and thus metabolize his own poisonous tissues.

Table 16.2   Upper Limits of Longevity of Chlorinated Hydrocarbons in Soils

Pesticide	Percent Remaining
	14 Years Later
Aldrin	40
Chlordane	40
Endrin	41
Heptachlor	16
BHC	10
Toxaphene	45
	15 Years Later
Aldrin	28
Dieldrin	31
	17 Years Later
DDT	39

Source: "Persistence of Chlorinated Hydrocarbon Insecticides in Soil," R. G. Nash and C. A. Woolson, *Science,* Vol. 157 (August 25, 1967), pp. 924–927.

facetiously, that each member of the board of directors of any corporation that discharges wastes into waterways be required to drink a glassful of their effluent at a public ceremony every six months. A less facetious suggestion is that board members be required to pay the corporation's fines from their own pockets.

Pesticide Pollution

Pesticides are biologically rather interesting substances. They have no known counterpart in the natural world, and most of them didn't even exist thirty years ago. Today, however, a metabolic product of DDT, called DDE, may be the most common and widely distributed man-made chemical on earth. It has been found in the tissues of living things from the polar regions to the remotest parts of the oceans, forests, and mountains. Although the permissible level of DDT in cow's milk, set by the U.S. Food and Drug Administration, is 0.05 parts per million, it often occurs in human milk in concentrations as high as 5 parts per million and in human fat at levels of more than 12 parts per million.

Pesticides, of course, are products that kill pests. But what is a pest? Biologically the term has no meaning. The Colorado potato beetle, for example, was never regarded as a pest until it made its way (as a result of human activity) to Europe, where it began to seriously interfere with potato production. Perhaps this points up the best defintion of a pest: it is usually something that competes with man.

It seems that the greatest pesticidal efforts have been directed at insects and much of it has been beneficial. The heavy application of DDT since World War II resulted in sharp decreases in malaria and yellow fever in certain areas of the world. But DDT and other chlorinated hydrocarbons have continued to be spread indiscriminately any place in which insect pests are found.

There are several risks involved in such use of insecticides. For example, most such products are unselective in their targets; they simply kill *all* the insects they contact. Many insects, of course, are beneficial, and as a group they form an important part of ecosystems. Also, chemical insecticides move easily through the environment and can permeate far larger areas than intended. Another particularly serious problem with pesticides is indicated in Table 16.2: many of them persist in the environment for long periods. In other words, the chemicals are very stable and it is difficult for natural processes to break them down to their harmless components.

The most persistent insecticide is DDT. Since it just doesn't break down very easily, it remains intact in the environment, poisoning every susceptible organism it encounters. And it does this for years. The persistence assumes major significance when you consider the history of heavy DDT use.

Persistent chlorinated hydrocarbons such as DDT, aldrin, dieldrin, and benzene hexachloride have another interesting feature. They not only

Box 16.6 Pest Control

Ranchers have persuaded their representatives in Washington that some kinds of animals were either preying on their stock or eating the stock's feed. They argued that their profits were being reduced. As a result the government sent men, employed at taxpayers' expense, to trap, shoot, or poison the designated offenders. These efforts have been aided by a bounty system that enables any citizen to get a piece of the action. As an example of one such effort, men are hired to stroll through prairie dog towns spreading poisoned grain. As the tiny prairie dogs lie in their death throes, they become easy prey for various predatory birds as well as the rare black-footed ferret. The poisons in the bodies of the prairie dogs are then transferred to other plains animals such as coyotes.

The government has also regularly employed men who place meat over a short-barreled device so that when a coyote or other animal tugs at the meat, cyanide is discharged by a shotgun blast into the animal's head and mouth. (The same economic reasoning undoubtedly contributed to recent slaughters of eagles by midwest ranchers shooting from private planes. Their targets included the bald eagle, our endangered national symbol.) Data to show the economic effects of such predators are not available. However, many conservation biologists feel the ranchers' attacks are unwarranted, but are, instead, attempts to gain favorable consideration when losses are reported for tax purposes.

NOW, THIS'N HERE IS THE BALD EAGLE, OUR NATIONAL BIRD AND SYMBOL — I WANT YOU
TO KILL EVERY ONE OF THE VARMINTS YOU SEE!

16.7 Insects are effective competitors with man for the crops that he plants. He is especially possessive of cultivated plants; any competitors are viewed as thieves and are dealt with swiftly and harshly. Here, the alfalfa on the left has been sprayed with insecticides while the one on the right has been left to the tender mercies of alfalfa weevils.

16.8 Once DDT enters the food chain it becomes increasingly concentrated as it moves along from one link to the next (opposite page). Whereas its concentration may be very low in such organisms as algae, fish that eat algae tend to store most of the DDT they eat in their tissues. Other fish that eat those fish, then, encounter an increased concentration of DDT in their food supply and, in turn, store most of it in their tissues. They eventually pass these boosted amounts along to their own predators. Each level, then, ends up with more DDT in its tissues than the species on which it feeds. The numbers in the figure indicate the parts per million of DDT in the tissues of various species that have been tested in the chain. Where would humans enter the system?

remain in the environment for long periods, but they also persist in the bodies of animals that ingest them. They are thus passed intact through the food chain. This is why human milk often has higher DDT levels than cow's milk. The cow eats the grass that has been sprayed, and the fat-soluble DDT moves into the cow's tissues. Then, when a human eats the cow, the DDT that has been accumulating for years in the cow's tissues is passed along to the human. If that person is fond of beef, the DDT level in his tissues will be quickly magnified. Other animals that occupy the top of a food pyramid may concentrate DDT even further.

The tendency of DDT to be magnified in food chains has been particularly disastrous for those birds that feed high on the food pyramid. Reproductive failures in peregrine falcons, the brown pelican, and the Bermuda petrel have been attributed to this phenomenon. The problem is that DDT interferes with the birds' ability to metabolize calcium. As a result, the birds lay eggs with shells so thin that they cannot support the weight of a nesting parent.

Manufacturers of DDT have tried to demonstrate its harmlessness to humans by publicly swallowing spoonfuls of the pesticide. The idea is an ad-man's dream, but it actually proves nothing. What we're concerned about are the effects of long-term, low-level exposure to the chemical, at such dosages that occur throughout the environment. There are a host of unanswered questions regarding DDT. What overall effect is the insecticide having on the complex life systems on the earth? How serious is the evidence that it inhibits photosynthesis? In a more immediate sense, what effect does it have on mammals, particularly humans? Are there alternatives to its use?

Much of the recent evidence of the effects of DDT on mammals is not encouraging. For example, it has been shown to increase the incidence of cancers, especially liver cancer, in mice. In rats DDT causes an increase in certain liver enzymes that render many drugs useless. It also induces an increase in the deposit of dextrose in the uterus of female rats. DDT can make rats sterile, and it has similar effects on the sex hormones of birds. In the seals of San Miguel Island, off California, high levels of DDT in the tissues of females are correlated with premature births and abnormal behavior of mothers toward their young.

We don't yet know whether these effects operate in humans, since there would probably be a problem in rounding up experimental subjects. However, certain alarming findings have come to light regarding the effects of DDT on humans. For example, concentrations of DDT, DDE, DDD (another breakdown product of DDT), and dieldrin were higher in the fat of patients who died of softening of the brain, brain hemorrhage, hypertension, cirrhosis of the liver, and cancers than in a group of patients who died of infectious diseases. Those with the highest DDT levels in their body were those who had used the chemical most freely at home. Human fat may be so high in DDT that it may be dangerous for some people to diet. Extreme dieting has produced peculiar neurological effects, ap-

Box 16.7 Insecticides

The majority of synthetic insecticides falls into two groups: *chlorinated hydrocarbons* and *organophosphates*. The chlorinated hydrocarbons include DDT (and its still-potent breakdown product DDE), benzene hexachloride (BHC), dieldrin, endrin, aldrin, chlordane, lindane, toxophene, and isodrin. DDT is by far the most studied of these. DDT is also rather typical for the group, although other compounds may be more or less toxic, water soluble, or persistent (hard for nature to break down). These chemicals somehow work on the central nervous system of insects causing excitability, convulsions, and death. They are soluble in fats and tend to collect in the fatty tissues of animals rather than being excreted. Chlorinated hydrocarbons are rather broad-spectrum poisons and affect a wide variety of animals in addition to insects, including vertebrates. They are also very stable (especially DDE), and mobile. The threat from chlorinated hydrocarbons is thus based on four properties: their affinity for living systems, their range of biological activity, their stability, and their mobility.

The organophosphates include parathion, malathion, diazinon, azondrin, TEPP, phosdrin, and others. These poisons are descended from a nerve gas called Tabun developed by the Nazis in World War II. Basically, they function as inhibitors of cholinesterase activity. Cholinesterase breaks down the neural transmitter substance, acetylcholine. The poison thus causes increased activity of the nervous system and the victim dies, twisting and jerking wildly. Organophosphates may be more immediately poisonous to vertebrates than chlorinated hydrocarbons, but they are unstable and nonpersistent and thus do not accumulate in food chains.

Box 16.8 Integrated Control of Insect Pests

It has been suggested by promoters of chlorinated hydrocarbons that all that stands between us and starvation because of insect-borne diseases are insecticides like DDT. Not so. We are aware of short-lived insecticides that bear very careful handling by humans because they also may kill mammals. And there are ways to reduce threats from insects that involve means other than chemical. It has been found that sterile males released in large numbers will virtually wipe out populations of some insects. The females can breed only once. If they mate with a sterile male, of course, no progeny are produced. If the population density should become so low that even potent individuals are unlikely to find each other, the population is doomed.

In the 1880s the cottony cushion scale threatened the developing California citrus industry. The growers introduced vedalia beetles, which attacked the scale and the problem was solved. In the 1940s, however, DDT was used on the citrus. It killed off the vedalia beetles and the scale insects blossomed again. When DDT use was discontinued and the beetle reintroduced, the scale insect was controlled once again. So the introduction of *very carefully* chosen predators may reduce pests.

Probably the system called "integrated control" is our best bet at present. Integrated control involves a multifaceted attack that simply keeps numbers of pests below levels where they pose a threat. For example, mosquitos might be controlled by draining swamps, introducing fish that eat mosquito larvae, spreading biodegradable chemicals on the surface of water to suffocate larvae, the judicious use of short-lived insecticides, and luring adults to traps. This way, no dangerous pesticides are used.

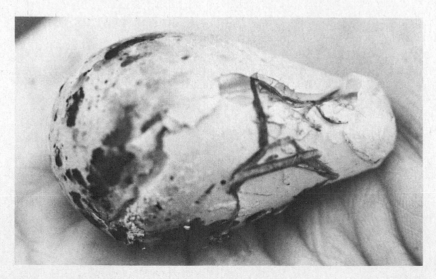

DDT causes birds to lay eggs with very thin shells that break under the weight of a nesting parent. This is a tern egg from Great Gull Island, New York.

parently a result of a person utilizing his own fat. In a bizarre sense, then, his fat may be poisoning him.

In summary, we know that the earth's living things are being exposed to long-term low dosages of chlorinated hydrocarbons. We don't know what the effects of such exposure are on living systems, or on individual species, including our own. Even in the face of such questions, DDT continues to be used in large quantities on a worldwide basis, each application being added to the still active residue of earlier years. Thus, whether we like it or not, we are part of a grand experiment. The results should prove interesting.

Biodegradable Pesticides

It would seem that the obvious solution to the long-term pesticide problem would be to switch to short-lived biodegradable pesticides. These have the obvious drawback of being deadly, in some cases to humans, while they are active. But what about after they are supposed to be decomposed into their harmless components? There are some recent disturbing findings that indicate that the harmless by-products may not be so harmless after all.

Swallowtail butterflies can be reared in the laboratory, the larvae being fed on parsley. After parsley was sprayed with a nonpersistent pesticide to rid it of mites, the plants were left twice as long as the manufacturer had suggested to allow its breakdown. The larvae then ate the parsley and formed pupae and then adults. But the adults, it turned out, were sterile! The problem lay in a minor malformation of the male sex organs. But those sex organs didn't even exist when the larvae ate the sprayed parsley.

Obviously, something is happening here that was not anticipated. Another way of saying it is, we don't really understand enough about the effects of pesticides. Do our sprayed foods have the potential of causing catastrophic changes in our offspring twenty years from now?

Radiation

Life on earth has always been exposed to the effects of ionizing radiation in one form or another. In fact, we live in a veritable sea of radiation. One source of radiation is the earth's crust, where radioactive substances undergo spontaneous nuclear disintegration, emitting both high-speed atomic particles and penetrating electromagnetic rays (similar to X rays). In addition, we are continually bombarded with radioactive products from extraterrestrial sources in the form of cosmic rays. Other radioactive substances, such as potassium 40, circulate through living systems. Such naturally occurring radiation is called *background radiation*. We do not know at present the degree to which background radiation is responsible for cancer or birth defects in the human population. We do know, however, that ionizing radiation causes mutations and cancer under experimental conditions.

A much more serious problem is the radiation level to which we are

Box 16.9 The Brown's Ferry Incident

On March 22, 1975, at Brown's Ferry, Alabama, workmen at the world's largest nuclear power installation where checking for air leaks between the reactor building and the main plant. They were using a candle, watching the flame to warn of any draught. However, the polyurethane used to seal off the two areas caught fire and the flames quickly spread to the myriad cables controlling the reactor's various functions. The resulting fiasco was unbelievable. It was 15 minutes before an alarm sounded, perhaps due to the fact that a metal plate covered the alarm control. Inside, lights were blinking randomly and all monitoring instruments failed. Furthermore, the fire extinguishers at the facility proved inadequate. The fire department at Athens, Alabama, 10 miles away was reached after a few wrong numbers. When the fire department finally arrived at the scene, the burning cable insulation had ignited other cables that had been placed near them despite warnings from federal investigators. None of the equipment or fire-retardant chemicals were effective and the hoses wouldn't reach the fire. Finally, after 5 hours the local fire department tried water and managed to stop the fire. Water was not used earlier because the local fire chief couldn't get permission from the plant authorities.

The coolant water over the core was now dangerously low, threatening a catastrophic meltdown. None of the facility pumps would work, but someone managed to rig a low pressure water pump and succeeded in cooling the core. The city most threatened was downwind Decatur, but on that day its air sampler was not working. In spite of the close call, those in charge of evacuation did not hear of the fire until two days later.

There had been no fewer than 207 safety inspections between 1966 and 1975 and the walls were covered with formal and assuring plaques and safety citations. We had been assured.

At present the nuclear industry admits that there are 30 unresolved technical issues surrounding nuclear power, and that there are often unexplained vibrations deep within the cores. The Brown's Ferry incident, by the way, was not defined as a "nuclear accident," because the problem was confined to wiring. The record remains unblemished, and air leaks are now checked with . . . feathers.

Box 16.10 "The Kyshtym Disaster"

"Superficially, everything was in order. There were trees and grass. But it was an empty land, like on the moon. There were no villages, no towns, only the chimneys of destroyed houses; no cultivated fields or pastures, no herds or people—nothing."

This is the description of Leo Tumerman, 78, a Russian-born physicist who had returned to Russia after settling in Israel. The West had heard rumors of what he described, but this was the first eyewitness account. What he had seen was believed to be the result of a nuclear accident that had shattered the Soviet heartland about twenty years ago.

No one knows what caused the problem but people still whispered about the terrible "Kyshtym disaster" in which the village of Sverdlovsk was destroyed. Tumerman was told the accident resulted from careless handling of nuclear wastes at a plant where plutonium was processed for hydrogen bombs. An exiled Soviet geneticist, Zhores Medvedev, said that buried wastes had exploded "like a volcano" and that strong winds had spread the radioactivity for hundreds of miles. Western scientists believed the accident was due to a malfunctioning reactor, possibly triggered by an earthquake.

Medvedev and other scientists from a Moscow radiation laboratory were hastily rushed into the area. They found hundreds dead and thousands more suffering from radiation poisoning. The Soviets had no official comment, but Tumerman told of a highway sign that warned travelers not to stop, and to drive the next 20 miles as fast as their cars would go.

exposed through our own activities. The use of radioactive substances has greatly increased in the last thirty or forty years, and now poses severe problems for us. More than 90 percent of man-made radiation exposure occurs in medical and dental uses, and there is some evidence that such use should be sharply curtailed. For example, health physicist Karl Z. Morgan has argued that unnecessary medical radiation may presently be causing 3,000 to 30,000 deaths annually in the United States alone.

We are also exposed to the results of fallout from nuclear weapons. In the United States, however, fallout presently accounts for only 1 or 2 percent of our environmental radiation. The rate of increase of radioactivity was rising sharply until the United States, Britain, and Russia stopped atmospheric testing in 1963. Of course, the Chinese and the French have continued surface testing of atomic weapons and we are all exposed to the radioactivity. Mark Twain would have new things to say about the French if he were here to witness their nuclear policies.

Nuclear Power Plants

In the United States an argument has been raging for the past few years over the risk involved in the proliferation of nuclear power plants. There can be no question that with our electrical power "needs" doubling every eight years, we cannot rely indefinitely on the earth's remaining fossil fuel supply. The question is, can we safely shift our reliance to nuclear fission power plants—considering the present state of our knowledge and technology?

It is not likely that an explosion of the type produced by atomic bombs can occur in the kinds of nuclear reactors being used today. But other types of tragedies are more likely. Radioactivity could be released into the environment from activities related to mining and processing the fuel, from the transportation and recycling of the fuel and from storage of the radioactive wastes. In addition, accidents can occur while the plants are in operation. Should the reactor core melt, or be disrupted by some event such as an earthquake or sabotage, radioactivity equivalent to that which would be produced by hundreds of bombs the size of those that devastated Hiroshima and Nagasaki could be released into the environment. Such accidents are said to be unlikely, but the extremity of any such disaster renders "odds" meaningless. Recently, two members of the same family in the Northeast were winners in a lottery. The odds against that were 5 billion to one, the same odds that nuclear power proponents were giving against a major disaster. (These odds, by the way, are recorded in the "Rasmussen Report," a paper that has been highly criticized by mathematicians as well as nuclear engineers for being based on countless unwarranted assumptions.)

Even on the basis of the calculated risk, insurance companies, which are presumably well advised in such matters, will not underwrite more than about 1 percent of the potential liability of such an accident. Instead, by act of Congress, the public would receive up to $478 million in public

16.9 Diagram of a nuclear power plant. Uranium bars comprise the fuel. As the decomposing uranium heats water, steam is produced, which turns turbines and operates electric generators. The fuel is highly reactive U^{235}, which is not present in concentrations great enough to cause a nuclear explosion. As a molecule of U^{235} degenerates (or explodes), its escaping neutrons are allowed to strike other U^{235} molecules after having been slowed by some sort of dense material. The molecules that are struck then become unstable and explode, producing a controlled chain reaction.

U^{238}, which is much more common, is very stable and hence cannot be used as a fuel. However, if it is struck by fast-moving neutrons, it is converted to highly reactive U^{239} or plutonium. Plutonium made in this way could provide our nuclear needs for another 60,000 years. But there are problems. The process takes place in breeder reactors, and these have yet to be perfected. Also it is a rather simple matter to make nuclear bombs from plutonium and it is estimated that, if the program progresses according to schedule, within 40 years there will be 100,000 shipments of nuclear material in the United States each year—a tempting target for terrorists. In mid-1977 the U.S. government admitted that it could not account for substantial amounts of nuclear fuel, but they added that they did not believe it had been stolen.

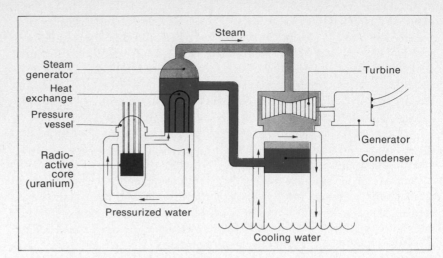

funds and assume the rest of the damages themselves. Figure *that* one out! A 1957 report by the Atomic Energy Commission concluded that a single major accident might cause 3,400 deaths at distances up to 15 miles and 43,000 injuries at distances up to 45 miles, with property damage up to $7 billion. That report is twenty years old. Today property is higher priced, populations are denser, and nuclear power plants are larger. Of course, these figures refer only to the immediate destruction. The residents of Nagasaki are still dying from the progressive effects of radiation damage incurred over thirty years ago. Also, today, there is another factor that was not operative in 1957. Today *I* own a house and *I* live near a nuclear power site! The issue is suddenly more than academic for millions of people.

Completely apart from the possibility of accidents, there is the unsolved problem of what to do with the radioactive wastes generated in the course of normal nuclear plant reactions. The problem of how to store nuclear wastes is a tough one since such wastes can only be rendered safe by the passage of time. The waste radioactivity is generated in the fuel system of the reactors because only a part of the fuel is fissionable, and for technical reasons not all of the fissionable elements are spent. Much of the spent fuel materials removed from the reactor core can be reused. However, some of the radioactive fuel in the spent elements cannot be recovered, and this material comprises the radioactive waste.

And then there is a problem with coolants. The coolant, whether it is water or gas, becomes radioactive as fission products manage to escape through the metal casing of the uranium. Water coolants may also become radioactive by a process called *neutron activation,* which occurs when one of the hydrogen ions in the water molecule is replaced by a tritium ion. The present practice is to dilute this radioactive water with plain water and then to dump the mixture onto the ground, where, of course, it can seep into the water table below. The enthusiastic developers of this process have even proposed that water with too high a radioactive level for human consump-

Box 16.11   Radiation

The nuclei of some atoms are unstable. Of the more than 320 isotopes that exist in nature, about 60 are unstable, or radioactive. In addition, man has created about 200 more. Such creations are called *radioisotopes*. The substances in which they exist are *radioactive*. Some of these unstable isotopes occur in natural systems; others are manufactured by man. When a radioisotope decays (or decomposes) certain particles (positively or negatively charged) and rays are emitted from its nucleus. The radioisotope is then changed from one element to another. The new atom may also be radioactive, or it may be stable.

Radioisotopes decay at predictable rates. These rates are usually expressed as the element's *half-life*. Half-life refers to the time that must expire in order for half the atoms in any amount of the isotope to decay to another kind of atom. Half-lives may be very long. The half-life of natural uranium (U_{238}) is 4,500,000,000 years! However, as a rule of thumb, the time it takes for a substance to become nonradioactive is considered at 20 half-lives.

When radioisotopes decay ("explode" may be a better word), the particles and rays may strike other atoms in the environment in such a way as to tear electrons away from them. Thus the process is called *ionizing radiation*. When living tissue is thus disrupted, the effects can range from reduced life expectancy through cancer, leukemia or other means, to instant death. There is *no* threshold below which radiation is genetically harmless. Also, the genetic effects of radiation are accumulative in the exposed individual and his offspring. A "snowball effect" operates so that an increased radiation level, which has little effect in the first few generations, may, if maintained, produce disasters, even extinction, in following generations.

Radioactivity of any substance can only be reduced by allowing the radioisotopes to decay naturally. The process cannot be hastened.

Potatoes subjected to varying doses of radiation. Clockwise from upper left, the first received none and sprouted and softened normally in response to its genetic signals. Increasing doses altered the genome until the one at lower far left failed to soften or sprout at all.

tion be used for such purposes as irrigating crops and watering livestock. A single course in general biology should dispel such foolish ideas soon enough. The frightening thing is that competent, highly-trained scientists have proposed other ideas that are potentially just as dangerous.

Highly radioactive liquid wastes are presently being stored in steel tanks at several sites around the country. It was recently found that 200,000 gallons of this waste had leaked from a tank farm holding 80 million gallons at Hanford, Washington! Because such wastes remain dangerous for thousands of years, there is a problem in building containers that will safely hold the material for that length of time. Certainly, steel holding-tanks are not the answer. The Atomic Energy Commission has suggested reducing high-level wastes to solid form and then storing these, perhaps permanently, in salt beds, deep underground. Technical problems and an increasingly skeptical public have stalled these plans for the present.

Heavy Metal Pollution

Believe it or not, lead-lined food containers, according to some historians, may have contributed to the fall of both the Greek and Roman civilizations. Because lead is noncorrodible and has a sweet taste, it was used to line their bronze vessels. However, lead can cause brain damage. Thus, it is believed, widespread lead poisoning may have caused mental debilitation in the citizenry of these great societies.

Today, we prefer our lead other ways, such as in our gasoline (so our high-compression cars don't knock), pesticides, paint, pipes, ceramics, glassware, and the solder in food cans. We're very familiar with the hazards of lead, but unleaded gasoline in mid-1971 accounted for only 5 percent of all gasoline sales. Part of the reason is that unleaded gasoline is from two to six cents higher in price per gallon. Since 1970, however, cars have become more tolerant of unleaded gas, and since 1975, all cars have been required to use it; that is, unless we relax our new regulations. The strictures on depositing lead into the environment should be strengthened because of the severe effects of lead poisoning. The symptoms include general apathy, weakness, loss of appetite, loss of coordination, and miscarriage.

Mercury provided the news media with a new toy recently as they projected the dangers of this metal into the public consciousness. The story involved finding high levels of mercury in the flesh of swordfish. Swordfish sales immediately plummetted, despite statements from the fishing industry that prodigious amounts of swordfish would have to be eaten for the mercury to pose any danger to humans. Because of the apparent threat, studies were immediately launched to see whether human activities had raised the mercury levels in the oceans. The findings indicated that man had not added significantly to natural levels of the oceans' mercury.

Mercury pollution in fresh-water bodies is a different story. It seems that man has sharply raised mercury levels in rivers, lakes, and streams. The Saskatchewan River, which runs through Edmonton, Alberta, has only half the amount of mercury before it enters the city that it has when it leaves. The usual sources of freshwater mercury are industrial processes (especially those involved in the production of plastics and caustic soda), pulp and paper mills, seed fungicides, and fossil fuels. The relatively small amount of mercury in fossil fuels is a threat because of the prodigious volume of such fuels that is burned.

Interestingly, mercury is relatively harmless in its metallic form. However, certain microorganisms present in water are able to convert it to methylmercury (CH_3Hg) and dimethylmercury ($(CH_3)_2Hg$), highly poisonous substances that are water soluble, and hence easily dispersed. The startling part of the story is that even if we immediately stopped dumping mercury into water on a worldwide basis, there may already be enough metallic mercury in the environment to pose a serious threat to later generations, as it continues to be slowly converted to the methylmercuries.

Mercury, like DDT, is amplified as it moves through the food chain, which accounts for the high mercury levels in the large fish at the top of the food pyramid. It is also likely to appear in higher concentrations in human populations that rely heavily on fish in their diet. The symptoms of severe mercury poisoning include blindness, deafness, loss of coordination, insanity, and death.

Other heavy metals are also increasing in our environment and are being viewed suspiciously, although we do not fully understand the implications of long-term, low-dosage exposure. Such metals include cadmium, chromium, arsenic, and nickel.

Noise Pollution

The real meaning of noise pollution is well known to anyone who has ever retreated in solitude to a remote area, to be alone, to think, only to have the silence rudely shattered by the blast of a trail bike. His eyes may turn to mere slits while he wistfully fondles an oak branch as the biker roars by, supremely confident in a sporting image, well cultivated by paunchy advertisers.

The same animosity is often described by people exposed to the raucous spattering of snowmobiles, power lawnmowers, or any number of other machines that seem to be designed with disturbance in mind. For sailors, nothing can ruin a day easier than sharing the water with a chromed and sparkle-painted powerboat. Then there are the "camper vehicles," those behemoths that may come replete with televisions and dishwashers, utterly dependent on outside power sources, and that increasingly disturb our national parks, requiring the development of more and more facilities to cater to their monumental needs. (You may detect faint personal prejudices on my part here and there in spite of my obviously intense efforts to be objective. After recently returning from a couple years living under sometimes primitive, but serene, conditions in Europe and the Near East I am newly sensitive to the sounds we seem to tolerate. Why are those inane TV commercials louder than the programs

16.10 Minamata is a small fishing village on the southern Japanese island of Kyushu. It is also the home of the Chisso chemical company whose founder is credited with saying, "Treat the workers like cows and horses." But they did worse. They polluted Minamata Bay so drastically that the beautiful sea was turned into a veritable sludge dump of industrial poisons. Years ago the people found that their fish catches were falling off and blamed the company, which eventually paid each fisherman a token amount and continued to pollute. Finally, in the early 1950s a mysterious disease began to take the citizens of the town. It was called the "strange disease" because no one knew what caused it. When anyone accused the company, Chisso would produce a gallery of "expert witnesses" who would faithfully exonerate it. It blocked every effort to stop its dumping. Once, when it began to dump in a new place, people there began to fall ill. Finally, one of its researchers could keep quiet no longer. The agent was mercury from the company's drains.

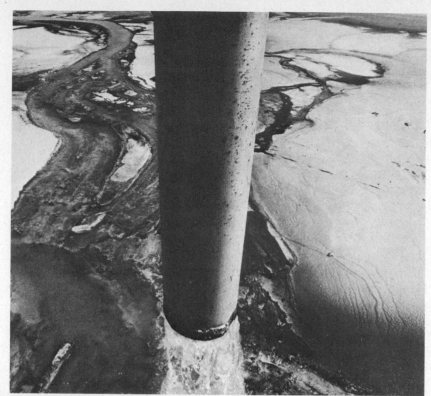

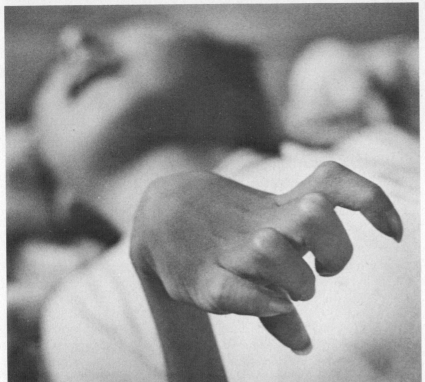

"Minamata Disease" begins with a tingling around the lips. Then numbness sets in and speech blurs. Finally blindness engulfs the victims even as their movements become jerky and uncoordinated and they shout uncontrollably. Many die. The famous essay photographer, W. Eugene Smith, went to Minamata to shoot these pictures of the victims' plight and at a demonstration received a broken back by the hands of thugs for his trouble. What he found were helpless people, cared for by their families, people who would never recover because they had eaten fish from Minamata Bay. Autopsies showed the brains of the victims were soft and spongy because the cells had deteriorated, the result of methyl-poisoning. The culprit was unveiled when a company researcher fed effluent to the now-famous "cat number 400" and it developed the disease. The company immediately censured the researcher and hid his findings. Chisso continued to pollute the water until 1968 when the mercury method of production became outmoded. But people continued to fall ill. Kumamoto University doctors have suggested that the number may reach 10,000.

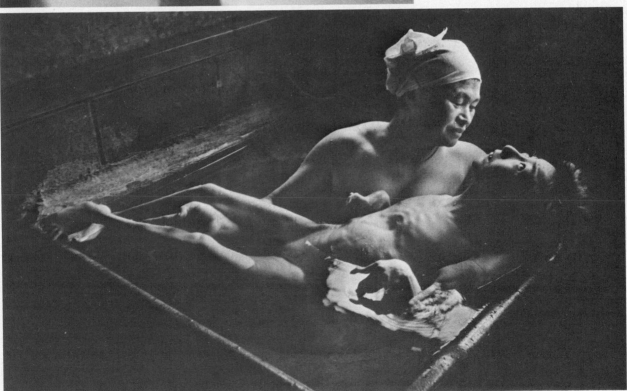

Table 16.3 Decibel Levels of Sounds

Noise Level	Decibels
Threshold of hearing	0
Normal breathing	10
Whispering	30
Home noises	45
Conversation	60
Food blender	80
Heavy automobile traffic	100
Jet aircraft taking off	120

themselves? Why do we allow them to force endless silly skits and songs upon us at irritating intensities? Why was a planned investigation into this problem a few years ago suddenly squelched?)

We need not enumerate other well-known major sources of noise, especially since you undoubtedly have your own list, but two deserve special mention. There is some evidence that the high amplification of rock music causes *permanent* hearing damage. Thus, the advent of hard rock in the '60s may have contributed to an entire generation of the hard of hearing. By the way, if you're prepared to bet that classical music is just as loud, you'll lose. Classical music just seems loud because it is characterized by sudden onsets of sound that, to the ear, translates as loudness, but without the disruptive effects. To hear for yourself, play a classical tape backwards sometime.

There is great scientific concern over the potential effects of supersonic transport planes (SSTs), should they finally find legislative approval for commercial use. Several powerful lobbying organizations quietly continue to work rather effectively in Washington in an effort to secure approval for the development and use of the SST in this country. The British-French version, the money-losing Concorde, already has permission to land at Dulles and is pressuring for permission to use New York's airports, but it seems at present that New Yorkers have had about enough.

Noise is usually measured in decibels. A 10-fold increase in the volume of a sound adds 10 decibel units. A 100-fold increase adds 20, and so on. Table 16.3 gives the decibel units of some familiar sounds.

You might also like to know that in addition to hearing loss, noise may cause several stress-related diseases such as hypertension and peptic ulcers. There is also growing evidence that noise may be associated with irreversible changes in the nervous system. Much of the evidence is circumstantial, and researchers are presently engaged in gathering and analyzing more specific data on the effects of noise.

HIDDEN DECISIONS

These, then, are a few of the problems we now face. You may wonder, why haven't we acted? Why have these situations been allowed to arise? If you're asking these sorts of questions, you may be at first gratified, and then horrified, to learn to what extent we *have* been addressing ourselves to these problems. Hard decisions *are* being made. The problem is, we are often not aware of their biological implications. Our lawmakers set policies that show disturbing tunnel vision. Often they are never really aware of what global wheels they are putting in motion. Let's consider a few examples.

The United States is proud of its system of public education. Every child in this country has the right, even the obligation, to be educated to some degree. The institutions exist and are funded. Education is obviously good, so what lawmaker wants to stand against it? However, public education has a side effect. It subtly encourages people to have children, for the cost of the education of their own children is borne by everyone, including childless couples. It is true that those childless couples might willingly forego an occasional weekend at the slopes to avoid having to live among the ignorant offspring of other couples, but nevertheless, their subsidizing of other people's children encourages the bearing of offspring.

How about our system of taxation? It was once cheaper to be married. Couples filing jointly were taxed less. By 1977, however, it was often cheaper to be single and just live together. Probably few people failed to get married because of increased taxation, but some couples did divorce and continue cohabiting to reduce taxation. If taxation policy can force a change in marriage customs, doesn't this constitute a governmental decision about how we will lead our lives?

Do tax deductions for home mortgages encourage building? Does home building affect forests? What are the effects on land use? On sewage? On water supplies? On road building and its handmaidens? What are the social effects of home ownership as opposed to apartment rental?

And then there are our "rapid depreciation schedules." This means that a company that provides automobiles for its traveling employees is encouraged to dump perfectly serviceable cars in order to purchase new ones. The encouragement lies in a system of tax write-offs. New cars depreciate faster and reduce the apparent earnings of the company. New car sales are thus boosted. More people are put to work building automobiles. More mountains are shredded, more sky is blackened, more waterways are poisoned, more lungs are damaged, more work days are missed, and the need for health care increases. The lawmakers drawing up "rapid depreciation schedules," however, were not thinking of blackened lungs.

How about our relative treatment of railroads and highways? Highways, of course, are largely publicly financed through taxation, while railroads have traditionally been privately financed. If railroads had had more public backing (with its attendant government control), perhaps passenger service would not be the farce it now is, with railroads apparently making people-carrying so inconvenient to the passengers that they prefer to travel by other means. The railroads are then in a position to ask regulatory agencies for permission to drop passenger runs in favor of much more lucrative cargo. What are the effects of people being forced to use higher-polluting and energy consuming cars on acreage-eating highways? What social patterns have developed from our powerful encouragement to use private means of transport? How have land-use policies been influenced by the automobile? What happens when farmland is paved?

The list of our hidden decisions is endless. What are the effects of building a dam and localizing a water supply? How do populations shift? Are the people who follow the life-giving water easier to police, or does their proximity make it easier for them to attack each other? What effect does water control have on farming policy? Do we really know enough to continue to attempt to harness our rivers? Can we believe the firm assurances of the chambers of commerce and the Army Corps of Engineers?

The point of all this is, not only *can* we make biological decisions, we *do* make biological decisions. And we do it almost casually, on a daily basis. The ignorance of the decision-makers does not lessen their effect. And no one would say that decisions should not be made. We have come too far to try to leave well enough alone. Not only *are* they being made, the decisions *must* be made. However, we must become aware of the far-reaching effects of our manipulations. In other words, our decisions must make biological sense. Perhaps the answer lies in educating our lawmakers. Or perhaps the answer lies in first educating ourselves and then electing lawmakers from among us who are fully aware of what they're doing and who can confidently expect the support of an enlightened public.

THE FUTURE

On the basis of the magnitude of some of our problems, you might be touched by traces of pessimism regarding our future. However, that spirit simply cannot be allowed to prevail. Some of our problems demand immediate, vigorous, and intelligent action, and they must be boldly addressed if we are going to buy the time to consider long-range solutions.

If we can deal symptomatically, through technology and stopgap efforts, with our most pressing problems, perhaps we can buy that time. However, we need to clearly differentiate those kinds of solutions from the long-range ones. For example, bolstering food production, at least with present methods, can only be considered stopgap. The population will eventually catch up. There is no question, in fact, that overreliance on technology at this point will ultimately lead to quite a different kind of world than the one that we could build if we shifted the *direction* of our development.

The ultimate answer does not lie in finding newer and better ways to do what we're doing. It can only be found, it seems, in a change in our behavior, a wholesale redirection of our course. Such shifts, of course, demand new thinking, new ways of looking at things, new values. This is the hard part. What if we thought through our options one more time and decided that welfare rolls must be drastically *increased* to pull people *out* of production, and that, in order to support people who choose not to work at all, we must all pay much *higher* taxes; that industrial management is somehow akin to prostitution; that acquirers without direct need are socially maladjusted; that a pleasant storyteller is a more socially valuable individual than a manufacturer; that patriots are those people who refuse to fight, or support fighting, on any soil other than their own. Spring those arguments on someone you know in order to see how entrenched some of our ideas are.

But aren't there things we can do that we know would be beneficial? Things that are more realistic? Something that we might be physically and psychologically prepared to do at present? Some changes, after all, would not greatly alter our life style. Some are even absurdly obvious. We could cut back on our use of disposable goods. We could reduce our reliance on the automobile by developing good mass-transit systems. We could decrease our use of oil and plastic. We could reduce our bizarre fascination with packaging. (Notice how many nonessential goods you end up with when you simply buy a shirt.) It is argued that such measures would reduce production and that would lead to unemployment. The answer is to put these people to work making things we really need.

So, what if people were hired to produce things we need rather than things that are used once and thrown away. Undoubtedly the initial reorganization and retooling would temporarily reduce profits. In the long run, however, we might benefit from changing from material-centered production to labor-centered services. People who work in automobile production, a trade that demands new steel from our earth, new plastics, new cloth, and immense amounts of energy, could switch to automobile maintenance. Other people could be put to work finding better ways for us to travel.

The new trend, then, might be toward services. We need mechanics, repairmen, doctors, lawyers, regional planners, plumbers, environmental architects, city planners, and even teachers. There is plenty of work to be done. And, although you may disagree, I think we need artists, entertainers, performers, and musicians. It may be worthwhile to shift a greater amount of our financial resources in their direction. One of the many lessons I learned from living in New York is that there are uncounted, virtually unknown artisans among us, people who truly excel at their trade. There are musicians who can leave us spellbound and writers who can tell stories that reflect our spirit as accurately as any mirror. Many of these people can barely keep body and soul together. Would we be willing to pay higher taxes to sponsor such people? The point is that perhaps we can come to place less importance on *things* and to place more importance on services. We might eventually come to elevate the value of sheer experience. If many of us felt this way, we might cultivate new Nureyevs, Picassos, Croces, Stravinskys, and Hendrixes among us.

One would hope that with some new insight, life could come to be more than a series of accumulative events. This is not to say that any new perspective or information will produce a utopian society. Perhaps that was possible at one time, but not now. Now we must engage in trade-offs, partly because there is no chance of pulling our population down to a reasonable size for many years. Driving a well-maintained 1957 automobile is not the same as having a new car. Cardboard comes from trees; toothpaste tubes without boxes are harder to stack. If we should de-emphasize production, we will have less goods. If we opt for subsidizing nonproducers, we will have less money.

Economist Kenneth Boulding has suggested that our problems of resources and pollution could be greatly helped by a single philosophical shift—from a "cowboy economy" to a "spaceman economy." The cowboy attitude involves a reckless, exploitive philosophy. The American cowboy could always assume that more resources were waiting just over the horizon. His charge was to press on, find those resources and leave the garbage behind in the untracked wilderness. It may have made sense—then. Human welfare was most likely to be improved through economic growth—*any* kind of growth, with no thought for the wastes. The important thing was to grow, develop, and surge onward to seek new ways to grow.

But that's over now. At least it should be. Now we've covered the entire land and exploited most of the resources that could be reached. Our numbers are so great that our efforts to grow threaten the system with collapse. Our waste is poisoning us, our animals, and our crops. There is no place to put it—no way to leave it. We are spacemen in cowboy hats. The sooner we realize that our planet is merely a spaceship with only limited resources to last us through our entire journey, the sooner we may be able to shed our cowboy mentality and see ourselves for what we are: travelers on the Spaceship Earth.

If you are reading this book, chances are that you are young and so you have the most to lose. The decisions will be particularly important for you, whether you make them yourself or allow them to be imposed on you. There are those who say that yours is the last generation that will live out its lifetime. Whether they are right or wrong depends on what you do, what you will allow and what you simply will not allow.

We have but a few more years of this odd interlude in history in which we will be able to ignore our biological imperatives, to fail to recognize that we are a recent upstart among the venerable and ancient species, and that our future is not insured. It is important that, as this period of borrowed time runs out, we become aware that our limitations are those of fragile cytoplasm, that we see the urgency of a steady-state global economy, an economy no longer based on a mortgaged future. What of the Global Debt we bequeath our posterity? We leave them as part of an enormous and growing population, with vast quantities of resources rendered useless or dangerous, and with chemical toxins permeating the globe. Will it be said by the survivors of our era that we stupidly and insensitively foreclosed their future and demeaned their existence? Will our children curse us? Or will they gratefully stand in admiration of our strength, dignity, and wisdom at a time when decisions were hard?

Appendix A
The Classification of Organisms

Living things may be classified a number of ways and no single way is right, the others being wrong. It finally gets down to whom one wishes to trust. This is not to say that classification is completely arbitrary. In some cases the divisions are obvious. If something has leaves it goes in the plant kingdom. If it can read it's an animal. But the problem immediately becomes apparent; not all plants have leaves and not all animals can read. So how much does one expand the plant and animal categories? At which point does an organism differ so much from others that it cannot be called one of them? Grackles are large, noisy birds, at least in Texas. Those in Puerto Rico are small versions with different voices. But there are those who study grackles who argue that they are the same species, that if they were to associate, they would interbreed. And interbreeding is one criterion that places two groups in the same species. Generally, if a group of similar animals strongly interacts to the extent that they interbreed and can produce healthy, viable offspring, they are said to be of the same species. However, those offspring must also be able to interbreed. Thus, although donkeys and mares produce mules, the mules are sterile, so the parents are put in different species. But even the interbreeding criterion sometimes breaks down. For example, there is a salamander in California whose range is rather long and narrow, and it curves around so that one end of the range overlaps the other end:

Species range

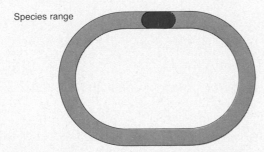

Whereas the salamanders interbreed along their range, they gradually change from one area to the next so that where the ends of the range overlap, the populations at those two ends can no longer interbreed.

These, then, are some of the problems with classification. The rules are not hard and fast. However, the disagreements generally lie at the lower end of the scale, among the genera and species. There is less disagreement about the levels of classification we will present here.

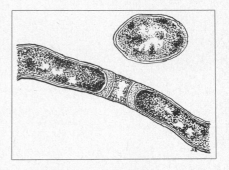

We will begin our discussion of classification with the simplest kingdom, the Monera. This system may soon have to be revised, however. On an earth only 4.6 billion years old, scientists have now discovered that a familiar bacterialike life form (see above) may be 3.5 billion years old. If so, this group, called the *archaebacteria* will be the oldest and simplest life form known. This type of organism lives in oxygen-free pockets of the earth in such places as cow stomachs and hot springs. It takes in carbon dioxide and hydrogen and releases methane (and so is also called a *methanogen*). Thus it could have survived on a planet in those remote days when oxygen was rare. It has long been assumed that all terrestrial life evolved along two distinct lines, one giving rise to the bacteria, the other to higher organisms. The peculiar ribosomal RNA of the archaebacteria suggests a third line of evolution. The discovery was announced in late 1977 and is sure to trigger heated debates among taxonomists.

Note: This system of classification is based closely upon that of R. H. Whittaker (in "New Concepts of Kingdoms of Organisms," *Science,* vol. 163, pp. 150–160, January 10, 1969).

KINGDOM MONERA

Prokaryotes (lack a nuclear envelope, plastids, mitochondria) and $9 + 2$ flagellae. Unicellular but may aggregate. Nutrition by absorption or autotrophism. Reproduction usually asexual (fission or budding), but some species may practice conjugation.

PHYLUM

Phylum Schizophyta Bacteria. Unicellular, prokaryotic (no nuclear membrane), reproduce usually by fission. Usually heterotrophic. About 3,000 species.

Phylum Cyanophyta Blue-green algae. Unicellular but may be colonial. Prokaryotic. Chlorophyll existing without chloroplasts. Usually autotrophic. Reproduce by fission. About 200 species.

KINGDOM PROTISTA

Unicellular eukaryotes (having a nuclear membrane), sometimes aggregated. Autotrophic or heterotrophic. Sexual and asexual reproduction. Nonmotile, or move by pseudopods or $9 + 2$ flagellae and cilia.

PHYLUM

Phylum Protozoa Microscopic, unicellular, or colonial heterotrophs. Reproduction usually by fission and conjugation. About 30,000 species.

CLASS

Class Mastigophora Protozoans with flagellae. Some symbiotic forms. Includes the African sleeping sickness organism, *Trypanosoma*.

Class Sarcodina Protozoans with pseudopods. Includes the amoebas. Soft cell membrane.

Class Ciliophora Protozoans with cilia. Includes the paramecia.

Class Sporozoa Parasitic protozoans. Usually nonmotile at some life stage. Includes the malarial parasite, *Plasmodium*.

PHYLUM

Phylum Chrysophyta Golden algae and diatoms. Autotrophs. Cell wall contains mainly pectin, sometimes with siliceous materials. 6,000 to 10,000 species.

CLASS

Class Bacillariophyceae Diatoms. Sometimes motile. 5,000 to 9,000 species.

Class Chrysophyceae Golden algae. Includes flagellated, amoeboid, and nonmotile forms. Some species with glassy cell wall. Over 1,000 species.

PHYLUM

Phylum Pyrrophyta Fire algae (golden-brown algae). Autotrophic with food stored as starch. Cell walls with cellulose. Mostly biflagellated.

CLASS

Class Dinophyceae Dinoflagellates. Lateral flagella. Very unusual reproductive mechanism via fission. Over 1,000 species.

Phylum Euglenophyta   Euglenoids. Usually autotrophic. Store food as paramylon, an unusual carbohydrate. A single terminal flagellum and contractile vacuole. Flexible cell wall. Reproduce by fission. About 450 species.

Phylum Gymnomycota   Slime molds. Heterotrophic amoeboids. Usually lack a cell wall. Form sporangia. Nutrition generally by ingestion.

Class Myxomycetes   Plasmodial slime molds. Multinucleate, creeping plasmodium. Form multinucleate sporangia with many spores. About 450 species.

Class Acrasiomycetes   Cellular slime molds. Separate amoebas that aggregate, eventually form a compound sporangium. About 26 species.

Class Protostelidomycetes   Recently discovered amoebas that may exist separately or as a mass. Each one forms a stalked sporangium with one or two spores. Over 12 species.

Eukaryotic unicellular or multinucleate organisms. Nuclei occur in an essentially continuous mycelium. Heterotrophic with nutrition by absorption. Often both sexual and asexual reproductive cycles. About 100,000 species.

Class Oomycetes   Mostly aquatic fungi with motile cells at some stage. Cell walls contain cellulose. Several hundred species.

Class Zygomycetes   Terrestrial fungi. Includes black bread mold. Septate hyphae only during formation of reproductive bodies. Cell walls mainly chitin. Several hundred species.

Class Ascomycetes   Terrestrial and aquatic fungi. Includes truffles. Septate hyphae, but septae are perforated. Cell walls mainly chitin. Sexual reproduction involves the formation of the ascus, in which meiosis takes place. Hyphae may be packed together into fruiting bodies. About 30,000 species.

Class Basidiomycetes   Terrestrial fungi, includes mushrooms. Septate but perforated hyphae. May form spores. Chitin in cell walls. Sexual reproduction involves basidia, in which meiosis takes place and where spores arise. About 25,000 species.

Class Fungi Imperfecti   Fungi similar to Ascomycetes but with no known sexual cycle. Perhaps some belong in other classes. One causes athlete's foot, another makes penicillin. About 25,000 species.

Class Lichens   Lichens. Fungi (mostly Ascomycetes), which occur symbiotically with unicellular algae. About 17,000 species.

KINGDOM PLANTAE

Multicellular and related unicellular eukaryotes. Nutrition principally by photosynthesis. Cell walls contain cellulose. Primarily fixed and non-motile. Alternating gametophytic and sporophytic generations (the former reduced in the vascular plants). Red and brown algae may have evolved separately from the rest.

Phylum Rhodophyta   Red algae. Primarily marine. Carbohydrate stored as a peculiar starch (floridian). No motile stages. Plant body unspecialized. About 4,000 species.

Phylum Phaeophyta   Brown algae. Multicellular marine plants. Carbohydrates stored as laminarin. Motile cells biflagellate. Strong tissue differentiation in some kelps. No roots, stems, leaves. About 1,100 species.

Phylum Chlorophyta   Green algae. Unicellular or multicellular. Carbohydrate reserve is starch. Motile cells have two terminal flagellae. Not well differentiated. At least 7,000 species.

Phylum Bryophyta   Moss, hornworts, and liverworts. Multicellular with photosynthesis and carbohydrate reserves similar to those of green algae. Gametangia with multicellular jacket of sterile cells. Motile, flagellated sperm. Complex development but often lacking conducting tissue. Most photosynthesis carried out by gametophyte. About 24,000 species.

CLASS

Class Hepaticae   Liverworts. Gametophytes, leafy or thallose (not differentiated into roots, stems and leaves). Simple sporophytes. About 9,000 species.

Class Antherocerotae   Hornworts. Thallose gametophytes. Sporophyte grow continuously from a basal meristem. Stomata on sporophyte. About 100 species.

Class Musci   Mosses. Leafy gametophytes. Sporophytes discharge spores in complex manner. Stomata on sporophyte. About 14,500 species.

PHYLUM

Phylum Tracheophyta   Vascular plants. Terrestrial with roots, stems, and leaves. Well-developed conducting tissue. Reduced gametophyte generation in most species.

SUBPHYLUM

Subphylum Lycophytina   Lycophytes. Spores identical (homosporous) or spores of two types (heterosporous). Extremely diverse. Motile sperm. About 1,000 species.

Subphylum Spenophytina   Horsetails. Homosporous. Vascular. Jointed stems. Elevated glassy ribs. Scalelike leaves. Motile sperm. About 24 species, all in same genus.

Subphylum Pterophytina Ferns, cone-bearers, and flowering plants. Highly diverse group with megaphyll (large, veiny leaf). About 260,000 species.

Class Filicineae Ferns. Mostly homosporous. Photosynthetic gametophyte, which is generally free-living. Free-swimming sperm. About 11,000 species.

Class Coniferinae Conifers (cone bearers). Seed plants. Simple leaves. Active cambial growth. Ovules not enclosed. Sperm not flagellated. About 550 species.

Class Cycadinae Cycads. Seed plants with pinnately compound, fernlike or palmlike leaves. Naked ovules. Flagellated sperm transported by a pollen tube. About 100 species.

Class Ginkgoinae Ginkgo. Seed plants with cambial growth and fanshaped leaves. Fleshy ovules not enclosed. Flagellated sperm are carried in a pollen tube. One species, *Ginkgo biloba,* survives today.

Class Angiospermae Flowering plants. Seed plants in which most species have ovule enclosed in a carpel. Seeds found within fruits. Diverse but flower-bearing and usually pollinated by insects. Gametophytes reduced. Complex fertilization. About 250,000 species.

Subclass Dicotyledonae Dicots. Flowers usually divided into four or five parts. Leaf venation may be netlike, palmate, or pinnate. Secondary growth. Vascular cambium common. Two cotyledons. About 190,000 species.

Subclass Monocotyledonae Monocots. Flowers usually in three parts. Usually parallel venation in leaves. No true secondary growth. One cotyledon. About 60,000 species.

Eukaryotic multicellular heterotrophic organisms. Nutrition usually by ingestion. Most are motile. Usually lack rigid cell walls. Considerable tissue differentiation. Primarily reproduce sexually. Perhaps 10 million species.

KINGDOM ANIMALIA

Phylum Porifera Sponges. Simple, multicellular, mainly marine animals. Stiff skeletons. Porous bodies. Contain choanocytes (collar cells). About 4,200 species.

Phylum Coelenterata Coelenterates. Radially symmetrical, two-layered bodies with jellylike cell matrix. Reproduction sexual or asexual. Contain cnidoblasts (stinging cells). All are aquatic and most are marine. About 11,000 species.

Class Hydrozoa Hydra and similar animals. Some colonial. Some with alternation of generations (sexual and asexual). Polyp form dominant.

Class Scyphozoa Marine jellyfish. Usually with alternation of generations. Medusa form dominant. Possess true muscles.

Class Anthozoa Sea anemones and colonial corals. No medusa stage.

Phylum Ctenophora Comb jellies and sea walnuts. Almost spherical, free-swimming. Translucent, gelatinous, delicately hued. Some are bioluminescent. Propelled by eight bands of cilia. About 80 species.

Phylum Platyhelminthes Flatworms. Bilaterally symmetrical. Three germ layers. One gut opening. No coelom or circulatory system. Hermaphroditic. Excretion by flame cells. About 15,000 species.

Class Turbellaria Planaria and other nonparasitic flatworms. Ciliated, carnivorous, with ocelli (eyespots).

Class Trematoda Flukes. Parasitic flatworms. No gut, absorbing nutrients through body wall.

Class Cestoidea Tapeworms. Parasitic flatworms. No gut, absorbing nutrients through body wall.

Phylum Rhynochocoela Proboscis worms. Nonparasitic. Tubelike gut. Protrusible proboscis with hooks for capturing prey. Simple circulatory and reproductive systems. About 600 species.

Phylum Nematoda Roundworms. Free-living and parasitic. Cylindrical, bilaterally symmetrical bodies. About 8,000 species.

Phylum Acanthocephala Spiny-headed worms. Parasitic. No digestive tract. Head with many hooks. About 300 species.

Phylum Chaetognatha Arrow worms. Free-swimming planktonic marine worms. Coelom present. Mouth with hooks on both sides. About 50 species.

Phylum Nematomorpha Horsehair worms. Slender, brown or black worms up to three feet long. Larvae parasitic in insects. Adults free-living. About 250 species.

Phylum Rotifera "Wheel animalcules." Microscopic, wormlike, or spherical. Complete digestive tract, flame cells in excretory tract. Circle of cilia on head suggests a wheel. Males degenerate or unknown. About 1,500 species.

Phylum Gastrotricha Microscopic wormlike animals. Propelled by longitudinal rows of cilia. About 140 species.

Phylum Bryozoa "Moss animals." Microscopic. Aquatic. U-shaped row of ciliated tentacles for feeding. Usually form branching colonies. Possess anus and coelom. Retain larvae in brood pouch. About 4,000 species.

Phylum Brachiopoda Lamp shells. Clamlike marine animals with hard dorsal and ventral shells. Fixed by one shell or a stalk as adults. Feed by ciliated tentacles. About 260 species.

Phylum Phoronidea Sedentary, marine, wormlike animals that secrete a leathery tube in which they live. About 15 species.

Phylum Annelida Segmented worms. Well-developed digestive tract. Coelom, circulatory system, nephridia, well-developed nervous system. About 8,800 species.

Class Archiannelida Small, simple marine worms. About 35 species.

Class Polychaeta Mainly marine worms. Distinct head with palps and tentacles. Bristled appendages. About 4,000 species.

Class Oligochaeta Freshwater, marine, and soil worms (including the earthworm). Poorly differentiated head. Few bristles, about 2,500 species.

Class Hirudinea Leeches. Posterior sucker, often an anterior sucker. Freshwater, marine, and terrestrial. Free-living or parasitic. About 300 species.

Phylum Mollusca Unsegmented animals with a head, a mantle, and a foot. Soft-bodied. Mostly aquatic. Some possess shells. Three-chambered heart. About 110,000 species.

Class Amphineura Chitons. Simplest molluscs. Elongated, flattened body with the mantle enclosing eight dorsal plates. About 700 species.

Class Pelecypoda Bivalve molluscs, including clams. No head. Hatchet-shaped foot. Rather sessile. About 15,000 species.

Class Scaphopoda Tooth shells (tusk shells). Marine with long conical shell. About 350 species.

Class Gastropoda Asymmetrical molluscs. Includes snails and slugs. Two pairs of tentacles on head. Often with spiral shell. About 80,000 species.

PHYLUM

CLASS

PHYLUM

CLASS

Class Cephalopoda "Head-foot" animals. Includes octopus and squid. Possess arms or tentacles. Parrotlike jaws. Well-developed eyes and nervous system. Shell external, internal, or absent. About 400 species.

Phylum Arthropoda Joint-legged animals. Segmented. The largest phylum. Hard exoskeleton. Reduced coelom. Dorsal brain, ventral nerve cord. Paired ganglia in each segment. About 765,000 species.

Class Merostomata Horseshoe crabs. Primitive arthropods. Aquatic. Book gills (surface area increased by pagelike extensions). About 5 species.

Class Crustacea Lobsters, crabs, shrimp, crayfish. Mostly aquatic with two pairs of antennae, one pair of mandibles and usually two pairs of maxillae. Appendages present on thoracic segment but possibly not on abdominal segment. About 25,000 species.

Class Arachnida Spiders, mites, and scorpions. Most are terrestrial, air-breathing. Usually four or five pairs of legs with the first pair used for grasping. Pincers or fangs in place of jaws and antennae. About 30,000 species.

Class Onychophora Simple, terrestrial, elongate arthropod. Many short unjointed pairs of legs. One genus, *Peripatus*. About 73 species.

Class Insecta Insects. Most are terrestrial and breathe through trachea. One pair of antennae, three pairs of legs. Body divided into head, thorax, abdomen. Most have two pairs of wings. About 700,000 species.

Class Chilopoda Centipedes. Many (15 to 173) body segments, each with one pair of jointed legs. About 2,000 species.

Class Diplopoda Millipedes. Abdomen with 20 to 100 segments, each with two pairs of legs. About 7,000 species.

Phylum Echinodermata Starfish and sea urchins. Adults radially symmetrical. Well-developed coelom. Calcareous endoskeleton. Water vascular system. Tube feet. Marine. About 6,000 species.

Class Crinoidea Sea lilies and feather stars. Sessile, attached by jointed stalk. Ten arms with slender lateral branches. Most are known from fossils.

Class Asteroideae Starfish. Five to fifty arms. Mouth on undersurface.

Class Ophiuroidea Brittle stars and serpent stars. Usually elongated with slender, flexible arms. Fast moving.

Class Ecinoidea Sea urchins and sand dollars. Skeletal plates form rigid outer shell with many movable spines.

Class Holothuroidea Sea cucumbers. Sausage or worm-shaped body.

Phylum Hemichordata Wormlike marine animals. Notochordlike structure in anterior end, gill slits, and a solid nerve cord. About 91 species.

Phylum Chordata Animals having a notochord at some stage, as well as a dorsal hollow nerve cord and pharyngeal gill slits. About 45,000 species.

Subphylum Tunicata Tunicates or ascidians. Marine. Adults saclike, sessile, often forming branching colonies. Feed by currents set up by cilia. Reduced nervous system. Gill slits. No notochord. Larvae active with well-developed nervous system and notochord. About 1,600 species.

Subphylum Cephalochordata Lancelets. Amphioxus and related forms. Fishlike with a permanent notochord extending the length of the body, a nerve cord, pharyngeal gill slits. No cartilage or bone. About 13 species.

Subphylum Vertebrata The vertebrates. Notochord is replaced by cartilage or bone. Vertebral column. Skull around large brain. Usually a tail present. About 43,090 species.

Class Agnatha Lampreys and hagfish. Eellike aquatic species. No limbs. Jawless, sucking mouths. No bones, fins, or scales.

Class Chondrichthyes Sharks and rays. Cartilagenous skeleton, complex reproductive organs. Scales present. Air bladders absent. Almost all are marine.

Class Osteichthyes Bony fish. Almost all freshwater fish. Usually an air bladder, rarely a lung. All classes of fish comprise about 23,000 species.

Class Amphibia Salamanders, frogs, and toads. Usually breathe by gills as larvae and by lungs as adults. Usually with double circulation (to respiratory organ and to body). Limbs present as legs. Eggs unprotected by shell or embryonic membranes. About 2,000 species.

Class Reptilia Snakes, turtles, lizards, and crocodilians. Breathe by lungs. Double circulation. Scales usually present. Legs present except in snakes and some lizards. Poikilothermic. Embryo protected by shell and embryonic membranes. About 5,000 species.

CLASS

PHYLUM

SUBPHYLUM

CLASS

Class Aves   Birds. Warm-blooded. Feathered. Double circulation. Wings. Embryo protected by shell and membranes. About 9,000 species.

Class Mammalia   Mammals. Warm-blooded. Hairy. Mothers give milk. Four limbs. Respiratory diaphragm. Lower jaw comprised of one pair of bones. Three bones in middle ear (hammer, anvil, stirrup). Seven vertebrae in neck. About 4,500 species.

SUBCLASS

Subclass Prototheria   Monotremes. Egg-laying mammals. Poor temperature regulation. Two species (duckbill and spiny anteater).

Subclass Metatheria   Marsupials. Includes kangaroos, opossums. Viviparous (live bearers). Placenta absent or poorly developed. Young born undeveloped and are carried in pouch. Mainly found in Australia.

Subclass Eutheria   Mammals with highly developed placenta (i.e., most mammals).

ORDER

Order Insectivora   Insect eaters. Shrews, moles, etc.

Order Edentata   Toothless mammals. Anteaters, armadillos, etc.

Order Rodentia   Rodents. Rats, mice, squirrels, etc.

Order Artiodactyla   Even-toed hoofed animals. Cattle, deer, etc.

Order Perissodactyla   Odd-toed hoofed animals. Horses, zebras, rhinoceroses, etc.

Order Proboscidea   Elephants.

Order Lagomorpha   Rabbits and hares.

Order Sirenia   Large aquatic mammals with finlike forelimbs. Hindlimbs absent. Sea cows, dugong, and manatee.

Order Carnivora   Carnivores. Cats, dogs, wolves, weasels, bears, etc.

Order Cetacea   Whales, dolphins, and porpoises.

Order Chiroptera   Bats.

Order Primates   Lemurs, monkeys, apes, and us, unless this book has fallen into the wrong hands.

Appendix B
Geologic Timetable

Era and Duration	Period	Epoch	Years Before Present (in millions)	Principal Points
Cenozoic 75 Million Years	Quaternary	Recent	(11 thousand)	Age of man; end of last ice age; warmer climate.
		Pleistocene	1	First human societies; large scale extinctions of plant and animal species; repeated glaciation.
	Tertiary	Pliocene	11	Appearance of man; volcanic activity; decline of forests; grasslands spreading.
		Miocene	25	Appearance of anthropoid apes; rapid evolution of mammals. Formation of Sierra Mountains.
		Oligocene	36	Appearance of most modern genera of mammals and monocotyledons; warmer climate.
		Eocene	54	Appearance of hoofed mammals and carnivores; heavy erosion of mountains.
		Paleocene	65	First placental mammals.
Mesozoic 165 Million Years	Cretaceous		135	Appearance of monocots; oak and maple forests; first modern mammals; beginning of extinction of dinosaurs. Formation of Andes, Alps, Himalayas, and Rocky Mountains.
	Jurassic		181	Appearance of birds and mammals; rapid evolution of dinosaurs; first flowering plants; shallow seas over much of Europe and North America.
	Triassic		220	Appearance of dinosaurs; gymnosperms dominant; extinction of seed ferns; continents rising to reveal deserts.
Paleozoic 360 Million Years	Permian		280	Widespread extinction of animals and plants; cooler, drier climates; widespread glaciation; mountains rising; atmospheric carbon dioxide and oxygen reduced.
	Pennsylvanian		310	Appearance of reptiles; amphibians dominant; insects common. Gymnosperms appear; vast forests; great life abundant. Climates mild; lowlying land; extensive swamps; formation of enormous coal deposits.
	Mississippian		355	Many sharks and amphibians; large scale trees and seed ferns; climate warm and humid.
	Devonian		405	Appearance of seed plants; ascendance of bony fishes; first amphibians; small seas; higher, drier lands; glaciations.
	Silurian		425	Atmospheric oxygen reaches second critical level. Explosive evolution of many forms of life over the land; first land plants and animals. Great continental seas; continents increasingly dry.
	Ordovician		500	Appearance of vertebrates, but invertebrates and algae dominant. Land largely submerged. Warm climates worldwide.

Era and Duration	Period	Epoch	Years Before Present (in millions)	Principal Points
Paleozoic 360 Million Years (*continued*)	Cambrian		600	Atmospheric oxygen reaches first critical level. Explosive evolution of life in the oceans; first abundant marine fossils formed; tribolites dominant; appearance of most phyla of invertebrates. Lowlying lands; climates mild.
Pre-Cambrian 2100 Million Years			2700	Life confined to shallow pools, fossil formation extremely rare. Volcanic activity, mountain building, erosion, and glaciation. Photosynthetic life.

Abundance of species through the ages

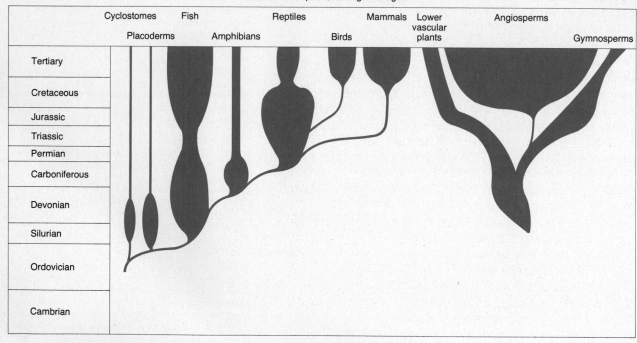

Appendix C
Evolution

As you know, we have discussed evolution throughout this book, beginning with those first coacervate droplets that were a little more organized than the rest, through the molecular mechanisms of genetic change and population genetics, to, finally, such esoteric discussions as the mechanisms for altruism. You should have a pretty good idea of how evolution operates by now, but perhaps a brief, concise review will help to organize your thoughts.

We are aware that members of populations vary—some populations being more diverse than others. Seagulls of the same age look alike to a human observer, but elephants don't. Elephants come in a variety of sizes and their ears and tusks may be quite distinctive, so they may be recognized individually. Even John Vanwinkle, who made the memorable statement, "You seen one alligator, you seen 'em all," would probably change his mind if he spent more time watching alligators, which isn't likely. The point is, populations do vary, and probably more than we realize, since the human observer is attuned to only certain kinds of features, such as facial appearance, size, and sound (features that are important in helping us to recognize each other).

At the same time, some populations vary more than others. And, as a general rule, it is assumed that high variability is a result of weak selection. If selection is strong for, say, a certain body size, then most animals will fall within a narrow range of body sizes. Those much larger or smaller than the optimum would either die or suffer reproductively, thus reducing the frequency of "large size" and "small size" genes in the population (Figure C.1A). If size is not particularly important to survival or reproduction, selection is said to be low and greater variation can be expected (Figure C.1B).

As long as the environmental conditions (and hence, selection) remain stable, the phenotype will tend to cluster, weakly or strongly, around a certain mean (*stabilizing selection*). But if the environment should change, the optimum phenotype may shift. For example, if the weather should grow colder, larger size might be selected for as a means of reducing the body's surface area in relation to its volume, a heat-saving device. (As the surface of a sphere is squared, its volume is cubed, if that helps.) So if larger size is advantageous, larger animals will tend to be more successful and will tend to leave more offspring. Thus, in time, the size of individuals

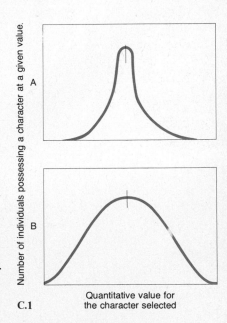

C.1

Number of individuals possessing a character at a given value.

A

B

Quantitative value for the character selected

in the population will increase. Such changes give rise to *directional selection* (Figure C.2).

If there is more than one ideal phenotype for a population, *disruptive selection* may occur, producing a bimodal curve. We see the result of disruptive selection in species in which one sex is larger than the other (if it is disadvantageous to be of intermediate size).

Such changes have been manipulating populations for ages. In some cases the selection has been too strong and the species were unable to adapt; they perished (Appendix B, Geologic Timetable). In a sense, then, all species are part of one grand experiment and, as the earth changes, sooner or later all are doomed to fail. Does this include our own? Could an unchronicled, slime-dominated world exist? Does life need people? Or are we simply destined to be one more experiment that failed.

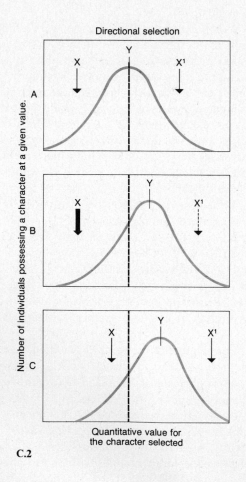

C.2

Appendix D
Metric Measurements

	Symbol	Fundamental unit	Quantity	Numerical unit	American measurement
Time		second (sec)			second
	msec		millisecond	0.001 sec	
	μsec		microsecond	0.000001 sec	
Length		meter (m)			39.4 inches
	km		kilometer	1,000 m	0.62137 miles
	cm		centimeter	0.01 m	0.3937 inches
	mm		millimeter	0.001 m	
	μm		micrometer	0.000001 m	
	nm		nanometer	0.000000001 m	
	Å		angstrom	0.0000000001 m	
Volume (liquids)		liter (l)			1.06 quarts
	ml		milliliter	0.001 liter	
	μl		microliter	0.000001 liter	
Volume (solids)		cubic meter (m^3)			1.308 cubic yards
	cm^3		cubic centimeter	0.000001 m^3	0.061 cubic inches
	mm^3		cubic millimeter	0.000000001 m^3	
Mass		gram (g)			0.035 ounces
	kg		kilogram	1,000 g	2.2 pounds
	mg		milligram	0.001 g	
	μg		microgram	0.000001 g	

Appendix E
Temperature Conversion Chart

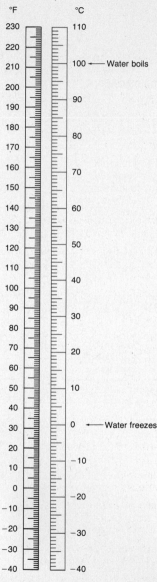

°F   °C

100 ← Water boils

0 ← Water freezes

To convert fahrenheit to
centigrade, use this formula:
$°C = \frac{5}{9}(°F - 32)$

To convert centigrade to
fahrenheit, use this formula:
$°F = \frac{9}{5}(°C + 32)$

Glossary

Abortion: The expulsion or removal of an embryo or fetus before it can survive on its own.

Abscission (ab·′sizh·ən): Dropping of leaves, flowers, stems or fruits after the growing season by developing an abscission layer of cells.

Acceptor: A molecule that unites chemically with hydrogen or electrons in the oxidation-reduction reactions of cells.

Acetylcholine (as′·ət′l·kō·lēn): A chemical agent that transmits neutral impulses across synapses from one neuron to another.

Acid: A substance that releases hydrogen ions into a solution. It has a sour taste, unites with bases to form salts.

Actin: The moveable filament of protein found in the myofibrils of muscle.

Action potential: Changes in the electrical potential across the membrane of a nerve cell that occur during nerve impulse transmission.

Active transport: The movement of solutes against a concentration gradient, usually requiring the expenditure of cell energy.

Adaptation: Evolving characteristics that make a population or individual better suited to its environment.

Adaptive radiation: The diversity of species that is attributed to subpopulations adapting to a unique set of environmental conditions.

Adenine (ad′·n·ēn′): One of the nitrogenous bases that are found in nucleic acids and in ATP.

Adenosine diphosphate (ə·den′·ə·s'n dī·fäs′·fāt): ADP—A product formed by the hydrolysis of ATP, Adenosine Triphosphate, with the release of energy.

Adenosine triphosphate (ə·den′·ə·s'n trī·fôs′·fāt): ATP—A compound that is found in all cells, which serves as an energy source as it is broken down into ADP plus phosphate with the release of energy.

Adrenal: A vertebrate endocrine gland. The outer area produces steroid hormones, the inner produces epinephrine.

Adrenaline (ə·dren′·′l·in): See epinephrine.

Aerobe (er′·ōb): An organism that uses oxygen in its respiration.

Afferent nerve: A nerve that directs impulses toward the central nervous system.

Afterbirth: The placenta with its associated membranes, which is expelled from the uterus at birth.

Alcohol: A carbon-hydrogen molecule that contains one or more OH groups.

Alkaline: Relating to substances that release hydroxyl ions in water and have a pH greater than 7.

Allele (ə·lēl′): One of two or more genes that produce a specific characteristic such as blood type, hair color, etc.

Alpha waves: A particular pattern of brain waves that are characteristic of a person who is awake, but at rest.

Alternation of generations: The alternation of haploid and diploid stages in the life cycle of a species.

Alveolus (al·vē′·ə·ləs): One of the tiny air sacs of the lung, in which carbon dioxide and oxygen are exchanged.

Amino acid: An organic molecule that contains at least one carboxy group, COOH, and one amino group, NH_2.

Amnion (am′·nē·ən): A fluid-filled sac that surrounds the embryo in the uterus, or in the embryonic egg of egglaying animals.

Amphibian: A class of vertebrates that lives part of the time in water and part on land.

Anaerobe (an·er′·ōb): An organism that does not utilize oxygen in its respiration.

Analogous: A term that describes structures that serve the same function but have different morphologies and embryonic origin.

Anaphase (an′·ə·fāz′): The third phase of mitosis in which the chromatids separate and move to opposite sides of the cell.

Androgen (an′·dra·jən): A male sex hormone that is secreted by the testicles.

Angiosperm: A flowering plant.

Animal pole: That end of an animal zygote that has relatively little yolk and experiences more rapid cell division.

Annual: A plant that completes its life cycle in a single season.

Anterior: Toward the head or front.

Anther (an′·thər): A floral structure at the tip of the stamen in which male spores are formed.

Anthropomorphism: Attribution of human characteristics to other animals.

Antibody: A globular protein produced in response to a foreign substance.

Antidiuretic hormone (ADH) (ant·i·dī·yù·′ret·ik): Hormone secreted by the hypothalamus that induces reabsorption of water in the kidneys.

Antigen (an′·tə·jən): Any substance that triggers an immune response.

Aorta: The large artery that, in humans, leaves the heart and supplies blood to all parts of the body except the lungs.

Apical meristem (ap′·ə·k'l mer′·ə·stem′): A region at the tip of a plant shoot that is marked by actively dividing cells.

Arteriole (är·tir′·ē·ōl): A small artery.

Asexual reproduction: Reproduction without the union of sex cells; the production of new individuals from one parent cell or organism.

Asymmetric: Lacking an orderly arrangement of parts.

Atomic number: The number assigned to a particular element, determined by the number of protons in its atoms.

Atomic weight: The average weight of the atom of an element. The unit of atomic weight is $\frac{1}{12}$ the mass of carbon 12.

Atrium (ā′·trē·əm): The smaller compartment of the heart that receives venus blood from the body or lungs.

Autonomic nervous system (ôt′·ə·näm′·ik): That part of the peripheral nervous system that regulates the internal environment by the control of glands and smooth muscles.

Autosome: Any chromosome other than a sex chromosome.

Autotroph: Self-feeder, able to manufacture food such as through photosynthesis.

Auxin: A general name for a plant hormone.

Avoidance behavior: That behavior that helps an organism escape or elude a stimulus.

Axon: A conducting fiber extending from a neuron that transmits impulses away from the cell body.

Bacteriophage (bak·tir′·ē·ə·fāj′): A virus that reproduces in bacterial cells.

Barbiturates: Sedative drugs that act as depressants on the central nervous system.

Bark: All the layers of a woody plant stem from the vascular cambium outward.

Basal body: A bulblike structure at the base of a cilium.

Basal metabolism: The metabolism of an organism at rest.

Base: Any chemical that releases hydroxyl (OH^-) ions in water, or joins with free protons. Bases leave a bitter taste and form salts when they react with acids.

Behaviorism: The psychological field that uses objective observation of behavior without reference to awareness on the part of the observed subject.

Benthic: Referring to the bottom of a lake or sea.

Bile: A yellow secretion of the vertebrate liver that emulsifies fats.

Bio-degradable: Capable of being oxidized or broken down by living organisms.

Biological clock: A mechanism or model that explains the cyclic or rhythmic changes of behavior in animals and plants.

Biological control: Using a predator or parasite of some pest organism in order to reduce the level of its population.

Biomass: The total weight of all the organisms in a prescribed area.

Biome (bī′·ōm): A geographical area that is characterized by relatively similar plant and animal representatives, for example, a tundra.

Biosphere: That portion of the earth that can support life.

Biotic community: Populations of plants and animals, sharing a region and interacting in the exchange of materials and energy.

Bladder: An organ that stores liquid such as bile or urine.

Blastocyst (blas′·tō·sist′): A blastula; embryonic stage when the cells form a hollow sphere.

Blastulation (blas′·choo·lā′·shən): The cellular activity of a zygote that results in the formation of a hollow sphere of cells, the blastula.

Bond energy: The energy stored in a chemical bond. When the bond is broken that energy is released.

Bowman's capsule: A curved sac at the beginning of the kidney nephric unit, which surrounds the glomerulus.

Brain stem: The lower portion of the brain that includes midbrain, the pons, and the medulla.

Breeder reactor: A nuclear reactor that generates highly radioactive plutonium from non-fissionable U^{238}.

Bronchiole (brän′·kē·ōl′): A small branch of the bronchi in the vertebrate lung.

Bronchus (brän′·kəs): A major branch of the trachea.

Budding: Asexual reproduction seen in yeasts and other organisms where smaller cells bud or grow from a parental cell.

Buffer: A substance that reduces changes in environmental conditions, including pH.

Bulbourethral gland (Cowper's gland): One of a pair of pea-sized glands, located below the prostate in mammalian males, that adds its secretions to the seminal fluid.

Calorie: A unit of heat needed to raise the temperature of one gram of water one degree centigrade.

Calvin cycle: A pathway of carbon dioxide that occurs during the dark phase of photosynthesis as carbohydrate is produced.

Cambium (kam′·bē·əm): A layer of meristematic tissue between the phloem and the xylem.

Cancer (kan′sər): An abnormal and uncontrolled cell growth occurring in plants and animals.

Cannabis (kan·na·bis): A plant of the genus cannabis; marijuana and hemp are included in this group.

Capillary: A thin-walled blood vessel branching from an artery and later joining with other capillaries to form veins.

Carbohydrate (kär′·bə·hī′·drāt): An organic substance containing carbon, hydrogen, and oxygen in the ratio of CH_2O. Includes sugars, starches, cellulose, etc.

Carbonic anhydrase (kär·bän′·ik an·hī′·drās): An enzyme in red blood cells that catalyzes the changing of carbon dioxide into carbonic acid and the reverse reaction.

Carcinogenic (kär′·sə·nō·jen′ik): Cancer producing.

Carpel: A floral structure that encloses the ovule of an angiosperm.

Cartilage: A connective tissue of the vertebrate skeleton.

Catalyst: A substance, such as an enzyme, that increases the rate of a reaction without entering into the reaction.

Cecum (sē′·kəm): A pouch between the small and large intestine in man. The appendix extends from the cecum.

Cell (sel): The basic unit of life. The simplest unit that can exist as an independent system.

Cell plate: A structure that appears along the equatorial plane as cytoplasm is divided in telophase of plant mitosis: it will form the cell wall separating daughter cells.

Cellulose (′sel·yə·lōs): An insoluble carbohydrate composed of glucose units. Forms cell walls in plants.

Central nervous system (CNS): The brain and spinal cord.

Centriole (sen′·trē·ōl′): A paired organelle near the nucleus of animal cells that functions in cell division.

Centromere (sen′·trə·mir′): A structure on the chromosome to which the spindle fiber attaches that divides when chromatids separate in meiosis and mitosis.

Centrosome (sen′·trə·sōm′): The area of specialized cytoplasm containing centriole in a cell that usually lies near the nucleus.

Cerebellum (ser′·ə·bel′·əm): The distinct, walnut-shaped structure at the back of the vertebrate brain; involved in coordination and equilibrium.

Cerebral cortex (kôr′•teks): The external layer of the forebrain or cerebrum.

Cerebrum (ser′•ə•brəm): In humans, the largest portion of the brain; coordinates voluntary action of muscles and many other activities.

Cervix (sur′viks): The narrow neck of the uterus that opens into the vagina.

Chain reaction: The phenomenon in which particles emitted from atomic nuclei collide with other nuclei causing an avalanche of similar events.

Chiasma (kī-az′•mə): A crossover area on the chromosome seen in prophase I of meiosis.

Chitin (′kīt•ən): A tough polysaccharide in the exoskeleton of arthropods, the surface of some other invertebrates and the cell walls of some fungi.

Chlorophyll (klôr′•ə•fil′): The light absorbing pigments of plants and some bacteria that trap light energy for photosynthesis.

Chlorophyll A: A type of chlorophyll common to all plants.

Chlorophyll B: A type of chlorophyll common in certain land plants and in green algae.

Chloroplast (klôr′•ə•plast′): A chlorophyll containing organelle that functions in photosynthesis in plant cells.

Chorion (kôr′•ē•än′): A membrane surrounding the embryo.

Chromatid (krō′•mə•tid): One of two duplicated strands in a chromosome.

Chromatin (krō′•mə•tin): Indistinct, granular chromosomal material seen at interphase.

Chromosome (krō′•mə•sōm′): Threadlike, condensed chromatin (DNA), visible at cell division only, occurring in pairs with specific numbers for each species.

Chromosome mapping: Plotting relative positions of genes along chromosomes.

Cilium (sil′•ē•əm): Hairlike cellular processes that beat rhythmically on the cell surface.

Circadian rhythm (sər•kā′•dē•ən): Pertaining to a daily cycle.

Class: A major taxonomic subdivision of a phylum consisting of subordinate groups known as orders.

Classical conditioning: Learning in which a condition stimulus presented together with

an unconditional stimulus initiate a reaction to the condition stimulus.

Cline (klīn): Small variations in genetic characters across the range of a species population.

Clitoris (klit′•ər•əs): An external organ of the female genitalia considered to be homologous to the penis in the male.

Clone (klōn): The descendents of a single cell, generally identical.

Codon (kō′•dän): A sequence of three nucleotide bases required to form one amino acid in a protein chain synthesized on the RNA messenger.

Coelenterata (si•′lent•ə•rət•ə): An animal phylum characterized by only one digestive opening, including the jellyfish.

Coelom (sē′•ləm): The cavity within the body wall and surrounding the gut in higher animals.

Coitus (kō′•it•əs): Sexual intercourse.

Colchicine (käl′•chə•sēn′): A meiotic poison that causes the doubling of chromosome numbers in plants.

Collar cell: Certain flagellated cells in sponges that create a current and ingest food particles from the water.

Collenchyma (kə•len′•ki•mə): A plant supportive tissue consisting of thick-walled, elongated cells.

Colloid (käl′•oid): A liquid containing fine particles of matter in permanent suspension.

Colon: The large intestine.

Community: The interacting organisms of a habitat.

Compound: A chemical substance composed of one kind of molecule.

Conditioned response: Any response given to a learned stimulus.

Conditioned stimulus: A stimulus that causes a response after training.

Congenital: Any condition present at birth.

Conspecific: Members of the same species.

Convergent evolution: A similarity in genetically different organisms resulting from their adaptation to similar habitats.

Copulation: Sexual intercourse or coitus.

Cork cambium: The perpetually growing tissue of a plant or stem that forms the bark.

Corolla (kə•räl′•ə): All the petals of a flower.

Corpus callosum (kôr′•pəs kə•lō′•səm): A connecting strip of nerve fibers that coordinates the activities of the cerebral hemispheres.

Corpus luteum (kôr′•pəs lōō′•tē•əm): The body formed by the ovarian follicle cells after the egg has been released.

Cortex: The outer layer of tissue. In plants the tissue between the epidermis and vascular tissue.

Cotyledon (kät′•′l•ēd′•′n): The seed leaf in the angiosperm embryo.

Covalent bond: A bond formed between two atoms as they share electrons.

Creatine (krē′•ə•tēn′): The amino acid part of creatine phosphate, important in muscle contraction.

Crista (kris′•tə): Inner membranes of the mitochondria.

Crop: A food storage sac located below the esophagus in birds, earthworms, and other animals.

Cross fertilization: (Fertilization of one flower by another.

Crossover: An exchange of chromosomal material between homologous chromosomes in prophase I of meiosis.

Cytochrome (sīt′•ə•krōm′): An iron-containing compound that functions in electron transport during oxidative respiration.

Cytoplasm (sit′•ə•plaz′m): The living matter in a cell, outside of the nucleus.

Cytosine (sīt′•ə•sēn′): One of the nitrogenous bases that are found in nucleic acids.

Dark reactions: The stage of photosynthesis that includes carbohydrate production.

Deciduous: Plants that seasonally shed their leaves.

Deme (dēm): An interbreeding population.

Dendrite: The short branched fiber of the neuron that receives nerve impulses.

Density dependent factor: Any influence on the growth rate of a population that is a function of the population size.

Deoxyribonuclease (dē•äk′•sə•rī′•bō•nōō•klē•ās): An enzyme that destroys DNA.

Deoxyribonucleic acid (dē•äk′•sə•rī′•bō•nōō•klē′•ik as′•id): DNA, a nucleic acid containing deoxyribose, which is found chiefly in the nucleus of cells. DNA is the genetic message, functioning in protein synthesis.

Depolarization: Reversal of the polarization of a resting neuron by the inward flow of sodium ions.

Diaphragm: The muscular dome below the lungs that functions in breathing. Also a contraceptive device that fits over the cervix.

Dicot (dī′·kät′): Abbreviation for dicotyledon, which is any of the higher plants that produces two seed leaves.

Differentiation: Cellular specialization during development.

Diffusion: The movement of particles from an area of greater concentration to lesser as a result of random movement.

Dihybrid (dī·hī′·brid): The result of a genetic cross in which two different traits are involved.

Diploidy (dip′·loi′·dē): Having the full number of chromosomes (2N).

Directional selection: In evolution, that selection that favors one or the other of two alleles in a deme.

Distal convoluted tubule: That region of the nephric unit that empties into the collecting tubule. It regulates the solutes present in the filtrate.

Dominant: In genetics, the allele of a pair that is expressed while the other is hidden or repressed.

Dorsal: Upward or toward the back.

Dorsal root ganglion: A cluster of receptive cell bodies that lies outside each segment of the spinal column.

Double bond: A covalent bond in which two pairs of electrons are shared between two atoms.

Double helix: A double spiral in which each strand is wound around the other.

Duodenum (doo′·ə·dē′·nəm): The first segment of the small intestine posterior to the stomach.

E. Coli *(Escherichia coli)*: A common rod-shaped intestinal bacteria of humans.

Ecology: The study of the interrelationship of the organism and its environment.

Ecosystem: A community of organisms interacting together and their physical environment.

Ectoderm (ek′·tə·durm′): The outer cell layer in an animal embryo that gives rise to skin, nervous tissue, etc.

Effector: A muscle or gland that responds to impulses from an efferent neuron.

Efferent nerve fiber: A nerve fiber that carries impulses from the central nervous system to an effector.

Egg: The female gamete.

Ejaculation: The forceful expulsion of semen from the penis.

Electrolyte: A solution of ions that exhibits electrical conductivity.

Electron: A minute, negatively charged particle of an atom, in motion about the nucleus.

Electron donor: A substance that yields electrons in an oxidation-reduction reaction.

Electron acceptor: A substance that receives electrons through reduction.

Electron transport: A series of reactions in which electrons are passed from one compound to the next in an energy-producing process.

Electronegative: A state of attraction to the negative electron by other atoms or molecules.

Electron shell: A theoretical band around an atomic nucleus where electrons of a certain energy level are found.

Element: A substance consisting of atoms of one kind only, such as iron, gold, sulfur, etc. There are 102 known elements.

Embryo: The developing stage of any organism.

Embryonic induction: An influence one group of embryonic cells has on another during development.

Emphysema (em′·fə·sē′·mə): Distension and rupturing of the alveoli of the lungs, interfering with respiration.

Endergonic (end·er·gon′·ik): Any chemical reaction characterized by the absorption of energy from an outside source.

Endocrine gland (en′·də·krin): A gland that releases hormones into the blood. Each hormone has specific target cells somewhere in the organism.

Endoderm (en′·də·durm′): The innermost cell layers of an embryo that differentiate into such organs as the digestive tract, liver, and pancreas.

Endogenous theory (en·däj′·ə·nəs): The theory that explains biological rhythm by assuming the existence of regulatory mechanisms known as biological clocks.

Endometrium (en′·dō·mē′·trē·əm): A temporary glandular lining of the uterus, in which the fertilized egg implants and develops.

Endoplasmic reticulum: Layers of membrane folded through the cytoplasm of a cell, forming complex inner surfaces, after being covered with ribosomes.

Endosperm (en′·də·spurm′): Nutritive tissue in the seeds that nourishes the developing embryonic sporophyte.

Entropy (en′·trə·pē): The opposite of "Free Energy." The degree of randomness in a system.

Enzyme (en′·zīm): An organic catalyst, usually a protein, capable of changing the rate of chemical reactions in cells.

Epidermis (ep′·ə·dur′·mis): The outermost cell layer of an organism.

Epididymus (ep′·ə·did′·i·məs): An oblong, convoluted tube, resting on the testes, which stores sperm, releasing them into the spermatic duct.

Epinephrine (ep′·ə·nef′·rin): Adrenaline; a hormone secreted by the adrenal medulla.

Epithelium (ep′·ə·thē′·lē·əm): Tissue covering internal and external surfaces.

Equatorial plate: The plane of the cell in which chromosomes arrange themselves as the cell begins metaphase.

Equilibrium: The point at which a chemical reaction proceeds in both directions at the same rate.

Esophagus: The muscular tube connecting the pharynx to the stomach.

Essential amino acid: An amino acid that is necessary to the metabolism of the organism.

Estrogen (es′·trə·jən): A female hormone; one of the hormones involved in the production of secondary sex characteristics and the menstrual cycle.

Estrous: The condition of being in the estrus period.

Estrus: The mating period of female mammals when they become sexually receptive.

Eucaryotic (yū·kar′·ē·o′·tic): Cells having a true nucleus and inner membrane structures.

Eutrophication (yoo·träf′·i·kā·shən): The aging process whereby a body of water supports increasing numbers of organisms, often causing lakes to become marshes and then terrestrial communities.

Evolution: The continuous genetic changes occurring in populations as a result of adapting to the environmental changes.

Excitation: Absorption of energy by electrons resulting in their movement to greater distances from the nucleus. Also, may refer to stimulation of a neuron.

Exergonic (ek'·ser·gon·ik): Any chemical reaction that releases energy.

Exogenous theory (ek·säj'·ə·nəs): The theory that says that biological rhythms are due to external geophysical factors.

Exoskeleton (ek'·sō·skel'·ə·t'n): An external skeleton, a hardened outside covering of arthropods such as crayfish and insects.

Exothermic reaction (ek'·sə·thur'·mik): A chemical reaction in which heat is released.

Extraembryonic membranes: Those membranes that lie outside the embryo proper but that contribute to its maintenance.

F_1 (first filial generation): The offspring of the parent generation.

F_2 (second filial generation): The offspring of the F_1 generation.

Fallopian tube: See oviduct.

Fat: A compound composed of three fatty acids linked to glycerol.

Fatty acids: Constituents of fats and oils. Long chains of carbon atoms with hydrogen attached, and ending with carboxyl groups.

Fermentation: Anaerobic respiration with the usual end product being alcohol and carbon dioxide.

Fertilization: The union of sperm and egg.

Fetus: The term used to describe the human embryo after eight weeks of development.

Fibril: Any tiny, threadlike structure within a cell.

Filtrates: Substances that have passed through a filter.

Fitness: A measure of reproductive success.

Flagellum (flə·jel'·əm): A threadlike organelle that extends from the cell's surface; it may beat with whiplike motions.

Flower: The reproductive structure of angiosperms.

Follicle (fäl'·i·k'l): A layer of cells surrounding the egg cell in the ovary.

Follicle stimulating hormone: FSH; A hormone produced by the anterior pituitary that stimulates the growth and maturation of eggs and sperm.

Food chain: Transfer of food energy from producers such as green plants to consumers such as herbivores and carnivores.

Food web: A scheme showing complex feeding relationships among the organisms in a community.

Forebrain: The anterior most subdivision of the vertebrate brain consisting of the cerebral hemispheres and certain anterior structures such as the thalamus.

Foreskin: A fold of skin at the end of the penis that is removed in circumcision.

Fossil fuel: Fuel for burning produced by the remains of organisms once alive, e.g. coal and natural gas.

Free energy: The energy that is available to do work.

Fruit: Layers of tissues produced by the ovary of an angiosperm.

Functional group: A group of atoms in an organic molecule that are free to participate in chemical reactions.

Gamete (gam'·ēt): A male or female sex cell with half the chromosomal material. Fertilization is the union of gametes, producing the full amount of hereditary material.

Gametogenesis (gə·mēt'·ə·jen'·ə·sis): The formation of gametes.

Gametophyte (gə·mēt'·ə·fīt'): The sexual generation of a plant in which the cells have half the normal chromosome numbers. It alternates with the sporophyte generation.

Ganglion (gan'·glē·ən): A group of nerve cell bodies, situated outside the central nervous system.

Gastrulation (gas'·troo·lā'·shən): The formation of the gastrula stage of an embryo. This is the stage in which the embryo acquires its three germ layers.

Gause's rule: The notion that two species with the same ecological requirements cannot occupy the same habitat indefinitely.

Gene: The portion or portions of DNA that produce a recognizable effect or trait, e.g. genes produce enzymes, color pigments, etc.

Gene frequency: The fractional distribution of alleles for a specific trait in a population, e.g. 0.3T + 0.7t — the gene frequency for tallness and shortness.

Gene pool: All the alleles in a population. (Also, a little kid living near the author).

Genetic code: The biochemical specification (DNA) for producing a particular genetic trait of the entire species' characteristics.

Genetic drift: Changes in the composition of a gene pool caused by the random loss of a particular gene through death or migration rather than by selection.

Genetic mapping: The determination of the position of genes on chromosomes, relative to each other, expressed as map units.

Genetic recombination: (1) the reorganizing of genetic material in the offspring, producing combinations present in neither parent; (2) the insertion of DNA from one organism into the genome of another, primarily bacteria. The result is that the recipient forms new products, directed by the DNA of the donor.

Genotype (jen'·ə·tīp'): The assortment of genes that make up the genetic characteristics of an organism.

Genus (jē'·nəs): A major subdivision of a family, consisting of one or more species.

Geotropism (jē·ät'·rə·piz'm): The influence of gravity upon growth or movement.

Germ cell: The sex cell; sperm or egg.

Germ layer: A layer of undifferentiated embryonic cells.

Gestation period (jes·tā'·shən): The duration of pregnancy.

Gibberellins (jib·ə·'rel·ən): A group of plant growth hormones.

Gill: The respiratory organ of certain aquatic organisms through which oxygen and carbon dioxide are exchanged with water.

Gizzard: A muscular sac in the digestive system that mechanically changes food by mashing it against sand particles; found in birds and some other animals where chewing is not possible.

Gland: A group of cells specialized for secretion.

Glands of Bartholin: Mucus glands of the vagina and labia majora whose secretions lubricate the vagina during intercourse.

Glans penis (glanz pē·nis): The end or caplike structure at the termination of the penis.

Glial cell: One of the numerous cells that fill the spaces between nerve cells.

Globular protein: A protein that has doubled back on itself several times so the structure is almost spherical.

Glomerulus (glö·mer′·yoo·ləs): A cluster of capillaries within the Bowman's capsule of the kidney.

Glucose: A six carbon sugar found in many fruits and animal tissues. The basic unit of starch and cellulose.

Glycerol (glis′·ər·ōl′): A three carbon alcohol molecule that combines with fatty acids to form fats and oils.

Glycogen (glī′·kə·jən): A complex chain of glucose units produced by animals.

Glycolysis (glī·käl′·ə·sis): The first phase of respiration where glucose is broken down with a small energy yield.

Golgi bodies (gōl′·jē): A secretory structure in some cells consisting of a complex system of folded membranes.

Gonad (gō′·nad): An ovary or testis.

Granum: A green granule within chloroplasts, appearing as stacks of thylakoids. Contains chlorophyll and carotenoids. The site of the light reaction.

Guanine (gwä′·nēn): One of the nitrogenous bases that are found in nucleic acids.

Guard cell: One of the two cells that surround the stoma in leaves or young stems of plants.

Gullet: The esophagus or food receiving tube in an animal.

Gyrus (jī′·rəs): A ridge formed by the convolutions of the cerebral cortex of the brain.

Habitat: The specific place where an organism lives.

Habituation (hah·bich·oo·ay·shun): A simple form of learning in which an organism's response to a stimulus diminishes.

Half-life: The time required for one half of a substance to disappear or decay.

Hallucinogenic (he·lōō′·sə·nə·jen·ic): The quality of producing hallucinations.

Haploid (hap′·loid): Having the half number of chromosomes per cell.

Hardy-Weinberg Law: A law in genetics that states that in a stable population the overall frequency of genes will not change in the absence of migration or other unstabling factors.

Hemisphere: Either of two symmetrical halves of the brain.

Hemoglobin (hē′·mə·glō′·bin): A protein-iron complex in red blood cells that binds with oxygen.

Hemophilia (hē·mə·fil′·ē·ə): A sex linked condition in which the blood does not clot normally. It is also called bleeder's disease.

Herbivore (hur′·bə·vôr′): A plant-eating animal.

Heredity: The transmission of traits from parents to offspring.

Hermaphrodite (hər·maf′·rə·dīt′): An individual who possesses both female and male reproductive organs.

Heterotroph (het′·ər·ə·träf′): An organism that depends on an external source of organic substances for its food.

Heterozygous (het′·ər·ō·zī′·gəs): A mixed genotype. Different alleles for a particular trait.

High energy bond: A bond that releases a relatively high amount of energy when broken; for example, the phosphate bonds in ATP.

Hindbrain: The most posterior of the three subdivisions of the vertebrate brain. It includes the pons, the medulla, and the cerebellum.

Homeostasis (hō′·mē·ō·stā′·sis): The tendency of organisms to maintain stability and consistency.

Homeotherm: An animal able to maintain relatively uniform internal temperatures, such as birds and mammals.

Homolog (häm′·ə·lôg′): One member of a pair of chromosomes.

Homozygous (hō·mō·zī′·gəs): Having identical alleles for a particular trait.

Hormone: A chemical messenger; a substance produced in the body that has a regulatory effect in another part of the body.

Host: An organism that supports a parasite.

Hybrid: Offspring resulting from the cross of parents with different characteristics or parents of different species.

Hydrocarbon: Any organic compound that contains simply carbon and hydrogen.

Hydrogenation: To combine with hydrogen.

Hydrogen bond: A weak attractive force between two molecules where one contains hydrogen.

Hydrolysis: The breakdown of a compound by the addition of water molecules.

Hydrophilic (hī′·drə·fil′·ik): Water loving, or having an attraction for water.

Hydrophobic (hī′·drə·fō′·bik): Water fearing, or repelling of water.

Hymen (hī′·mən): A membrane that may partially block the opening of the vagina.

Hypnotic drugs: Drugs that act to bring about depression and sleep.

Hypothalamus (hī′·pə·thal′·ə·məs): The region that lies in front of the thalamus in the forebrain. It regulates the internal environment, pituitary secretions, and some of the basic drives.

Imprinting: A type of learning that occurs in very specific periods early in life and is generally not reversible.

Independent assortment: A law developed by Mendel that states that the distribution of one gene to the offspring is not affected by the distribution of another gene. This permits recombination or the mixture of genetic material to the offspring.

Induction: The influence of cell differentiation or development by neighboring cells.

Inert: Chemically showing a lack of activity.

Innate releasing mechanism (IRM): An ethological term that refers to a genetic mechanism that produces a specific behavioral event when triggered by a particular stimulus from the environment.

Inorganic: Pertaining to chemical substances not produced by cells and not containing carbon.

Instinct (in′·stinkt): Any inherent, unlearned behavioral pattern, which is functional the first time it happens and can occur in animals reared in total isolation.

Insulin: A hormone produced in the pancreas. Lowers blood sugar concentration.

Intercalated disc (in·tur′·kə·lā′·ted): The junction that lies between the subunits of cardiac muscle fibers.

Interneuron (in′·tər·noor′·än): A nerve cell that links sensory cells and motor cells.

Interphase: The period between two mitotic events in the same cell.

Intrauterine device (IUD) (in′·trə·yōot′·ər·in): A plastic device, sometimes containing copper, that is inserted into the uterus as a means of preventing conception.

Ion: An atom or molecule that has gained or lost one or more electrons.

Ionic bond: An attractive force between atoms that have different charges as a result of the transfer of electrons between them.

Ionization: The separation of a substance into its ions when in solution.

Isosmotic (ī'·säs·mät'·ik): A solution with the same osmotic properties as a cell or another solution.

Isotope: An atom with a lighter or heavier nucleus than the average; may exhibit radioactivity.

Jejunum (ji·jōō'·nəm): The area of the small intestine that lies between the duodenum and the ilium.

Karyotype ('kar·ē·ə·tīp): The physical appearance (including size, shape, and number) of the chromosomes in a cell.

Kidney: The organ that removes soluble waste products thus maintaining the composition and concentration of solutes in the blood.

Kilocalorie (kil'·ə·kal'·ər·ē): One thousand calories or the amount of heat necessary to raise the temperature of one thousand grams of water one degree.

Kingdom: Any of five categories in which organisms are usually placed.

Krebs citric acid cycle: The oxidative phase of respiration in aerobic reactions where the fuel is broken down to CO_2 and H_2O.

Labia majum (lā'·bē·ə mā'·jəm): The large conspicuous outer folds of skin forming the perimeter of the female reproductive organs.

Labia minum (lā'·bē·ə mī'·nəm): The smaller inner folds of skin that border the vagina.

Lamella: A layer, or flattened sheet.

Larva: An immature animal with an appearance different from the adult.

Laterite: A hardened tropical soil, high in iron and aluminum. Often formed when jungles are cleared.

Leucocyte (lōō'·kə·sīt'): Blood cells that aid in resisting infections.

Leucoplast (lōō'·kə·plast'): A colorless body in plant cells that stores proteins, lipids, and starch.

Linkage: The condition of genes existing on the same chromosome.

Lipase (lī'·pās): A pancreatic enzyme that digests lipids.

Lipid (lib'·id): A type of fatty organic compound.

Liver: A large organ in higher animals that secretes bile, stores glycogen, and carries out other essential chemical activities.

Locus: The position of a gene on a chromosome.

Loop of Henle: The prominent U-shaped loop in the renal tubule of the mammalian kidney.

Luteal phase (lōōt'·ē·əl): The latter part of the menstrual cycle in which the corpus luteum secretes hormones that prepare the uterus for the fertilized egg.

Luteinizing hormone: LH-A pituitary hormone that causes ovulation and stimulates hormone production in the corpus luteum.

Lymph: The clear yellowish fluid or plasma found in the lymphatic system.

Lymph node: A clump of cells in the lymphatic system that produces lymphocytes and acts as a filter for lymph.

Lymphocyte (lim'·fə·sīt'): Any of a variety of white blood cells that are formed in the lymph nodes.

Lysosome: A membrane-bounded organelle that forms or isolates disruptive enzymes.

Macromolecule: A very large molecule.

Macronucleus (mak'·rō·nōō'·klē·əs): The larger of two types of nuclei found in certain protozoans.

Mantle: The shell-secreting organ in mollusks.

Marine: Pertaining to salt water.

Mass number: The total number of neutrons and protons in the nucleus of an atom.

Mechanistic theory: The theory that the processes of life are based on the same physical and chemical laws that apply to the nonliving phenomena.

Medulla (mi·dul'·ə): The region of the hindbrain that appears to be an expanded portion of the spinal cord. It controls many activities of the internal organs.

Medusa: The free-swimming stage of many coelenterates.

Meiosis (mī·ō'sis): The process of cell division in which the chromosome number is halved during the formation of gametes.

Melanin (mel'·ə·nin): Coloring pigments in the hair and skin of man, other animals, and some plants.

Menstrual cycle: The events in the human female that produce a mature egg and prepares the uterus for receiving the zygote.

Meristem (mer'·ə·stem'): Undifferentiated embryonic tissue in plants.

Mesoderm: One of the three embryonic germ layers. It gives rise to connective tissue, muscle, the circulatory system, and much of the reproductive and excretory systems.

Mesophyll (mes'·ə·fil'): The parenchyma tissue of a leaf.

Messenger RNA (mRNA): RNA that carries the genetic code to the ribosomes where it is translated into protein production.

Metabolism: The total chemical processes of life, including energy storage, release, and transfer.

Metamorphosis (met'·ə·môr'·fə·sis): Any morphological change such as the transformation of an insect larva into an adult.

Metaphase (met'·ə·fāz'): The second phase in mitosis during which the chromosomes line up at the center of the cell along the equator of the spindle.

Micron: One thousandth of a millimeter.

Micronucleus (mī'·krō·nōō'·klē·əs): The smaller of two types of nuclei in certain protozoa.

Microtubule (mī·krō): A microscopic tube or hollow cylinder of protein filaments.

Midbrain: The central of the three subdivisions of the vertebrate brain.

Middle lamella: Layer between adjacent cell walls.

Millimicron: One thousandth of a micron.

Mitochondrion (mīt'·ə·kän'·drē·ən): An organelle in the cytoplasm that is sometimes called the powerhouse of the cell because it functions in energy producing processes.

Mitosis: Nuclear division resulting in exact chromosome duplication.

Mitotic apparatus: Any cell part involved in the process of mitosis.

Molecule: Smallest unit of a compound, composed of two or more atoms.

Monocot (män'·ə·kät'): Brief form of monocotyledon. Any plant that produces one seed leaf such as grasses or palms.

Monosaccharide (män·ə·'sak·ə·rīd): A simple sugar, such as glucose or fructose.

Mons veneris (mänz' ven'·ər·is): A fatty mound of tissue covering the pubic bone in the human female.

Morula (môr'·yoo·lə): A stage of the embryo in which the cells are arranged in a solid mass.

Motor neuron: A nerve cell that carries impulses from the central nervous system to a muscle or gland.

Mutation: Any change in a gene that results in a permanent genetic change in the individual; a deletion or addition to the basic code of the DNA strand.

Mutualism (myōō'·choo·wəl·iz'm): A symbiotic relationship in which both organisms benefit in some way.

Myelin sheath: The membranes of Schwann cells. Surrounds the axons of some vertebrate neurons.

Myofibril (mī'·ə·fī'·brəl: The slender protein thread in the skeletal muscle that acts in contraction of the muscle.

Myofilament (mī'·ə·fil'·ə·mənt): A filament that makes up the myofibril in the skeletal muscle cell.

Myosin (mī'·ə·sin): The nonmoving protein of the myofibril of muscle.

NADP or NADPH: Nicotinamide Adenine Dinucleotide Phosphate. A molecule that functions in the electron transport system of plant chloroplasts.

Narcotic: Any agent that produces stupor or insensibility.

Natality: A measurement of births in a population.

Natural selection: The differential survival and reproduction of certain individuals in a population. In the same process, other organisms less suited to the environment are eliminated.

Nephridium (ni·'frid·ē·əm): A kind of invertebrate excretory organ.

Nephron (nef'·rän): The functional unit of the kidney.

Nerve: A bundle of neurons and their accompanying connective tissue.

Nerve impulse: An electrochemical signal in which an impulse is transmitted from one neuron to another neuron or to other cells.

Neural groove (noor·al): An ingrowth of the neural plate of a vertebrate embryo; it eventually forms the spinal cord in mammals.

Neural plate: A thick plate formed by rapid cell division over the notochord on the dorsal side of the embryo; it eventually forms the spinal cord in vertebrates.

Neural tube: The early brain and spinal cord of a vertebrate.

Neuron: A nerve cell.

Neutron: An atomic nuclear particle that has about the same mass as a proton, but is electrically neutral.

Niche (nich): All aspects of the biological and physical environment that relate to the activities of an organism.

Nitrogen cycle: Worldwide circulation of nitrogen atoms.

Nitrogenous base (nī·träj'·ə·nəs): The purine or pyrimidine constituent of nucleotides and nucleic acids (adenine, guanine, cytosine, thymine, uracil).

Nocturnal animal: An animal that is active mainly at night.

Node of Ranvier: A depression that separates Schwann cells that lie along neurons.

Nondisjunction: The failure of homologous chromosomes to separate in meiosis.

Notochord (nōt'·ə·kôrd'): A rodlike arrangement of cartilage cells that form the chief supporting structure in the chordate body. In vertebrates it is present only in the embryonic stage.

Nuclear membrane: The outer selectively permeable membrane of the cell nucleus.

Nuclear pore: Tiny pores in the nuclear membrane of cells.

Nuclease (nōō'·klē·ās'): Any enzyme that acts to split nucleic acids.

Nucleic acids (nōō·klē'·ik): Any of a number of acids that carry genetic information. There are two kinds: RNA and DNA.

Nucleolus (nōō·klē'·ə·ləs): An organelle that lies within the nucleus of the cell. It is high in RNA and protein. It forms ribosomal RNA.

Nucleotide (nōō'·klē·ə·tīd'): A molecule made up of phosphate, a 5-carbon sugar, and a purine or pyrimidine base.

Nucleus: An organelle within a cell bounded by a double membrane that contains most

of the DNA of the cell. Also, in the atom, a positively charged mass composed of neutrons and protons.

Occipital lobe (äk·sip'·ə·t'l lōb): One of the four major lobes of the brain. It is involved in the reception and processing of visual information.

Omnivore (äm'·nə·vôr'): An animal that eats both plant and animal matter.

Oocyte (ō'·ə·sīt'): A cell that undergoes meiosis to form an egg cell.

Oogenesis (ō'·ə·jen'·ə·sis): The process of egg formation.

Ootid (ō'·ə·tid'): One of four cells formed by meiosis. In animals, one will develop into an egg cell.

Organelle (ôr'·gə·nel'): A specific structure within a cell, usually bounded by a membrane.

Operant conditioning: A form of learning in which an animal performs an act to receive rewards.

Orbital: The approximate and hypothetical path on which an electron moves around the nucleus.

Order: A taxonomic subdivision of a class in plants and animals that may consist of several families.

Organic: Pertaining to chemical compounds that are normally synthesized by living organisms that contain carbon, hydrogen, and oxygen.

Orgasm: The final stage or climax of sexual intercourse, usually followed by a period of relaxation.

Orientation: An organism's means of determining its location in respect to some reference point in its environment.

Osmosis (äs·mō'·sis): The tendency of water to move across a membrane from an area of lower solute concentration to one that has a higher solute concentration.

Ovary: The female gonad in which the eggs are formed.

Oviduct (ō'·vi·dukt'): The tube opening near the ovary and ending in the uterus in most animals.

Ovulation (ō'·vyə·lā'·shən): The sudden release of a mature egg from the ovary.

Ovum: The egg or female reproductive cell.

Oxidation (äk'·sə·dā'·shən): The loss of electrons in chemical reactions; or the combination of oxygen with another substance.

Palisade cells: A layer of elongated, chloroplast containing cells that lie under the epidermis of the leaf in higher plants.

Pancreas (pan'•krē•əs): A gland that is located behind the stomach in mammals that secretes digestive enzymes.

Parasite: An organism that lives in or on another organism and causes some harm to its host as it derives food from it.

Parasympathetic nervous system (par'•ə•sim'•pə•thet'•ik): One of the divisions of the autonomic nervous system.

Parenchyma (pə•ren'•ki•mə): Thin walled, unspecialized plant-storage cells.

Parietal lobe (pə•rī'•ə•t'l): One of the four major lobes of the cortex. The detection of body position and sensory input are the major functions.

Passive transport: Movement of substance through the cell membrane without cellular energy (ATP) involved.

Penis: The male sex organ.

Peptide (pep'•tīd): A compound comprised of two or more amino acids.

Perennial: A plant that persists, in whole or in part, from year to year.

Peripheral nervous system (pə•rif'•ər•əl): One of the two divisions of the vertebrate nervous system. It includes the nervous tissue outside the central nervous system.

Petiole (pet'•i•ōl'): The stalk by which a leaf is attached to the stem.

pH: The relative concentration of hydrogen ions in a solution. A measurement of acidity, neutrality being 7, and the range from 0 to 14, lower numbers denoting greater acidity.

Pharynx (far'•inks): The organ below the mouth that is the common respiratory and digestive structure leading to the esophagus and larynx.

Phenotype (fē'•nə•tīp'): The observable genetic characteristics of any organism.

Pheromone (fer'•ə•mōn'): A substance that is secreted by an organism that changes the behavior in another organism of the same species.

Phloem (flō'•em): The vascular tissues in plants, consisting of sieve tubes and companion cells and specializing in nutrient transport.

Phospholipid (fäs'•fō•lip'•id): A fat like substance of the cell membrane, which contains phosphorus, fatty acids, glycerol, and a nitrogenous base.

Phosphorylation (fös'•fər•ə•lā'•shən): The addition of phosphate to an organic molecule such as ADP, e.g. ADP + Phosphate → ATP.

Photon: A unit of light energy.

Photoperiodism (fōt'•ō•pir'•ē•əd•iz'm): The rhythmic response of organisms to cyclic variations in light and darkness.

Photosynthesis (fōt'•ə•sin'•thə•sis): The process in which plants convert carbon dioxide and water into carbohydrates through the use of solar energy.

Phototropism (fō•tät'•rə•piz'm): A group response of plants to the presence of light.

Phylum (fī'•lem): The major primary subdivision of a kingdom. It is composed of classes.

Pigment: Any colored substance.

Pistil: The seed-bearing part of a flower.

Pituitary gland (pi•tōō'•ə•ter'•ē): A tiny endocrine gland at the base of the brain.

Placenta: Embryonic membranes in the uterus of pregnant mammals through which the embryo transports wastes and receives essential nutrients.

Plankton: Organism that drifts in a body of water; usually very small or microscopic.

Plasma: The liquid portion of the blood.

Plasmid: Small circles of bacterial DNA that exist outside the single large circular chromosome.

Plastids (plas'•tid): Small bodies in plant cells.

Pleiotropy (plī'•ä•trə•pē): Multiple effects of a single gene.

Polar body: Small, nonfunctional egg formed during meiosis.

Pollen: The male microspores of flowering plants consisting of reproductive cells in a protective case.

Polygenic inheritance: Determination of a trait due to the action of multiple genes.

Polyp: The sessile stage of coelenterates.

Polypeptide (päl'•i•pep'•tīd): A group of two or more amino acids; a protein fragment.

Pons (pänz): A part or extension of the medulla; it contains nerve cells that relate to the spinal motor nerves.

Population: An interbreeding group of organisms. Usually a subset of a species.

Posterior: Pertaining to the rear end or, in humans, the back area.

Primary consumers: Animals that feed on plants; herbivores.

Primary structure: The amino acid sequence of proteins.

Primitive streak: An elongated group of cells in the early embryo that constitutes the axis of the body.

Procaryotic (prō•karee•ah•tic): Organisms that constitute the base of the food chain, mainly photosynthetic plants.

Progesterone (prō•jes'•tə•rōn'): An ovarian hormone, produced by the corpus luteum, assists in the preparation of the uterus for implantation of a fertilized egg.

Prophase (prō'•fāz): The first stage in mitosis.

Prostaglandin: A kind of fatty acid hormone found first in semen. May play a role in fertilization.

Prostate gland: A gland of the male reproduction system, situated at the base of the urethra; secretes a fluid that is added to the semen.

Protein: A specific chain of amino acids. Essential in producing cell structures and a variety of enzymes.

Proton: An atomic nuclear particle that has a mass similar to a neutron, but with a positive charge.

Protoplasm: The living colloidal material that makes up cells.

Pseudopodium (sōō'•də•pō'•dē•əm): Literally, a false foot; a temporary projection from the cell used in amoeboid cells for feeding and movement.

Psychedelic (sī'•kə•del'•ik): Related to relaxation; mind-freeing, accompanied by bizarre thought patterns and behavior.

Pulmonary: Pertaining to the lungs. In vertebrates pulmonary arteries carry blood to the lungs, pulmonary veins carry oxygenated blood to the heart.

Pupa ('pyü•pə): A nonmotile, nonfeeding, sometimes encapsulated stage between larval and adult stages in some insects.

Purine: Nitrogenous bases of DNA and RNA, containing two nitrogen rings—adenine and guanine.

Pyrimidine (pi•rim'•ə•dēn'): Nitrogenous bases of DNA and RNA, containing one nitrogen ring—thymine, cytosine, and uracil.

Pyruvate (pī·rōō′·vāt): Pyruvic acid; a molecule formed at the end of glycolysis.

Radial symmetry: A body plan consisting of parts arranged around a longitudinal axis. Any line through the axis will divide the organism in roughly equal parts.

Radiant energy: Sunlight energy.

Radioactivity: Particles or energy emitted from radioactive elements as they decay.

Radula (raj′·oo·lə): The rough surfaced tongue or feeding organ found in many mollusks.

Receptor: A sensory structure that detects a particular form of stimulus in the environment such as light or sound.

Recessive: The allele that is not expressed in the phenotype of an organism.

Reduction: Adding electrons to an atom or molecule.

Reduction division: Meiosis.

Reflex: An involuntary response to a stimulus.

Reflex arc: A simple nervous pathway that involves a sensory cell, a connecting neuron and a muscle cell.

Refractory period: The recovery period following a nerve impulse during which the nerve cannot initiate a new impulse.

Regeneration: Regrowth of a lost part or tissue.

Reinforcement: In behavior conditioning, a process that increases the likelihood of a given response.

Renal (rē′·n′l): Referring to the kidney.

Replication: The process of duplicating DNA prior to cell division.

Respiration: Cellular transfer of energy in fuels to ATP energy.

Resting potential: The relative charges on both sides of a nerve fiber in the unstimulated state.

Reticular formation: A nerve cell mass in the medulla that functions as an arousal mechanism.

Retina: The layer of light sensitive cells in the vertebrate eye.

RH factor: A type of antigen in the blood.

Ribonucleic acid (rī′·bō·nōō·klē′·ik): A nucleic acid containing ribose sugar and the nitrogen base, uracil. Functions in protein synthesis.

Ribosomal RNA (rRNA): A type of RNA molecule found in the ribosomes; its actions are not clearly known, but it is believed to function in protein synthesis.

Ribosome (rī′·bə·sōm′): A grainy structure found in great number in cells. Important in protein synthesis.

Ribulose (rī′·bū·lōs): A five carbon sugar produced in the Calvin cycle during photosynthesis.

Ribulose diphosphate (rī′·bū·lōs dī′·fos·fāt): A form of ribulose that acts as the carbon dioxide acceptor in the Calvin cycle of photosynthesis.

RNA polymerase: An enzyme that aids in the formation of RNA along the DNA template.

Root cap: A group of cells covering the growing tip of a root.

Salt: A product of the reaction between a base and an acid.

Sarcolemma (sar·koh·lem·ah): The surrounding membrane of a striated muscle fiber.

Schwann cell: A cell that helps to form the sheath around the peripheral nerve fibers.

Scrotum: The sac surrounding the testis.

Secondary consumers: Animals that feed primarily on herbivores.

Secondary sex characteristic: The physical development that occurs during puberty, specific for each sex.

Second filial generation (F₂) (fil′·ē·əl): The offspring produced by the mating of two individuals of the first filial generation.

Secretion: A cellular product, usually released into a duct or organ to serve a function.

Seed: Formed by fertilization of a mature ovule in seed plants.

Segregation: Separation of two alleles during meiosis.

Self-fertilization: A sexual process in plants where gametes from the same plant combine.

Semen: The sperm-containing fluid ejaculated by the male.

Seminal duct: The tube that carries semen from the epididymus to the ejaculatory duct. It is also called a vas deferens.

Seminal vesicle: A saclike part of the male reproductive duct in which sperm are stored.

Senescence (sə·nes′·'ns): Gradual deterioration caused by age.

Sensory neuron: A neuron that receives impulses; an afferent neuron.

Sepal (sē′·p'l): The leaflike floral parts surrounding the base of the petals.

Sessile (ses′·il): Pertaining to organisms that do not move about.

Sex chromosome: Chromosomes associated with the determination of sex. In man, the 23rd pair.

Sex-linked: A trait determined by a gene located on a sex chromosome.

Sickle cell anemia: A hereditary disease caused by a form of hemoglobin that is defective in its nature and causes sickle-shaped red blood cells.

Sieve cell: A specialized phloem cell through which nutrients move in the vascular plant.

Sieve tube: A tube formed by joining sieve cells end to end.

Sociobiology: The study of the impact of evolution on social behavior.

Solar energy: Any radiant energy produced by the sun.

Solute (säl′·yōōt): Any substance dissolved in a solvent.

Solvent: A fluid in which substances may dissolve forming solutions.

Soma (sō′·mə): Pertaining to the body.

Somatic cells: Usually diploid (2n) cells composing body tissues. Does not form gametes.

Somite (sō′·mīt): One of the paired mesadermal segments along the neural axis of an embryo.

Species: The major subdivision of a genus. Individuals of a species are able to breed among themselves under natural conditions.

Sperm: A mature male gamete.

Spermatid (spur′·mə·tid): A haploid cell resulting from meiotic divisions of sperm-forming cells. Will become a sperm cell.

Spermatogenesis (spər·mat′·əjen′·ə·sis): The formation of sperm cells.

Sphincter (sfink′·tər): A ring of muscles that controls the movement of materials through passages, e.g. sphincters at either end of the stomach.

Spinal cord: The complex band of neurons that runs through the spinal column of vertebrates to the brain.

Spindle: A structure formed in mitosis and meiosis that consists of fine threads radiating from the centrosomes along which the chromosomes appear to move.

Spiracle: An external opening to the respiratory system of terrestrial arthropods.

Spleen: An organ, containing many blood vessels, that is situated near the stomach in man. It forms certain white blood cells and destroys worn out red blood cells. It also acts as a reservoir for red blood cells.

Spontaneous generation: The notion that living organisms are derived from nonliving matter.

Sporangium (spə·′ran·jē·əm): A hollow structure in which spores are formed.

Spore: An asexual reproductive cell that can develop into an adult.

Sporophyte (spôr′·ə·fīt′): The spore producing, diploid generation in plants. It alternates with the sexual or gametophyte generation.

Stamen: The male organ of the flower that bears pollen.

Starch: A long chain of glucose molecules, formed in plants for glucose storage.

Stereotyped (ster′·ē·ə·tīpt′): Predictable or unchanging.

Steroid (stir′·oid): The type of lipid consisting mainly of fatty acids attached to complex alcohols.

Stigma: The female part of a flower that receives the pollen.

Stoma (stō′·mə): A small opening in a leaf through which gases may pass.

Stroma: Matrix of a chloroplast in which the grana are imbedded.

Style: The portion of the pistil that connects the stigma with the ovary in flowers.

Subnormal releaser: An artificial stimulus that is more effective than the natural stimulus in releasing a particular behavioral pattern.

Subphylum: A group of related organisms within a phylum.

Substrate (sub′·strāt): A substance that is acted upon by an enzyme.

Sulcus (sul′·kəs): A convolution of the cerebral cortex.

Surface tension: A cohesive force among surface molecules that resists rupture.

Symbiosis (sim′·bī·ō′·sis): The association between two organisms of two different species when at least one organism benefits in the association.

Sympathetic nervous system: A subdivision of the autonomic nervous system that increases energy expenditure and prepares the body for emergency situations.

Synapse (si·naps′): The space between neurons, across which impulses are chemically transmitted.

Systole (sis′·tə·lē′): The contraction of the heart.

Taxonomy: The science of classifying life.

Telophase (tel′·ə·fāz′): The final stage of mitosis that includes total separation of chromosomes, cytoplasmic division, and the return of the interphase nucleus.

Template: A pattern or mold.

Temporal lobe: One of the four lobes in the human brain; it contains auditory and visual centers.

Territory: Any defended area.

Testis: The male gonad where sperm and sex hormones are produced.

Testosterone (tes·täs′·tə·rōn′): A male hormone, produced in the testes, important in the sex drive and producing secondary sex characteristics.

Tetanus: A state of continuous muscle contraction.

Thalamus (thal′·ə·məs): A cluster of neural cells in the posterior forebrain; it is involved in auditory and visual functions.

Thermal energy: Energy due to the motion of molecules.

Threshold: The intensity level necessary to create a response to a stimulus.

Thylakoid: A saclike, membranous structure in the chloroplasts. Stacks of these form grana.

Thymine (thī′·mēn): One of the nitrogenous bases of DNA.

Thymus gland (thī′·məs): An organ lying behind the breast bone that grows smaller as an individual ages; it functions in immune responses.

Thyroxin (thī·räk′·sin): A thyroid hormone that functions in metabolism.

Tissue: A group of cells organized to perform the same function.

Trachea (trā′·kē·ə): The tube that carries air from the larynx to the bronchi in land vertebrates; in insects a complex system of branched tubules carrying air to the interior.

Tracheid (trā′·kē·id): A type of xylem cell in plants adapted for conduction and support.

Transfer RNA: Lightweight nucleic acid molecules that identify with specific amino acids during protein synthesis.

Trophic levels (träf′·ik): The various feeding levels in a food chain, including producers and consumers.

Trophoblast (träf′·ə·blast′): The thin wall side of a blastocyst that forms the chorion when implantation occurs in mammals.

Tropism: A growth that is a response to a specific external stimulus.

Tubal ligation: Female sterilization by cutting and tying the fallopian tube.

Tumescence (tōō·mes′·′ns): The swelling of an organ or part.

Turgor: Stiffness or rigidity produced by fluid pressure.

Turner's Syndrome: An abnormality due to the lack of one X chromosome; it is characterized by retarded growth and sexual development.

Umbilical cord: Blood vessels that transport substances between the placenta and the embryo.

Unconditioned response: A natural response to an unfamiliar stimulus.

Unconditioned stimulus: An unfamiliar stimulus that causes response.

Unit membrane: The basic structure of all cellular membranes.

Uracil (yoor′·ə·sil): One of the nitrogenous bases of RNA.

Ureter (yoo·rēt′·ər): The tube that conducts urine from the kidney to the bladder in higher animals.

Urethra (yoo·rē′·thrə): The tube that conducts urine from the bladder to the outside of the body in higher animals.

Uterus (yōōt′·ər·əs): In female animals, the reproductive tract in which the fetus is contained.

Vacuole: A space within a cell, bounded by a membrane.

Vagina: The female copulatory organ and birth canal.

Vas deferens: The mammalian tube carrying sperm from the testes to the urethra.

Vascular tissue: Any tissue that contains vessels through which fluids are passed.

Vasectomy (vas·ek′·tə·mē): Male sterilization by cutting and tying the seminal ducts.

Vein: A vessel that carries blood toward the heart.

Vena cava: A large vein carrying blood to the right atrium of the heart.

Ventral: Referring to the undersurface or belly of an animal.

Villus (vil′·əs): A small protrusion of the intestinal wall, greatly increasing the absorbing surface of that organ.

Virulent: Capable of causing diseases.

Virus: An infectious, submicroscopic parasite that consists of a RNA or DNA core with a protein coat.

Viscera (vis′·ər·ə): A term that pertains to the internal abdominal organs of an animal. Sometimes called the guts.

Vitalism (vīt′·'l·iz′m): The notion that life has unique mystical properties that are distinct from those ascribed by chemical and physical laws.

Vitamin: A coenzyme. Any of a group of unrelated organic substances that are essential for normal metabolism.

Vulva: The external part of the female genitals.

White matter: The outer layer of the spinal cord in which myelinated fibers are found.

Xanthophyll (zan′·thə·fil): The yellow, red, or orange pigments of green plants.

X chromosome: A sex chromosome. Females carry two X chromosomes while males have one.

Xylem (zī′·ləm): Water conducting and supporting tissue in vascular plants.

Y chromosome: A sex chromosome found only in males.

Yolk: The nutrient portion of the egg.

Yolk sac: An embryonic tissue that assists in the nourishment of the embryo in some fishes, reptiles, and birds, and in most mammals.

Zooplankton: The nonphotosynthetic plankton.

Z scheme: A series of photochemical events centered around photosynthesis that results in the net movement of electrons from water to reduced NADP.

Zygote (zī′·gōt): A fertilized egg in plants or animals.

Suggested Readings

Chapter 1: Yesterday and Today

Darwin, C. R. 1962. *The voyage of the* Beagle. Garden City, N.Y.: Doubleday.

Dobzhansky, T. 1962. *Mankind evolving.* New Haven, Conn.: Yale University Press.

Eiseley, L. 1956. Charles Darwin. *Scientific American* 194(2):62–72.

———. 1958. *Darwin's century.* Garden City, N.Y.: Doubleday.

———. 1960. *The firmament of time.* New York: Atheneum.

Hamilton, T. H. 1967. *Process and pattern in evolution.* New York: Macmillan.

Howell, F. 1965. *Early man.* New York: Time-Life.

Mayr, E. 1970. *Populations, species and evolution.* Cambridge, Mass.: Harvard University Press.

Moorehead, A. 1969. *Darwin and the* Beagle. New York: Harper & Row.

Porter, E. 1971. *Galápagos.* New York: Ballantine.

Smith, H. W. 1961. *From fish to philosopher.* Garden City, N.Y.: Doubleday.

Chapter 2: The Beginnings

Baker, J. J. W., and Allen, G. E. 1968. *Hypothesis, prediction and implication in biology.* Reading, Mass.: Addison-Wesley.

Barghoorn, E. S. 1971. The oldest fossils. *Scientific American* 224(5):30–42.

Conant, J. B. 1951. *Science and common sense.* New Haven, Conn.: Yale Univ. Press.

Henshaw, P. S. 1971. *This side of yesterday—extinction or utopia?* New York: Wiley.

Jastrow, R. 1967. *Red giants and white dwarfs.* New York: Harper & Row.

Jorgensen, J. J., ed. 1970. Biology and culture in modern perspective. *Readings from Sci. Amer.* San Francisco: Freeman.

Koestler, A. 1972. *The roots of coincidence.* New York: Random House.

Kummel, B. 1961. *History of the earth.* San Francisco: Freeman.

Miller, S. L. 1955. Production of some organic compounds under possible primitive earth conditions. *Journal of the American Chemical Society* 77:2351–2361.

Monod, J. 1972. *Chance and necessity: an essay on the natural philosophy of modern biology.* New York: Knopf.

Oparin, A. I. 1938. *Origin of Life.* New York: Dover.

Wald, G. 1955. The origin of life. In *The physics and chemistry of life,* ed. of *Scientific American.* New York: Simon and Schuster.

Chapter 3: Molecules

Asimov, I. 1954. *The chemicals of life.* New York: Abelard-Schuman.

———. 1962. *Life and energy.* N.Y.: Avon.

Baker, J. J. W., and Allen, G. E. 1965. *Matter, energy and life.* Reading, Mass.: Addison-Wesley.

Barry, J. M., and Barry, E. M. 1969. *An introduction to the structure of biological molecules.* Englewood Cliffs, N.J.: Prentice-Hall.

Bernhard, S. A. 1968. *The structure and function of enzymes.* New York: Benjamin.

Farago, P., and Lagnado, J. 1972. *Life in action.* New York: Vintage.

Neurath, H. 1964. Protein-digesting enzymes. *Scientific American* 211(6):68–79.

Roberts, J. D. 1957. Organic chemical reactions. *Scientific American* 197(5):117–126.

Steiner, R. F., and Edelhock, H. 1965. *Molecules and life.* Princeton, N.J.: Van Nostrand.

Chapter 4: Energy

Amerine, M. A. 1964. Wine. *Scientific American* 211(2):46–56.

Asimov, I. 1968. *Photosynthesis.* New York: Basic Books.

Fogg, G. E. 1968. *Photosynthesis.* New York: American Elsevier.

Goldsby, R. A. 1967. *Cells and energy.* New York: Macmillan.

Lehninger, A. L. 1961. How cells transform energy. *Scientific American* 205(3):62–82.

———. 1965. *Bioenergetics, the molecular basis of biological energy transformations.* New York: Benjamin.

Levine, R. P. 1969. The mechanism of photosynthesis. *Scientific American* 221(6):58–70.

Margaria, R. 1972. The sources of muscular energy. *Scientific American* 226(3):84–91.

McGilvery, R. W. 1970. *Biochemistry.* Philadelphia: W. B. Saunders.

Rabinowitch, E. I., and Govindjee, I. 1965. The role of chlorophyll in photosynthesis. *Scientific American* 213(1):74–83.

Chapter 5: Cells

Crick, F. H. C. 1966. The genetic code: III. *Scientific American* 215(4):55–62.

Dupraw, E. J. 1968. *Cell and molecular biology.* New York: Academic Press.

———. 1970. *DNA and chromosomes.* New York: Holt, Rinehart and Winston.

Giese, A. C. 1968. *Cell physiology.* Philadelphia: Saunders.

Goldstein, L., ed. 1966. *Cell biology.* Dubuque, Iowa: William C. Brown.

Holter, J. 1961. How things get into cells. *Scientific American* 205(9):167–180.

Ingram, V. M. 1958. How do genes act? *Scientific American* 198(1):68–74.

Kennedy, D., ed. 1965. The living cell. *Readings from Sci. Amer.* San Francisco: Freeman.

Koller, P. C. 1971. *Chromosomes and genes.* New York: Norton.

Ledbetter, M. C. and K. Porter. 1970. *Introduction to the fine structure of plant cells.* New York: Springer.

Lowey, A. G., and Siekevitz, P. 1969. *Cell structure and function.* New York: Holt, Rinehart and Winston.

McElroy, W. D. 1971. *Cell physiology and biochemistry.* Englewood Cliffs, N.J.: Prentice-Hall.

Palade, G. 1966. *In Cell Biology.* Ed. by L. Goldstein. Dubuque, Iowa: W. C. Brown.

Porter, K., and Bonneville, M. A. 1968. *An introduction to the fine structure of cells and tissues.* Philadelphia: Lea and Febiger.

Robertson, J. D. 1962. The membrane of the living cell. *Scientific American* 206(4):64–82.

Swanson, C. 1964. *The cell.* Englewood Cliffs, N.J.: Prentice-Hall.

Watson, J. D. 1968. *The double helix.* New York: Atheneum.

———. 1970. *Molecular biology of the gene.* New York: Benjamin.

Woese, C. R. 1967. *The genetic code, the molecular basis for genetic expression.* New York: Harper and Row.

Chapter 6: Inheritance

Bonner, D. M., and Mills, S. E. 1964. *Heredity.* Englewood Cliffs, N.J.: Prentice-Hall.

Boyd, W. C. 1963, Genetics and the human race. *Science* 140:3571.

Crow, J. F., and Kimura, M. 1970. *An introduction to population genetics theory.* New York: Harper and Row.

Dobzhansky, T. 1963. Evolutionary and population genetics. *Science* 142:3596.

Eckhardt, R. B. 1972. Population genetics and human origins. *Scientific American* 226(1):94–102.

Goldsby, R. A. 1971. *Race and races.* New York: Macmillan.

Gray, G. W. 1951. Sickle-cell anemia. *Scientific American* 185(2):56–59.

Johnson, W. H., and Steere, W. C., eds. 1962. *This is life.* New York: Holt, Rinehart and Winston.

Levine, R. P. 1968. *Genetics.* New York: Holt, Rinehart and Winston.

McKusick, V. A. 1965. The royal hemophilia. *Scientific American* 213(2):88–95.

Scriven, M. 1959. Explanation and prediction in evolutionary theory. *Science* 130:477–482.

Srb, A., Owen, R., and Edgar, R. 1965. *General genetics.* San Francisco: Freeman.

Chapter 7: Reproduction

Allen, R. D. 1959. The moment of fertilization. *Scientific American* 201(1):124–134.

Asdell, S. A. 1964. *Patterns of mammalian reproduction.* Ithaca, N.Y.: Cornell University Press.

Bell, P. R., and Woodcock, C. L. F. 1968. *The diversity of green plants.* Reading, Mass.: Addison-Wesley.

Bullough, W. S. 1961. *Vertebrate reproductive cycles.* New York: Barnes & Noble.

Corner, E. J. H. 1968. *The life of plants.* New York: New American Library.

Csapo, A. 1958. Progesterone. *Scientific American* 198(4):40–46.

Frazer, J. F. D. 1959. *The sexual cycles of vertebrates.* London: Hutchinson University Library.

Hardin, G. 1970. *Birth control.* New York: Pegasus.

Katchadowrian, H., and Lunde, D. T. 1972. *Fundamentals of human sexuality.* New York: Holt, Rinehart and Winston.

Masters, W., and Johnson, V. 1966. *Human sexual response.* Boston: Little, Brown.

Michelmore, S. 1965. *Sexual reproduction.* Garden City, N.Y.: Natural History Press.

Peel, J., and Potts, M. 1969. *Textbook of contraceptive practices.* New York: Cambridge University Press.

Wilmoth, J. H. 1967. *Biology of invertebrata.* Englewood Cliffs, N.J.: Prentice-Hall.

Chapter 8: Development

Bonner, J. T. 1963. *Morphogenesis.* New York: Atheneum.

Fischberg, M., and Blackler, A. W. 1961. How cells specialize. *Scientific American* 205(3):124–140.

Flickinger, R. A., ed. 1966. *Developmental biology.* Dubuque, Iowa: Brown.

Grobstein, C. 1964. Cytodifferentiation and its controls. *Science* 143:643.

Odell, W. D., and Moyer, D. L. 1971. *Physiology of reproduction.* St. Louis: Mosby.

Rugh, R., and Shettles, L. B. 1971. *From conception to birth: the drama of life's beginnings.* New York: Harper and Row.

Smith, C. A. 1963. The first breath. *Scientific American* 209(4):27–35.

Spratt, N. T., Jr. 1964. *Introduction to cell differentiation.* New York: Van Nostrand-Reinhold.

Sussman, M. 1960. *Animal growth and development.* Englewood Cliffs, N.J.: Prentice-Hall.

Tanner, J. M., Taylor, G. R., and the ed. of *Life.* 1965. *Growth.* New York: Time-Life.

Waddington, C. H. 1966. *Principles of development and differentiation.* New York: Macmillan.

Wilmoth, J. H. 1967. *Biology of invertebrata.* Englewood Cliffs, N.J.: Prentice-Hall.

Chapter 9: Homeostasis: Controlling the Internal Environment

Macey, R. I. 1968. *Human physiology.* Englewood Cliffs, N.J.: Prentice-Hall.

Schmidt-Nielsen, K. 1964. *Animal physiology,* 2d ed. Englewood Cliffs, N.J.: Prentice-Hall.

———. 1964. *Desert animals.* New York: Oxford University Press.

———. 1972. *How animals work.* New York: Cambridge University Press.

Chapter 10: Systems and Their Control

Asimov, I. 1964. *The human body: its structure and operation.* New York: Signet.

Beck, W. S. 1971. *Human design: molecular, cellular and systematic physiology.* New York: Harcourt Brace Jovanovich.

Best, C. H., and Taylor, N. B. 1968. *The living body.* New York: Holt, Rinehart and Winston.

Currey, J. 1970. *Animal skeletons.* New York: St. Martin's Press.

Guyton, A. C. 1967. *Function of the human body.* Philadelphia: Saunders.

Hoyle, G. 1970. How is muscle turned on and off? *Scientific American* 222(4):72–82.

Huxley, H. E. 1965. The mechanism of muscular contraction. *Scientific American* 213(6):18–27.

Morton, J. E. 1967. *Guts.* New York: St. Martin's Press.

Ramsay, J. A. 1968. *Physiological approach to the lower animals.* New York: Cambridge University Press.

Chapter 11: Nerves and Hormones

Asimov, I. 1963. *The human brain: its capacities and functions.* New York: Signet.

Baker, P. F. 1966. The nerve axon. *Scientific American* 214(3):74–82.

Case, J. 1966. *Sensory mechanisms.* New York: Macmillan.

Clegg, P. C., and Clegg, A. C. 1968. *Hormones, cells and organisms.* Stanford, Calif.: Stanford University Press.

Davidson, E. H. 1965. Hormones and genes. *Scientific American* 212(6):36–45.

Day, R. H. 1971. *Perception.* Dubuque, Iowa: Brown.

Eccles, J. C. 1963. *The physiology of synapses.* New York: Academic Press.

Frye, B. E. 1967. *Hormonal control in vertebrates.* New York: Macmillan.

Greene, R. 1970. *Human hormones.* New York: McGraw-Hill.

Griffin, D. R., and Novick, A. 1970. *Animal structure and function.* New York: Holt, Rinehart and Winston.

Heimer, L. 1971. Pathways in the brain. *Scientific American* 225(1):48–60.

Kimble, D. P. 1973. *Psychology as a biological science.* Santa Monica, Calif.: Goodyear.

McCashland, B. W. 1968. *Animal coordinating mechanisms.* Dubuque, Iowa: Brown.

Riedman, S. R. 1962. *Our hormones and how they work.* New York: Collier.

Snider, R. S. 1958. The cerebellum. *Scientific American* 199(2):84–90.

Whalen, R. E. 1967. *Hormones and behavior.* New York: Van Nostrand-Reinhold.

Chapter 12: Behavior

Eibl-Ebesfeldt, I. 1970. *Ethology.* New York: Holt, Rinehart and Winston.

Hinde, R. 1970. *Animal behavior.* New York: McGraw-Hill.

Jolly, A. 1972. *The evolution of primate behavior.* New York: Macmillan.

Klopfer, P. H., and Hailman, J. P. 1967. *An introduction to animal behavior: ethology's first century.* Englewood Cliffs, N.J.: Prentice-Hall.

————. 1973. *Behavioral aspects of ecology.* Englewood Cliffs, N.J.: Prentice-Hall.

Manning, A. 1972. *An introduction to animal behavior.* Reading, Mass.: Addison-Wesley.

Marler, P., and Hamilton, W. 1966. *Mechanisms of animal behavior.* New York: Wiley.

Morris, D. 1967. *The naked ape.* New York: Dell.

Tiger, L. 1969. *Men in groups.* New York: Random House.

Tinbergen, N. 1951. *The study of instinct.* Oxford: Oxford University Press.

————. 1953. *Social behavior in animals.* London: Methuen.

Wallace, R. A. 1978. *The ecology and evolution of animal behavior,* 2d ed. Santa Monica, Calif.: Goodyear.

Wilson, E. O. 1975. *Sociobiology.* Cambridge, Mass.: Belknap.

Chapter 13: Community and Competition

Cornell, J., Mertz, D., and Murdoch, W. eds. 1970. *Readings in ecology and ecological genetics.* New York: Harper & Row.

Emlen, J. M. 1973. *Ecology: An evolutionary approach.* Reading, Mass.: Addison-Wesley.

Hazen, W. 1970, 1975. *Readings in population and community ecology,* 1st and 2d eds. Philadelphia: Saunders.

Kendeigh, S. C. 1974. *Ecology.* Englewood Cliffs, N.J.: Prentice-Hall.

Kormondy, E. 1969. *Concepts of ecology.* Englewood Cliffs, N.J.: Prentice-Hall.

Pianka, E. 1974. *Evolutionary ecology.* New York: Harper & Row.

McNaughton, S. J., and Wolf, L. L. 1973. *General ecology.* New York: Holt, Rinehart and Winston.

Moen, A. 1973. *Wildlife ecology.* San Francisco: Freeman.

Ricklefs, R. 1973. *Ecology.* Newton, Mass.: Chiron.

Vernberg, F. J., and Vernberg, W. B. 1970. *The animal and the environment.* New York: Holt, Rinehart and Winston.

Chapter 14: Populations

Appleman, B. 1965. *The silent explosion.* Boston, Mass.: Beacon.

Calahan, D. 1972. Ethics and population limitation. *Science* 175:487–492.

Calhoun, J. R. 1962. Population density and social pathology. *Scientific American* 206(2):139–148.

Croat, T. 1972. The role of overpopulation and agricultural methods in the destruction of tropical ecosystems. *BioScience* 22:465–467.

Ehrlich, P. 1968. *The population bomb.* New York: Ballantine.

————, and Ehrlich, A. 1970. *Population, resources, environment.* San Francisco: Freeman.

Hulett, H. 1970. Optimum world population. *BioScience* 20:160–161.

Langer, W. 1972. Checks on population growth: 1750–1850. *Scientific American* 226(2):92–99.

Meadows, D., et al. 1972. *Limits to growth.* New York: Universe.

Westing, A. 1971. Ecoside in Indochina. *Natural History* 80(3):56–61.

Chapter 15: Resources and Energy

Boerma, A. 1970. A world agricultural plan. *Scientific American* 223(2):54–69.

Carson, R. 1962. *Silent Spring.* Boston: Houghton Mifflin.

Commoner, B. 1966. *Science and survival.* New York: Viking Press.

————. 1971. *The closing circle.* New York: Knopf.

Hardin, G. 1968. The tragedy of the commons. *Science* 162:1243–1248.

————. 1972. *Exploring new ethics for survival.* New York: Viking.

Rienow, R., and Rienow, L. 1967. *Moment in the sun.* New York: Ballantine.

Shepard, P., and McKinley, D., eds. 1969. *The subversive science: essays toward an ecology of man.* Boston: Houghton Mifflin.

Chapter 16: Bioethics and Environment

Bresler, J. 1966. *Human ecology: collected readings.* Reading, Mass.: Addison-Wesley.

Clement, R. 1970. Pesticide do's and don't's. *Audubon* 72(2):50–51.

Lappe, F. 1971. *Diet for a small planet.* New York: Ballantine.

Miller, G. T. 1975. *Living in the environment: Concepts, problems, and alternatives.* Belmont, Calif.: Wadsworth.

Murdoch, W. 1975. *Environment,* 2d ed. Stamford, Conn.: Sinauer.

Wagar, J. A. 1970. Growth versus the quality of life. *Science* 168:1179–1184.

Photo Acknowledgements

First endpaper, Nick Pavloff
Second endpaper, Nick Pavloff
Third endpaper, Wayne Miller, Magnum
Page i, Donald McCullin, Magnum
Page ii, Nick Pavloff
Page iii, Bruce Davidson, Magnum
Page iv, Bill Eppridge, LIFE MAGAZINE © Time, Inc.
Page v, Michael and Barbara Reed, Earth Scenes
Page vi, Paul Caponigro
Page x (left) Radio Times Hulton Picture Library
Page xi (left) Redrawn from photo by Edward Leigh, Cambridge, England, courtesy John Kendrew, in American Institute of Biological Sciences, *Biological Science: An Inquiry into Life* (New York: Harcourt, Brace, World, 1963); (center) Walter Dawn, National Audubon Society, from Photo Researchers
Page xii (left) Radio Times Hulton Picture Library; (right) Bevilacqua–Cedri
Page xiii (center) Runk/Schoenberger, Grant Heilman Photography; (right) Radio Times Hulton Picture Library
Page xv (left) Cornell Capa, Magnum; (right) David Seymour, Magnum
Page xvi, Elliott Erwitt, Magnum
Page xvii, Dr. K. J. Carlson, Ardea
Page xviii (left) Lou de la Torre, BBM; (right) Nick Pavloff
Page xix Grant Heilman Photography
Page xx By kind permission of the President and Council of the Royal College of Surgeons of England

Chapter 1

1.1 Photographie Bulloz
1.2 Photographie Bulloz
Page 4 (top) Radio Times Hulton Picture Library; (bottom) The Granger Collection
Page 5 (top left) Courtesy of the Darwin Museum; (top right) Radio Times Hulton Picture Library; (bottom) Radio Times Hulton Picture Library
Page 6 (left) Elaine F. Keenan; (right) T. A. de Roy, Bruce Coleman, Inc.
Page 7 (top left) Bruce Coleman, Inc.; (top right) Hal Harrison, Grant Heilman Photography; (bottom) Hal Harrison, Grant Heilman Photography
Page 8 (top left) Bruce Coleman Inc.; (top right) Larry Keenan, Jr.; (bottom left) Michael and Barbara Reed, Animals Animals; (bottom right) Larry Keenan, Jr.
Page 9 (left) Brian Hawkes, N.H.P.A.; (right) Tui de Roy, Bruce Coleman, Inc.
Page 10 Larry Keenan, Jr.
Page 11 (top) Larry Keenan, Jr.; (bottom left) Alan Root, Bruce Coleman, Inc.; (bottom right) George H. Harrison, Grant Heilman Photography
Page 12 (left) Larry Keenan, Jr.; (top right) Kenneth Fink, Ardea; (bottom right) Elaine F. Keenan
1.3 Radio Times Hulton Picture Library
1.4 Brown Brothers
1.5 Western History Research Center, University of Wyoming

Page 17 Utah State Historical Society
1.6 Photographie Bulloz
1.7 Burndy Library, Norwalk,
Connecticut

Chapter 2

Page 22 Peter Menzel, Stock, Boston
2.4 (top left) Courtesy Carolina
Biological Supply Company; (top
right) Leonard Lee Rue III, Bruce
Coleman; (bottom left) Runk/
Schoenberger, Grant Heilman
Photography; (bottom right) Ansel
Adams, Magnum
2.6 Paul S. Conklin
2.7 H. Charles Laun, National
Audubon Society, from Photo
Researchers
2.8 Michael Bertan, Van Cleve
Photography; (inset) Ansel Adams,
Magnum

Chapter 3

Page 40 Erich Hartman, Magnum
3.2 Jack Dermid, National Audubon
Society, from Photo Researchers
Page 47 Burk Uzzle, Magnum
3.11 Eleutherian Mills Historical
Library
3.17 Stephen Dalton, N.H.P.A.
Page 64 Popperfoto
3.23 Redrawn from photo by Edward
Leigh, Cambridge, England, Courtesy
John Kendrew, in American Institute
of Biological Sciences, *Biological
Science: An Inquiry into Life* (New
York: Harcourt, Brace, World, 1963)

Chapter 4

Page 66 Ansel Adams, Magnum
4.2 Walter Dawn, National Audubon
Society, from Photo Researchers
Page 71 Science Software Systems
4.9 Robert H. Wright, National
Audubon Society, from Photo
Researchers

Chapter 5

Page 82 Runk/Schoenberger, Grant
Heilman Photography

Page 85 (top left) Richard H. Falk,
University of California, Davis; (top
right) Daniel Branton, Harvard
University; (bottom) Terence Reeve,
courtesy Cambridge Instrument
Company
5.5 (A) Runk/Schoenberger, Grant
Heilman Photography; (B) Courtesy
Carolina Biological Supply
Company; (C) Runk/Schoenberger,
Grant Heilman Photography;
(D) Runk/Schoenberger, Grant
Heilman Photography; (E) Courtesy
Carolina Biological Supply
Company; (F) Science Software
Systems
5.6 (left) Science Software Systems
5.7 (left) Science Software Systems
5.8 (center) Science Software Systems
5.9 (left) Courtesy Carolina
Biological Supply Company
5.10 (top) Science Software Systems
5.11 (top) Courtesy Carolina
Biological Supply Company
5.12 (top) Courtesy Carolina
Biological Supply Company
5.13 (top) Science Software Systems
5.14 (top) Science Software Systems
5.15 (top) Courtesy Carolina
Biological Supply Company
5.16 (top) Science Software Systems
5.19 Michigan Department of
Natural Resources
5.23 Photographs courtesy Carolina
Biological Supply Company
5.25 (top) Keystone Press Agency;
(bottom) Popperfoto
5.26 Photographs courtesy Carolina
Biological Supply Company
5.32 (top left) Harry S. Truman
Library; (top center) Harry S.
Truman Library; (top right) Harry
S. Truman Library; (center) *St. Louis
Globe Democrat*, courtesy Harry S.
Truman Library; (bottom) UPI
Page 122 (top) Pfizer Inc.; (bottom)
Courtesy Carolina Biological Supply
Company

Chapter 6

Page 124 Elliott Erwitt, Magnum
6.1 Radio Times Hulton Picture
Library
6.5 (top) Keystone Press Agency;
(bottom) Wide World Photos

6.9 Science Software Systems
Page 136 Gersheim Collection, Humanities Research Center, The University of Texas, Houston
6.11 The National Foundation, March of Dimes
6.15 Adapted from *Biological Science* by William T. Keeton; illustrated by Paula Desanto Bensadoun; © 1967 by W. W. Norton & Co., Inc.
6.18 Radio Times Hulton Picture Library
Page 144 Courtesy of Philips Electronic Instruments, Inc.
6.19 Wide World Photos
Page 148 Office of Public Affairs, University of Alabama
6.20 Redrawn by permission from TIME, The Weekly Newsmagazine; copyright © 1977, Time, Inc.
6.21 National Institutes of Health
Page 150 University of California, San Francisco

Chapter 7

Page 152 Ylla, Photo Researchers
7.1 Leonard Lee Rue III, National Audubon Society, from Photo Researchers
7.6 Grant Heilman Photography
7.11 Courtesy Carolina Biological Supply Company
7.13 Redrawn from Joan Rahn, *Seeing What Plants Do* (New York: Atheneum, 1972)
7.21 From Sheldon J. Segal, "The Physiology of Human Reproduction"; copyright © by *Scientific American*, Inc.; all rights reserved
7.22 Adapted from Leslie Brainard Arey, *Developmental Anatomy*, 7th ed. (Philadelphia: Saunders, 1965); after Dickinson

Chapter 8

Page 186 Popperfoto
Page 197 Charles Harbutt, Magnum
Page 204 (left) Michael Hayman, Photo Researchers; (right) Courtesy Rugh and Shettles: *From Conception to Birth: The Drama of Lifes*

Beginnings, 1971
Page 205 (all) Bevilacqua-Cedri
Page 206 (left) Bevilacqua-Cedri Courtesy Rugh and Shettles: *From Conception to Birth: The Drama of Lifes Beginnings,* 1971; (right) Bevilacqua-Cedri
Page 207 Bevilacqua-Cedri
Page 208 Bevilacqua-Cedri
Page 209 (all) Bevilacqua-Cedri
Page 210 Bevilacqua-Cedri
Page 211 (left) Bevilacqua-Cedri; (right) Courtesy Rugh and Shettles: *From Conception to Birth: The Drama of Lifes Beginnings,* 1971
Page 212 (all) Jean Villain-Cedri

Chapter 9

Page 218 Ian Berry, Magnum
9.2 From Robert I. Macey, *Human Physiology;* © reprinted by permission of Prentice-Hall, Englewood Cliffs, N.J.
9.3 Steinhart Aquarium, Tom McHugh, Photo Researchers
9.4 E. Stanford, Photo Researchers
Page 227 Jill Freedman, Magnum

Chapter 10

Page 234 Pamela R. Schuyler, Stock, Boston
10.2 John H. Gerard, National Audubon Society, from Photo Researchers
10.3 (top) Courtesy Carolina Biological Supply Company; (bottom) Runk/Schoenberger, Grant Heilman Photography
10.10 (left) Courtesy Carolina Biological Supply Company; (center) Grant Heilman Photography; (right) Robert J. Munn, University of California, Davis
10.12 The Metropolitan Museum of Art, Gift of Charles Bregler, 1941
10.15 Runk/Schoenberger, Grant Heilman Photography
10.20 Runk/Schoenberger, Grant Heilman Photography
10.23 Grant Heilman Photography
Page 256 (top) Courtesy Carolina Biological Supply Company

Page 261 Bernard Pierre Wolff, Magnum

Chapter 11

Page 272 Paul Caponigro
11.4a (left) Peter Gowland; (right) Courtesy of Arnold Schwarzennegger
11.4b Radio Times Hulton Picture Library
11.6 Michigan State University
Page 293 Department of HEW, Public Health Service Food and Drug Administration
11.20 Dan DeWilde

Chapter 12

Page 312 F. Roche, Animals Animals
12.2 Wide World Photos
12.3 Adapted from N. Tinbergen, *The Study of Instinct* (Oxford: Clarendon Press)
12.4 Adapted from N. Tinbergen, *The Study of Instinct* (Oxford: Clarendon Press)
12.6 From Doty and Bosma, 1956, *Journal of Neurophysiology,* 19:44–60
12.9 (top left) National Library of Medicine, Bethesda
12.10 Will Rapport
12.12 From Brown, American Astronautical Society, *Advances in the Astronautical Sciences,* 1964, 17:29–39
Page 332 (left) Grant Heilman Photography; (right) Jan Kopec, Photo Trends
Page 333 (left) C. C. Lockwood, Bruce Coleman, Inc.; (right) Gary Griffen, Animals Animals
Page 334 (top) Patricia Caulfield; (bottom) Breck P. Kent, Animals Animals
Page 335 Michael Morcombe, N.H.P.A.
Page 336 (top) Frank Woehr, Photo Trends; (bottom) Peter Johnson, N.H.P.A.
Page 337 (top) Michael Godfrey; (bottom) Jane Burton, Bruce Coleman, Inc.
Page 338 (left) Michael Godfrey;

(right) Larry Keenan, Jr.
Page 339 (top) Bruno Barbey, Magnum; (bottom) Anthony Bannister, N.H.P.A.
Page 340 (left) Annan Photo Features; (top right) Ivan Polunin, N.H.P.A.; (bottom) J. Dermid, Bruce Coleman, Inc.
12.15 I.B.&S. Bottomley, Ardea
12.16 From P. Marler, in *Darwin's Biological Work,* ed. by P. R. Bell, Cambridge University Press, 1959
12.17 Ron Garrison, San Diego Zoo Photo
12.18 From P. Marler, "Developments in the Study of Animal Communication," in *Darwin's Biological Work,* ed. by P. R. Bell (Cambridge University Press, 1959), pp. 150–206
12.19 From P. Marler, "Developments in the Study of Animal Communication," in *Darwin's Biological Work,* ed. by P. R. Bell (Cambridge University Press, 1959), pp. 150–206
12.20 Stephen Dalton, N.H.P.A.
12.21 (right) A. A. Francescorie, National Audubon Society, from Photo Researchers; (left) Allan D. Cruickshank, National Audubon Society, from Photo Researchers
12.22 From Etkin, *Social Behavior from Fish to Man* (Chicago: University of Chicago Press, 1967)
12.23 George Holton, Photo Researchers
Page 350 (left) Russ Kinne, Photo Researchers; (center) Toni Angermayer, Photo Researchers; (right) Christa Armstrong, Rapho/Photo Researchers
Page 351 Louis Lemieux, Rapho/Photo Researchers
12.24 Lynwood Chace, National Audubon Society, from Photo Researchers
12.25 San Diego Zoo Photos
12.26 Grant Heilman Photography
12.27 After Walther Jahrbuch der Gesellshaft von Opel-Rreigehege Tierforshung, 1961
12.28 Russ Kinne, Photo Researchers
12.29 After Siebenaler and Caldwell, "Cooperation Among Dolphins,"

Journal of Mammology, 1956
12.30 M. W. F. Tweedie, National
Audubon Society, from Photo
Researchers
12.31 Keystone Press Agency
Page 363 American Cancer Society
12.34 Charles Gatewood, Magnum
12.35 Leonard Freed, Magnum

Chapter 13

Page 372 Laurence Pringle, Photo
Researchers
13.1 From MacArthur, "Population
Ecology of Some Warblers of N.E.
Coniferous Forests," *Ecology,*
39:599–618
Page 377 NASA
Page 381 EPA
Page 384 (left) John Hodder; (right)
Paul Caponigro
Page 385 (left) Nick Pavloff; (top
right) Jack Dermid, National
Audubon Society, from Photo
Researchers; (bottom right) Grant
Heilman Photography
Page 386 (top) Nick Pavloff;
(bottom) Keith Gunnar, National
Audubon Society, from Photo
Researchers
Page 387 (top) Michael Muckley,
Photophile; (bottom left) Laurence
Pringle, Photo Researchers; (bottom
right) William Dicker, Photophile
Page 388 (left) F. Erize, Bruce
Coleman, Inc.; (right) Naval
Photographic Center
Page 389 (top left) Jen and Des
Bartlett, Bruce Coleman, Inc.;
(bottom) Brian Hawks, N.H.P.A.;
(top right) B. Ruth, Bruce Coleman,
Inc.
Page 390 (top left) Joy Spurr, Bruce
Coleman, Inc.; (bottom left) J.
Couffer, Bruce Coleman, Inc.; (right)
James Tallon, N.H.P.A.
Page 391 (left) Bruce Coleman, Inc.;
(right) Grant Heilman Photography
Page 392 (top) Hiroji Kubota,
Magnum; (bottom left) Michael
Godfrey; (bottom right) Stephen
Dalton, N.H.P.A.
Page 393 (top) Jen and Des Bartlett,
Bruce Coleman, Inc.; (bottom) Frank

Woehr, Photo Trends
Page 394 (left) K. W. Fink, Bruce
Coleman, Inc.; (top right) Michael
Godfrey; (bottom right) Wardene
Weisser, Ardea
Page 395 (top left) Alan Pitcairn,
Grant Heilman Photography;
(bottom left) Mike and Barbara
Reed, Earth Scenes; (right) Alan
Blank, Bruce Coleman, Inc.
Page 396 (left; both) Roy D. Mackay,
N.H.P.A.; (right) Ivan Polunin,
N.H.P.A.
Page 397 (left) E. H. Roo, N.H.P.A.;
(right) Marilyn Silverstone, Magnum
13.8 (left) Howard E. Uible, Photo
Researchers; (right) Sonja Bullaty,
National Audubon Society, from
Photo Researchers
13.9 Tom McHugh, National
Audubon Society, from Photo
Researchers
Page 414 George H. Harrison, Grant
Heilman Photography
Page 415 Alexander Lowry, Photo
Researchers
Page 416 Marc and Evelyne
Bernheim, courtesy of the
Rockefeller Foundation

Chapter 14

Page 418 Inger McCabe, Rapho/
Photo Researchers
14.1 Australian Information Service
14.8 (top) Leonard Lee Rue III,
Bruce Coleman, Inc.; (bottom)
Stephen Dalton, N.H.P.A.
14.9 Reprinted by permission of the
Chicago Tribune-New York News
Syndicate, Inc.
14.10 Carson Baldwin © Animals
Animals
14.11 Wide World Photos
14.12 From G. F. Gause, *The
Struggle for Existence* (Baltimore:
Williams and Wilkens, 1934)
Page 435 Rex Weyler, Courtesy of
Greenpeace
14.13 Wallace Kirkland, LIFE
Magazine © Time, Inc.
14.14 From *N.Y. Zoological Society
Newsletter,* November 1968; reprinted
by permission; 1975 figures from

ECO, representatives from ecology groups to the 1977 IWC meeting in Canberra, Australia
14.15 Andrew M. Anderson, N.H.P.A.
14.17 From H. T. Odum, "Trophic Structure and Productivity of Silver Springs, Florida," *Ecological Monographs,* 1957, 27:55–112
14.18 From Annabelle Desmond, "How Many People Have Ever Lived on Earth?" *Population Bulletin,* vol. 18, no. 1, Washington, D.C.: Population Reference Bureau, February, 1962
Page 449 Cornell Capa, Magnum
14.19 From Annabelle Desmond and Judy K. Morris, "The Story of Mauritius—From DoDo to Stork," *Population Bulletin,* vol. 18, no. 5, Washington, D.C.: Population Reference Bureau, August, 1962
14.20 From *Human Ecology: Problems and Solutions,* by Paul R. Ehrlich, Anne H. Ehrlich, and John P. Holdren; W. H. Freeman and Company; copyright © 1973
14.21 Adapted from *Newsweek* Magazine; copyright © 1976 by Newsweek, Inc.; all rights reserved; reprinted by permission
14.22 Adapted from *Newsweek* Magazine; copyright © 1976 by Newsweek, Inc.; all rights reserved; reprinted by permission
14.23 (top) Film Study Center, Peabody Museum, Harvard University; (bottom) SFC Jack H. Yamaguchi, U.S. Army Photo
Page 457 F. Roche, Animals Animals
14.24 Cornell Capa, Magnum
14.25 W. Eugene Smith
Page 465 Burk Uzzle, Magnum

Chapter 15

Page 468 Peter Menzel, Stock, Boston
15.1 (left) UPI; (right) Marilyn Silverstone, Magnum
15.2 From *Human Ecology: Problems and Solutions,* by Paul R. Ehrlich, Anne H. Ehrlich, and John P. Holdren; W. H. Freeman and Company; copyright © 1973

Page 473 Donald McCullin, Magnum
15.3 Cooperative Extension, University of California, Davis
Page 475 Rene Burri, Magnum
Page 477 George Harrison, Grant Heilman Photography
15.4 From S. J. Holt, "The Food Resources of the Ocean"; copyright © by Scientific American, Inc.; all rights reserved
15.5 S. J. Holt, "The Food Resources of the Ocean"; copyright © by Scientific American, Inc.; all rights reserved
Page 480 G. Mendoza, WFP/FAO
15.6 Peter Vilms, Jeroboam
Page 487 Earl Dotter
Pages 488–489 Arthur Bilstein, Photophile
15.8 Burk Uzzle, Magnum
15.9 Wide World Photos
15.10 (top left) ERDA; (top right) United Nations; (bottom left) W. J. Bailey, Bureau of Reclamation; (bottom right) David Seymour, Magnum

Chapter 16

Page 496 Ansel Adams, Magnum
16.2 UPI
Page 501 Horst Schäfer, Photo Trends
16.3 Wide World Photos
Page 504 Elliott Erwitt, Magnum
16.4 John Hendry Jr., National Audubon Society, from Photo Researchers
Page 509 EPA
Page 510 (left) Ake Lindau, Ardea; (right) Carson Baldwin, Jr. © Animals Animals
16.6 Wayne Miller, Magnum
Page 513 Pat Oliphant, copyright © 1971, *The Denver Post,* reprinted with permission, Los Angeles Times Syndicate
16.7 Grant Heilman Photography
Page 516 Mary M. Thacher, Photo Researchers
Page 521 Brookhaven National Laboratory
16.10 W. Eugene Smith
Page 531 Michael Hayman, Photo Researchers

Index

A

Abductors, 244
Abert squirrel, 409f
Abiotic control, of population, 432–433
Abortion, 183–184
Absorption, of food in digestive system, 267. *See also* Digestive system.
Acetylcholine, 294, 295f, 296, 308
Acetylcholinesterase, 295f, 296
Acetyl–CoA, 78, 81
Acid. *See also specific acids.*
defined, 54b
on pH scale, 54b
Acid group, in amino acids, 62
Acromegaly, 275
ACTH, 277, 279
Actin, 246f, 247
Action potential, 291, 292f
Action-specific energy, 321
Active transport, 88
in digestive system, 267
in kidney, 230, 232
Adaptation, 98, 156, 158. *See also* Speciation.
to heat, 223–225
Adaptive radiation, 405–413
Adductors, 244
Adenine, 70, 72
in DNA, 112
in formation of pyruvic acid, 75
structure of, 115
Adenosine diphosphate (ADP)
formation, 68
in light reaction, 72
in muscle contraction, 247
in photosynthesis, 74f
Adenosine monophosphate (cyclic AMP), 277
Adenosine triphosphate (ATP), 75, 277, 291

composition of, 70
as energy source, 31, 68, 88
in electron-transport chain, 80–81
formation of, 80
in glycolysis, 75–78
in Krebs cycle, 78–80
in muscle contraction, 247
in photosynthesis, 72, 74
production of, 91
Adenyl cyclase, 277
ADH, 233
ADP. *See* Adenosine diphosphate.
ATP. *See* Adenosine triphosphate.
Adrenal gland, 255, 277, 279. *See also* Endocrine system.
Adrenaline, 225, 279, 308
Advanced traits, 284
Aerobic organisms, 91
respiration of, 76–77, 78
Afferent neuron, 290
Afterbirth, 214. *See also* Placenta.
Agar, 280
Age profiles, 450–451
Aggression, 349, 352–356. *See also* Aggressive behavior, Animal behavior, Fighting.
among animals, 457b
defined, 349
as instinct, 355
in territorial species, 438
Aggressive behavior, 457b
in birds, 320–321
in fish, 321
Agriculture
in early human populations, 443–444
as human food supply, 445
Air pollution, 498–505
effects of, 501b
prospects for, 503, 505
sources of, 498–499. *See also individual pollutants.*
Alanine, 117

Alaska oil pipeline, 402b, 486
Albinism, 143, 145
Alcohol, 365–367
effects of, 366b
and elimination of wastes, 230–232
and pregnancy, 197
Aldosterone, 277
Aldrin, 512, 516b
Algae, 30, 31
blooms, 476
circulation in, 252
in freshwater bodies, 379, 380
in hot water, 222
Allantois, 194
Allele, 129, 131, 135, 138. *See also* Genes.
frequency of, 142
Alternative generations, development of, 156, 158
Alternation of generations, 155–156. *See also* Asexual reproduction.
Altruism, 358–360
among social insects, 360
and effect on reproduction, 359–360
in related groups, 360
Alveolus, 253f
Amazon basin, 472
Amino acid
composition of, 62
configuration of, 62
as energy source, 228
in formation of peptides, 63–64
and genetic code, 116–117
in protein structure, 65
as result of Miller's experiments, 25
Amino group, 51
Ammonia (NH_3)
in early atmosphere, 23
elimination of, 228
in Miller's experiments, 25
structure of, 228
Amnion, 194

Amniotic fluid, 195
Amoeba, 87, 98, 255
    digestive system in, 265
    excretory system in, 225
Amphetamines, 369
Amphibian
    and body temperature, 222
    heart, 257
Amytal, 369
Anaerobic organisms, respiration in,
    75–76
Anaphase
    in meiosis, 108–109, 111, 138–139
    in mitosis, 99, 103, 102, 110
Anderson, Jack, 461
Animal behavior
    aggressive, 349
    copulatory, 169
Animal cell, 100f, 102. *See also* Cell.
    meiosis in, 103–107
    mitosis in, 99–107
Animal pole, 188
    in frog egg, 189f
Annelids, 248
    circulatory systems in, 255
Annual, 376
Antagonistic muscles, 244
Anterior lobe, of pituitary, 275
Anther, 159
Antidiuretic hormone (ADH), 230
Anus, 265
Aorta, 257
Apgar test series, 214
Apical meristem, 161
Apodeme, 249
Appendicular skeleton, 238f
Appetitive behavior, 320
Archaeopteryx, 224f
Archenteron, 189f
Ardrey, Robert, 361
Arginine, 117
Argon, 45
    in early atmosphere, 23
Arms Control and Disarmament
    Agency (ACDA), 464
Army Corps of Engineers, 384f, 511
Arterial system, 262f. *See also*
    Circulatory system.
Artery, 255
Arteriole, 255
Arthropods, 248
    circulatory system in, 255
Artificial selection, 16–17, 125
Asexual reproduction, 153–155. *See*

also Alternation of generations,
    Sporophyte generation.
Asparagine, 117
Aspartic acid, 117
Aswan High Dam, 475b
Atmosphere, early, 23
Atom(s), 42–44
    defined, 42
    inert, 45
    properties of, 45–47
    structure of, 43
Atomic Energy Commission, 520
Atomic number, 43
Atrium, 257
Autonomic learning, 308–309. *See*
    also Learning.
Autonomic nervous system, 306–308.
    See also Nervous system.
Autosome, 134
Autotroph, 29–31
Autotrophy, 31. *See also*
    Photosynthesis.
Auxin, 279–280
Axial skeleton, 238f, 239
Axon, 288, 295f

B

Baboons, 332f
    social bonds among, 169
    visual signals among, 342–343, 344f
Background radiation, 517
Bacteria, and depletion of oxygen in
    freshwater, 379, 380f
Balanced equations, 53
Ball-and-socket joints, 241
Banded sea snake, 409f
Banff National Park, 391f
Banyon trees, 338f
Baobab tree, 393f
Barbiturates, 369
Barr bodies, 134, 135
Base
    defined, 54b
    on pH scale, 54b
Beagle, 1–2, 14–15
Bean, germination and development
    of, 165f
Bearded dragon (Amphibolurus
    barbatus), 335f
Beetle, development of, 275f
Behavior. *See also* Aggression,
    Aggressive behavior.

in birds, 319
early study of, 313
and environment, 320–321
as result of natural selection, 361
stereotyped, 408
Behavioral hierarchy, 321, 323
Benzedrine, 369
Benzene hexachloride, 512
"Bergmann's rule," 336f
Biceps brachii, 243
Biological control, dual nature of, 330
Biology, beginnings of, as a science, 3,
    13–14
Biome, 376–378, 388–397. *See also*
    types.
Biotic control, of population, 433–439
Biotic potential, 421
Birch, L. C., 374
Bird
    Adaptation to niche, 401, 403–404
    aggressive behavior among, 345
    courting behavior of, 319
    development of, 194–196
    feeding behavior, 321
    feeding patterns, 320
    feeding zones, 374, 375f
    fighting behavior among, 320
    and flight learning, 317, 326, 328
    individual recognition, 348
    instinctive behavior, 315–316
    and migration, 331
    reproduction rate of, 426–428
    song learning in, 328
    sound signals of, 345
    spectrograms of, 345f
    sun-compass orientation of, 341
Birth, human, 212f, 213–214
Birth control pill, 180–181
Birth rates, 447b, 449b
Bitterman, M. E., 326
Blagden, Charles, 219
Blastocoel, 189f
Blastocyst, in human egg, 189f
Blastospore, 188, 189f, 194
Blastula, 189f
    defined, 188
    human, 199
Blastulation, 188
Blest, David, 440
Blond ring dove, courting behavior
    of, 319
Blood, 256b
Blood cells, 237
Blood vessels, 226. *See also*

Circulatory system.
of humans, 262f, 263
Blue-footed boobies (*Sula Nebouxii*), 12f
Bobcat, 36f
Body temperature, 220
in humans, 220
in mammals, 222–223
Bombykol, 346
Bonding, 48–52
Bone. See Skeletal systems.
Botulism, 293b
Boulding, Kenneth, 529
Bowman's capsule, 230, 231f, 232f
Brackish water, 379
Brain
fish, 284f
human, 299–305
mammal, 284f
reptile, 284f
Brazil, 472
"Breeder reactor," 494
Bronchote, 253f
Brontosaurus, 283
Brown, Frank A., 330
Brown's experiments, 330–331
Brown's Ferry incident, 518b
Budding, 154
"Buffer," 54
Buffon, George-Louis Leclerc, de, 13
Bulbourethra. *See* Cowper's gland.
Burundi, 461–462
Bush, Guy, 476
Butane, 50

C

Caesarian section, 213
Caffeine, 363
Calcium salts, in skeletal systems, 237
California ground squirrels, 394f
Caloric consumption, 471f
Calvin cycle, 74
Cambium. *See* Lateral meristems.
Camel, and water storage, 227
Canaliculi, 237
Cancer, 120, 122, 123
Cancer research, 151
Cannabis sativa. See Marijuana.
Capillary, 255. *See also* Circulatory system.
Carbohydrate, 57–60
composition of, 57

as energy source, 57–58
Carbon
and methane formation, 50
properties of, 49–50
Carbon chains, 51
Carbon dioxide (CO_2), 81
in blood, 257
as by-product of ATP formation, 80
in cells, 88
in dark reaction, 73
in early atmosphere, 23
in Krebs cycle, 78–80
in photosynthesis, 31, 70
as product of anaerobic respiration, 75
in respiration, 252
Carbonic acid, in blood, 257
Carbonic anhydrase, 257
Carbon monoxide, as air pollutant, 499
Carbon rings, 51
Carboxyl group (COOH), 51
in fats, 60–61
Cardiac muscle, 241, 247
Cardinal, 392f
Cardio-pulmonary resuscitation (CPR), 261
Carnivores, and energy pyramid, 444
Carotenoids, 94
Carpals, 240
Carrying capacity, 421–422
Carter, President, 462, 491f
Catalysis, 56
Catalyst, 56
Cattle, 16f, 17f,
as food source, 478
Cell(s)
components of, 88–99, 100–101. *See also individual components.*
conducting, 89f
eucaryotic, 84, 86
of heart muscle, 80, 89f
procaryotic, 84, 86
of skeletal muscle, 242–243
of spinal cord, 89f
Squamus epithelium, 89f
variety of, 89f
Cell biology, problems in, 120–123
Cell body, of neuron, 288, 305
Cell control, 84
Cell development, as reversible process, 153
Cell differentiation, 192

Cell division, 120, 122, 147, 161. *See also* Cloning.
of chicken embryo, 188, 189f
and cytokinins, 281
early, 188
in frog embryo, 188, 189f
and gibberellin, 280
of human embryo, 188, 189, 198
Cell elongation, 280
"Cell garbage," 225
Cell membrane, 97, 100, 101
composition of, 90–91
of fish gills, 250
movement of O_2 and CO_2 across, 235
permeability of, 91
of plant, 88, 90
Cell replication, 99, 102–120
"Cell sap," 94
Cell specialization, 83–84, 119, 153, 161, 163–164, 191
Cellular movement, 84, 86–88
Cellular respiration, 91
Cellulose, 59, 88
Cephalic ganglia, 283
Central nervous system
evolution of, 281–283
of humans, 296–309
Centriole, 95, 96, 100, 101
Centromere, 108–109
Cerebellum, 284
of humans, 299
Cerebral cortex, function maps of, 304f
Cerebrum, 284, 302
Cervical cap, 180
Cervix, 170f, 171
Channel Islands, 387f
Chaparral, 384f
Character displacement, 407
Cheetah, 332f
Chemical communication, 345–347
Chemical effluents, as water pollutant, 507
"Chemical messengers." *See* Hormone.
Chemical notation, 49, 53, 57
Chemical reactions, defined, 52–53
Chestnut tree, 438
Chicken egg, development of, 189f
Child mortality, 430t
Chimpanzees
reproduction rate of, 423, 425
sexual behavior in, 169

Chitin, 60, 236
Chiton, respiration in, 249–250
Chlordane, 516b
Chloride, in formation of
   hydrochloric acid (HC1), 53
Chlorinated hydrocarbons, 516b
   longevity of, in soils, 512t
Chlorine (C1), 46
   in neuron cell, 291
   in table salt, 48
   as water additive, 507
Chlorophyll, 29, 86
   in photosynthesis, 31
   structure of, 71b
Chloroplast, 80, 94, 101
CHNOPS, 43
Choanocytes, 236f
Cholesterol, 62
Chorioallantoic membrane, 195
Chorion, 194
Chorionic villi, 199
Chromatid, 102, 139
Chromatin, 98, 102
Chromatin net, 100, 101
Chromatography, 64b
Chromosome, 86, 102, 131
   composition of, 98
   gene assortment on, 138
   as gene carriers, 133
   homologs, 138–139
   human, 135
   in meiosis, 108–109
   in mitosis, 110–111
   sex, 134
   single-stranded, 108–109
Chromosome mapping, 139, 140
Chromosome mutation, 140–141
Chromosome number
   in meiosis, 103, 110–111
   in mitosis, 110–111
Cilium (pl.: Cilia), 95–96
Circadian rhythm, 329–330
Circulatory systems, 252–263
   annelid, 255
   arthropod, 255
   coelenterate, 255
   flatworm, 255
   human, 257–263
   mollusk, 255
Classical conditioning, 323, 324–325.
   See also Learning.
Classification, of living organisms, 13,
   254b, 533–542
Clavicles, 240

Clean Air Act, 505
Clear-cutting, 477b
Clitoris, 167, 170f, 171
Cloaca, 265
Cloning, 147, 150
Closed circulatory system, 255
Cloud, Preston, 483
Coacervate droplet
   defined, 26
   growth of, 26–27
   internal structure of, 27–28
   physical properties of, 26–27
   reproduction of, 27–28
Coastal areas, 398–400. See also
   Rocky areas, Mud flat, Sandy
   beach.
Cocaine, 367–368
Coconut milk, 281
Coconuts, 340f
Codon, 117, 140
Coelenterate, 248
   circulatory system in, 255
Coenzyme A (CoA), 78, 79, 80, 81
Coitus interruptus, 177, 179
Colanic acid, 148
Cold-blooded animals, 222–223
Collagen, 237
Collenchyma, 163, 164f
Colonization, 403–405, 410
Color blindness, 135, 137
Commodities
   as nonrenewable resources, 482
   and warfare, 454, 456, 458–462, 463
Communication, 342–347
Communities, 379
Comparative psychology, 313–314
Competition, 352, 404–405, 434, 514
   as cause of war, 454
   in population balance, 438–439
Comorant (Nannopterum harrisi), 8f
Concentration gradient, 86, 88
Conception, in humans, 174. See also
   Fertilization.
Condensation, 58
   in formation of fats, 61
Condoms, 177, 179
Coniferous forest, 391f
Conifers, 159
Connecting tubules, 230
Connective tissue, 243
   types of, 237
Connector neuron, 297
Consummatory behavior, 320
Continental shelves, 398

Contraception, 174. See also Birth
   control.
   developments in, 184–185
   methods of, 177–185, 187t
Contractile fibers, 241. See Muscle,
   types of.
Contractile systems, 247–249
Contractile vacuole, 94, 225
Control mechanisms, of reproduction
   rate, 422
Copernicus, 2f, 3
Copulation, 166. See also Sexual
   intercourse, Reproduction.
   and bonding in humans, 169
   functions of, 169, 171
Cooperation, 356–358
Cork cambium, 162
Corn blight, 476
Corpus callosm, 305
Corpus luteum, 173
Cortex
   of adrenal gland, 277
   in developing plant, 161
Cortisone, 277
Cotyledon, 161, 164, 165f
Covalent bond, 49–50
Cowper's gland, 168f, 169
Cranial nerves, 309, 310f
Cranium, 239, 242f
Crayfish, 236f, 337f
Cretinism, 275
Crick, Francis, 107
Crista (pl.: Cristae), 91
Crop, 265
Crossover, 107
   percent, 139
   of chromosomes, 108–109
CS-M3, 474f
Curette, 183
Curtiss, Roy, III, 148b
Cuyahoga River, 507
Cypress swamp, 385f
Cysteine, 117
Cytokinins, 281
Cytology, 83
Cytoplams, 75, 87
Cytosine, 112
   structure, 115

D

Danielli, J. F., 90
Dark reaction, 70, 73, 75. See also

Calvin Cycle, Photosynthesis.
Darwin, Charles, 1, 4–20, 125, 126, 279, 315, 405, 422
Darwin, Erasmus, 13, 14, 315
Darwin's finches, 15–16, 405–406. *See also* Speciation.
Darwin, Francis, 279
Darwin's plant experiments, 280
Darwin's theory of evolution, 14–15
Daughter cells, 110–111
Da Vinci, Leonardo, 146b
DDT, 296, 432, 512, 514, 516b. *See also* Pesticides, Insecticides, Pollution.
DDE, 512, 514, 516b. *See also* Pesticides, Insecticides, Pollution.
Death. *See also* Child mortality, Mortality.
  from malnutrition, 470
  in population change, 439–441
  from starvation, 470
Death rates, 447b, 449b
Decibel levels of sounds, 526t. *See also* Noise pollution.
Dehydration, 58
Delerium tremens (DTs), 367
Deletion, of chromosome, 140
Dendrites, 288
Density-dependent influences, on populations, 433–434
Density-independent influences, on populations, 432
Deoxyribonucleic acid. *See* DNA.
Depolarization, of neurons, 291
Depressors, 244
Descent with modification, 17
Desalination, 485b
Desert, 376, 377f, 378, 395f
Desert lizard, 223f
Desert locust, 60f
Development
  animal, 161
  of brain, and evolution, 288
  of chick embryo, 196f
  of embryo, 193–194
  of human embryo, 200f
  plant, 161–164
  sequence of, 192
Dexedrine, 369
Diaphragm, as birth control method, 179, 180f
Dieldrin, 512, 514, 516b
Diffusion, 86, 267
  in kidney, 230

Digestive system, 263–267
  amoeba, 265, 266f
  earthworm, 266f
  flatworm, 266f
  human, 267–271
  hydra, 266f
  salamander, 266f
Dihybrids, 129
Dilation and curettage (D and C), 183
Dilators, 244
Dimethylmercury $((CH_3)_2Hg)$, 523
Dinosaurs, 223–225
Disaccharide, 58
Disarmament, 464
Discrimination, in classical conditioning, 325
Disease, as population control, 439
Ditran, 396
DNA
  and cytokinins, 281
  and genetic mutation, 140
  mutation, of, 149
  in protein synthesis, 119–120
  replication, 110–111, 113, 114
  structure of, 107
  in viruses, 122
Dog
  consummatory behavior in, 320
  cooperation among, 357–358
Dolphins, cooperation among, 357
Domestic animals, as food source, 476, 478
Dominance, principle of, 126, 127
Dominance hierarchy, 349t
Dominant trait, 127
  in humans, 133
Dorsal nerve root, 297
Double helix, 107
Down's syndrome, 137
Drones, 346
Drosophila melanogaster, 103, 133, 134, 138, 139, 140
Drought, 434f
Drugs. *See drug names.*
"Drumstick," in white blood cells, 134–135
Dumont, René, 470
Duplication, of chromosome, 140
Dutch elm disease, 438
Duve, Christian, de, 93

E

Early civilizations, and nature of

world, 3
Earth, as a young planet, 23–25
Earthworm
  digestive system of, 265
  excretory system of, 225
  nervous system of, 283
Ecology, 37–39
Ecosystems, 378–379
Ectoderm, 194
  of gastrula, 188
Effector, 290
Efferent neuron, 290
Egg, 173f, 174. *See also* Gamete, Ovule.
  human, 171
  sizes of, 187
  types of, 187
Eibl-Eibesfeldt, Irenäus, 328
Eisenhower, Dwight, 460–461
Ejaculation, in human male, 167, 174
Electrical energy, 291
Electrical potential, 291
Electron acceptor, 80
Electron microscope, 83, 85b
Electrons, 43
Electron shell, 44
Electron transfer. *See* Oxidation, Reduction.
Electron-transport chain, 78, 80–81
Element,
  arrangement of, 43
  defined, 42
  designation of, 42
  half-life, 521b
  number of, 42
Elephants, 16, 422
El Salvador, 459
Endergonic energy change, 57
Endocrine glands, 273
  and their hormones, 278t
Endocrine system
  of humans, 274f, 275–279
Endoderm
  of chicken, 194
  of gastrula, 188
Endogenous control, 329
Endometrium, 170f, 171
Endoplasmic reticulum (ER), 95, 97, 100, 101
Endoskeletons, 237
Endosperm, in plant zygote, 159
Endrin, 516b
Energy
  from light, 70

and metabolism, 444
sources, 490t
sources of, on young earth, 25
types of, 490–495
Energy of activation, 53, 55–56
Energy changes, 57
Energy pyramid, 444–445
Energy transmission, 295f
Entropy, 67, 273
Environment, and behavior, 320–321
Environment. *See* Land environment,
Water environment.
Environmental Protection Agency,
505, 511
Environmental resistance, 421
Enzymes, 56–57, 147
manufacture of, 116–120
proteolytic, 199
role of, 116, 119
Epicotyl, 164, 165f
Epidermis, in developing plant, 161
Epididymis, 168f, 169
Epilepsy, 296
Epinephrine, 277
Epiphyte, 9f, 340f
Equatorial plate, 106
Erastosthenes, 3
Erythrocytes, 256
Escherichia coli (E. coli), 148, 149,
151. *See also* Recombinant DNA.
Esophagus, 265
Estivation, 222
Estrogen, 171, 185, 277
Estrus, 166
Ethane, 50
Ethanol, 75
Ethology, defined, 313
Ethyl alcohol. *See* Alcohol.
Eucaryotic cell, 84, 86, 91
Euglena, 96
European robins, visual
communication among, 342
and aggressive behavior, 320
Eutrophy, 476. *See also* Lake Erie.
Everglades cougar, 334f
Evolution, 545–546
of central nervous system, 281–283
defined, 142
and learning ability, 326
principles of, and scientific
community, 19–20
of vertebrate brain, 286–288
and warfare, 456, 458
Evolutionary tree, 287f

Excitatory transmitter substances, 296
Excretory system, 225–233
amoeboid, 225
human, 230–233
Exergonic energy change, 57
Experimental techniques, 33–34
Experiments, on plants, 126–133, 280.
See also Darwin's experiment,
Mendel's experiment.
Exogenous control, 329
Exoskeleton, 60, 236f, 237, 249. *See
also* Chitin.
Extensor muscle, 244
Extinction
and humans, 413, 417
of species, 411–413
Extinction rate, on islands, 401–403
Eyes, development of, in humans, 203

F

FAD, 80–81
Fallopian tube, 170f, 171, 174. *See
also* Oviduct.
Fascia, 243
Fats. *See also* Lipids.
composition of, 60–61
in digestive system, 267
saturated, 61
structures of, 61
unsaturated, 61
Fatty acids, 60–61
Fawn, 333f
Fear, as part of aggression, 349
Feedback system, 220f, 221f, 219–220,
278–279
Female reproductive system, 170f, 171
Ferns, 156f
alternation of generation in, 156
life cycle of, 157f
Fertility rate, 451–452
as predictor of human
population, 451
in United States, 452f
Fertilization, 103
animal, 165–174
defined, 159
human, 173
plant, 159–164
Fetus, 201, 203f, 210f–211f
mother's influence on, 197
Fever, and body temperature, 225
Fighting. *See also* Aggression, Animal

Behavior.
among animals, 352–355
stylized, 353
Fish
aggressive behavior in, 355f
brain of, 284f
breeding behavior of, 321, 322f
as food source, 478–479, 481
freshwater, and water
temperature, 222
in Great Lakes, 380f
heart in, 257
respiration in, 250
saltwater, and water
temperature, 222
world catch, 479f
Fissure of Rolando, 303
Fissure of Sylvius, 303
Fixed action pattern, 316, 319
Flagellum (pl.: Flagellae), 86, 95–96
Flatworms
circulatory system in, 255
digestive system in, 265
regeneration of, 191
respiration in, 249
nervous system in, 283
Flexor muscle, 244
Floating ribs, 240
Fluid matrix, 91
Follicle-stimulating hormone (FSH),
171, 172f, 173, 428
Food, 470–482
distribution of, in world, 470–472
production of, 472, 474
sources of, 476–482
Food chain, mercury in, 523. *See also*
Food web
Food pyramid, in ocean, 382, 383f
Food vacuole, 265
Food web, 411, 412
"Force of life," 13
Foreplay, 166–167
Fossil fuels, as energy source, 492, 494
Foxes, 436b
Franklin, Rosalind, 107
Free radicals, 51
Fresh water, 379
Freshwater bodies, 379, 385f
Frigate bird, 12f
Fringilla coelebs, 343f
Frog, 187–194
development, 187–194
insemination in, 166f
nervous system of, 283

orienting behavior in, 316f
skeleton of, 239f
Frontal lobe, 303
Fructose, structure of, 58
Fruit fly. See *Drosophila Melanogaster.*
Functional groups, 51
Fusion, of molecules, 25–26

G

Galapagos Islands, 4f–12f, 15–16, 405
Galapagos finches. See Darwin's finches.
Galapagos flamingo (*Phoenicopterus*), 12f
Galapagos penguin (*Spheniscus mendiculus*), 8f
Galileo, 3
Gamete, 126, 129, 140, 155f, 159
Gametophytic generation, 155. See also Alternation of generation.
Gamma rays, 25
Ganglion, 283, 309
Gastric cavity, 265. See also Digestive system.
Gastrula, 188, 189f
Gastrulation, 188, 190–191
Gene, 98, 129, 134, 135. See also Alleles.
arrangement of, 139
assortment of, 138
and behavior, 352–353
exchange, 407
linkage, 138–139. See also Crossover.
Generalization, in classical conditioning, 325
Genetic code, 116, 117. See also DNA.
Genetic engineering, 147–149
Genetic isolation, 409
Genetic mutation, 140
Genetic variation, 476
Genetics
classical, 131–139
Mendelian, 125–133
population, 141–143, 145, 147
Genotype, defined, 128
Geothermal energy, 492
Germ layers, 188, 190f
Germinal spot, 194
Giant, 275

Giant tree cactus (*Opuntia*), 8f
Giant water lilly, 339f
Gibberellin, 280–281
Gila monster, 395f
Giraffe, 13–14, 17–19
Gizzard, 265
Glans penis, 168f, 169
Glial cell, 290
Glomerulus, 230, 231f
Glucose, 220
in ATP synthesis, 80–81
as product of photosynthesis, 31, 70
structure of, 58
Glutamine, 117
Glyceraldehyde, 58
Glycerol, in fats, 60
Glycine, 117
Glycogen, 59
Glycolysis, 75–78, 81
Goats, 10f
as food source, 476
Golden plover, 342f
Golgi body, 92–93, 100, 101
Golgi, Camillo, 92
Gonads, 103, 277
Goodall, Jane, 33, 169, 423
Gordon, J. E., 470
"Graded display," 343f
Grassland, 376, 377f, 393f
"Gray matter." See Cerebrum.
Great Lakes, 379, 380f
Greenpeace Foundation, 435
"Green revolution," 474. See also Food, sources of.
Greylag goose, nesting behavior, 316f
Group, 288
Growth rate, of human population, 445–450
Guanine,
in DNA, 112
structure of, 115
Guard cells, 89
Gulf Stream, 382. See also Ocean currents.
Gulls, 426, 427f

H

Habit, 363
Habitat, 373
Habituation, 323–324
Haeckel, E. E., 175
Haldane, J. B. S., 360

Hamilton, W. D., 360
Hangover. See Alcohol, effects of.
Haploid cells, 108–109
Haploid gametophyte, 157f
Hardin, Garrett, 478
Hardy, G. W., 142, 143
Hardy-Weinberg principle, 142–143, 145
Hashish, 363, 364. See also Marijuana.
Haversian canals, 237
Heart, 257–263
amphibian, 257
mammal, 259
reptile, 259
Heart attack, 261b
Heart rate, 259
Heat
and chemical reactions, 222
as product of metabolism, 223
and protein, 222
as water pollutant, 508, 511
Heat exchange, in human body, 223–224
Heavy metal pollution, 522–523
Heinroth, Oscar, 288, 315
Helium, 45
atom, 43
composition of, 43
in early atmosphere, 23
Hemispheres, of human brain, 302–303, 305
Hemoglobin, 65, 114, 256
Hemophilia, 135, 136b
Herbivores, in energy pryamid, 444
Heritable trait, 17–19
Heritability of change, 14
Heroin, 367
Heterotrophs, 29–31
Heterozygous, 129
Hierarchical model of behavior, 355
Hierarchies, behavior of animals in, 439
Highway system, U.S., 504b
Hinge joint, 242f, 241
Histidine, 117
Hoffman, K., 341
Hogs, as food source, 478
Homeostasis, 219–225
defined, 219
in human body, 220
temperature regulation, 220–225
Homeotherm. See Warm-blooded animals.
Homolog, 131, 133

Homozygous plants, 128
Honduras, 459
Honeybees,
    and chemical communication, 346
    cooperation among, 358
Hormone. *See also individual hormone*
    names.
    defined, 171, 273
    experiments with, 273–274
    and insect development, 275f
    and menstrual cycle, 172f, 173
    and neural control, 279
    in plants, 279–281
Hooke, Robert, 83
Host, 436
Housefly (*Musca domestica*), 423
Human
    aggression, 355–356
    brain, 299–305
    caloric requirements, 470
    circulatory system, 257–263
    development of, 196–215
    digestive system, 267–271
    early populations, 441–444
    embryo, 206–209f
    endocrine system, 275–279
    estimated average life span, 428
    excretory system, 230–233
    and extinction, 413, 417
    heart, 260f
    impact of, on environment, 432
    kidney, 230–233
    lymphatic system, 263–267
    musculature, 244, 245f
    nervous system, 296–309
    populations, 441–450
        doubling times of, 446t
    reproduction rate of, 428–431
    reproductive system, 169–177
    respiratory system, 253f
    skeletal structure, 238, 239–241
Hybrid, 126
Hydra
    digestive system of, 268
    nervous system of, 281, 282f
Hydrocarbon, 23
    as air pollutants, 500
    defined, 50
    functional groups with, 51
Hydrochloric acid, (HC1), 53
Hydroelectric energy, 491
Hydrogen (H_2)
    atom, 43
    in early atmosphere, 23

in glycolysis, 75
in hydrochloric acid formation, 53
in methane formation, 50
in Miller's experiments, 25
in water formation, 49, 53
Hydrogen bond, 51–52
Hydrogen power. *See* pH.
Hydroxyl groups (OH), 51, 60
Hydrolysis, 58, 265
Hyenas, 457b
Hymen, 171
Hypertonic medium, 228
Hypertonic urine, formation of in
    humans, 232
Hypocotyl, 165f
Hypothesis, defined, 33
Hypothalamus, 225, 233, 275, 279,
    299, 301–302
Hypotonic medium, 226

I

Impalas, 353f
Inclusion bodies, 86. *See also*
    individual names.
Independent assortment, principle of,
    126–127, 129–131
Individual recognition, 348–349
Indoleacetic acid (IAA), 280
Induction
    of eye lens, 194
    as regulatory factor, 193–194
    of spinal cord, 193f
Infection, and body heat, 225
Infectious hepatitis, 507
Inferior vena cava, 259
Inhibitory synapse, 296
Inhibitory transmitter substances, 296
Innate behavioral patterns, 313
Innate releasing mechanisms (IRMs),
    320–321
Inner mass cell, 199
Inorganic phosphate (P_i), 68
Insects
    chemical communication
        among, 346
    cooperation among, 358
    respiratory system of, 251–252
    exoskeleton of, 60
Insecticides, 516b
    effects of human nervous
        system, 296
Instinct, 314–323

theory of,
    and learning, 326–329
Instinctive pattern, 316
Insulin, 64b, 150b
Intercourse. *See* Sexual intercourse.
Interdependency of species, 413f
Intermediate inheritance, 130–131
Interphase
    in meiosis, 111
    in mitosis, 94, 104, 110
Interspecific cooperation, 357
Intestine, 265. *See also* Digestive
    system.
Intrauterine device (IUD), 180
Inversion, of chromosome, 140
Invertebrates, hormonal function
    in, 273
"Involuntary" muscle, 241–242
Ions
    and behavior, 47b
    defined, 44
    structure, 44
Ionic bicarbonate, 257
Ionic bond, 48–49
Ionizing radiation, 521b
Intelligence, and maternal
    nutrition, 213
IR-8, 474
Island life, 403–405
    biotic factors, 401, 403
    physical factors, 400–401
Islands, colonization of, 400–401. *See
    also* Colonization, Speciation.
Isodrin, 516b
Isoleucine, 117
Isotopes, 43

J

Jackals, 457b
Johnson, V., 169
Joints, types of, 240–241. *See also*
    Skeletal system.

K

Kaibob squirrel, 409f
Kangaroo rat, 81f, 227
Kidney, 226. *See also*
    Excretory system.
Kinetic energy, 55, 67
Kinetin, 281

Kissinger, Henry, 461
Kittiwake, 427f
Klinefelter's syndrome, 137
Krebs cycle, 75, 78–80, 81
Krebs, Sir Hans, 78
Kurosawa, E., 280
Kwashiorkor, 470, 473b.
Kyshtym disaster, 518b

L

Labia majum, 170f, 171
Labia minum, 167, 170f, 171
Labor, 213. *See also* Birth.
Lactic acid, 76
Lake Erie, 381b, 498
Lamarck, Jean Baptiste, de, 13
Lamella, (pl.: lamellae), 250f
Land
  cultivation of, 472
  environment, 374, 376–379
Land iguana (*Conolothus subcristatus*), 7f
Lateral buds, in plant, 161, 163f
Lateral meristems, 162
Laterite, 474
Latin America, population growth in, 459f
Latka-Volterra oscillation, 434f
Learning, 323–326
  and instinct, 326–329
Lehrman, Daniel, 319
Lemmings, 389f
Leopard frog, orienting movements, 316f
Leucine, 117
Leucocyte, 256
Levators, 244
Lichens, 339
Life expectancy, 429
Ligaments, 239
Ligase, 149
"Lightening," 213. *See also* Birth, Labor.
Light microscope, 85, 162
Light reaction, 70, 72–73, 74. *See also* Photosynthesis.
Lignin, 90
Lindane, 516b
Linneaus, Carolus, 13, 15f
Lion, 30, 31
  and aggression, 349
Lipid, 60–62. *See also* Fats.

Liver, 220
Lizards, 15
  and body temperature, 222
Loop of Henle, 230, 231f, 232
Lorenz, Konrad, 313, 314f, 315, 321, 361
Lungs, 251. *See also* Respiratory system.
Luteinizing hormone (LH), 171, 173, 428
Lyell, Charles, 14–15
Lymph, 263
Lymph capillaries, 263
Lymphatic system, 263
  human, 264f, 265
Lymphocytes, 263
Lysergic acid diethylamide (LSD), 369
Lysine, 117
Lysosomes, 93, 100

M

MacArthur, Robert H., 374
Magnesium, 46
Malathion, 296
Male reproductive system, 168f
Malthus, Thomas, 16
Mammal
  brain of, 284f
  defined, 165
  heart of, 259
  respiratory system of, 251–252
Mangrove tree, 340f
Marijuana, 90, 363–364
Marine iguana (*Amblyrhynchus cristatus*), 7f
Marrow, 237
"Master gland," 275. *See also* Pituitary gland.
Master, W., 169
Mate recognition, 348
Matrix, 237
Mauritius, 449, 450
Mayr, Ernest, 403
Mechanism, 20–21
Medulla
  of adrenal gland, 277
  of brain, 284, 299
Meiosis, 103, 106–107, 133, 139
  in alternation of generations, 155
  in animal cells, 110–111
  chromosomes in, 103, 106

compared to mitosis, 107
  defined, 103
  in ovaries, 107
  phases in, 106, 108–109
  in plant cells, 108–109
  and population variation, 158
  in testes, 106–107
Memory, 309–310
Mendel's experiment, 126–133
Mendel, Gregor Johann, 126–131
Menses, 172f
Menstrual cycle, 171–173
Menstruation, defined, 171
Merced River, 384f
Mercury, as pollutant, 522
Mercury poisoning, 524b–525b
Mescaline, 369
Mesenchyme, 193, 236f
Mesoderm, 194
  of gastrula, 188
  in human embryo, 199
Messenger RNA. *See* RNA.
Metabolic rate, 225
Metabolic waste, 225
Metacarpals, 240
Metaphase
  in meiosis, 108–109, 111, 138
  in mitosis, 99, 102, 104–105, 110
Metatarsal, 240
Methadone, 367
Methane (CH_4), 50
  in early atmosphere, 23
  in Miller's experiments, 25
  as a source of organic compounds, 26
Methedrine, 369
Methionine, 117
Methylmercury (CH_3Hg), 523
Methyl poisoning. *See* Mercury poisoning.
Metric measurements, 547
Mew gull, 427f
Microhabitat, 373
Microorganisms, as agents of decay, 26
Microvilli, 267
Midlobe, of pituitary, 275
Midget. *See* Pituitary dwarf.
Milgau antelope, 354f
Miller, N. E., 308
Miller, Stanley L., 25–26
Minanata, Japan, 524b–525b
Miracle crops, 474, 476. *See also* "Green revolution."

Miscarriage, 214–215
Mitochondrion (pl.: Mitochondria), 78, 80, 91–92, 100, 101, 247
Mitosis, 99, 102–103, 104–105, 147
   in animal cells, 102, 110–111
   continuous, 102
   in plant cells, 161
Mole, defined, 54
Molecular makeup of chemical signals, 346
Molecules
   stability of, in early atmosphere, 26
   structure of, 23–25
Mollusks, circulatory system in, 251
Mongolism. *See* Down's syndrome.
Monosaccharide, 59
Mons veneris, 170f, 171
Morgan, Karl Z., 519
Morgan, Thomas Hunt, 133–134, 139
Morgan's experiments, 133–134
"Morning-after pill," 185
Morphine, 367
Morris, Desmond, 361
Mortality, and population control, 431–439
Mosquito, 249
Moth, 440
"Motivation," 319
Movement, types of, 244
Mud flat, 386b,
   effects of human development of, 399–400
Multicellularity, 235
Muscle
   comparison of types, 247t
   contraction of, 244, 246–247
   types, 241–243
Muscular dystrophy, 135
Mutation, 133, 140–141, 149, 158–159
Myelin sheath, 288
Myofibril, 244
Myofilaments, 247
Myosin, 246f, 247

N

NAD, 80–81
NADP$_{ox}$, 72, 74, 75
NADP$_{red}$, 72, 74, 75
Natural reserves, world, 484f
Natural selection, 16–17, 141, 431, 476
Negative feedback, 219–220, 221f

Nematods, 248
Nembutal, 369
Nephric units, 230
Nephridium (pl.: Nephridia), 226
Neon, 45
Nerve, 283, 290
Nerve impulse, 290–292, 294–296
Nervous system, 281–288
   earthworm, 282f
   frog, 282f
   human, 296–309
   hydra, 281, 282f
   planarian, 282f
Neural folds, 193
Neural stage, in egg development, 189f
Neural regulation, 308
Neuron, 281, 288–292, 294–296, 297f
Neurophysiology, 314
Neutron, 43
Neutron activation, 520
Newton, Isaac, 3, 14f
New York City, 509b
Niche
   competition for, 376
   defined, 373
   expanded, 403
Nicotiana tabacum. See Tobacco.
Nitrates
   effects on hemoglobin, 507
   as water pollutant, 507
Nitrogen (N_2)
   in early atmosphere, 23
   elimination, in amino acid breakdown, 228
Nitrogen oxides, as air pollutants, 499–500
Nitrogenous waste, 228, 257
Nitroglycerin, 55f
Nixon, Richard, 460, 462
Nodes, in plant stems, 161
Noise pollution, 523, 526
Nonrenewable resources, 482–490
Norepinephrine, 277
North American warblers, 374, 375f
Notochord, role of, 194
"Nuclear accident," 518b. *See also* Kyshym disaster, Brown's Ferry incident.
Nuclear fission, 494
Nuclear fusion, 495
Nuclear membrane, 86, 97, 100, 101
Nuclear power, 494
Nuclear power plants, 519–522

Nucleic acid, in coacervate droplets, 28
Nucleolus, 99, 100, 101
Nucleotide
   and amino acid synthesis, 140
   in coacervate droplets, 28
   defined, 112
Nucleus, 43, 96, 97–98, 100, 101
Nutrition, and pregnancy, 213. *See also Starvation, Kwashiorkor.*

O

Ocean bottom, life at, 398
Ocean currents, 382
Oceans, 379, 383
   as food source, 478
   as islands, 409, 411
Octopus, 285b
Oil companies, 486
Oil spills, 432, 433, 510b
Olecranon process (elbow), 244
Olfaction, 286
Olympic National Park, 386f
Ona Indians, 5f
Oparin, A. I., 23, 26
Open circulatory system, 255
Operant conditioning, 325–326. *See also* Learning.
Opiates. 367. *See also drug names.*
Opium, 367
Organic compounds, synthesis of, 25–26
Organophosphates, 516b
Organs, defined, 235. *See also individual organ names.*
Orgasm (climax), 167
Orientation, 331, 341
Orientation cage, 341f
Orienting movements, 316
Origin, of muscle, 243
Osmosis, 86–88, 226
Ostia, 255
Out-group aggression, 457b. *See also* Aggression, Animal behavior.
Ovaries, 170, 171, 277
Overpopulation, 16. *See also entries under* Population.
Oviduct, 170f, 171, 182, 198. *See also* Fallopian tube.
Ovule, 159, 161. *See also* Egg, Gamete.
Oxaloacetic acid, 78

Oxidation, 44–45, 55, of
    of glucose, 75
Oxygen (O_2), 31–32, 42, 81 as agent
        of decay, 26
    in ATP synthesis, 80
    in blood, 256
    in cells, 88
    composition of, 43
    in early atmosphere, 23
    isotopes of, 43
    in photosynthesis, 31, 68
    properties of, 46
    in respiration, 249, 252
    source of, 382
    in water formation, 49, 53
Ozone (O_3), 32

P

Pair bonding (mating), 350
Panama Canal, 409
Parasite, 436
Parasitism, 436, 438
Parasympathetic nerve, 306, 307f
Parathion, 296
Parenchyma, 163, 165f
Parietal lobe, 303
Particulate matter, as air pollutant,
    500, 502
Pauling, Linus, 107
Pavlov, Ivan, 324
Pavlov's experiment, 324–325
Peacock, 350
Pectoral girdle, 240
Pelvic girdle, 240
Pelvis, 242f
Penfield, Wilder, 309
Penfield's mapping, 309–311
Penguins, 336f, 338f
    individual recognition among, 349
Penis, 167, 168f, 169
Peptide, 63
Peregrine falcon, hunting behavior in,
    317–339
Pereira, Dennis, 148b
Peripheral nervous system, 305
Permafrost, 378
Pest control, 513b, 516b
Pesticide pollution, 512–517
Pesticides
    biodegradable, 517
    and food chain, 514–515
Peterson, Roger Tory, 420

Peyote, 369
pH, 54
Pharynx, 265
Phenotype, 141
    defined, 129
Phenylalanine, 117, 119, 147
Phenylketonuria (PKU), 147
Pheromones, 346
Phloem, 162
Phosphoglyceraldehyde (PGAL), 73
Phospholipid, 62
Photosynthesis, 31–32, 68–75. *See
    also* Dark reaction, Light
    reaction.
Phytoplankton, 382
Pituitary dwarf, 275
Pituitary gland, 171, 225, 230, 274f,
    275, 279
Placenta, 199, 214. *See also*
    Afterbirth.
Plant
    alternation of generations, 155–156
    cells, 88–89
    cortex of, 161
    cotyledons, 161, 164, 165f
    development in, 161–165
    experiments with, 126–133, 280
    flowering, 160f, 161
    hormones, 279–281
    leaf, 165f
    movement in, 329
    reproduction in, 153–161
    stem, 164f
    vascular system, 252
Plant community, in biomes, 376
Plasma, 256, 263
Plasma membrane. *See* Cell
    membrane.
Plasmids, 149
Plastids, 93–94
Platelets, 256
Platinum, 56
Platyhelminthes, 248
Plutonium (U^{239}), 520f
Poikilotherms. *See* Cold-blooded
    animals.
Polar bear, 388f
Polar bodies, 107
Polar ice cap, 388f
Pollen, 159
Pollen germination, 281
Pollination, 159
Pollutant, 497. *See also individual
    pollutants.*

Pollution, 491f, 497–526. *See also
    types of pollution.*
Polysaccharide, 58
Pons, of human brain, 299
Population. *See also* Birth rate,
    Death rate.
    changes in, 419–422, 448, 449
    control, 431–439
    crash, 421, 422f
    defined, 141
    geometric progression of, 16
    growth, 420–421, 447
    roles, in early human, 443
    structure, in U.S., 453f
    world, 448t
Population recognition, 348
Porpoises, cooperation among, 357
Positive feedback, 220, 221f
Posterior lobe, or pituitary gland, 275
Potassium, 40, 517
    movement into cells, 88
    in neuron, 291
Potatoes, radiation of, 521b
Potential energy, 67
Poultry, as food source, 478
Predation, 434–436
Prefrontal area, of human brain, 303
Pregnancy, 197b
    as altruism, 359
Prenatal influences, 197. *See also*
    Pregnancy.
Pressure gradient, in kidney
    tubule, 232b
Primary consumer, 444
Primitive traits, 284
Procaryotic cells, 84, 86
Progesterone, 173, 277
Progestin, 180, 184–185
Proline, 117
Pronghorn antelope, 153f, 393f
Pronators, 244
Propagation, 215–217
Propane, structure of, 50
Prophase
    in meiosis, 108–109, 111, 139
    in mitosis, 99, 102, 104–105, 110
Prostaglandins, 185, 277
Prostate gland, 168f, 169
Protein, 56, 62–65
    deficiency and child development,
    denatured, 222
    structure of, 65
    synthesis, 117–120
    world consumption, 471f

Protons, 43
Protozoa, 96, 147
    cloning in, 147
    excretory system in, 225
Psychedelics, 369–370
Psilocybin, 369
Pubic symphysis, 242f
Public education, in U.S., 526–527
Pulmonary arteries, 259
Punnett squares, 129, 142
Pupfish, 222f
Purple pitcher plant, 338f
Purine, 112
Pyruvic acid, 75, 78, 81
Pyrimidine, 112

Q

Queen bee, 346
Quickening, 196

R

Rabbit test, 197
Radial symmetry, 248
Radiation, 68, 140, 517–522
Radioactive wastes, 520
Radioactivity, 519
Radioisotope, 521b
Radius, 243
Radula, 250
"Rasmussen Report," 519
Ratio
    of dominant and recessive traits,
        128, 129, 130
    gene, 142
Rats, 416b
    and chemical communication, 346
    effects of DDT on, 514
Rattlesnake, 353f
Reagan, Ronald, 414
Recapitulation, theory of, 175b
Receptor, 290, 297
Recessive traits, 127
    in humans, 133
Recombinant DNA
    controversy, 149–151
    and social responsibility, 35
Red blood cells, and sickle-cell
    anemia, 144
Red-cheeked parrot, 396f
Red marrow, 237

"Red tides," 476
Reduction, 45
Red wolf, 333f
Redwoods, 414b–415b
Reflex arc, 296–298
Refuse Act of 1899, 511
Regeneration, 154
Regulation. *See also* Homeostasis.
    of body functions, 273
    of body temperature, 220–225
    of cell development, 191
    hormonal, 273
    of water and salt balances, 229
Reinforcement, and reward, 327b
Releasers, 321–322
Repolarization, of neuron, 291
Reproduction
    and bonding, 166
    controversy, 215–217
    flatworm, 153, 154f
    human, 169–177
    hydra, 154
Reproduction rate, 422–431
Reproductive behavior, and
        population varition, 158
Reproduction cycles, 155–156. *See
    also* Alternation of generations.
Reptile
    and body temperature, 222
    brain, 284f
    heart, 259
Resolution, in microscopes, 85
Respiration
    in anaerobic organisms, 75–76
    cellular, 75–81
    in chick embryo, 195
    external, 249–252
Response threshold, 319
Resting potential, 291
Reticular system, 299–300
Reversible reaction, 53
Ribulose diphosphate, (RuDP), 73, 74
Ribonucleic acid. *See* RNA.
Ribose, 70
Ribosomes, 95, 100, 101
    function of, in protein
        synthesis, 119
RNA, 99
    composition of, 116
    manufacture of, 115, 116
    and memory, 310
    messenger (mRNA),
        production of, 277
        in protein synthesis, 117–120

    ribosomal RNA, 119
    transfer (tRNA), 119
    in viruses, 122
Rocky coasts, 398. *See also* Coastal
    areas.
Roberts, Lamar, 310
Rosier, Bernard, 470
Rough ER. *See* Endoplasmic
    reticulum.
Rhesus monkeys, copulation of, 169
Rhythm method, 178–179
Rhythms, 329–331

S

Sac fungus (*Endothia parasitia*),
    436, 438
Saliva, 267
Sally lightfoot crabs (*Grapsus
    grapsus*), 11f
Salticid spider, 348
Salt marsh 387f. *See also* Coastal
    areas.
Salt water, 379
Saltwater animals, and
    elimination, 228
Sand shark, respiratory system
    of, 251f
Sandy beach, 387f, 399. *See also*
    Coastal areas.
Sanger, Frederick, 64b
Santa Barbara Channel, 486
"Scale of nature," 3
Scapulas, 240
Schleidin, Matthias, Jakob, 83
Sclerenchyma, 163, 164f
Schlesinger, James, 491f
Schwann cell, 288
Schwann, Theodor, 83
Science
    and ideas, 32–35
    and religion, 3
    and social responsibility, 35–39
Scientific knowledge, 3
Scientific process, 33–35
Scrimshaw, N. A., 470
Scrotum, 168f, 169
Sea anemones, 337f
Seabirds, and salt removal glands, 227
Seconal, 369
Secondary sex characteristics, 277
Sedative-hypnotics. *See* Barbiturates.
Seed germination, 159–160, 161, 165f

"Seed leaves." *See* Cotyledons.
Segregation
  of genes, 130
  principle, 126, 127–129
Semen, 174
Seminal vesicle, 186f, 169
Seminiferous tubules, 168f, 169
Senses, 283
Sequencing, 192
Serine, 117
Serynl, 369
Sewage, 505–507, 509b
Sex chromosome. *See* X chromosome, Y chromosome.
Sex, determination of, 133–134
Sex inheritance, 134–137
Sex linkage, in humans, 134–137
Sexual dimorphism, 350b–351b
Sexual intercourse, in humans, 166–171
Sexual reproduction
  adaptiveness of, 158
  advantages of, 158–159
  as a response to stress, 156, 158
Sickle-cell anemia, 144b, 145
Sierra Club, 414
Sinuses, 255
Sinusoids, 255
Shale oil, 488b–489b
Sharks, and elimination, 228
Sheep, as food source, 476, 478
Shepherd's purse (*Capsalla*), 162f
Short-term visual signals, 343
Skeletal muscles, 243–244, 247
Skeletal systems, 236–237
Skinner, B. F., 325, 327b
Skinner box, 325
"Sleep movement," in plants, 329
Small intestine, 267. *See also* Digestive system.
Smith, W. Eugene, 524b–525b
Smooth ER. *See* Endoplasmic reticulum.
Smooth muscle, 241, 247
Snowy owls, 389f
Sociobiology, 361b
Sodium
  ionization of, 46
  ions in nervous system, 291
  properties of, 46
  "pump," 88, 291
  reabsorption of, in kidney, 277
Sodium chloride, 46
  ionic bonding of, 48–49

Solar energy, 492. *See also* Sun.
Somatotropin, 275
Somite, 195, 199
Song sparrows, sexual dimorphism in, 350
Sound communication, 343–345
Sound spectrogram, 345
Southeast Asia, 459–462
Spanish moss, 340f
Speciation, 407, 409, 410. *See also* Character displacement.
Species recognition, 347–348
Sperm, human, 174. *See also* Gamete.
Spermicides, 180
Sphincter, 167, 244, 267
Spicule, 236f
Spinal cord, 296–298, 309
  development of, 193–194
  in humans, 199
Spindle, 102
Sponge
  cells of, 236f
  skeleton of, 236
Sporangium (pl.: sporangia), 156, 159
Spore, 154
Sporophyte, 157f
Sporophyte generation, 155, 159
Sporulation, 154–155
Squirrel, and nut opening behavior, 328
Stamen, 161
Starch, 59
Starlings, orientation in, 331
Starvation, 473b. *See also* Kwashiorkor, Protein deficiency.
Sterilization, 181–183
Sternum, 239
Steroid, 62, 277
Stevia rebaudiana, 185
"Stimulus control," in operant conditioning, 325
Stigma, 159, 161
Stomach, 265, 267. *See also* Digestive system.
Stomata, 165f
Stress, as regulatory factor, 192–193
Striations, 243
Strip mining, 487b
Sturtevant, A. H., 139
Sucrose, structure of, 58
Sudden death, 261
Sulfur oxides, as air pollutants, 500
Sun, as energy source, 25, 68
Sunlight, in photosynthesis, 31

Superior vena cava, 259
Supersonic transport planes (SSTs), 526
Supinators, 244
Sutton, Walter S., 133
Suture lines, 241, 242f
Swallowing, as consummatory behavior, 320
Sweating, as regulatory function, 220
Swift (*Apus apus*), 426
Sympathetic nervous system, 306, 307b, 308
Synapse, 290, 294–296, 297f
Systems
  defined, 235
  supportive, 235–241

T

Taiga, 378, 390f
Tapeworm
  life cycle of, 424f
  reproduction rate of, 423
Tarsals, 240
Taxation, 527
Telophase,
  in meiosis, 108–109
  in mitosis, 99, 102, 104–105
Temperature
  and living organisms, 222
  regulation of internal, 220
Temperature conversion, 548
Temperature inversions, as air pollutant, 502–503
Temporal lobe, 303
Temperate deciduous forest, 376, 377f, 392f
Tendons, 239, 243
Terminal bud, 161, 163f
Territorial species, behavior in, 438
Tetrads, 108–109
Tetrahydrocannabinols (THC), 364
Testes, 168f, 169, 277
Testosterone, 277
Thalamus, 299–300
Thalidomide, 197
Thermodynamics, laws of, 67
Theory, defined, 32–33
Thompson's gazelle, 30, 31
Threonine, 117
Threshold value, 294
Thymine, 112, 115, 148
Thyroid, 225, 275

Thyroxin, 275
Tinbergen, Niko, 313, 314f, 321, 355
Tissue, defined, 235
Tobacco, 362–363
Tortoise, 15
  Geochelone elephantopus, 9f
  Testudo elephantopus, 8f
Toxophene, 516b
Trachea, 253f
Trade winds, 382
Transverse commissures, 283
Triage, 480b
Tricepts brachii, 244
Trophoblast, 199
Tropical deciduous forest, 397f
Tropical rainforest, 378, 396f
Tropic hormones, 275
Tropic responses, 248
Truman, Harry, 121
Tryptoplan, 117
Tubal ligation, 182
Tube-within-a-tube structure, 194
Tuinal, 369
Tules, 385f
Tumescence, 167
Tundra, 378, 389f
Turner, Admiral Stansfield, 462
Turtles, and salt removal glands, 227.
  See also Tortoise.
Tyrosine, 117

U

Ultraviolet rays, 25, 32
United Kingdom, 450, 451
United Nations Food and Agriculture
  Organization, 470, 478
Unit membrane. *See* Cell membrane.
Uracil, 16
  structure of, 115
Uranium, 520f
Urea, 228
Ureter, 230, 231f
Urethra, 168f, 169, 170f, 171, 230
Uric acid, 228
Urinary bladder, 230
U.S. Food and Drug Administration,
  185, 512
U.S. imports, 486t. *See also*
  Commodities.
U.S. Public Health Service, 511
Uterus, 170f, 171, 198–199

V

Vacuole, 94–95, 100, 101
Vacuum behavior, 321
Vacuum curettage, 183
Vagina, 167, 170f, 171
Valine, 117
Values, 465b
van Overbeek, J., 281
Vascular cambium, 162
Vascular tissue, in developing
  plant, 161
Vas deferens (sperm duct), 168f, 169
Vasectomy, 180, 181f, 182
Vegetal pole, 188, 189f
Vegetative fission, 153–154
Vein, 257
Venous system, 262
Ventilation (breathing), 251–252
Ventral nerve root, 297
Ventricle, 257
Venules, 257
Vernix caseosa, 202
Vertebrae, 239, 240f
Vertebrate
  brain, 283–288
  circulatory system, 255–257
  endocrine system, 278
  heart, 258
  development of, 191f
Victoria plant. *See* Giant water lilly.
Villi, 267
Viperfish, 398f
Virus, 122b
Visual communication, 342–343
Vitalism, 20–21
Vultures, 336f
Vulva, 171

W

Wallace, Alfred Russell, 19
War, 454–456, 458–466
  causes of, 462–463
  effect on population, 456
  and evolution, 456, 458
Warm-blooded animals, 223–225
Water
  and ATP formation, 80
  in early atmosphere, 23
  elimination of, 227b
  environment, 379–382
  in Miller's experiment, 25
  in photosynthesis, 31
  as resource, 482
  as source of oxygen, 32
  U.S. consumption of, 482, 483f
Water pollution, 505–512
Water Pollution Control Act, 511
Watson, James, 107
Waxes, 61
Weaver ants (*Oecophylla
  smargdina*), 358f
Weinberg, W., 142
Went, Frits W., 280
Whales, 435, 436, 437
White-crowned sparrows, and song
  learning, 328
White-footed mouse (*Peromyseus
  leucopus*), 334f
Whitman, C. O., 315
Wind, as energy source, 492
Wolf, 390f, 457b
Woodpecker, 131, 403–404, 426
  species recognition, 347f
Woodpecker finch, 11f

X

X chromosome, 134, 135, 137
X ray. *See* Radiation.
Xylem, 162

Y

Y chromosome, 134, 137
Yaks, cooperation among, 357
Yeast, 75, 8l, respiration of, 75–76
Yellow marrow, 237
Yellowstone National Park, 391f
Yolk
  as food store, 187
  plug, 188

Z

Z reaction. *See* Light reaction.
Zero population growth (ZPG),
  452–454
Zooplankton, 382
Zygote, 159, 187, 189
  bird, 194
  human, 174, 198–199

Short-eared Owl
Asio flammeus

Common Raven
Corvus corax

Long-tailed Weasel
Mustela frenata

Tule
Scirpus robustus

Lasthenia
Lasthenia glabrata

Silverweed
Potentilla Egedei

Sea Fig
Mesembryanthemum chilense

Clapper Rail*
Rallus longirostris

California Ground Squirrel
Citellus beecheyi

Fiddler Crab
Uca crenulata

Pacific Razor Clam
Siliqua patula

Mudsucker
Gillichthys mirabilis